Lecture Notes in Computer Science 5665

Commenced Publication in 1973
Founding and Former Series Editors:
Gerhard Goos, Juris Hartmanis, and Jan van Leeuwen

Orr Dunkelman (Ed.)

Fast
Software Encryption

16th International Workshop, FSE 2009
Leuven, Belgium, February 22-25, 2009
Revised Selected Papers

 Springer

Volume Editor

Orr Dunkelman
École Normale Supérieure
Département d'Informatique
45 rue d'Ulm, 75230 Paris CEDEX 05, France
E-mail: orr.dunkelman@ens.fr

Library of Congress Control Number: 2009931058

CR Subject Classification (1998): E.3, I.1, E.2, D.4.6, K.6.5

LNCS Sublibrary: SL 4 – Security and Cryptology

ISSN   0302-9743
ISBN-10   3-642-03316-4 Springer Berlin Heidelberg New York
ISBN-13   978-3-642-03316-2 Springer Berlin Heidelberg New York

springer.com

© International Association for Cryptologic Research 2009
Printed in Germany

Typesetting: Camera-ready by author, data conversion by Scientific Publishing Services, Chennai, India
Printed on acid-free paper   SPIN: 12731466   06/3180   5 4 3 2 1 0

Preface

Fast Software Encryption 2009 was the 16th in a series of workshops on symmetric key cryptography. Starting from 2002, it is sponsored by the International Association for Cryptologic Research (IACR). FSE 2009 was held in Leuven, Belgium, after previous venues held in Cambridge, UK (1993, 1996), Leuven, Belgium (1994, 2002), Haifa, Israel (1997), Paris, France (1998, 2005), Rome, Italy (1999), New York, USA (2000), Yokohama, Japan (2001), Lund, Sweden (2003), New Delhi, India (2004), Graz, Austria (2006), Luxembourg, Luxembourg (2007), and Lausanne, Switzerland (2008).

The workshop's main topic is symmetric key cryptography, including the design of fast and secure symmetric key primitives, such as block ciphers, stream ciphers, hash functions, message authentication codes, modes of operation and iteration, as well as the theoretical foundations of these primitives.

This year, 76 papers were submitted to FSE including a large portion of papers on hash functions, following the NIST SHA-3 competition, whose workshop was held just after FSE in the same location. From the 76 papers, 24 were accepted for presentation. It is my pleasure to thank all the authors of all submissions for the high-quality research, which is the base for the scientific value of the workshop. The review process was thorough (each submission received the attention of at least three reviewers), and at the end, besides the accepted papers, the Committee decided that the merits of the paper "Blockcipher-Based Hashing Revisited" entitled the authors to receive the best paper award. I wish to thank all Committee members and the referees for their hard and dedicated work.

The workshop also featured two invited talks. The first was given by Shay Gueron about "Intel's New AES Instructions for Enhanced Performance and Security" and the second was given by Matt Robshaw about "Looking Back at the eSTREAM Project." Along the presentation of the papers and the invited talks, the traditional rump session was organized and chaired by Dan J. Bernstein.

I would like to thank Thomas Baignères for the iChair review management software, which facilitated a smooth and easy review process, and Shai Halevi for the Web Submission and Review Software for dealing with the proceedings.

A special thanks is due to the organizing team. The COSIC team from Katholieke Universiteit Leuven, headed by Program Chair Bart Preneel, did a wonderful job in hosting the workshop. The warm welcome that awaited more than 200 delegates from all over the world was unblemished. The support given to the FSE 2009 workshop by the sponsors Katholieke Universiteit Leuven, PriceWaterhouseCoppers, and Oberthur technologies is also gratefully acknowledged.

May 2009   Orr Dunkelman

Fast Software Encryption 2009

Leuven, Belgium, February 22–25, 2009

Sponsored by the
International Association for Cryptologic Research (IACR)

Program and General Chairs

Program Chair	Orr Dunkelman
	École Normale Supérieure, France
General Chair	Bart Preneel
	Katholieke Universiteit Leuven, Belgium

Program Committee

Steve Babbage	Vodafone Group R&D, UK
Alex Biryukov	University of Luxembourg, Luxembourg
Dan J. Bernstein	University of Illinois at Chicago, USA
Joan Daemen	STMicroelectronics, Belgium
Christophe De Cannière	École Normale Supérieure, France
	and Katholieke Universiteit Leuven, Belgium
Orr Dunkelman (Chair)	École Normale Supérieure, France
Henri Gilbert	Orange Labs, France
Louis Granboulan	EADS Innovation Works, France
Helena Handschuh	Spansion, France
Tetsu Iwata	Nagoya University, Japan
Nathan Keller	Hebrew University, Israel
Stefan Lucks	Bauhaus-University Weimar, Germany
Mitsuru Matsui	Mitsubishi Electric, Japan
Willi Meier	FHNW, Switzerland
Kaisa Nyberg	Helsinki University of Technology
	and NOKIA, Finland
Raphael Phan	Loughborough University, UK
Bart Preneel	Katholieke Universiteit Leuven, Belgium
Håvard Raddum	University of Bergen, Norway
Christian Rechberger	Graz University of Technology, Austria
Thomas Ristenpart	UC San Diego, USA
Greg Rose	Qualcomm, Australia
Serge Vaudenay	EPFL, Switzerland
Yiqun Lisa Yin	Independent Consultant, USA

Referees

Elena Andreeva
Kazumaro Aoki
Frederik Armknecht
Jean-Philippe Aumasson
Guido Bertoni
Olivier Billet
Billy Brumley
Rafik Chaabouni
Donghoon Chang
Joo Yeon Cho
Shanshan Duan
Baha Dundar
Ewan Fleischmann
Chistian Forler
Pasqualina Fragneto
Benedikt Gierlichs
Michael Gorski
Jian Guo
Risto Hakala
Miia Hermelin
Shoichi Hirose
Michael Hutter
Sebastiaan Indesteege
Kimmo Järvinen
Pascal Junod
Charanjit Jutla
Liam Keliher
Shahram Khazaei
Dmitry Khovratovich
Jongsung Kim
Matthias Krause

Mario Lamberger
Changhoon Lee
David McGrew
Florian Mendel
Nicky Mouha
Jorge Nakahara Jr.
Maria Naya-Plasencia
Ivica Nikolić
Khaled Ouafi
Matthew Parker
Sylvain Pasini
Chris Peikert
Thomas Peyrin
Thomas Roche
Martin Schläffer
Yannick Seurin
Zhijie Shi
Thomas Shrimpton
Hervé Sibert
Dirk Stegemann
Daisuke Suzuki
Stefano Tessaro
Stefan Tillich
Elena Trichina
Gilles Van Assche
Martin Vuagnoux
Ralf-Philipp Weinmann
Bo-Yin Yang
Scott Yilek
Erik Zenner
Fan Zhang

Sponsors

Katholieke Universiteit Leuven, Belgium
PriceWaterhouseCoppers, Belgium
Oberthur technologies, Belgium

Table of Contents

Block Ciphers Analysis

Hash Functions Analysis II

Block Ciphers

Theory of Symmetric Key

Message Authentication Codes

Cube Testers and Key Recovery Attacks on Reduced-Round MD6 and Trivium

Jean-Philippe Aumasson[1,*], Itai Dinur[2], Willi Meier[1,**], and Adi Shamir[2]

[1] FHNW, Windisch, Switzerland
[2] Computer Science Department, The Weizmann Institute, Rehovot, Israel

Abstract. CRYPTO 2008 saw the introduction of the hash function MD6 and of cube attacks, a type of algebraic attack applicable to cryptographic functions having a low-degree algebraic normal form over GF(2). This paper applies cube attacks to reduced round MD6, finding the full 128-bit key of a 14-round MD6 with complexity 2^{22} (which takes less than a minute on a single PC). This is the best key recovery attack announced so far for MD6. We then introduce a new class of attacks called cube testers, based on efficient property-testing algorithms, and apply them to MD6 and to the stream cipher Trivium. Unlike the standard cube attacks, cube testers detect nonrandom behavior rather than performing key extraction, but they can also attack cryptographic schemes described by nonrandom polynomials of relatively high degree. Applied to MD6, cube testers detect nonrandomness over 18 rounds in 2^{17} complexity; applied to a slightly modified version of the MD6 compression function, they can distinguish 66 rounds from random in 2^{24} complexity. Cube testers give distinguishers on Trivium reduced to 790 rounds from random with 2^{30} complexity and detect nonrandomness over 885 rounds in 2^{27}, improving on the original 767-round cube attack.

1 Introduction

1.1 Cube Attacks

Cube attacks [29, 9] are a new type of algebraic cryptanalysis that exploit implicit low-degree equations in cryptographic algorithms. Cube attacks only require black box access to the target primitive, and were successfully applied to reduced versions of the stream cipher Trivium [6] in [9]. Roughly speaking, a cryptographic function is vulnerable to cube attacks if its implicit algebraic normal form over GF(2) has degree at most d, provided that 2^d computations of the function is feasible. Cube attacks recover a secret key through queries to a *black box polynomial* with *tweakable public variables* (e.g. chosen plaintext or IV bits), followed by solving a linear system of equations in the secret key variables. A one time preprocessing phase is required to determine which queries should be made to the black box during the on-line phase of the attack. Low-degree implicit equations were previously exploited in [11, 27, 21, 10] to construct

---

* Supported by the Swiss National Science Foundation, project no. 113329.
** Supported by GEBERT RÜF STIFTUNG, project no. GRS-069/07.

O. Dunkelman (Ed.): FSE 2009, LNCS 5665, pp. 1–22, 2009.

distinguishers, and in [32, 12, 15] for key recovery. Cube attacks are related to saturation attacks [17] and to high order differential cryptanalysis [16].

Basics. Let $\mathcal{F}_n$ be the set of all functions mapping $\{0, 1\}^n$ to $\{0, 1\}$, $n > 0$, and let $f \in \mathcal{F}_n$. The *algebraic normal form* (ANF) of f is the polynomial p over GF(2) in variables $x_1, \ldots, x_n$ such that evaluating p on $x \in \{0, 1\}^n$ is equivalent to computing $f(x)$, and such that it is of the form[1]

$$\sum_{i=0}^{2^n - 1} a_i \cdot x_1^{i_1} x_2^{i_2} \cdots x_{n-1}^{i_{n-1}} x_n^{i_n}$$

for some $(a_0, \ldots, a_{2^n - 1}) \in \{0, 1\}^{2^n}$, and where i_j denotes the j-th digit of the binary encoding of i (and so the sum spans all monomials in $x_1, \ldots, x_n$). A key observation regarding cube attacks is that for any function $f : \{0, 1\}^n \mapsto \{0, 1\}$, the sum (XOR) of all entries in the truth table

$$\sum_{x \in \{0,1\}^n} f(x)$$

equals the coefficient of the highest degree monomial $x_1 \cdots x_n$ in the algebraic normal form (ANF) of f. For example, let $n = 4$ and f be defined as

$$f(x_1, x_2, x_3, x_4) = x_1 + x_1 x_2 x_3 + x_1 x_2 x_4 + x_3 .$$

Then summing $f(x_1, x_2, x_3, x_4)$ over all 16 distinct inputs makes all monomials vanish and yields zero, i.e. the coefficient of the monomial $x_1 x_2 x_3 x_4$. Instead, cube attacks sum over a *subset* of the inputs; for example summing over the four possible values of (x_1, x_2) gives

$$f(0, 0, x_3, x_4) + f(0, 1, x_3, x_4) + f(1, 0, x_3, x_4) + f(1, 1, x_3, x_4) = x_3 + x_4 ,$$

where $(x_3 + x_4)$ is the polynomial that multiplies $x_1 x_2$ in f:

$$f(x_1, x_2, x_3, x_4) = x_1 + x_1 x_2 (x_3 + x_4) + x_3 .$$

Generalizing, given an index set $I \subsetneq \{1, \ldots, n\}$, any function in $\mathcal{F}_n$ can be represented algebraically under the form

$$f(x_1, \ldots, x_n) = t_I \cdot p(\cdots) + q(x_1, \ldots, x_n)$$

where t_I is the monomial containing all the x_i's with $i \in I$, p is a polynomial that has no variable in common with t_I, and such that no monomial in the polynomial q contains t_I (that is, we factored f by the monomial t_I). Summing f over the *cube* t_I for other variables fixed, one gets

$$\sum_I t_I \cdot p(\cdots) + q(x_1, \ldots, x_n) = \sum_I t_I \cdot p(\cdots) = p(\cdots),$$

---

[1] The ANF of any $f \in \mathcal{F}_n$ has degree at most n, since $x_i^d = x_i$, for $x_i \in$ GF(2), $d > 0$.

that is, the evaluation of p for the chosen fixed variables. Following the terminology of [9], p is called the *superpoly* of I in f. A *cube* t_I is called a *maxterm* if and only if its superpoly p has degree 1 (i.e., is linear but not a constant). The polynomial f is called the *master polynomial*.

Given access to a cryptographic function with public and secret variables, the attacker has to recover the secret key variables. Key recovery is achieved in two steps, a preprocessing and an online phase, which are described below.

Preprocessing. One first finds sufficiently many maxterms t_I of the master polynomial. For each maxterm, one computes the coefficients of the secret variables in the symbolic representation of the linear superpoly p. That is, one *reconstructs* the ANF of the superpoly of each t_I. Reconstruction is achieved via probabilistic linearity tests [5], to check that a superpoly is linear, and to identify which variables it contains. The maxterms and superpolys are not key-dependent, thus they need to be computed only once per master polynomial.

The main challenge of the cube attack is to find maxterms. We propose the following simple preprocessing heuristic: one randomly chooses a subset I of k public variables. Thereafter one uses a linearity test to check whether p is linear. If the subset I is too small, the corresponding superpoly p is likely to be a nonlinear function in the secret variables, and in this case the attacker adds a public variable to I and repeats the process. If I is too large, the sum will be a constant function, and in this case he drops one of the public variables from I and repeats the process. The correct choice of I is the borderline between these cases, and if it does not exist the attacker retries with a different initial I.

Online Phase. Once sufficiently many maxterms and the ANF of their superpolys are found, preprocessing is finished and one performs the online phase. Now the secret variables are fixed: one evaluates the superpoly's p by summing $f(x)$ over all the values of the corresponding maxterm, and gets as a result a linear combination of the key bits (because the superpolys are linear). The public variables that are not in the maxterm should be set to a fixed value, and to the same value as set in the preprocessing phase.

Assuming that the degree of the master polynomial is d, each sum requires at most 2^{d-1} evaluations of the derived polynomials (which the attacker obtains via a chosen plaintext attack). Once enough linear superpolys are found, the key can be recovered by simple linear algebra techniques.

1.2   MD6

Rivest presented the hash function MD6 [24, 25] as a candidate for NIST's hash competition[2]. MD6 shows originality in both its operation mode—a parametrized quadtree [7]—and its compression function, which repeats hundreds of times a simple combination of XOR's, AND's and shift operations: the r-round compression function of MD6 takes as input an array $A_0, \ldots, A_{88}$ of 64-bit words, recursively computes $A_{89}, \ldots, A_{16r+88}$, and outputs the 16 words $A_{16r+73}, \ldots, A_{16r+88}$:

---

[2] See http://www.nist.gov/hash-competition

for $i = 89, \ldots, 16r + 88$
$\quad x \leftarrow S_i \oplus A_{i-17} \oplus A_{i-89} \oplus (A_{i-18} \wedge A_{i-21}) \oplus (A_{i-31} \wedge A_{i-67})$
$\quad x \leftarrow x \oplus (x \gg r_i)$
$\quad A_i \leftarrow x \oplus (x \ll \ell_i)$
return $A_{16r+73, \ldots, 16r+88}$

A *step* is one iteration of the above loop, a *round* is a sequence of 16 steps. The values S_i, r_i, and ℓ_i are step-dependent constants (see Appendix A). MD6 generates the input words $A_0, \ldots, A_{88}$ as follows:

1. $A_0, \ldots, A_{14}$ contain constants (fractional part of $\sqrt{6}$; 960 bits)
2. $A_{15}, \ldots, A_{22}$ contain a key (512 bits)
3. A_{23}, A_{24} contain parameters (key length, root bit, digest size, etc.; 128 bits)
4. $A_{25}, \ldots, A_{88}$ contain the data to be compressed (message block or chain value; 4096 bits)

The proposed instances of MD6 perform at least 80 rounds (1280 steps) and at most 168 (2688 steps). Resistance to "standard" differential attacks for collision finding is proven for up to 12 rounds. The designers of MD6 could break at most 12 rounds with high complexity using SAT-solvers.

The compression function of MD6 can be seen as a device composed of 64 non-linear feedback shift registers (NFSR's) and a linear combiner: during a step the 64 NFSR's are clocked in parallel, then linearly combined. The AND operators ($\wedge$) progressively increase nonlinearity, and the shift operators provide wordwise diffusion. This representation will make our attacks easier to understand.

1.3 Trivium

The stream cipher Trivium was designed by De Cannière and Preneel [6] and submitted as a candidate to the eSTREAM project in 2005. Trivium was eventually chosen as one of the four hardware ciphers in the eSTREAM portofolio[3]. Reduced variants of Trivium underwent several attacks [23, 19, 20, 31, 32, 10, 12, 22], including cube attacks [9].

Trivium takes as input a 80-bit key and a 80-bit IV, and produces a keystream after 1152 rounds of initialization. Each round corresponds to clocking three feedback shift registers, each one having a quadratic feedback polynomial. The best result on Trivium is a cube attack [9] on a reduced version with 767 initialization rounds instead of 1152.

1.4 The Contributions of This Paper

First we apply cube attacks to keyed versions of the compression function of MD6. The MD6 team managed to break up to 12 rounds using a high complexity attack based on SAT solvers. In this paper we show how to break the same 12 round version and recover the full 128-bit key with trivial complexity using

---

[3] See http://www.ecrypt.eu.org/stream/

Table 1. Summary of the best known attacks on MD6 and Trivium ("$\sqrt{}$" designates the present paper)

#Rounds	Time	Attack	Authors
		MD6	
12	hours	inversion	[25]
14	2^{22}	key recovery	$\sqrt{}$
18	2^{17}	nonrandomness	$\sqrt{}$
66*	2^{24}	nonrandomness	$\sqrt{}$
		Trivium	
736	2^{33}	distinguisher	[10]
736$^\diamond$	2^{30}	key-recovery	[9]
767$^\diamond$	2^{36}	key-recovery	[9]
772	2^{24}	distinguisher	$\sqrt{}$
785	2^{27}	distinguisher	$\sqrt{}$
790	2^{30}	distinguisher	$\sqrt{}$
842	2^{24}	nonrandomness	$\sqrt{}$
885	2^{27}	nonrandomness	$\sqrt{}$

*: for a modified version where $S_i = 0$.
$^\diamond$: cost excluding precomputation.

a cube attack, even under the assumption that the attacker does not know anything about its design (i.e., assuming that the algorithm had not been published and treating the function as a black box polynomial). By exploiting the known internal structure of the function, we can improve the attack and recover the 128-bit key of a 14-round MD6 function in about 2^{22} operations, which take less than a minute on a single PC. This is the best key recovery attack announced so far on MD6.

Then we introduce the new notion of *cube tester*, which combines the cube attack with efficient property-testers, and can be used to mount distinguishers or to detect nonrandomness in cryptographic primitives. Cube testers are flexible attacks that are adaptable to the primitive attacked. Some cube testers don't require the function attacked to have a low degree, but just to satisfy some testable property with significantly higher (or lower) probability than a random function. To the best of our knowledge, this is one of the first explicit applications of property-testing to cryptanalysis.

Applying cube testers to MD6, we can detect nonrandomness in reduced versions with up to 18 rounds in just 2^{17} time. In a variant of MD6 in which all the step constants S_i are zero, we could detect nonrandomness up to 66 rounds using 2^{24} time. Applied to Trivium, cube testers give distinguishers on up to 790 in time 2^{30}, and detect nonrandomness on up to 885 rounds in 2^{27}. Table 1 summarizes our results on MD6 and Trivium, comparing them with the previous attacks .

As Table 1 shows, all our announced complexities are quite low, and presumably much better results can be obtained if we allow a complexity bound of 2^{50}

(which is currently practical on a large network of PC's) or even 2^{80} (which may become practical in the future). However, it is very difficult to estimate the performance of cube attacks on larger versions without actually finding the best choice of cube variables, and thus our limited experimental resources allowed us to discover only low complexity attacks. On the other hand, all our announced attacks are fully tested and verified, whereas other types of algebraic attacks are often based on the conjectured independence of huge systems of linear equations, which is impossible to verify in a realistic amount of time.

2   Key Recovery on MD6

2.1   Method

We describe the attack on reduced-round variants of a basic keyed version of the MD6 compression function. The compression function of the basic MD6 keyed version we tested uses a key of 128 bits, and outputs 5 words. Initially, we used the basic cube attack techniques that treat the compression function as a black box, and were able to efficiently recover the key for up to 12 rounds. We then used the knowledge of the internal structure of the MD6 compression function to improve on these results. The main idea of the improved attack is to choose the public variables in the cube that we sum over so that they do not mix with the key in the initial mixing rounds. In addition, the public variables that do not belong to the cube are assigned predefined constant values that limit the diffusion of the private variables and the cube public variables in the MD6 array for as many rounds as possible. This reduces the degree of the polynomials describing the output bits as functions in the private variables and the cube public variables, improving the performance of the cube attack.

   The improved attack is based on the observation that in the feedback function, A_i depends on A_{i-17}, A_{i-89}, A_{i-18}, A_{i-31} and A_{i-67}. However, since A_{i-18} is ANDed with A_{i-21}, the dependency of A_i on A_{i-18} can be eliminated regardless of its value, by zeroing A_{i-21} (assuming the value of A_{i-21} can be controlled by the attacker). Similarly, dependencies on A_{i-21}, A_{i-31} or A_{i-67} can be eliminated by setting the corresponding ANDed word to zero. On the other hand, removing the linear dependencies of A_i on A_{i-17} or A_{i-89} is not possible if their value is unknown (e.g. for private variables), and even if their values are known (e.g. for public variables), the elimination introduces another dependency, which may contribute to the diffusion of the cube public variables (for example it is possible to remove the dependency of A_i on A_{i-89} by setting A_{i-17} to the same value, introducing the dependency of A_{i-17} on A_{i-89}).

   These observations lead to the conclusion that the attacker can limit the diffusion of the private variables by removing as many quadratic dependencies of the array variables on the private variables as possible. The basic MD6 keyed version that we tested uses a 2-word (128-bit) key, which is initially placed in A_{15} and A_{16}. Note that the MD6 mode of operation dedicates a specific part of the input to the key in words $A_{15}, \ldots, A_{22}$ (512 bits in total).

Table 2 describes the diffusion of the key into the MD6 compression function array up to step 189 (the index of the first outputted word is 89).

In contrast to the predefined private variable indexes, the attacker can choose the indexes of the cube public variables, and improve the complexity of the attack by choosing such cube public variables that diffuse linearly to the MD6 array only at the later stages of the mixing process. Quadratic dependencies of an array word on cube public variables can be eliminated if the attacker can control the value of the array word that is ANDed with the array word containing the cube public variables. It is easy to verify that the public variable word that is XORed back to the MD6 array at the latest stage of the mixing process is A_{71}, which is XORed in step 160 to A_{160}. Thus, the array word with index 71 and words with index just under 71, seem to be a good choice for the cube public variables. Exceptions are A_{68} and A_{69} which are mixed with the key in steps 135 and 136 and should be zeroed. We tested several cubes, and the best preprocessing results were obtained by choosing cube indexes from A_{65}. One of the reason that A_{65} gives better results than several other words (e.g. A_{71}) is that it is ANDed with just 2 words before it is XORed again into the array in step 154, whereas A_{71} is ANDed with 4 words before step 170. This gives the attacker more freedom to choose the values of the fixed public variables, and limit the diffusion of the private and cube public variables for more rounds. Table 3 describes the diffusion of A_{65} into the MD6 compression function array up to step 185 (the index of the first outputted word is 89).

2.2   Results

We were able to prevent non-linear mixing of the cube public variables and the private variables for more than 6 MD6 compression function rounds. This was made possible by zeroing all the MD6 array words whose indexes are listed in the third column of Table 2 and Table 3 (ignoring the special "L" values). As described in the previous section, we set the values of several of the 63 attacker controlled words, excluding A_{65} (from which the cube public variables were chosen), to predefined constants that zero the words specified in the third column. Public variables whose value does not affect the values of the listed MD6 array words were set to zero. We were not able to limit the diffusion of the cube public variables and the private variables as much when all the cube public variable indexes were chosen from words other than A_{65}.

We describe the cube attack results on the keyed MD6 version. The results were obtained by running the preprocessing phase of the cube attack with the special parameters describes above. We found many dense maxterms for 13-round MD6, with associated cubes of size 5. Each of the maxterms passed at least 100 linearity tests, thus the maxterm equations are likely to be correct for most keys. During the online phase, the attacker evaluates the superpolys by summing over the cubes of size 5. This requires a total of about 2^{12} chosen IVs. The total complexity of the attack is thus no more than 2^{12}.

We were able to find many constant superpolys for 14 rounds of MD6, with associated cubes of size 7. However, summing on cubes of size 6 gives superpolys

Table 2. Diffusion of the private variables into the MD6 compression function array in the initial mixing steps. The third column specifies the MD6 array index of the word that is ANDed with the key-dependent array word index in the step number specified by the first column. The output of step i is inserted into A_i. If the key-dependent array word is diffused linearly, then L is written instead. Note that once a dependency of an MD6 array word on the private variables can be eliminated, it does not appear any more as key-dependent (i.e. we assume that this dependency is eliminated by the attacker).

Step	Key-dependent array index	ANDed index
104	15	L
105	16	L
121	104	L
122	105	L
122	104	101
123	105	102
125	104	107
126	105	108
135	104	68
136	105	69
138	121	L
139	122	L
139	121	118
140	122	119
142	121	124
143	122	125
152	121	85
153	122	86
155	138	L
156	139	L
156	138	135
157	139	136
159	138	141
160	139	142
169	138	102
170	139	103
171	104	140
172	105	141
172	155	L
173	156	L
173	155	152
174	156	153
176	155	158
177	156	159
186	155	119
187	156	120
187	121	157
188	122	158

Table 3. Diffusion of A_{65} into the MD6 compression function array in the initial mixing rounds (if the key-dependent array word is diffused linearly, then L is written instead)

Step	A_{65}-dependent array index	Multiplicand index
96	65	29
132	65	101
154	65	L
171	154	L
172	154	151
175	154	157
185	154	118

of high degree in the key bits. In order to further eliminate most (but not all) high degree terms from the superpolys obtained by summing on cubes of size 6, we added more public variable indexes from words other than A_{65}. The best results were obtained by choosing the remaining indexes from A_{32}, A_{33}, A_{49} and A_{50} (which are directly XORed with key bits in steps 121, 122, 138 and 139). Using this approach, we found many dense maxterms for 14-round MD6, with associated cubes of size 15. Some of these results are listed in Table 5 (Appendix A), many more linearly independent maxterms can be easily obtained by choosing other cube indexes from the same words listed in Table 5. During the online phase, the attacker evaluates the superpolys by summing over the cubes of size 15. This requires a total of about 2^{22} chosen IVs. The total complexity of the attack is thus no more than 2^{22}. In fact every IV gives many maxterms, so the required total of chosen IVs is lower than 2^{22}, and the total complexity of the attack is less than 2^{22}.

We were able to find many constant superpolys for 15 rounds of MD6, with associated cubes of size 14. We were not able to find low degree superpolys for 15-round MD6. However, it seems likely that low degree equation for 15-round MD6 can be obtained using approaches similar to the one we used to recover the key for 14-round MD6. Hence we believe that cube attacks can efficiently recover the key for 15-round MD6. Furthermore, we believe that cube key recovery attacks will remain faster than exhaustive search for 18-19 MD6 rounds.

3   Cube Testers

3.1   Definitions

Recall that $\mathcal{F}_n$ denotes the set of all functions mapping $\{0,1\}^n$ to $\{0,1\}$, $n > 0$. For a given n, a *random function* is a random element of $\mathcal{F}_n$ (we have $|\mathcal{F}_n| = 2^{2^n}$). In the ANF of a random function, each monomial (and in particular, the highest degree monomial $x_1 \cdots x_n$) appears with probability $1/2$, hence a random function has maximal degree of n with probability $1/2$. Similarly, it has degree $(n-2)$ or less with probability $1/2^{n+1}$. Note that the explicit description of a random function can be directly expressed as a circuit with, in average, 2^{n-1}

gates (AND and XOR), or as a string of 2^n bits where each bit is the coefficient of a monomial (encoding the truth table also requires 2^n bits, but hides the algebraic structure).

Informally, a *distinguisher* for a family $\mathcal{F} \subsetneq \mathcal{F}_n$ is a procedure that, given a function f randomly sampled from $\mathcal{F}^\star \in \{\mathcal{F}, \mathcal{F}_n\}$, efficiently determines which one of these two families was chosen as $\mathcal{F}^\star$. A family $\mathcal{F}$ is *pseudorandom* if and only if there exists no efficient distinguisher for it. In practice, e.g. for hash functions or ciphers, a family of functions is defined by a k-bit parameter of the function, randomly chosen and unknown to the adversary, and the function is considered broken (or, at least, "nonrandom") if there exists a distinguisher making significantly less than 2^k queries to the function. Note that a distinguisher that runs in exponential time in the key may be considered as "efficient" in practice, e.g. 2^{k-10}.

We would like to stress the terminology difference between a *distinguisher* and the more general detection of *pseudorandomness*, when speaking about cryptographic algorithms; the former denotes a distinguisher (as defined above) where the parameter of the family of functions is the cipher's *key*, and thus can't be modified by the adversary through its queries; the latter considers part of the key as a public input, and assumes as secret an arbitrary subset of the input (including the input bits that are normally public, like IV bits). The detection of nonrandomness thus does not necessarily correspond to a realistic scenario. Note that related-key attacks are captured by neither one of those scenarios.

To distinguish $\mathcal{F} \subsetneq \mathcal{F}_n$ from $\mathcal{F}_n$, cube testers partition the set of public variables $\{x_1, \ldots, x_n\}$ into two complementary subsets:

- *cube variables* (CV)
- *superpoly variables* (SV)

We illustrate these notions with the example from §1.1: recall that, given

$$f(x_1, x_2, x_3, x_4) = x_1 + x_1 x_2 x_3 + x_1 x_2 x_4 + x_3 ,$$

we considered the *cube* $x_1 x_2$ and called $(x_3 + x_4)$ its *superpoly*, because

$$f(x_1, x_2, x_3, x_4) = x_1 + x_1 x_2 (x_3 + x_4) + x_3 .$$

Here the cube variables (CV) are x_1 and x_2, and the superpoly variables (SV) are x_3 and x_4. Therefore, by setting a value to x_3 and x_4, e.g. $x_3 = 0$, $x_4 = 1$, one can compute $(x_3 + x_4) = 1$ by summing $f(x_1, x_2, x_3, x_4)$ for all possibles choices of (x_1, x_2). Note that it is not required for a SV to actually appear in the superpoly of the maxterm. For example, if $f(x_1, x_2, x_3, x_4) = x_1 + x_1 x_2 x_3$, then the superpoly of $x_1 x_2$ is x_3, but the SV's are both x_3 and x_4.

Remark. When f is, for example, a hash function, not all inputs should be considered as variables, and not all Boolean components should be considered as outputs, for the sake of efficiency. For example if f maps 1024 bits to 256 bits, one may choose 20 CV and 10 SV and set a fixed value to the other inputs. These fixed inputs determine the coefficient of each monomial in the ANF with CV and SV as variables. This is similar to the preprocessing phase of key-recovery cube attacks, where one has access to all the input variables. Finally, for the sake of efficiency, one may only evaluate the superpolys for 32 of the 256 Boolean components of the output.

3.2   Examples

Cube testers distinguish a family of functions from random functions by testing a property of the superpoly for a specific choice of CV and SV. This section introduces this idea with simple examples. Consider

$$f(x_1, x_2, x_3, x_4) = x_1 + x_1 x_2 x_3 + x_1 x_2 x_4 + x_3$$

and suppose we choose CV x_3 and x_4 and SV x_1 and x_2, and evaluate the superpoly of $x_3 x_4$:

$$f(x_1, x_2, 0, 0) + f(x_1, x_2, 0, 1) + f(x_1, x_2, 1, 0) + f(x_1, x_2, 1, 1) = 0 ,$$

This yields zero for any $(x_1, x_2) \in \{0, 1\}^2$, i.e. the superpoly of $x_3 x_4$ is zero, i.e. none of the monomials $x_3 x_4$, $x_1 x_3 x_4$, $x_2 x_3 x_4$, or $x_1 x_2 x_3 x_4$ appears in f. In comparison, in a random function the superpoly of $x_3 x_4$ is null with probability only $1/16$, which suggests that f was not chosen at random (indeed, we chose it particularly sparse, for clarity). Generalizing the idea, one can deterministically test whether the superpoly of a given maxterm is constant, and return "random function" if and only if the superpoly is not constant. This is similar to the test used in [10].

Let $f \in \mathcal{F}_n$, $n > 4$. We present a probabilistic test that detects the presence of monomials of the form $x_1 x_2 x_3 x_i \ldots x_j$ (e.g. $x_1 x_2 x_3$, $x_1 x_2 x_3 x_n$, etc.):

1. choose a random value of $(x_4, \ldots, x_n) \in \{0, 1\}^{n-4}$
2. sum $f(x_1, \ldots, x_n)$ over all values of (x_1, x_2, x_3), to get

$$\sum_{(x_1, x_2, x_3) \in \{0,1\}^3} f(x_1, \ldots, x_n) = p(x_4, \ldots, x_n)$$

where p is a polynomial such that

$$f(x_1, \ldots, x_n) = x_1 x_2 x_3 \cdot p(x_4, \ldots, x_n) + q(x_1, \ldots, x_n)$$

where the polynomial q contains no monomial with $x_1 x_2 x_3$ as a factor in its ANF
3. repeat the two previous steps N times, recording the values of $p(x_4, \ldots, x_n)$

If f were a random function, it would contain at least one monomial of the form $x_1 x_2 x_3 x_i \ldots x_j$ with high probability; hence, for a large enough number of repetitions N, one would record at least one nonzero $p(x_4, \ldots, x_n)$ with high probability. However, if no monomial of the form $x_1 x_2 x_3 x_i \ldots x_j$ appears in the ANF, $p(x_4, \ldots, x_n)$ always evaluates to zero.

3.3   Building on Property Testers

Cube testers combine an efficient property tester on the superpoly, which is viewed either as a polynomial or as a mapping, with a statistical decision rule. This section gives a general informal definition of cube testers, starting with basic definitions. A *family tester* for a family of functions $\mathcal{F}$ takes as input a function f of same domain $\mathcal{D}$ and tests if f is close to $\mathcal{F}$, with respect to a bound ϵ on the distance

$$\delta(f, \mathcal{F}) = \min_{g \in \mathcal{F}} \frac{|\{x \in \mathcal{D}, f(x) \neq g(x)\}|}{|\mathcal{D}|} .$$

The tester accepts if $\delta(f, \mathcal{F}) = 0$, rejects with high probability if f and $\mathcal{F}$ are not ϵ-close, and behaves arbitrarily otherwise. Such a test captures the notion of property-testing, when a property is defined by belonging to a family of functions $\mathcal{P}$; a *property tester* is thus a family tester for a property $\mathcal{P}$.

Suppose one wishes to distinguish a family $\mathcal{F} \subsetneq \mathcal{F}_n$ from $\mathcal{F}_n$, i.e., given a random $f \in \mathcal{F}^\star$, to determine whether $\mathcal{F}^\star$ is $\mathcal{F}$ or $\mathcal{F}_n$ (for example, in Trivium, $\mathcal{F}$ may be a superpoly with respect to CV and SV in the IV bits, such that each $f \in \mathcal{F}$ is computed with a distinct key). Then if $\mathcal{F}$ is efficiently testable (see [26,14]), then one can use directly a family tester for $\mathcal{F}$ on f to distinguish it from a random function.

Cube testers detect nonrandomness by applying property testers to super-polys: informally, as soon as a superpoly has some "unexpected" property (that is, is anormally structured) it is identified as nonrandom. Given a testable property $\mathcal{P} \subsetneq \mathcal{F}_n$, cube testers run a tester for $\mathcal{P}$ on the superpoly function f, and use a statistical decision rule to return either "random" or "nonrandom". The decision rule depends on the probabilities $|\mathcal{P}|/|\mathcal{F}_n|$ and $|\mathcal{P} \cap \mathcal{F}|/|\mathcal{F}|$ and on a margin of error chosen by the attacker. Roughly speaking, a family $\mathcal{F}$ will be distinguishable from $\mathcal{F}_n$ using the property $\mathcal{P}$ if

$$\left| \frac{|\mathcal{P}|}{|\mathcal{F}_n|} - \frac{|\mathcal{P} \cap \mathcal{F}|}{|\mathcal{F}|} \right|$$

is non-negligible. That is, the tester will determine whether f is significantly closer to $\mathcal{P}$ than a random function. Note that the dichotomy between structure (e.g. testable properties) and randomness has been studied in [30].

3.4   Examples of Testable Properties

Below, we give examples of efficiently testable properties of the superpoly, which can be used to build cube testers (see [14] for a general characterization of efficiently testable properties). We let C be the size of CV, and S be the size of SV; the complexity is given as the number of evaluations of the tested function f. Note that each query of the tester to the superpoly requires 2^C queries to the target cryptographic function. The complexity of any property tester is thus, even in the best case, exponential in the number of CV.

Balance. A random function is expected to contain as many zeroes as ones in its truth table. Superpolys that have a strongly unbalanced truth table can thus be distinguished from random polynomials, by testing whether it evaluates as often to one as to zero, either deterministically (by evaluating the superpoly for each possible input), or probabilistically (over some random subset of the SV). For example, if CV are $x_1, \ldots, x_C$ and SV are $x_{C+1}, \ldots, x_n$, the deterministic balance test is

1. $c \leftarrow 0$
2. **for** all values of $(x_{C+1}, \ldots, x_n)$
3.   compute

$$p(x_{C+1}, \ldots, x_n) = \sum_{(x_1, \ldots, x_C)} f(x_1, \ldots, x_n) \in \{0, 1\}$$

4. $c \leftarrow c + p(x_{C+1}, \ldots, x_n)$
5. **return** $\mathcal{D}(c) \in \{0, 1\}$

where $\mathcal{D}$ is some decision rule. A probabilistic version of the test makes $N < 2^S$ iterations, for random distinct values of $(x_{C+1}, \ldots, x_n)$. Complexity is respectively 2^n and $N \cdot 2^C$.

Constantness. A particular case of balance test considers the "constantness" property, i.e. whether the superpoly defines a constant function; that is, it detects either that f has maximal degree strictly less than C (null superpoly), or that f has maximal degree exactly C (superpoly equals the constant 1), or that f has degree strictly greater than C (non-constant superpoly). This is equivalent to the maximal degree monomial test used in [10], used to detect nonrandomness in 736-round Trivium.

Low Degree. A random superpoly has degree at least $(S - 1)$ with high probability. Cryptographic functions that rely on a low-degree function, however, are likely to have superpolys of low degree. Because it closely relates to probabilistically checkable proofs and to error-correcting codes, low-degree testing has been well studied; the most relevant results to our concerns are the tests for Boolean functions in [1, 28]. The test by Alon et al. [1], for a given degree d, queries the function at about $d \cdot 4^d$ points and always accepts if the ANF of the function has degree at most k, otherwise it rejects with some bounded error probability. Note that, contrary to the method of ANF reconstruction (exponential in S), the complexity of this algorithm is *independent of the number of variables*. Hence, cube testers based on this low-degree test have complexity which is independent of the number of SV's.

Presence of Linear Variables. This is a particular case of the low-degree test, for degree $d = 1$ and a single variable. Indeed, the ANF of a random function contains a given variable in at least one monomial of degree at least two with probability close to 1. One can thus test whether a given superpoly variable appears only linearly in the superpoly, e.g. for x_1 using the following test similar to that introduced in [5]:

1. pick random $(x_2, \ldots, x_S)$
2. **if** $p(0, x_2, \ldots, x_S) = p(1, x_2, \ldots, x_S)$
3.   **return** nonlinear
4. repeat steps 1 to 3 N times
5. **return** linear

This test answers correctly with probability about $1 - 2^{-N}$, and computes $N \cdot 2^{C+1}$ times the function f. If, say, a stream cipher is shown to have an IV bit linear with respect to a set of CV in the IV, independently of the choice of the key, then it directly gives a distinguisher.

Presence of Neutral Variables. Dually to the above linearity test, one can test whether a SV is neutral in the superpoly, that is, whether it appears in at least one monomial. For example, the following algorithm tests the neutrality of x_1, for $N \le 2^{S-1}$:

1. pick random $(x_2, \ldots, x_S)$
2. **if** $p(0, x_2, \ldots, x_S) \neq p(1, x_2, \ldots, x_S)$
3.   **return** not neutral
4. repeat steps 1 to 3 N times
5. **return** neutral

This test answers correctly with probability about $1 - 2^{-N}$ and runs in time $N \cdot 2^C$. For example, if x_1, x_2, x_3 are the CV and x_4, x_5, x_6 the SV, then x_6 is neutral with respect to $x_1 x_2 x_3$ if the superpoly $p(x_4, x_5, x_6)$ satisfies $p(x_4, x_5, 0) = p(x_4, x_5, 1)$ for all values of (x_4, x_5). A similar test was implicitly used in [12], via the computation of a *neutrality measure*.

Remarks. Except low degree and constantness, the above properties do not require the superpoly to have a low degree to be tested. For example if the maxterm $x_1 x_2$ has the degree-5 superpoly

$$x_3 x_5 x_6 + x_3 x_5 x_6 x_7 x_8 + x_5 x_8 + x_9 ,$$

then one can distinguish this superpoly from a random one either by detecting the linearity of x_9 or the neutrality of x_4, with a cost independent on the degree. In comparison, the cube tester suggested in [9] required the degree to be bounded by d such that 2^d is feasible.

Note that the cost of detecting the property during the preprocessing is larger than the cost of the on-line phase of the attack, given the knowledge of the property. For example, testing that x_1 is a neutral variable requires about $N \cdot 2^C$ queries to the function, but once this property is known, 2^C queries are sufficient to distinguish the function from a random one with high probability.

Finally, note that tests based on the nonrandom distribution of the monomials [11, 27, 21] are not captured by our definition of cube testers, which focus on high-degree terms. Although, in principle, there exist cases where the former tests would succeed while cube testers would fail, in practice a weak distribution of lower-degree monomials rarely comes with a good distribution of high-degree ones, as results in [10] and of ourselves suggest.

4   Cube Testers on MD6

We use cube testers to detect nonrandom properties in reduced-round versions of the MD6 compression function, which maps the 64-bit words $A_0, \ldots, A_{88}$ to $A_{16r+73}, \ldots, A_{16r+88}$, with r the number of rounds. From the compression function $f : \{0,1\}^{64 \times 89} \mapsto \{0,1\}^{64 \times 16}$, our testers consider families of functions $\{f_m\}$ where a random $f_i : \{0,1\}^{64 \times 89-k} \mapsto \{0,1\}^{64 \times 16}$ has k input bits set to a random k-bit string. The attacker can thus query f_i, for a randomly chosen key i, on $(64 \times 89 - k)$-bit inputs.

The key observations leading to our improved attacks on MD6 are that:

1. input words appear either linearly (as A_{i-89} or A_{i-17}) or nonlinearly (as $A_{18}, A_{21}, A_{31},$ or A_{67}) within a step
2. words $A_0, \ldots, A_{21}$ are input once, $A_{22}, \ldots, A_{57}$ are input twice, $A_{58}, \ldots, A_{67}$ are input three times, A_{68}, A_{69}, A_{70} four times, A_{71} five times, and $A_{72}, \ldots,$ A_{88} six times
3. all input words appear linearly at least once ($A_0, \ldots, A_{71}$), and at most twice ($A_{72}, \ldots, A_{88}$)
4. A_{57} is the last word input (at step 124, i.e. after 2 rounds plus 3 steps)
5. A_{71} is the last word input linearly (at step 160, i.e. after 4 rounds plus 7 steps)
6. differences in a word input nonlinearly are "absorbed" if the second operand is zero (e.g. $A_{i-18} \wedge A_{i-21} = 0$ if A_{i-18} is zero, for any value of A_{i-21})

Based on the above observations, the first attack (A) makes only black-box queries to the function. The second attack (B) can be seen as a kind of related-key attack, and is more complex and more powerful. Our best attacks, in terms of efficiency and number of rounds broken, were obtained by testing the *balance* of superpolys.

4.1   Attack A

This attack considers CV, SV, and secret bits in A_{71}: the MSB's of A_{71} contain the CV, the LSB's contain the 30 secret bits, and the 4 bits "in the middle" are the SV. The other bits in A_{71} are set to zero. To minimize the density and the degree of the ANF, we set $A_i = S_i$ for $i = 0, \ldots, 57$ in order to eliminate the constants S_i from the expressions, and set $A_i = 0$ for $i = 58, \ldots, 88$ in order to eliminate the quadratic terms by "absorbing" the nonzero $A_{22}, \ldots, A_{57}$ through AND's with zero values.

The attack exploits the fact that A_{71} is the last word input linearly. We set initial conditions on the message such that modifications in A_{71} are only effective at step 160, and so CV and SV are only introduced (linearly) at step 160: in order to absorb A_{71} before step 160, one needs $A_{68} = A_{74} = A_{35} = A_{107} = 0$, respectively for steps 89, 92, 102, and 138.

Given the setup above, the attack evaluates the balance of the superpoly for each of the 1024 output components, in order to identify superpolys that are constant for a large majority of inputs (SV). These superpolys may be either constants, or unbalanced nonlinear functions. Results for reduced and modified MD6 are given in subsequent sections.

4.2   Attack B

This attack considers CV, SV, and secret bits in A_{54}, at the same positions as in Attack A. Other input words are set by default to S_i for $A_0, \ldots, A_{47}$, and to zero otherwise.

The attack exploits the fact that A_{54} and A_{71} are input linearly only once, and that both directly interact with A_{143}. We set initial conditions on the message such that CV and SV are only effective at step 232. Here are the details of this attack:

- step 143: input variables are transfered linearly to A_{143}
- step 160: A_{143} is input linearly; to cancel it, and thus to avoid the introduction of the CV and SV in the ANF, one needs $A_{71} = S_{160} \oplus A_{143}$
- step 92: A_{71} is input nonlinearly; to cancel it, in order to make A_{138} independent of A_{143}, we need $A_{74} = 0$
- step 138: A_{71} is input nonlinearly; to cancel it, one needs $A_{107} = 0$
- step 161: A_{143} is input nonlinearly; to cancel it, one needs $A_{140} = 0$
- step 164: A_{143} is input nonlinearly; to cancel it, one needs $A_{146} = 0$
- step 174: A_{143} is input nonlinearly; to cancel it, one needs $A_{107} = 0$ (as for step 138)
- step 210: A_{143} is input nonlinearly; to cancel it, one needs $A_{179} = 0$
- step 232: A_{143} is input linearly, and introduces the CV and SV linearly into the ANF

To satisfy the above conditions, one has to choose suitable values of A_1, A_{18}, A_{51}, A_{57}, A_{74}. These values are constants that do not depend on the input in A_{54}.

Given the setup above, the attack evaluates the balance of the superpoly for each of the 1024 output components, in order to identify superpolys that are constant for large majority of inputs (SV). Results for reduced and modified MD6 are given in §4.3.

4.3   Results

In this subsection we report the results we obtained by applying attacks A and B to reduced versions of MD6, and to a modified version of MD6 that sets all the constants S_i to zero. Recall that by using C CV's, the complexity of the attack is about 2^C computations of the function. We report results for attacks using at most 20 CV (i.e. doable in less than a minute on a single PC):

- with *attack A*, we observed strong imbalance after 15 rounds, using 19 CV. More precisely, the Boolean components corresponding to the output bits in A_{317} and A_{325} all have (almost) constant superpoly. When all the S_i constants are set to 0, we observed that all the outputs in A_{1039} and A_{1047} have (almost) constant superpoly, i.e. we can break 60 rounds of this modified MD6 version using only 14 CV's.

– with *attack B*, we observed strong imbalance after 18 rounds, using 17 CV's. The Boolean components corresponding to the output bits in A_{368} and A_{376} all have (almost) constant superpoly. When $S_i = 0$, using 10 CV's, one finds that all outputs in A_{1114} and A_{1122} have (almost) constant superpoly, i.e. one breaks 65 rounds. Pushing the attack further, one can detect nonrandomness after 66 rounds, using 24 CV's.

The difference of results between the original MD6 and the modified case in which $S_i = 0$ comes from the fact that a zero S_i makes it possible to keep a sparse state during many rounds, whereas a nonzero S_i forces the introduction of nonzero bits in the early steps, thereby quickly increasing the density of the implicit polynomials, which indirectly facilitates the creation of high degree monomials.

5   Cube Testers on Trivium

Observations in [9, Tables 1,2,3] suggest nonrandomness properties detectable in time about 2^{12} after 684 rounds, in time 2^{24} after 747 rounds, and in time 2^{30} after 774 rounds. However, a distinguisher cannot be directly derived because the SV used are in the key, and thus cannot be chosen by the attacker in an attack where the key is fixed.

5.1   Setup

We consider families of functions defined by the secret key of the cipher, and where the IV corresponds to public variables. We first used the 23-variable index sets identified in [8, Table 2]; even though we have not tested all entries, we obtained the best results using the IV bits (starting from zero)

$$\{3, 4, 6, 9, 13, 17, 18, 21, 26, 28, 32, 34, 37, 41, 47, 49, 52, 58, 59, 65, 70, 76, 78\} \ .$$

For this choice of CV, we choose 5 SV, either

– in the IV, at positions $0, 1, 2, 35, 44$ (to have a distinguisher), or
– in the key, at positions $0, 1, 2, 3, 4$ (to detect nonrandomness)

For experiments with 30 CV, we use another index set discovered in [8]:

$$\{1, 3, 6, 12, 14, 18, 22, 23, 24, 26, 30, 32, 33, 35, 36, 39, 40, 44, 47, 49, 50, 53, 59, 60, 61, 66, 68, 69, 72, 75\} \ .$$

IV bits that are neither CV nor SV are set to zero, in order to minimize the degree and the density of the polynomials generated during the first few initialization steps. Contrary to MD6, we obtain the best results on Trivium by testing the presence of *neutral variables*. We look for neutral variables either for a random key, or for the special case of the zero key, which is significantly weaker with respect to cube testers.

In addition to the cubes identified in [8, Table 2], we were able to further improve the results by applying cube testers on carefully chosen cubes, where the indexes are uniformly spread (the distance between neighbors is at least 2). These cubes exploit the internal structure of Trivium, where non linear operations are only performed on consecutive cells. The best results were obtained using the cubes below:

$$\{0, 3, 6, 9, 12, 15, 18, 21, 24, 27, 33, 36, 39, 42, 45, 48, 51, 60, 63, 66, 69, 72, 75, 79\}$$
$$\{0, 3, 6, 9, 12, 15, 18, 21, 24, 27, 30, 33, 36, 39, 42, 45, 48, 51, 54, 57, 60, 63, 66, 69, 72, 75, 79\}$$
$$\{0, 2, 4, 6, 8, 10, 12, 14, 16, 18, 21, 24, 27, 30, 33, 36, 39, 42, 45, 48, 51, 54, 57, 60, 63, 66, 69, 72, 75, 79\}$$

5.2   Results

We obtained the following results, by testing the *neutrality* of the SV in the superpoly:

- with 23 CV, and SV in the IV, we found a distinguisher on up to 749 rounds (runtime 2^{23}); SV 0, 1, 2, and 3 are neutral after 749 initialization rounds. Using the zero key, neutral variables are observed after 755 rounds (SV 0, 1 are neutral).
- with 23 CV, and SV in the key, we observed nonrandomness after 758 initialization rounds (SV 1, 2, 3 are neutral). Using the zero key, nonrandomness was observed after 761 rounds (SV 0 is neutral).
- with 30 CV, and SV in the key, we observed nonrandomness after 772 initialization rounds (SV 0, 2, 4 are neutral). Using the zero key, nonrandomness was observed after 782 rounds (SV 2, 3, 4 are neutral).

With the the new chosen cubes we obtain the following results:

- with 24 CV, we observe that the resultant superpoly after 772 initialization rounds is constant, hence we found a distinguisher on up to 772 rounds. Using the neutrality test, for the zero key, we detected nonrandomness over up to 842 rounds (the 4 first key bits are neutral).
- with 27 CV, we observe that the resultant superpoly after 785 initialization rounds is constant, hence we found a distinguisher on up to 785 rounds. Using the neutrality test, for the zero key, we detected nonrandomness over up to 885 rounds (bits 0, 3, and 4 of the key are neutral).
- with 30 CV, we observe that the resultant superpoly after 790 initialization rounds is constant, hence we found a distinguisher for Trivium with up to 790 rounds.

Better results are obtained when the SV's are in the key, not the IV; this is because the initialization algorithm of Trivium puts the key and the IV into two different registers, which make dependency between bits in a same register stronger than between bits in different registers.

In comparison, [10], testing the *constantness* of the superpoly, reached 736 rounds with 33 CV. The observations in [8], obtained by testing the *linearity* of SV in the key, lead to detectable nonrandomness on 748 rounds with 23 CV, and on 771 rounds with 30 CV.

6   Conclusions

We applied cube attacks to the reduced-round MD6 compression function, and could recover a full 128-bit key on 14-round MD6 with a very practical complexity of 2^{22} evaluations. This outperforms all the attacks obtained by the designers of MD6.

Then we introduced the notion of cube tester, based on cube attacks and on property-testers for Boolean functions. Cube testers can be used to mount distinguishers or to simply detect nonrandomness in cryptographic algorithms. Cube testers do not require large precomputations, and can even work for high degree polynomials (provided they have some "unexpected" testable property).

Using cube testers, we detected nonrandomness properties after 18 rounds of the MD6 compression function (the proposed instances have at least 80 rounds). Based on observations in [9], we extended the attacks on Trivium a few more rounds, giving experimentally verified attacks on reduced variants with up to 790 rounds, and detection of nonrandomness on 885 rounds (against 1152 in the full version, and 771 for the best previous attack).

Our results leave several issues open:

1. So far cube attacks have resulted from empirical observations, so that one could only assess the existence of feasible attacks. However, if one could upper-bound the degree of some Boolean component (e.g. of MD6 or Trivium) after a higher number of rounds, then one could predict the existence of observable nonrandomness (and one may build distinguishers based on low-degree tests [1]). The problem is closely related to that of bounding the degree of a nonlinear recursive Boolean sequence which, to the best of our knowledge, has remained unsolved.

2. Low-degree tests may be used for purposes other than detecting nonrandomness. For example, key-recovery cube attacks may be optimized by exploiting low-degree tests, to discover low-degree superpolys, and then reconstruct them. Also, low-degree tests for general fields [13] may be applicable to hash functions based on multivariate systems [4], which remain unbroken over fields larger than $GF(2)$ [2].

3. Our attacks on MD6 detect nonrandomness of reduced versions of the compression function, and even recover a 128-bit key. It would be interesting to extend these attacks to a more realistic scenario, e.g. that would be applicable to the MD6 operation mode, and/or to recover larger keys.

4. One may investigate the existence of cube testers on other primitives that are based on low-degree functions, like RadioGatún, Panama, the stream cipher MICKEY, and on the SHA-3 submissions ESSENCE [18], and Keccak [3]. We propose to use cube attacks and cube testers as a benchmark for evaluating the algebraic strength of primitives based on a low-degree component, and as a reference for choosing the number of rounds. Our preliminary results on Grain-128 outperform all previous attacks, but will be reported later since they are still work in progress.

References

1. Alon, N., Kaufman, T., Krivelevich, M., Litsyn, S., Ron, D.: Testing low-degree polynomials over GF(2). In: Arora, S., Jansen, K., Rolim, J.D.P., Sahai, A. (eds.) RANDOM 2003 and APPROX 2003. LNCS, vol. 2764, pp. 188–199. Springer, Heidelberg (2003)
2. Aumasson, J.-P., Meier, W.: Analysis of multivariate hash functions. In: Nam, K.-H., Rhee, G. (eds.) ICISC 2007. LNCS, vol. 4817, pp. 309–323. Springer, Heidelberg (2007)
3. Bertoni, G., Daemen, J., Peeters, M., Van Assche, G.: Keccak specifications. Submission to NIST 2008 (2008), http://keccak.noekeon.org/
4. Billet, O., Robshaw, M.J.B., Peyrin, T.: On building hash functions from multivariate quadratic equations. In: Pieprzyk, J., Ghodosi, H., Dawson, E. (eds.) ACISP 2007. LNCS, vol. 4586, pp. 82–95. Springer, Heidelberg (2007)
5. Blum, M., Luby, M., Rubinfeld, R.: Self-testing/correcting with applications to numerical problems. In: STOC, pp. 73–83. ACM, New York (1990)
6. De Cannière, C., Preneel, B.: Trivium. In: Robshaw, M.J.B., Billet, O. (eds.) New Stream Cipher Designs. LNCS, vol. 4986, pp. 244–266. Springer, Heidelberg (2008)
7. Crutchfield, C.Y.: Security proofs for the MD6 hash function mode of operation. Master's thesis, Massachusetts Institute of Technology (2008)
8. Dinur, I., Shamir, A.: Cube attacks on tweakable black box polynomials. IACR ePrint Archive, Report 2008/385, version 20080914:160327 (2008), http://eprint.iacr.org/2008/385
9. Dinur, I., Shamir, A.: Cube attacks on tweakable black box polynomials. In: Joux, A. (ed.) EUROCRYPT 2009. LNCS, vol. 5479, pp. 278–299. Springer, Heidelberg (2009); see also [8]
10. Englund, H., Johansson, T., Turan, M.S.: A framework for chosen IV statistical analysis of stream ciphers. In: Srinathan, K., Rangan, C.P., Yung, M. (eds.) INDOCRYPT 2007. LNCS, vol. 4859, pp. 268–281. Springer, Heidelberg (2007)
11. Filiol, E.: A new statistical testing for symmetric ciphers and hash functions. In: Deng, R.H., Qing, S., Bao, F., Zhou, J. (eds.) ICICS 2002. LNCS, vol. 2513, pp. 342–353. Springer, Heidelberg (2002)
12. Fischer, S., Khazaei, S., Meier, W.: Chosen IV statistical analysis for key recovery attacks on stream ciphers. In: Vaudenay, S. (ed.) AFRICACRYPT 2008. LNCS, vol. 5023, pp. 236–245. Springer, Heidelberg (2008)
13. Kaufman, T., Ron, D.: Testing polynomials over general fields. In: FOCS, pp. 413–422. IEEE Computer Society, Los Alamitos (2004)
14. Kaufman, T., Sudan, M.: Algebraic property testing: the role of invariance. In: Ladner, R.E., Dwork, C. (eds.) STOC, pp. 403–412. ACM, New York (2008)
15. Khazaei, S., Meier, W.: New directions in cryptanalysis of self-synchronizing stream ciphers. In: Chowdhury, D.R., Rijmen, V., Das, A. (eds.) INDOCRYPT 2008. LNCS, vol. 5365, pp. 15–26. Springer, Heidelberg (2008)
16. Knudsen, L.R.: Truncated and higher order differentials. In: Preneel, B. (ed.) FSE 1994. LNCS, vol. 1008, pp. 196–211. Springer, Heidelberg (1995)
17. Lucks, S.: The saturation attack - a bait for Twofish. In: Matsui, M. (ed.) FSE 2001. LNCS, vol. 2355, pp. 1–15. Springer, Heidelberg (2001)
18. Martin, J.W.: ESSENCE: A candidate hashing algorithm for the NIST competition. Submission to NIST (2008)
19. Maximov, A., Biryukov, A.: Two trivial attacks on Trivium. In: Adams, C.M., Miri, A., Wiener, M.J. (eds.) SAC 2007. LNCS, vol. 4876, pp. 36–55. Springer, Heidelberg (2007)

20. McDonald, C., Charnes, C., Pieprzyk, J.: Attacking Bivium with MiniSat. eS-TREAM, ECRYPT Stream Cipher Project, Report 2007/040 (2007)
21. O'Neil, S.: Algebraic structure defectoscopy. IACR ePrint Archive, Report 2007/378 (2007), http://eprint.iacr.org/2007/378
22. Pasalic, E.: Transforming chosen iv attack into a key differential attack: how to break TRIVIUM and similar designs. IACR ePrint Archive, Report 2008/443 (2008), http://eprint.iacr.org/2008/443
23. Raddum, H.: Cryptanalytic results on Trivium. eSTREAM, ECRYPT Stream Cipher Project, Report 2005/001 (2006)
24. Rivest, R.L.: The MD6 hash function. Invited talk at CRYPTO 2008 (2008), http://people.csail.mit.edu/rivest/
25. Rivest, R.L., Agre, B., Bailey, D.V., Crutchfield, C., Dodis, Y., Fleming, K.E., Khan, A., Krishnamurthy, J., Lin, Y., Reyzin, L., Shen, E., Sukha, J., Sutherland, D., Tromer, E., Yin, Y.L.: The MD6 hash function – a proposal to NIST for SHA-3, http://groups.csail.mit.edu/cis/md6/
26. Rubinfeld, R., Sudan, M.: Robust characterizations of polynomials with applications to program testing. SIAM J. Comput. 25(2), 252–271 (1996)
27. Saarinen, M.-J.O.: Chosen-IV statistical attacks on eStream ciphers. In: Malek, M., Fernández-Medina, E., Hernando, J. (eds.) SECRYPT, pp. 260–266. INSTICC Press (2006)
28. Samorodnitsky, A.: Low-degree tests at large distances. In: Johnson, D.S., Feige, U. (eds.) STOC, pp. 506–515. ACM, New York (2007)
29. Shamir, A.: How to solve it: New techniques in algebraic cryptanalysis. Invited talk at CRYPTO 2008 (2008)
30. Tao, T.: The dichotomy between structure and randomness, arithmetic progressions, and the primes. In: International Congress of Mathematicians, pp. 581–608. European Mathematical Society (2006)
31. Turan, M.S., Kara, O.: Linear approximations for 2-round Trivium. eSTREAM, ECRYPT Stream Cipher Project, Report 2007/008 (2007)
32. Vielhaber, M.: Breaking ONE.FIVIUM by AIDA an algebraic IV differential attack. IACR ePrint Archive, Report 2007/413 (2007), http://eprint.iacr.org/2007/413

A   Details on MD6

The word S_i is a round-dependent constant: during the first round (i.e., the first 16 steps) $S_i = $ 0123456789abcdef, then at each new round it is updated as

$$S_i \leftarrow (S_0 \ll 1) \oplus (S_0 \gg 63) \oplus (S_{i-1} \wedge \text{7311c2812425cfa}).$$

The shift distances r_i and ℓ_i are step-dependent constants, see Table 4.

Table 4. Distances of the shift operators used in MD6, as function of the step index within a round

Step	0	1	2	3	4	5	6	7	8	9	10	11	12	13	14	15
r_i	10	5	13	10	11	12	2	7	14	15	7	13	11	7	6	12
ℓ_i	11	24	9	16	15	9	27	15	6	2	29	8	15	5	31	9

The number of rounds r depends on the digest size: for a d-bit digest, MD6 makes $40 + d/4$ rounds.

Table 5. Examples of maxterm equations for 14-round MD6, with respect specified cube are listed

Maxterm equation	Output index
$A_{15}^0 + A_{15}^1 + A_{15}^3 + A_{15}^4 + A_{15}^6 + A_{15}^8 + A_{15}^9 + A_{15}^{14} + A_{15}^{20} + A_{15}^{21}$ $+A_{15}^{22} + A_{15}^{28} + A_{15}^{28} + A_{15}^{32} + A_{15}^{37} + A_{15}^{38} + A_{15}^{40} + A_{15}^{41} + A_{15}^{43} + A_{15}^{44}$ $+A_{15}^{47} + A_{15}^{48} + A_{15}^{49} + A_{15}^{50} + A_{15}^{56} + A_{15}^{58} + A_{15}^{60} + A_{15}^{61} + A_{15}^{62} + A_{15}^{63}$ $+A_{16}^1 + A_{16}^2 + A_{16}^3 + A_{16}^4 + A_{16}^5 + A_{16}^{10} + A_{16}^{11} + A_{16}^{12} + A_{16}^{13} + A_{16}^{15}$ $+A_{16}^{16} + A_{16}^{17} + A_{16}^{19} + A_{16}^{21} + A_{16}^{22} + A_{16}^{24} + A_{16}^{25} + A_{16}^{27} + A_{16}^{28} + A_{16}^{29}$ $+A_{16}^{31} + A_{16}^{32} + A_{16}^{36} + A_{16}^{37} + A_{16}^{38} + A_{16}^{39} + A_{16}^{43} + A_{16}^{44} + A_{16}^{48} + A_{16}^{49}$ $+A_{16}^{50} + A_{16}^{52} + A_{16}^{53} + A_{16}^{55} + A_{16}^{57} + A_{16}^{60} + A_{16}^{61} + A_{16}^{63} + A_{16}^8 + 1$	O_0^0
$A_{15}^0 + A_{15}^1 + A_{15}^3 + A_{15}^6 + A_{15}^8 + A_{15}^{10} + A_{15}^{11} + A_{15}^{14} + A_{15}^{16} + A_{15}^{21}$ $+A_{15}^{22} + A_{15}^{27} + A_{15}^{28} + A_{15}^{32} + A_{15}^{34} + A_{15}^{35} + A_{15}^{36} + A_{15}^{37} + A_{15}^{44} + A_{15}^{45}$ $+A_{15}^{48} + A_{15}^{50} + A_{15}^{54} + A_{15}^{55} + A_{15}^{57} + A_{15}^{58} + A_{15}^{59} + A_{15}^{60} + A_{15}^{63} + A_{16}^0$ $+A_{16}^2 + A_{16}^5 + A_{16}^6 + A_{16}^7 + A_{16}^9 + A_{16}^{10} + A_{16}^{11} + A_{16}^{13} + A_{16}^{15} + A_{16}^{16}$ $+A_{16}^{18} + A_{16}^{19} + A_{16}^{20} + A_{16}^{21} + A_{16}^{23} + A_{16}^{30} + A_{16}^{35} + A_{16}^{36} + A_{16}^{39} + A_{16}^{42}$ $+A_{16}^{43} + A_{16}^{44} + A_{16}^{47} + A_{16}^{48} + A_{16}^{49} + A_{16}^{50} + A_{16}^{51} + A_{16}^{53} + A_{16}^{59} + A_{16}^{61}$ $+A_{16}^{50} + A_{16}^{51} + A_{16}^{53} + A_{16}^{59} + A_{16}^{61} + 1$	O_0^1

An Efficient State Recovery Attack on X-FCSR-256

Paul Stankovski, Martin Hell, and Thomas Johansson

Dept. of Electrical and Information Technology, Lund University,
P.O. Box 118, 221 00 Lund, Sweden

Abstract. We describe a state recovery attack on the X-FCSR-256 stream cipher of total complexity at most $2^{57.6}$. This complexity is achievable by requiring $2^{49.3}$ output blocks with an amortized calculation effort of at most $2^{8.3}$ table lookups per output block using no more than 2^{33} table entries of precomputational storage.

Keywords: stream cipher, FCSR, X-FCSR, cryptanalysis, state recovery.

1 Introduction

A common building block in stream ciphers is the Linear Feedback Shift Register (LFSR). The bit sequence produced by an LFSR has several cryptographically interesting properties, such as long period, low autocorrelation and balancedness. LFSRs are inherently linear, so additional building blocks are needed in order to introduce nonlinearity. A Feedback with Carry Shift Register (FCSR) is an alternative construction, similar to an LFSR, but with a distinguishing feature, namely that the update of the register is in itself nonlinear. The idea of using FCSRs to generate sequences for cryptographic applications was initially proposed by Klapper and Goresky in [8].

Recently, we have seen several new constructions based on the concept of FCSRs. The class of F-FCSRs, Filtered FCSRs, was proposed by Arnault and Berger in [1]. These constructions were cryptanalyzed in [7], using a weakness in the initialization function. Also a time/memory tradeoff attack was demonstrated in the same paper.

Another similar construction targeting hardware environments is F-FCSR-H, which was submitted to the eSTREAM project [4]. F-FCSR-H was later updated to F-FCSR-H v2 because of a weakness demonstrated in [6]. F-FCSR-H v2 was one of the four ciphers targeting hardware that were selected for the final portfolio at the end of the eSTREAM project. Inspired by the success, Arnault, Berger, Lauradoux and Minier presented a new construction at Indocrypt 2007, now targeting software implementations. It is named X-FCSR [3]. The main idea was to use two FCSRs instead of one, and to also include an additional nonlinear extraction function inspired by the Rijndael round function. Adding this would allow more output bits per register update and thus increase throughput significantly. Two versions, X-FCSR-256 and X-FCSR-128,

O. Dunkelman (Ed.): FSE 2009, LNCS 5665, pp. 23–37, 2009.

were defined producing 256 and 128 bits per register update, respectively. According to the specification X-FCSR-256 runs at 6.5 cycles/byte and X-FCSR-128 runs at 8.2 cycles/byte. As this is comparable to the fastest known stream ciphers, it makes them very interesting in software environments. For the security of X-FCSR-256 and X-FCSR-128 we note that there have been no published attacks faster than exhaustive key search.

In [5] a new property inside the FCSR was discovered, namely that the update was sometimes temporarily linear for a number of clocks. This resulted in a very efficient attack on F-FCSR-H v2 and led to its removal from the eSTREAM portfolio.

In this paper we present a state recovery attack on X-FCSR-256. We use the observation in [5]. The fact that two registers are used, together with the extraction function, makes it impossible to immediately use this observation to break the cipher. However, several additional non-trivial observations will allow a successful cryptanalysis. The keystream is produced using state variables 16 time instances apart. By considering consecutive output blocks, and assuming that the update is linear, we are able to partly remove the dependency of several state variables. A careful analysis of the extraction function then allows us to treat parts of the state independently and brute force these parts separately, leading to an efficient state recovery attack. It is shown that the state can be recovered using $2^{49.3}$ keystream output blocks and a computational complexity of $2^{8.3}$ table lookups per output block. Note that table lookup operations are *much* cheaper than testing a single key.

The paper is organized as follows. In Section 2 we give an overview of the FCSR construction in general and the X-FCSR-256 stream cipher in particular. In Section 3 we describe the different parts of the attack. Each part of the attack is described in a separate subsection and in order to simplify the description we will deliberately base the attack on assumptions and methods that are not optimal for the cryptanalyst. Then, additional observations and more efficient algorithms are discussed in Section 4, leading to a more efficient attack. Finally, some concluding remarks are given in Section 5.

2 Background

This section will review the necessary prerequisites for understanding the details of the attack. FCSRs are presented separately as they are used as core components of the X-FCSR-256 stream cipher. The X-FCSR-256 stream cipher itself is outlined in sufficient detail for understanding the presented attack. For remaining details, the reader is referred to the specification [3].

2.1 Recalling the FCSR Automaton

An FCSR is a device that computes the binary expansion of a 2-adic number p/q, where p and q are some integers, with q odd. For simplicity one may assume that $q < 0 < p < |q|$. Following the notation from [2], the size n of the FCSR is the bitlength of $|q|$ less one. In stream ciphers, p usually depends on the secret key and the IV, and q is a public parameter. The choice of q induces a number of FCSR properties, the most important one being that it completely determines the length of the period T of the keystream.

The FCSR automaton as described in [2] efficiently implements generation of a 2-adic expansion sequence. It contains two registers, a main register M and a carries register C. The main register M contains n cells. Let $M = (m_{n-1}, m_{n-2}, \ldots, m_1, m_0)$ and associate M to the integer $M = \sum_{i=0}^{n-1} m_i \cdot 2^i$.

Let the binary representation of the positive integer $d = (1 + |q|)/2$ be given by $d = \sum_{i=0}^{n-1} d_i \cdot 2^i$. The carries register contains l active cells, $l + 1$ being the number of nonzero binary digits d_i in d. The active carry cells are the ones in the interval $0 \le i \le n - 2$ for which $d_i = 1$, and d_{n-1} must always be set.

Write the carries register as $C = (c_{n-2}, c_{n-3}, \ldots, c_1, c_0)$ and associate it to the integer $C = \sum_{i=0}^{n-2} c_i \cdot 2^i$. Note that l of the bits in C are active and the remaining ones are set to zero.

Representing the integer p as $\sum_{i=0}^{n-1} p_i \cdot 2^i$ where $p_i \in \{0, 1\}$, the 2-adic expansion of the number p/q is computed by the automaton given in Figure 1.

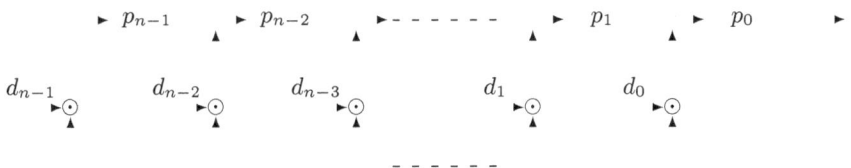

Fig. 1. Automaton computing the 2-adic expansion of p/q

The automaton is referred to as the Galois representation and it is very similar to the Galois representation of an LFSR. For all defined variables we also introduce a time index t, letting $M(t)$ and $C(t)$ denote the content of M and C at time t, respectively.

The addition with carry operation, denoted $\boxplus$ in Figure 1, has a one bit memory, the carry. It operates on three inputs in total, two external inputs and the carry bit. It outputs the XOR of the external inputs and sets the new carry value to one if and only if the integer sum of all three inputs is two or three.

In Figure 2 we specifically illustrate (following [2]) the case $q = -347$, which gives us $d = 174 = (10101110)_{binary}$. The X-FCSR family of stream ciphers uses two FCSR automatons at the core of their construction. For the purposes of this paper it is sufficient to recall the FCSR automaton as implemented in Figure 1 and Figure 2.

The FCSR automaton has n bits of memory in the main register and l bits in the carries register for a total of $n + l$ bits. If (M, C) is our *state*, then many states are equivalent in the sense that starting in equivalent states will produce the same output. As the period is $|q| - 1 \approx 2^n$, the number of states equivalent to a given state is in the order of 2^l.

$C(t)$	0	0	c_5	0	c_3	c_2	c_1	0
			▲		▲	▲	▲	
$M(t)$	►m_7	►m_6	►m_5	►m_4	►m_3	►m_2	►m_1	►m_0 ►
d	1	0	1	0	1	1	1	0

Fig. 2. Example of an FCSR

2.2 Brief Summary of X-FCSR-256 Prerequisites

X-FCSR-256 admits a secret key of 128-bit length and a public initialization vector (IV) of bitlength ranging from 64 to 128 as input. The core of the X-FCSR stream cipher consists of two 256-bit FCSRs with main registers Y and Z which are clocked in opposite directions.

$$Y(t) = (y_{t+255}, \ldots, y_{t+2}, y_{t+1}, y_t), \text{ clocked } \rightarrow$$
$$Z(t) = (z_{t-255}, \ldots, z_{t-2}, z_{t-1}, z_t), \text{ clocked } \leftarrow$$

X-FCSR combines Y and Z to form a 256-bit block $X(t)$ at each discrete time instance t according to

$$X(t) = Y(t) \oplus Z(t),$$

where $\oplus$ denotes bitwise XOR, so that

$$X(0) = (y_{255} \oplus z_{-255}, \ldots, y_2 \oplus z_{-2}, y_1 \oplus z_{-1}, y_0 \oplus z_0)$$
$$X(1) = (y_{256} \oplus z_{-254}, \ldots, y_3 \oplus z_{-1}, y_2 \oplus z_0, y_1 \oplus z_1)$$
$$X(2) = (y_{257} \oplus z_{-253}, \ldots, y_4 \oplus z_0, y_3 \oplus z_1, y_2 \oplus z_2)$$
$$\ldots$$

Further define

$$W(t) = round_{256}(X(t)) = mix(sr(sl(X(t)))), \tag{1}$$

where sl, sr and mix mimic the general structure of the AES round function;

sl is an s-box function applied at byte level,

sr is a row-shifting function operating on bytes,

mix is a column mixing function operating on bytes.

The round functions operate on a 256-bit input, as defined in (1). The general idea behind the round function operations becomes apparent if one considers how the functions operate on the 256-bit input when it is viewed as a 4×8 matrix $\mathbf{A}$ at byte level. Let the byte entries of $\mathbf{A}$ be denoted $a_{i,j}$ with $0 \leq i \leq 3$ and $0 \leq j \leq 7$.

The first transformation layer consists of an S-box function sl applied at byte level. The chosen S-box has a number of attractive properties that are described in [3].

The second operation shifts the rows of $\mathbf{A}$, and sr is identical to the row shifting operation of Rijndael. sr shifts (i.e., rotates) each row of $\mathbf{A}$ to the left at byte level, shifting the first, second, third and fourth rows 0, 1, 3 and 4 bytes respectively.

The purpose of the third operation, mix, is to mix the columns of $\mathbf{A}$. This is also done at byte level according to

$$mix_{256} \begin{pmatrix} a_{0,j} \\ a_{1,j} \\ a_{2,j} \\ a_{3,j} \end{pmatrix} = \begin{pmatrix} a_{3,j} \oplus a_{0,j} \oplus a_{1,j} \\ a_{0,j} \oplus a_{1,j} \oplus a_{2,j} \\ a_{1,j} \oplus a_{2,j} \oplus a_{3,j} \\ a_{2,j} \oplus a_{3,j} \oplus a_{0,j} \end{pmatrix}$$

for every column j of $\mathbf{A}$.

Note that sl, sr and mix are all both invertible and byte oriented. Finally, the 256 bits of keystream that are output at time t are given by

$$out(t) = X(t) \oplus W(t - 16). \tag{2}$$

This last equation introduces a time delay of 16 time units. The first block of keystream is produced at $t = 0$ and the key schedule takes care of defining $W(t)$ for $t < 0$.

3   Describing the Attack

3.1   Idea of Attack

A conceptual basis for understanding the attack is obtained by dividing it into the four parts listed below. Each part has been attributed its own section.

- LFSRization of FCSRs
- Combining Output Blocks
- Analytical Unwinding
- Brute-forcing the State

In Section 3.2 we describe a trick we call "LFSRization of FCSRs". We explain how an observation in [5] can be used to allow treating FCSRs as LFSRs. There is a price to pay for introducing this simplification, of course, but the penalty is not as severe as one may expect.

We observe that we can combine a number of consecutive output blocks to effectively remove most of the dependency on $X(t)$ introduced in (2). The LFSRization process works in our favor here as it provides a linear relationship between FCSR variables. Output block combination is explored in Section 3.3.

Once a suitable combination Q of output blocks is defined, state recovery is the next step. This is done in two parts. In Section 3.4 we explain how to work with Q analytically to transform its constituent parts into something that will get us closer to the state representation. As it turns out, we can do quite a bit here. Finally, having exhausted the analytical options available to us, we bring in the computational artillery and do the remaining job by brute-force. We find that the state can be divided into several almost independent parts and perform exhaustive search on each part separately. This is described in Section 3.5.

3.2   LFSRization of FCSRs

As mentioned above, an observation in [5] provides a way of justifying the validity in treating FCSRs as LFSRs, and does so at a very reasonable cost. We call this process LFSRization of FCSRs, or simply LFSRization when there is no confusion as to what is being treated as an LFSR. There are two parts to the process, a flush phase and a linearity phase.

The observation is simply that a zero feedback bit causes the contents of the carry registers to change in a very predictable way. Adopting a statistical view and assuming independent events is helpful here. Assuming a zero feedback bit, carry registers containing zeros will not change, they will remain zero. The carry registers containing ones are a different matter, though. A 'one' bit will change to a zero bit with probability $\frac{1}{2}$. In essence this means that one single zero feedback bit will cut the number of ones in the carry registers roughly in half.

The natural continuation of this observation is that a sufficient amount of consecutive zero feedback bits will eventually flush the carry registers so that they contain only zeros. On average, roughly half of the carry registers contain ones to start with, so an FCSR with N active carry registers requires roughly $\lg \frac{N}{2} + 1$ zero feedback bits to flush the 'ones' away with probability $\frac{1}{2}$. By expected value we therefore require roughly $\lg \frac{N}{2} + 2$ zero feedback bits to flush a register completely. For X-FCSR-256 we have $N = 210$, indicating that we need no more than nine zero feedback bits to flush a register.

After the flush phase, a register is ready to act as an LFSR. In order to take advantage of this state we need to maintain a linearity phase in which we keep having zero feedback bits fed for a sufficiently long duration of time. As we will see from upcoming arguments, we will in principle require the linearity property for two separate sets of six consecutive zero feedback bits, with the two sets being sixteen time units apart. We will need the FCSRs to act as LFSRs during this time, so our base requirement consists of two smaller LFSRizations, each requiring roughly $9 + 6$ bits for flush and linearity phase respectively. The probability of the two smaller LFSRizations occurring in *both* registers Y and Z simultaneously is $2^{-4(9+6)} = 2^{-60}$. In other words, our particular LFSRization condition appears once in about 2^{60} output blocks.

A real life deviation from the theoretical flush reasoning was noted in [5]. We cannot flush the carry register entirely as the last active carry bit will tend to one instead of zero. As further noted in [5], flushing all but the last carry bit does not cause a problem in practice. Consider the linearized FCSR in Figure 3, it produces a maximal number of zero feedback bits for an FCSR of its size.

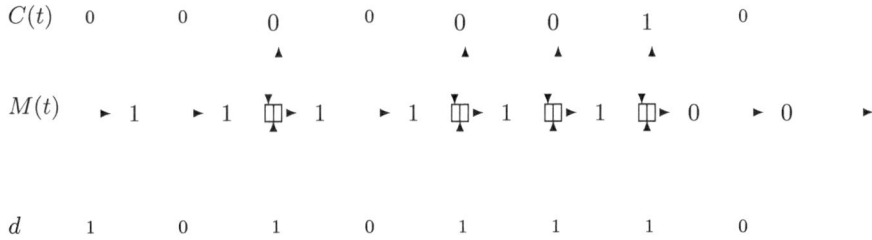

Fig. 3. Maximally linearized FCSR

In simulations and analytical work we must compensate for this effect, of course, but the theoretical reasoning to follow remains valid as we allow ourselves to treat FCSRs as simple LFSRs. The interested reader is referred to [5] for details on this part.

Furthermore, assumptions of independence are not entirely realistic. Although the theoretical reasoning above is included mainly for reasons of completeness, simulations show that we are not far from the truth, effectively providing some degree of validation for the theory. Our simulations show that we have $2^{28.7}$ for the Y register and $2^{27.5}$ for Z for a total of at most $2^{56.2}$ expected output blocks for LFSRization in X-FCSR as we require it.

Our requirements for the basic attack are as follows. At some specific time instance t we require the carry registers of X and Y to be completely flushed except for the last bit. Here we also require the tails of the main registers to contain the bit sequence 111100 as in Figure 3 to guarantee at least six consecutive zero feedback bits. At $t + 16$ we require this precise set-up to appear once again. In each set, the first five zero feedback bits are needed to ensure that the main registers are linear. The last remaining zero feedback bit is used only to facilitate equation solving in the state recovery part, as it guarantees that the last carry bit remains set.

To be fair and accurate we will use the simulation values, which puts us at

$$COST_{LFSRization} \leq 2^{56.2}$$

for the basic attack. Later, in Section 4.2, we will see how we can reduce the requirements to only four consecutive zero feedback bits per set for a complexity of

$$COST_{LFSRization} \leq 2^{49.3}.$$

3.3   Combining Output Blocks

The principal reason for combining consecutive output blocks is to obtain a set of data that is easier to analyze and work with, ultimately leading to a less complicated way to reconstruct the cipher state. Remember that we now treat the two FCSRs as LFSRs with the properties given in Section 3.2.

The main observation is that the modest and regular clocking of the two main registers provides us with the following equality:

$$X(t) \oplus [X(t+1) \ll 1] \oplus [X(t+1) \gg 1] \oplus X(t+2) = (\star, 0, 0, \ldots, 0, \star) \quad (3)$$

The shifting operations $\ll$ and $\gg$ on the left hand side denote shifting of the corresponding 256-bit block left and right, respectively. From this point onward we discard bits that fall over the edge of the 256 bit blocks, and we do so without loss of generality or other such severe penalties. The right hand side is then the zero vector[1], with the possible exception of the first and last bits which are undetermined (and denoted $\star$). Define

$$OUT(t) = out(t) \oplus [out(t+1) \ll 1] \oplus [out(t+1) \gg 1] \oplus out(t+2) \quad (4)$$

in the corresponding way. We have

$$OUT(t) =$$
$$X(t) \oplus [X(t+1) \ll 1] \oplus [X(t+1) \gg 1] \oplus X(t+2) \oplus$$
$$W(t-16) \oplus [W(t-15) \ll 1] \oplus [W(t-15) \gg 1] \oplus W(t-14)$$
$$=$$
$$(\star, 0, 0, \ldots, 0, \star) \oplus$$
$$W(t-16) \oplus [W(t-15) \ll 1] \oplus [W(t-15) \gg 1] \oplus W(t-14)$$
$$\approx$$
$$W(t-16) \oplus [W(t-15) \ll 1] \oplus [W(t-15) \gg 1] \oplus W(t-14), \quad (5)$$

where $\approx$ denotes bitwise equality except for the first and last bit. This expression allows us to relate keystream bits to bits inside the generator that are just a few time instances apart. This will turn out to be very useful when recovering the state of the FCSRs. In order to further unwind equation (5) we need to take a closer look at the constituent parts of W, namely the round function operations sl, sr and mix.

3.4   Analytical Unwinding

Reviewing the round function operations from Section 2.2, recall that all of the operations are invertible and byte oriented. We can also see that the operations mix, sr and their inverses are linear over $\oplus$, such that

$$mix(A \oplus B) = mix(A) \oplus mix(B),$$
$$sr(A \oplus B) = sr(A) \oplus sr(B).$$

Obviously, sl does not harbor the linear property. So, in order to unwind (5) as much as possible, we would now ideally like to apply mix^{-1} and sr^{-1} in that order. Let us begin with focusing on the mix operation.

---

[1] Recall that we ignore the effects of the last carry bit being one instead of zero, as explained in Section 3.2. The arguments below are valid as long as adjustments are made accordingly.

The linearity of mix over $\oplus$ is the first ingredient we need as it allows us to apply mix^{-1} to each of the W terms separately. The shifting does cause us some problems, however, since

$$mix^{-1}\left(W(t) \ll 1\right) \neq mix^{-1}\left(W(t)\right) \ll 1.$$

Therefore mix^{-1} cannot be applied directly in this way, but realizing that mix^{-1} is a byte-oriented operation, it is clear that the equality holds if one restricts comparison to every bit position except the first and last bit of every byte. This is easy to realize if one considers the origin and destination byte of the six middlemost bits as mix^{-1} is applied. One single bit shift does not affect the destination byte of these bits. Furthermore, the peripheral bits that are shifted out of their byte position are mapped to another peripheral bit position. We therefore have

$$
\begin{aligned}
mix^{-1}\left(OUT(t)\right) \cong\ & sr\left(sl\left(X(t-16)\right)\right) \oplus \\
& \left[\ sr\left(sl\left(X(t-15)\right)\right) \ll 1\ \right] \oplus \\
& \left[\ sr\left(sl\left(X(t-15)\right)\right) \gg 1\ \right] \oplus \\
& sr\left(sl\left(X(t-14)\right)\right),
\end{aligned}
$$

where $\cong$ denotes equality with respect to the six middlemost bits of each byte. The same arguments apply to sr^{-1}, so we define $Q(t) = sr^{-1}\left(mix^{-1}\left(OUT(t)\right)\right)$ to obtain

$$
\begin{aligned}
Q(t) \cong\ & sl\left(X(t-16)\right) \oplus \\
& \left[\ sl\left(X(t-15)\right) \ll 1\ \right] \oplus \\
& \left[\ sl\left(X(t-15)\right) \gg 1\ \right] \oplus \\
& sl\left(X(t-14)\right).
\end{aligned}
\tag{6}
$$

Loosely put, we can essentially bypass the effects of the mix and sr operations by ignoring the peripheral bits of each byte.

We have combined consecutive keystream blocks $out(t)$ into Q in hope of Q being easier to analyze than $out(t)$. Since the ultimate goal is to map $out(t)$ to Y and Z, we don't have very far to go now. As our expression for Q involves only X and sl, let's see how and at what cost we can brute-force Q and solve for Y and Z.

3.5   Brute-Forcing the State

The brute-forcing part can most easily be understood by focusing on one specific byte position in $Q(t)$. Given the, say, seventh byte in $Q(t)$, how can we *uniquely* reconstruct the relevant parts of Y and Z? Let us first figure out which bits one needs from $Y(t-16)$ and $Z(t-16)$ in order to be able to calculate the given byte in $Q(t)$. Note that this step is possible only because of the LFSRization described in Section 3.2.

Have another look at the first part of expression (6): $sl\left(X(t-16)\right)$. Since sl is an S-box function that operates on bytes, we need to know the full corresponding byte from $X(t-16)$. Those eight bits are derived from eight bits in each of Y and Z, totaling 16 bits, as shown in the left column of Figure 4 below.

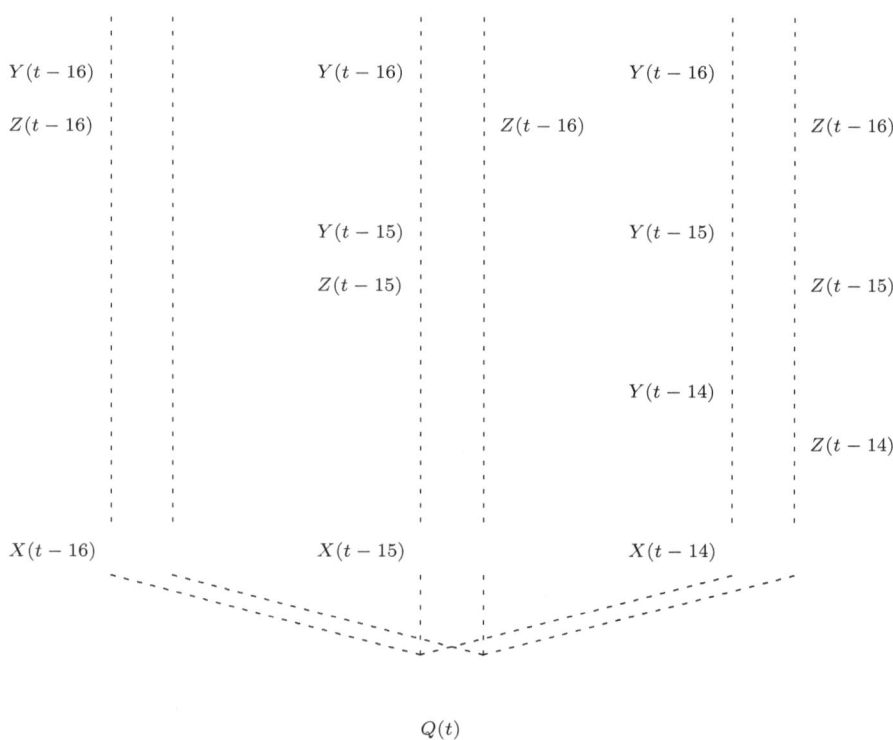

$Q(t)$

Fig. 4. Bit usage for one byte in $Q(t)$

The next parts of (6) involves $sl(X(t-15))$. The same reasoning applies here, we need to know the full corresponding byte of $X(t-15)$ in order to be able to calculate this S-box value. But, since the main registers act like LFSRs, most of the bits we need from Y and Z for $X(t-16)$ have already been employed for $X(t-16)$ previously. Since the two main registers are clocked only one step at each time instance, only two more bits are needed, one from Y and one from Z. This is illustrated by the middle column of Figure 4 below. We count 18 bits in Y and Z so far.

In the same vein, two more bits are needed from Y and Z to calculate $sl(X(t-14))$, illustrated in the remaining part of Figure 4. This brings the total up to 20 bits. All in all, for one byte position in $Q(t)$ we have total bit usage as shown in Figure 5.

So, 10 bits in $Y(t-16)$ and 10 bits in $Z(t-16)$ is what we require to be able to calculate one specific byte position in $Q(t)$. By restricting our attention to the six middlemost bits of each byte in Q we accomplish two objectives; we effectively reduce the number of unknown bits we are dealing with in Y and Z,

Fig. 5. Bit usage in $Q(t)$

and we simplify the expression for calculating the byte in Q by safely reducing the effects of the shifting operation. Specifically, shifting one bit left or right does not bring neighboring bytes into play.

Focusing on one single byte position gives us six equations, one for each of the six middlemost bits, and 20 unsolved variables, one for each bit position in Y and Z. This amounts to an underdetermined system, of course, but we can easily add more equations by having a look at the same byte position in $Q(t+1)$. The six middle bits of that byte give us six new equations at the cost of introducing a few new variables. To see how many, we must perform the analysis for $Q(t+1)$ corresponding to Figure 4. The total bit usage for one byte position in $Q(t+1)$ in terms of bits in $Y(t-16)$ and $Z(t-16)$ is given in Figure 6.

Fig. 6. Bit usage in $Q(t+1)$

From this we see that the six new equations have the downside of introducing two new variables. In total we therefore have 12 equations and 22 variables after including $Q(t+1)$. The system is still underdetermined, so we can add $Q(t+2)$ as well. This brings us to 18 equations and 24 variables, and so on. Adding $Q(t+3)$ provides 24 equations for 26 variables, but at this level we will obtain a resulting system that provides hope of being fully determined as we may also reduce the number of variables by reusing already determined values as we scan Q byte by byte from one end to the other to solve for bits in Y and Z. The corresponding bit usage for our four consecutive Q's in terms of bits in $Y(t-16)$ and $Z(t-16)$ is illustrated in Figure 7 below.

$Y(t-16)$

$Z(t-16)$

$Q(i)$

Fig. 7. Total bit usage for $Q(i)$, $t \leq i \leq t+3$

When brute-forcing one byte position in Q we essentially solve for 26 bits. If we scan Q from left to right, solving the corresponding system for each byte, we can reuse quite many of these bits. Instead of solving for 26, we need only solve for 16 as the remaining 10 have already been determined. This is illustrated in Figure 8. Reusing bits in this way works fine for all byte positions except the first one. For the first byte position we don't have any prior solution to lean back on, but we can use the LFSRization assumption. We have already assumed that we have 'zero' feedback bits coming in and these are valid to use when solving the system. The system for the first byte contains 21 unsolved variables, so the 24 equations do indeed provide a fully determined system.

$Y(t-16)$

$Z(t-16)$

$Q(t)$

Fig. 8. Reusing bits when solving for $Q(t)$

Employing bit reuse, the total cost for the brute-forcing part becomes

$$COST_{brute-force} \leq 2^{21} + 31 \times 2^{16} < 2^{22}.$$

This calculation is a little bit idealized, however, since we in practice do obtain multiple solutions in some cases. These occur sometimes because the peripheral bits in the system appear in only one or two of the equations, causing false solutions. These are easy to spot, though, as the succeeding equation system will generally be unsolvable as we attempt to reuse 10 of the bits from the false solution. And since the false solutions do not appear in abundance, we do not compensate for this complexity wise.

This concludes the basic attack, in which we have assumed availability of four separate sets of six consecutive zero feedback bits as described in Section 3.2.

4   Improving the Attack

4.1   Precomputation

We can reduce the workload of the brute-force part almost entirely using pre-computation. A precomputation table for solving the first byte system would require 2^{24} entries[2] as we have the 24 bits from the four Q's as input to resolve 21 bits. For succeeding byte positions we may limit the number of Q's to three, which provides 18 equations for the 16 unsolved variables. Adding the already determined 8 bits to the formula, we can see that a table with $2^{18+8} = 2^{26}$ entries will suffice. In this context we consider these tables to be of reasonable size.

The total amortized cost for attempting to solve for the entire state is then given by considering the relative frequencies of table lookups per byte position. Using table lookups as unit, we have

$$COST_{brute-force} < 1 + \frac{1}{8}\left(1 + \frac{1}{4} + \frac{1}{4^2} + \ldots\right) = \frac{7}{6}$$

using no more than 2^{27} table entries worth of storage.

4.2   Lowering the Required Keystream

In the basic attack we assumed existence of four separate sets of six consecutive zero feedback bits, as explained in Section 3.2. Our next improvement is to reduce the required keystream by loosening the above requirement to only four consecutive zero feedback bits in each set and increasing the calculation effort correspondingly.

To shine some light upon some of the details involved in this process, consider equation (5) once more. The purpose of the second of the two sets of zero feedback bits is to make way for the X's to cancel out properly according to equation (3). A 'one' feedback bit in the second set prohibits the X's from canceling out entirely. We can cope with this anomaly by compensating for such a non-null aggregate of the X's in equation (5). The important issue is that we are in control of the resulting changes.

The first set of zero feedback bits govern the composition of the W's. With zero feedback bits all the way we obtain a well defined system when solving for the first byte position in Q. If the fifth feedback bit is a 'one' the system changes somewhat, but it is still as well defined as before. Here, too, we are in

---

[2] The storage is trivially realized using on average at most two 4-byte words per entry.

control of the resulting changes. Our increase in computational effort consists of constructing and using the corresponding tables for the resulting systems, so that we can solve the resulting system regardless of these last bit values.

Without the sixth and last zero feedback bit in each set we would not know if the last remaining carry bit has ultimately been nullified or not. Our basic attack assumptions allow us to easily figure out the value of the last carry bit. We may remove the requirement of the sixth zero feedback bit in each set if we instead solve all the 16 similar but essentially different resulting variants of the system. In principle, we can allow creation of 16 new tables, one for each system, for a total workload increase factor of 16. Therefore, storage requirements increase to 2^{28} table entries for the first byte position systems but remain at at most 2^{26} for the succeeding byte position systems for a total of 2^{29} table entries. Note that no specialized tables for the last byte position system are needed because of the symmetry in the systems for the first and last byte positions.

The corresponding arguments are valid when removing the requirement of the fifth zero feedback bit. The fifth feedback bit from two of the sets affect the system of the first byte position for an increase in storage and computation of a factor of at most 16, again. Storage requirements increase to 2^{32} table entries for the first byte position systems and remain at at most 2^{26} for succeeding byte position systems. All in all, we can solve the entire system for all cases using only

$$COST_{brute-force} < 2^{4+4} \times \frac{7}{6} < 2^{8.3}$$

table lookups into at most 2^{33} table entries of storage. The interested reader is referred to [5], in which a similar situation is discussed.

In practice, the $COST_{LFSRization}$ part tells us how many keystream blocks we need to analyze before we can find a favorable situation that allows the brute-force method to go all the way to recovering the state. The $COST_{brute-force}$ part is payed by performing that many calculations for each analyzed keystream block. To summarize, we have

$$COST = COST_{LFSRization} \times COST_{brute-force} < 2^{49.3+8.3} = 2^{57.6}$$

using no more than 2^{33} table entries worth of precomputational storage.

5   Concluding Remarks

It is clear that the design of the X-FCSR stream cipher family is not sufficiently secure. Depending on one's inclination, it is possible to attribute this insufficiency to the modest clocking of the two FCSRs, the size or number of FCSRs, how they are combined, the complexity of the round function or some other issue. All of these factors are parts of the whole, but the key insight, however, is that it is important not to rely on the non-linear property of FCSRs too heavily. The LFSRization process shows that it is relatively cheap to linearize FCSRs, the cost being roughly logarithmic in the size of active carry registers.

More details on the last improvements and a more in-depth exposé of the effects of the last carry bit on system solving are available in the full version of this paper. There we also exploit the symmetry situation of requiring several consecutive 'one' feedback bits for an additional reduction in required keystream.

Let us end with a note on applicability to X-FCSR-128. The basic attack presented here works for X-FCSR-128 as well, but the resulting complexity is much less impressive. The LFSRization process is identical for both variants of X-FCSR, as is the analytical unwinding. Enter round functions. The two registers are 256 bits in size in both cases, but X-FCSR-128 "folds" the contents of the registers to produce a 128-bit result, implying that more bits are condensed into one byte position of Q as analyzed in Section 3.5. This affects cost in a negative way, actually making the attack more expensive for X-FCSR-128. We estimate that at least twelve consecutive Q's are needed for a fully determined first byte system. This leads to a guesstimated expected value of about 2^{74} output blocks for the attack to come through in the basic setting, each output block requiring roughly one table lookup into a storage of at most 2^{72} table cells.

References

1. Arnault, F., Berger, T.: F-FCSR: Design of a new class of stream ciphers. In: Gilbert, H., Handschuh, H. (eds.) FSE 2005. LNCS, vol. 3557, pp. 83–97. Springer, Heidelberg (2005)
2. Arnault, F., Berger, T., Lauradoux, C.: Update on F-FCSR stream cipher. eSTREAM, ECRYPT Stream Cipher Project, Report 2006/025 (2006),
   http://www.ecrypt.eu.org/stream
3. Arnault, F., Berger, T.P., Lauradoux, C., Minier, M.: X-FCSR - a new software oriented stream cipher based upon FCSRs. In: Srinathan, K., Pandu Rangan, C., Yung, M. (eds.) INDOCRYPT 2007. LNCS, vol. 4859, pp. 341–350. Springer, Heidelberg (2007)
4. ECRYPT. eSTREAM: ECRYPT Stream Cipher Project, IST-2002-507932,
   http://www.ecrypt.eu.org/stream/
5. Hell, M., Johansson, T.: Breaking the F-FCSR-H stream cipher in real time. In: Pieprzyk, J. (ed.) ASIACRYPT 2008. LNCS, vol. 5350, pp. 557–569. Springer, Heidelberg (2008)
6. Jaulmes, E., Muller, F.: Cryptanalysis of ECRYPT candidates F-FCSR-8 and F-FCSR-H. eSTREAM, ECRYPT Stream Cipher Project, Report 2005/046 (2005),
   http://www.ecrypt.eu.org/stream
7. Jaulmes, E., Muller, F.: Cryptanalysis of the F-FCSR stream cipher family. In: Preneel, B., Tavares, S. (eds.) SAC 2005. LNCS, vol. 3897, pp. 20–35. Springer, Heidelberg (2006)
8. Klapper, A., Goresky, M.: 2-adic shift registers. In: Anderson, R.J. (ed.) FSE 1993. LNCS, vol. 809, pp. 174–178. Springer, Heidelberg (1994)

Key Collisions of the RC4 Stream Cipher

Mitsuru Matsui

Information Technology R&D Center
Mitsubishi Electric Corporation
5-1-1 Ofuna Kamakura Kanagawa 247-8501, Japan
Matsui.Mitsuru@ab.MitsubishiElectric.co.jp

Abstract. This paper studies "colliding keys" of RC4 that create the same initial state and hence generate the same pseudo-random byte stream. It is easy to see that RC4 has colliding keys when its key size is very large, but it was unknown whether such key collisions exist for shorter key sizes. We present a new state transition sequence of the key scheduling algorithm for a related key pair of an arbitrary fixed length that can lead to key collisions and show as an example a 24-byte colliding key pair. We also demonstrate that it is very likely that RC4 has a colliding key pair even if its key size is less than 20 bytes. This result is remarkable in that the number of possible initial states of RC4 reaches $256! \approx 2^{1684}$. In addition we present a 20-byte near-colliding key pair whose 256-byte initial state arrays differ at only two byte positions.

1 Introduction

The RC4 stream cipher is one of the most widely-used real-world cryptosystems. Since its development in 1980's, RC4 has been used in various software applications and standard protocols such as Microsoft Office, Secure Socket Layer (SSL), Wired Equivalent Privacy (WEP). The architecture of RC4 is extremely simple and its encryption speed is remarkably fast. It has been suitable for not only PC applications but also embedding environments.

RC4 accepts a secret key whose length is 1 to 256 bytes, where a typical key size is 5 bytes (due to an old export regulation), 13 bytes (in the WEP encryption) or 16 bytes. It consists of two parts; the key scheduling algorithm (KSA) and the pseudo-random generating algorithm (PRGA). The KSA creates the 256-byte initial state from the secret key, and the PRGA generates pseudo-random byte stream from the initial state. Either of the algorithms can be described unambiguously in only a few lines. Due to this simplicity and wide applicability, a vast amount of efforts have been made for cryptanalysing RC4 since its specification was made public on Internet in 1994 [1].

As early as one year after the specification of RC4 was publicly available, it was pointed out by Roos [10] that the initial few bytes of its output stream are strongly correlated with the key. This observation was later extended in various ways. Mantin and Shamir [8] mounted a distinguishing attack of RC4 under the "strong distinguisher scenario" that an attacker can obtain many

O. Dunkelman (Ed.): FSE 2009, LNCS 5665, pp. 38–50, 2009.

short output streams using random and unrelated keys. Paul and Preneel [9] successfully demonstrated a distinguisher that requires a total of 2^{25} output bytes (2 bytes/key × 2^{24} keys).

For cryptanalysis of a real-world application using RC4, Fluhrer, Mantin and Shamir [3] showed an attack of WEP for the first time in 2001. They found that partial information of the key of RC4 gives non-negligible information of its output bytes. This fact was efficiently used for mounting a passive attack, since the initial vector of WEP is packet-variable and embedded in the key of RC4. It was recently improved by Klein [6], and finally in 2007 Tews, Weinmann and Pyskin [11] and Vaudenay and Vuagnoux [12] independently demonstrated a very practical key recovery attack of the 104-bit WEP.

For another attacks, Fluhrer and McGrew [4] presented a distinguishing attack of RC4 using $2^{30.6}$ output bytes generated by a single key, and Mantin [7] successfully reduced the output stream size required for the successful attack down to $2^{26.5}$ bytes and also showed how to predict output bits. More recently Biham and Carmeli [2] concentrated on the key scheduling algorithm of RC4 and discussed how to recover a secret key from a given internal state.

This paper studies another type of weakness of the key scheduling algorithm of RC4; that is, existence of secret keys that create the same initial state and hence generate the same pseudo-random byte stream, which we call "colliding keys". It had been already pointed out by Grosul and Wallach [5] in 2000 that RC4 has related-key key pairs that generate substantially similar hundred output bytes when the key size is close to the full 256 bytes. In this paper we explore much stronger key collisions in a shorter key size.

Since the total number of possible initial states of RC4 reaches $256! \approx 2^{1684}$, RC4 must have colliding keys if its key size exceeds $\lfloor 1684/8 \rfloor = 210$ bytes. Moreover, due to the birthday paradox, it is not surprising that RC4 has colliding keys when its key size is $\lfloor (1684/2)/8 \rfloor = 105$ (or more) bytes. However, it was unknown whether colliding keys exist in a shorter key size. The contribution of this paper is to give a positive answer to this problem.

In this paper, we begin by demonstrating a specific example of a colliding 64-byte key pair, whose internal 256-byte state arrays differ at most two byte positions in any step i ($0 \leq i \leq 255$) of the key scheduling algorithm. Then by generalizing this example, we show a state transition pattern that is applicable to a key pair of an arbitrary fixed length and estimate the probability that such pattern actually takes place for randomly given key pairs with a fixed difference.

We have confirmed that our probability estimation mostly agrees with computer experimental results when the key size is around 32 bytes or more. We also demonstrate that it is very likely that RC4 has a colliding key pair even when its key size is much shorter, say 20 bytes, while the minimal key size that has colliding keys is still an open problem.

We further extend our observation to a near-colliding key pair; that is, a key pair whose initial states differ at exactly two positions. In the same way as the key collision case, we show a state transition pattern of a key pair of an arbitrary length that can lead to a near-collision and analyze the probability that the pattern actually takes place. Finally we illustrate our (near-)colliding key pair search algorithm, which has successfully found a 24-byte colliding key pair and a 20-byte near-colliding key pair.

2   The RC4 Stream Cipher

RC4 is a stream cipher with a secret key whose length is 1 to 256 bytes. We define k and K as the target key size in byte and the k-byte secret key, respectively, and K[0]...K[k-1] denote k key bytes. For arbitrary i, K[i] means K[i mod k]. RC4 consists of the key scheduling algorithm (KSA), which creates the 256-byte initial state array S[0]...S[255] from the secret key, and the pseudo-random generating algorithm (PRGA), which generates byte sequence Z[0]...Z[L-1] of arbitrary length L from the initial state.

This paper discusses colliding key pairs of RC4 that create the same initial state. Hence only the key scheduling algorithm, described below in the syntax of C language, is relevant. When necessary, we use notations S1/S2 and j1/j2 for the state array and the state index for the first and second key K1/K2, respectively. The goal of this paper is to find, or show a strong evidence of existence of, key pair K1 and K2 such that the corresponding state arrays S1 and S2 are completely same at the end of the key scheduling algorithm.

We define "the distance of a key pair at step i" as the number of distinct bytes between S1 and S2 at the bottom (i.e. after the swap) of the i-th iteration in the state randomization loop. If the distance of key pair K1 and K2 is 0 at step 255, then they are a colliding key pair. This paper will deal with key pairs whose distance is at most 2 at any step i ($0 \leq i \leq 255$).

```
[The Key Scheduling Algorithm of RC4]
    /* State Initialization */
    for(i=0; i<256; i++){
        S[i] = i;
    }
    /* State Randomization */
    j=0;   /* Index j */
    for(i=0; i<256; i++){
        /* Step i */
        j = (j + S[i] + K[i % k]) & 0xff;
        SWAP(S[i], S[j]);
    }

[The Pseudo-Random Generator Algorithm of RC4]
    i = 0;
    j = 0;
    for(n=0; n<L; n++){
        i = (i + 1) & 0xff;
        j = (j + S[i]) & 0xff;
        SWAP(S[i], S[j]);
        Z[n] = S[(S[i] + S[j]) & 0xff];
    }
```

3   An Example: How It Works

In this section we demonstrate a specific example of a colliding key pair and explain how its collision is created in a step-by-step fashion. In fact, this simple example contains all tricks that we need for finding colliding key pairs of an arbitrary length in later sections. The following is a 64-byte key pair, written in hexadecimal form, which differs at only one byte position, K1[2] $\neq$ K2[2]. These two keys create the same initial state, and hence they are cryptographically indistinguishable.

```
K1 = 45 3d 7d 3d c9 45 57 12 00 00 00 00 00 00 00 00
     00 00 00 00 00 00 00 00 00 00 00 00 00 00 00 00
     00 00 00 00 00 00 00 00 00 00 00 00 00 00 00 00
     00 00 00 00 00 00 00 00 00 00 00 00 00 00 00 00

K2 = 45 3d 7e 3d c9 45 57 12 00 00 00 00 00 00 00 00
     00 00 00 00 00 00 00 00 00 00 00 00 00 00 00 00
     00 00 00 00 00 00 00 00 00 00 00 00 00 00 00 00
     00 00 00 00 00 00 00 00 00 00 00 00 00 00 00 00
```

Table 1 shows internal values in the state randomization loop for each key, where "@XX" denotes an address of the state array. For example, at the end of step i=02 (i.e. after the swap), j1=02, j2=03 and the state arrays differ at exactly two positions 02 and 03. At these positions, S1[02]=02, S2[02]=03, S1[03]=03 and S2[03]=02.

Table 1. The State Transition Pattern of the 64-byte Key Pair

i	K1 K2	j1 j2	differences between S1 and S2	
00	45 45	45 45	S1=S2	
01	3d 3d	83 83	S1=S2	
02	7d 7e	02 03	@02(S1=02 S2=03)	@03(S1=03 S2=02)
03	3d 3d	42 42	@02(S1=02 S2=03)	@42(S1=03 S2=02)
04-40			@02(S1=02 S2=03)	@42(S1=03 S2=02)
41	3d 3d	02 02	@41(S1=02 S2=03)	@42(S1=03 S2=02)
42	7d 7e	82 82	@41(S1=02 S2=03)	@82(S1=03 S2=02)
43-80			@41(S1=02 S2=03)	@82(S1=03 S2=02)
81	3d 3d	41 41	@81(S1=02 S2=03)	@82(S1=03 S2=02)
82	7d 7e	c1 c1	@81(S1=02 S2=03)	@c1(S1=03 S2=02)
83-bf			@81(S1=02 S2=03)	@c1(S1=03 S2=02)
c0	45 45	81 81	@c0(S1=02 S2=03)	@c1(S1=03 S2=02)
c1	3d 3d	c1 c0	S1=S2	
c2	7d 7e	54 54	S1=S2	
c3-ff			S1=S2	

Now let us take a look at this transition sequence more closely to see why/how the distance of this key pair remains 0 or 2 at any step.

Table 2. The State Transition Details of the 64-bit Key Pair

i

00,01 K1[i]=K2[i]. Hence S1=S2 at these steps.

 02   K1[i]=K2[i]-1 (1st time). Hence j1=j2-1.
       Also since i=j1, S1 and S2 differ at j1=02 and j2=03 only.
       (If i≠j1, S1 and S2 differ at three positions.)

 03   K1[i]=K2[i] and S1[i]=S2[i]+1. Hence j1=j2 again.
       Also since j1=j2=42, S1/S2[i] is swapped with S1/S2[42].
       Now S1 and S2 differ at 02 and 42. (Note that K1[42]=K2[42]-1.)

04-40 K1[i]=K2[i]. Hence j1=j2.
       Also since j1(=j2)≠02 and j1(=j2)≠42, S1 and S2 differ at 02 and 42.

 41   Since j1=j2=02, S1/S2[i] is swapped with S1/S2[02].
       Now S1 and S2 differ at 41 and 42.

 42   K1[i]=K2[i]-1 (2nd time). Since S1[i]=S2[i]+1, j1=j2.
       Also since j1=j2=82, S1/S2[i] is swapped with S1/S2[82].
       Now S1 and S2 differ at 41 and 82. (Note that K1[82]=K2[82]-1.)

43-80 K1[i]=K2[i]. Hence j1=j2.
       Also since j1(=j2)≠41 and j1(=j2)≠82, S1 and S2 differ at 41 and 82.

 81   Since j1=j2=41, S1/S2[i] is swapped with S1/S2[41].
       Now S1 and S2 differ at 81 and 82.

 82   K1[i]=K2[i]-1 (3nd time). Since S1[i]=S2[i]+1, j1=j2.
       Also since j1=j2=c1, S1/S2[i] is swapped with S1/S2[c1].
       Now S1 and S2 differ at 81 and c1.
       (Note that this time the swapped address is c1, not c2).

83-bf K1[i]=K2[i]. Hence j1=j2.
       Also since j1(=j2)≠81 and j1(=j2)≠c1, S1 and S2 differ at 81 and c1.

 c0   Since j1=81, S1/S2[i] is swapped with S1/S2[81].
       Now S1 and S2 differ at c0 and c1.

 c1   Since j1=c1 and j2=c0, the differences between S1/S2 disappear.
       Now S1 is the same as S2.

 c2   K1[i]=K2[i]-1 (4th time). Since j1=j2+1 at the previous step,
       now j1=j2. Hence S1=S2.

c3-ff K1[i]=K2[i]. Hence j1=j2 and S1=S2.

We have K1[i]=K2[i]-1 four times. For i=02, this difference is absorbed at the next step because it causes j1=j2-1 while S1[3]=S2[3]+1. Note that the relation K1[2]=K2[2]-1 is essential for this example. For i=42 and i=82, j1 remains the same as j2 because S1[42]=S2[42]+1 and S1[82]=S2[82]+1, respectively.

The last time i=c2 is a bit tricky; at two steps before i=c0, S1 and S2 differ at c0 and c1. Moreover at i=c1, j1=c0 and j2=c1 and hence the differences of S1 and S2 disappear at this step. The remaining difference is between j1 and j2, which is finally cancelled at i=c2 due to the relation K1[c2]=K2[c2]-1.

4   General Collision Sequence

In this section we extend the sequence shown in the previous section to a key pair of an arbitrary length and give a general transition pattern that can lead to the same initial state. We also estimate the probability that the key collision actually occurs for randomly given key pairs with a fixed difference.

We here consider k-byte key pair K1 and K2 such that K1[i]=K2[i]-1 if i=d ($0{\le}d{<}k$) and K1[i]=K2[i] otherwise. We also define n as $\lfloor(256+k-1-d)/k\rfloor$. Then K1[i]$\neq$K2[i] takes place exactly n times in the state randomization loop.

Table 3 illustrates details of the state transition pattern for given k and d with estimated probability, where "-" allows any address (a "don't-care" value) and "x" at step i=d+(n-1)k-2 is the value such that S1[x]=d and S2[x]=d+1 at step i=d+(n-1)k-3.

Table 3. The State Transition Pattern of a k-byte Colliding Key Pair

Step	Internal State Values	Estimated Prob.
(a) i=0...d-1	j1(=j2)$\neq$ d, j1(=j2)$\neq$d+1	$(254/256)^d$ or
	S1=S2	$(254/256)^{d-1}$
(b) i=d	j1=d, j2=d+1	1/256
	@d(S1=d S2=d+1)	
	@d+1(S1=d+1 S2=d)	
(c) i=d+1	j1=j2=d+k	1/256
	@d(S1=d S2=d+1)	
	@d+k(S1=d+1 S2=d)	
(d) i=d+2...d+k-1	j1(=j2)$\neq$d+k	$(255/256)^{k-2}$
	@-(S1=d S2=d+1)	
	@d+k(S1=d+1 S2=d)	
Repeat steps (e) and (f) for m=1...n-3.		
(e) i=d+mk	j1=j2=d+(m+1)k	1/256
	@-(S1=d S2=d+1)	
	@d+(m+1)k(S1=d+1 S2=d)	
(f) i=d+mk+1...	j1(=j2)$\neq$d+(m+1)k	$(255/256)^{k-2}$
d+(m+1)k-1	@-(S1=d S2=d+1)	
	@d+(m+1)k(S1=d+1 S2=d)	
(g) i=d+(n-2)k	j1=j2=d+(n-1)k-1	1/256
	@-(S1=d S2=d+1)	
	@d+(n-1)k-1(S1=d+1 S2=d)	
(h) i=d+(n-2)k+1...	j1(=j2)$\neq$d+(n-1)k-1	$(255/256)^{k-4}$
d+(n-1)k-3	@-(S1=d S2=d+1)	
	@d+(n-1)k-1(S1=d+1 S2=d)	
(i) i=d+(n-1)k-2	j1=j2=x	1/256
	@d+(n-1)k-2(S1=d S2=d+1)	
	@d+(n-1)k-1(S1=d+1 S2=d)	
(j) i=d+(n-1)k-1	j1=d+(n-1)k-1, j2=d+(n-1)k-2	1/256
(k) i=d+(n-1)k...255	S1=S2	1

The probability that event (a) takes place is $(254/256)^{d-1}$ only if d=k-1, and $(254/256)^d$ otherwise, because K1[0]=K2[0]=0 when d=k-1 (also see below). Now assuming that the state index j is uniformly random at all steps, the probability that this transition sequence actually takes place for randomly given key pairs with the fixed difference is

$$ColProb(k,d) = \begin{cases} (254/256)^d(255/256)^{(n-1)(k-2)-2}(1/256)^{n+2} & (d \neq k-1) \\ (254/256)^{d-1}(255/256)^{(n-1)(k-2)-2}(1/256)^{n+2} & (d = k-1) \end{cases} .$$

This probability is actually very close to $1/e(1/256)^{n+2}$, which depends on n only. Of course j's are not uniformly random in practice, and therefore this probability estimation is not necessarily correct. However this (very intuitive) assumption mostly agrees with our computer experimental results when the length of the key is around 32 bytes or more.

The following is one of the 43-byte colliding key pairs that we found (43 is the minimal k such that $\lfloor 256/k \rfloor$=5), and its experimentally observed conditional probability of each event. These keys differ at the last byte, hence d=42. Note that successful events (a) and (b) effectively determine one-byte information of the key. In other words, K[0]...K[d-1] uniquely determines K[d]. Moreover for meeting events (b) and (c), we must have K1[0]=K2[0]=0, more generally K1[d+1]=K2[d+1]=k-d-1. We hence do not have to "wait for" events (b) and (c) in searching for key collisions.

Table 4 shows that our experimental results, where we obtained two 43-byte colliding key pairs from $2^{41.5}$ candidate pairs, perfectly agree with our probability estimation.

Table 4. Experimental Results of Finding 43-byte Colliding Key Pairs

Event	Estimated Prob.	Measured Prob.
(a)	$0.725010=(254/256)^{41}$	$0.725010 = 2272363208729/3134252384256$
(d)	$0.851743=(255/256)^{41}$	$0.851638 = 1935231636873/2272363208729$
(e-1)	$0.003906=1/256$	$0.004120 = 7973306038/1935231636873$
(f-1)	$0.851743=(255/256)^{41}$	$0.851706 = 6790914484/7973306038$
(e-2)	$0.003906=1/256$	$0.003933 = 26707884/6790914484$
(f-2)	$0.851743=(255/256)^{41}$	$0.851605 = 22744564/26707884$
(g)	$0.003906=1/256$	$0.003891 = 88498/22744564$
(h)	$0.858437=(255/256)^{39}$	$0.858336 = 75961/88498$
(i)	$0.003906=1/256$	$0.003976 = 302/75961$
(j)	$0.003906=1/256$	$0.006623 = 2/302$

```
K1 = 00 6d 41 8b 95 46 07 a4 87 8d 69 d7 bc bc c4 70
     4a 3b ed 94 34 50 04 68 4d 4f 2e 30 c1 6e 20 a8
     bf 80 b6 ae df ae 43 56 0a 80 e7


K2 = 00 6d 41 8b 95 46 07 a4 87 8d 69 d7 bc bc c4 70
     4a 3b ed 94 34 50 04 68 4d 4f 2e 30 c1 6e 20 a8
     bf 80 b6 ae df ae 43 56 0a 80 e8
```

Also, assuming again that $ColProb(k,d)$ is correct for any k and d, the expected number of k-byte colliding key pairs out of a total of 2^{8k} keys is

$$ColPairs(k) = 2^{8k} \times \sum_{d=0}^{k-1} ColProb(k,d).$$

Table 5 is a list of $log_2(ColPairs(k))$ for $k=17\ldots64$. This table clearly shows that key collisions can exist in much shorter key length. On the other hand, $ColProb(k,d)$ does not always agree with our experimental results when k is small, say 30 or less, because probabilistic dependency of j's cannot be ignored in such range. It seems that more detailed probability analysis is needed for accurately estimating the density of colliding key pairs in a small key size.

Table 5. List of $log_2(ColPairs(k))$ for $k=17\ldots64$

k	Pairs	k	Pairs	k	Pairs	k	Pairs	k	Pairs
15	-	25	106.9	35	211.2	45	306.4	55	394.9
16	-	26	120.7	36	219.6	46	314.9	56	403.2
17	2.7	27	130.5	37	232.3	47	323.3	57	411.5
18	18.5	28	139.2	38	242.0	48	331.6	58	419.7
19	34.0	29	153.0	39	250.7	49	339.9	59	427.9
20	48.7	30	162.5	40	259.2	50	348.1	60	436.1
21	58.8	31	171.2	41	267.6	51	356.3	61	444.2
22	73.7	32	179.7	42	275.9	52	368.7	62	452.4
23	83.0	33	193.7	43	287.7	53	377.8	63	460.5
24	97.7	34	202.6	44	297.6	54	386.4	64	468.6

5 Near-Collision Sequence

This section gives another extension of section 2, a near-colliding key pair whose initial state arrays $S1$ and $S2$ differ at exactly two positions. We use the same notations as in the previous section.

Table 6 shows the details of our state transition pattern that can lead to a near-collision, which is the same as the collision case except the last part. "-" allows any address and "x" is any value equal to or less than $d+(n-1)k$. Note that if x exceeds $d+(n-1)k$, the distance between $S1$ and $S2$ exceeds two in (h).

The expected probability that this transition pattern actually takes place is

$NearColProb(k,d) =$
$$\begin{cases} (254/256)^d(255/256)^{(n-1)(k-2)}(1/256)^n(d+(n-1)k+1)/256 & (d \neq k-1) \\ (254/256)^{d-1}(255/256)^{(n-1)(k-2)}(1/256)^n(d+(n-1)k+1)/256(d=k-1) \end{cases}$$

This probability is actually very close to $1/e(1/256)^n$, which depends on n only, and hence roughly it holds that $NearColProb(k,d) = 2^{16}ColProb(k,d)$.

Table 6. The State Transition Pattern of a k-byte Near-Colliding Key Pair

Step	Internal State Values	Approx. Prob.
(a) i=0...d-1	j1(=j2)≠d, j1(=j2)≠d+1 S1=S2	$(254/256)^d$ or $(254/256)^{d-1}$
(b) i=d	j1=d, j2=d+1 @d(S1=d S2=d+1) @d+1(S1=d+1 S2=d)	1/256
(c) i=d+1	j1=j2=d+k @d(S1=d S2=d+1) @d+k(S1=d+1 S2=d)	1/256
(d) i=d+2...d+k-1	j1(=j2)≠d+k @-(S1=d S2=d+1) @d+k(S1=d+1 S2=d)	$(255/256)^{k-2}$

Repeat steps (e) and (f) for m=1..n-2

(e) i=d+mk	j1=j2=d+(m+1)k @-(S1=d S2=d+1) @d+(m+1)k(S1=d+1 S2=d)	1/256
(f) i=d+mk+1... d+(m+1)k-1	j1(=j2)≠d+(m+1)k @-(S1=d S2=d+1) @d+(m+1)k(S1=d+1 S2=d)	$(255/256)^{k-2}$
(g) i=d+(n-1)k	j1=x, j2=x @-(S1=d S2=d+1) @x(S1=d+1 S2=d)	(d+(n-1)k+1)/256
(h) i=d+(n-1)k+1...255	@-(S1=d S2=d+1) @-(S1=d+1 S2=d)	1

The following is a 33-byte near-colliding key pair that we found (33 is the minimal k such that $\lfloor 256/k \rfloor = 7$), and its experimentally observed conditional probability of each event. These two keys differ at the last byte, hence d=32.

```
K1 = 00 3d 3f 08 4f cd d8 f1 11 8c 83 80 1e 7f 5b c3
     d9 60 e2 c8 22 88 3c bc 56 2c 22 d2 b3 d9 ab d9 41


K2 = 00 3d 3f 08 4f cd d8 f1 11 8c 83 80 1e 7f 5b c3
     d9 60 e2 c8 22 88 3c bc 56 2c 22 d2 b3 d9 ab d9 42
```

Table 7 shows that our experimental results, where we obtained four 33-byte near-colliding key pairs from $2^{44.2}$ candidates, mostly agree with our probability estimation. Now the expected number of k-byte near-colliding key pairs out of a total of 2^{8k} keys is

$$NearColPairs(k) = 2^{8k} \times \sum_{d=0}^{k-1} NearColProb(k, d).$$

Table 8 is a list of $log_2(NearColPairs(k))$ for k=16...64. This table clearly shows that near-key collisions can also exist in much shorter key length. However,

Table 7. Experimental Results of Finding 33-byte Near-colliding Key Pairs

Event	Estimated Prob.	Measured Prob.
(a)	$0.784163 = (254/256)^{31}$	$0.784163 = 15852958662942/20216411062272$
(d)	$0.885741 = (255/256)^{31}$	$0.885471 = 14037333620475/15852958662942$
(e-1)	$0.003906 = 1/256$	$0.001946 = 27312181761/14037333620475$
(f-1)	$0.885741 = (255/256)^{31}$	$0.885555 = 24186440984/27312181761$
(e-2)	$0.003906 = 1/256$	$0.003954 = 95628579/24186440984$
(f-2)	$0.885741 = (255/256)^{31}$	$0.885499 = 84679053/95628579$
(e-3)	$0.003906 = 1/256$	$0.003930 = 332809/84679053$
(f-3)	$0.885741 = (255/256)^{31}$	$0.886160 = 885741/332809$
(e-4)	$0.003906 = 1/256$	$0.003774 = 1113/294922$
(f-4)	$0.882281 = (255/256)^{31}$	$0.881402 = 981/1113$
(e-5)	$0.003906 = 1/256$	$0.005097 = 5/981$
(f-5)	$0.882281 = (255/256)^{31}$	$0.800000 = 4/5$
(g)	$0.902344 = 231/256$	$1.000000 = 4/4$

Table 8. List of $log_2(NearColPairs(k))$ for k=16...64

k	Pairs	k	Pairs	k	Pairs	k	Pairs	k	Pairs
15	-	25	122.8	35	227.1	45	322.2	55	410.6
16	2.7	26	136.5	36	235.5	46	330.7	56	418.9
17	18.7	27	146.3	37	248.1	47	339.1	57	427.2
18	34.5	28	155.1	38	257.8	48	347.4	58	435.4
19	49.9	29	168.8	39	266.5	49	355.7	59	443.6
20	64.6	30	178.3	40	275.0	50	363.9	60	451.8
21	74.7	31	187.1	41	283.4	51	372.1	61	460.0
22	89.6	32	195.5	42	291.7	52	384.4	62	468.1
23	98.9	33	209.5	43	303.5	53	393.5	63	476.3
24	113.6	34	218.5	44	313.4	54	402.1	64	484.4

again, more detailed probabilistic analysis is needed for accurately estimating the density of near-colliding key pairs in a small key size.

6   Faster Collision Search

In previous sections we searched for colliding and near-colliding key pairs in a very simple fashion, — checking transition patterns step-by-step and if fails, restarting the search with another random candidate —, whose primary purpose was to confirm theoretical claims experimentally. In this section we explore a faster method for finding (near-)colliding key pairs for smaller key sizes.

We here try to find (near-)colliding key pairs by checking distance between two keys at each step but not checking our transition patters. We now define MaxColStep(K1,K2) as maximal step i such that distance between K1 and K2 is at most two at all steps up to step i. For a colliding or near-colliding key pair, `MaxColStep(K1,K2)=255`. Also for given key K and $0 \leq x,y<256$, we define a slightly modified key K<x,y> as

```
K<x,y>[x]   = K[x]   +y
K<x,y>[x+1] = K[x+1]-y
K<x,y>[i]   = K[i]         if i is not x or x+1.
```

It is naturally expected that the given key K and its modified keys K<x,y> are likely to create similar initial states. Hence when the check fails for K1 and K2, we can try K1<x,y> and K2<x,y>, which are likely to be a better pair, instead of rewinding and restarting the search with another random candidate. Which pair is better (i.e. closer to an actual (near-)collision) can be measured by the `MaxColStep` function. These observations lead to the following simple recursive search algorithm,

Collision Search Algorithm: Generate a random key pair K1 and K2 which differ at position d by one. Set `K1[d+1]=K2[d+1]=k-d-1` (see section 4) and call `Search(K1,K2)`. Repeat this until a (near-)colliding key pair is found:

```
Search(K1,K2)
S = MaxColStep(K1,K2)
if S = 255 then stop (found a (near-)collision!) or return (to find more)
MaxS = max_{x,y} MaxColStep(K1<x,y>,K2<x,y>)
if MaxS ≤ S then return
C = 0
For all x and y, do the following:
if MaxColStep(K1<x,y>,K2<x,y>) = MaxS
call Search(K1<x,y>,K2<x,y>)
C = C + 1
if C = MaxC then return
endif
return
```

where x runs from 0 to 255 except d and d+1 (for not changing `K1[d+1]`/K2 `[d+1]`), and y runs from 1 to 255. This algorithm finds mostly near-colliding key pairs, but may also find colliding key pairs (if we are lucky enough or patient enough). In fact, using this algorithm we reached a 24-byte colliding key pair and a 20-byte near-colliding key pair as follows:

```
[24-byte Colliding Key Pair]

K1 = 00 42 CE D3 DF DD B6 9D 41 3D BD 3A B1 16 5A 33
        ED A2 CD 1F E2 8C 01 76

K2 = 00 42 CE D3 DF DD B6 9D 41 3D BD 3A B1 16 5A 33
        ED A2 CD 1F E2 8C 01 77

[20-byte Near Colliding Key Pair]
```

```
K1 = 00 73 2F 6A 01 37 89 C5 15 49 9A 55 98 54 D7 53 4E F6 4F DC

K2 = 00 73 2F 6A 01 37 89 C5 15 49 9A 55 98 54 D7 53 4E F6 4F DD
```

MaxC is a pre-defined value. In our experiments, the search worked efficiently when MaxC is around 10, and the maximal depth of recursive calls was less than 20. Obviously this algorithm has a room for improvement. For instance, it can pick up the same near-colliding key pair twice or more; that is, the search contains some redundancy. Also another evaluation function, instead of MaxColStep, or another key modification is a possibility. Studying a better collision search method seems an interesting future topic.

7   Concluding Remarks

This paper explored key collisions of the RC4 stream cipher. We presented a 24-byte colliding key pair and a 20-byte near-colliding key pair, and demonstrated that our probabilistic analysis well agrees with experimental results when the key size is around 32 bytes. It seems now very likely that RC4 has colliding keys in even smaller key length, say less than 20 bytes.

While tables 5 and 8 suggest an existence of 17-byte colliding key pairs and 16-byte near-colliding key pairs, respectively, further research is needed for more accurately estimating $ColProb(k, d)$ and $NearColProb(k, d)$ in such small k. In fact, we have already seen that the observed probability of event (e-1) in table 7 was much smaller than the expected probability $1/256$. This kind of phenomenons (much larger or smaller than our estimation) frequently appears when k is small.

It might not be very easy to derive a simple formula of $ColProb(k, d)$ and $NearColProb(k, d)$ applicable to all k and d. As far as we know, this is the first paper that went deep into key collisions of RC4. We hope that our observation will lead to further study of this direction.

Acknowledgements

The author would like to thank an anonymous reviewer for suggesting the fast search algorithm, which has been further improved by the author and included in section 6. Another anonymous reviewer has pointed out that $ColProb(k, d)$ and $NearColProb(k, d)$ can approximate to $1/e(1/256)^{n+2}$ and $1/e(1/256)^{n}$, respectively, which makes things much clearer. The author is grateful to all reviewers for their valuable and constructive comments.

References

1. Anonymous: RC4 Source Code. CypherPunks mailing list (September 9, 1994), http://cypherpunks.venona.com/date/1994/09/msg00304.html, http://groups.google.com/group/sci.crypt/msg/10a300c9d21afca0

2. Biham, E., Carmeli, Y.: Efficient Reconstruction of RC4 Keys from Internal States. In: Nyberg, K. (ed.) FSE 2008. LNCS, vol. 5086, pp. 270–288. Springer, Heidelberg (2008)

3. Fluhrer, S., Mantin, I., Shamir, A.: Weaknesses in the Key Scheduling Algorithm of RC4. In: Vaudenay, S., Youssef, A.M. (eds.) SAC 2001. LNCS, vol. 2259, pp. 1–24. Springer, Heidelberg (2001)

4. Fluhrer, S., McGrew, D.: Statistical Analysis of the Alleged RC4 Keystream Generator. In: Schneier, B. (ed.) FSE 2000. LNCS, vol. 1978, pp. 19–30. Springer, Heidelberg (2000)

5. Grosul, A.L., Wallach, D.S.: A Related-Key Cryptanalysis of RC4. Technical Report TR-00-358, Department of Computer Science, Rice University (2000), http://cohesion.rice.edu/engineering/computerscience/tr/ TR_Download.cfm?SDID=126

6. Klein, A.: Attacks on the RC4 Stream Cipher. Designs, Codes and Cryptography 48(3), 269–286 (2008)

7. Mantin, I.: Predicting and Distinguishing Attacks on RC4 Keystream Generator. In: Cramer, R. (ed.) EUROCRYPT 2005. LNCS, vol. 3494, pp. 491–506. Springer, Heidelberg (2005)

8. Mantin, I., Shamir, A.: A Practical Attack on Broadcast RC4. In: Matsui, M. (ed.) FSE 2001. LNCS, vol. 2355, pp. 152–164. Springer, Heidelberg (2001)

9. Paul, S., Preneel, B.: A New Weakness in the RC4 Keystream Generator and an Approach to Improve Security of the Cipher. In: Roy, B., Meier, W. (eds.) FSE 2004. LNCS, vol. 3017, pp. 245–259. Springer, Heidelberg (2004)

10. Roos, A.: A Class of Weak Keys in the RC4 Stream Cipher (1995), http://marcel.wanda.ch/Archive/WeakKeys

11. Tews, E., Weinmann, R.P., Pyshkin, A.: Breaking 104 Bit WEP in Less than 60 Seconds. In: Kim, S., Yung, M., Lee, H.-W. (eds.) WISA 2007. LNCS, vol. 4867, pp. 188–202. Springer, Heidelberg (2007)

12. Vaudenay, S., Vuagnoux, M.: Passive-Only Key Recovery Attacks on RC4. In: Adams, C., Miri, A., Wiener, M. (eds.) SAC 2007. LNCS, vol. 4876, pp. 344–359. Springer, Heidelberg (2007)

Intel's New AES Instructions for Enhanced Performance and Security

Shay Gueron[1,2]

[1] Intel Corporation, Mobility Group, Israel Development Center, Haifa, Israel
[2] University of Haifa, Faculty of Science, Department of Mathematics, Haifa, Israel

Abstract. The Advanced Encryption Standard (AES) is the Federal Information Processing Standard for symmetric encryption. It is widely believed to be secure and efficient, and is therefore broadly accepted as the standard for both government and industry applications. If fact, almost any new protocol requiring symmetric encryption supports AES, and many existing systems that were originally designed with other symmetric encryption algorithms are being converted to AES. Given the popularity of AES and its expected long term importance, improving AES performance and security has significant benefits for the PC client and server platforms. To this end, Intel is introducing a new set of instructions into the next generation of its processors, starting from 2009. The new architecture has six instructions: four instructions (AESENC, AESENCLAST, AESDEC, and AESDELAST) facilitate high performance AES encryption and decryption, and the other two (AESIMC and AESKEYGENASSIST) support the AES key expansion. Together, these instructions provide full hardware support for AES, offering high performance, enhanced security, and a great deal of software usage flexibility, and are therefore useful for a wide range of cryptographic applications. The AES instructions can support AES encryption and decryption with each one of the standard key lengths (128, 192, and 256 bits), using the standard block size of 128 bits. They can also be used for all other block sizes of the general RIJNDAEL cipher. The instructions are well suited to all common uses of AES, including bulk encryption/decryption using cipher modes such as ECB, CBC and CTR, data authentication using CBC-MACs (e.g., CMAC), random number generation using algorithms such as CTR-DRBG, and authenticated encryption using modes such as GCM. Beyond improving performance, the AES instructions provide important security benefits. Since the instructions run in data independent time and do not use table lookups, they help eliminating the major timing and cache-based attacks that threaten table-lookup based software implementations of AES. In addition, these instructions make AES simple to implement, with reduced code size. This helps reducing the risk of inadvertent introduction of security flaws, such as difficult-to-detect side channel leaks. This paper provides an overview of the new AES instructions and how they can be used for achieving high performance and secure AES processing. Some special usage models of this architecture are also described.

O. Dunkelman (Ed.): FSE 2009, LNCS 5665, pp. 51–66, 2009.

Keywords: Advanced Encryption Standard, computer architecture, new instructions set.

1   Introduction

The Advanced Encryption Standard (AES), defined in 2001 by NIST [11]. (FIPS197 hereafter), is considered the state of the art in symmetric encryption, and a crucial ingredient for security and privacy applications. Rising requirements for high encryption/decryption bandwidths that have minimal impact on the user experience, increase the value of a high throughput AES solution for commodity processors. One example is disk encryption applications, such as Microsoft's Vista BitLocker [10], where due to increased volume size and disks speed, software encryption overhead may become a bottleneck for both the client and the server platforms.

The security of AES execution is an additional consideration added to the PC environment due to increased awareness to recent side channel attacks on AES software that uses lookup tables (e.g., [13]). Mitigation techniques significantly degrade the resulting performance, therefore making a hardware based AES solution even more advantageous.

Intel offers a comprehensive hardware solution for AES, introducing six new instructions to its processors, starting from the processor called "Westmere".

This paper describes the instructions, how they can be used efficiently and flexibly, and explains some of the benefits of this particular AES architecture.

2   Intel's AES Architecture

2.1   Preliminaries and Notations

Hereafter, we use the terminology of FIPS197, which details of all transformations, flows for encryption/decryption and key expansion that define AES.

We point out some subtlety related to the notation conventions. FIPS197 defines AES in terms of bytes. However, the algorithm is described using a text convention where hexadecimal strings are written with the low-memory byte on the left, and the high-memory byte on the right (this convention is analogous to writing integers in a "Big Endian" convention). This text convention determines the way in which the test vectors are written, and the description of some of the transformations. On the other hand, Intel's Architecture convention is the opposite: hexadecimal strings are written with the low-memory byte on the right and the high-memory byte on the left (analogous to writing integers in a "Little Endian" convention). Of course, store/load processor operations are consistent with the way that the AES instructions operate (i.e., using these instructions does not require any byte reversal). For reference, we provide here an example for all of the eight AES transformations, expressed in the "Little Endian" convention which is used on Intel's processors.

SubBytes(73744765635354655d5b56727b746f5d) =
8f92a04dfbed204d4c39b1402192a84c

MixColumns(627a6f6644b109c82b18330a81c3b3e5) =
7b5b54657374566563746f725d53475d

ShiftRows(7b5b54657374566563746f725d53475d) =
73744765635354655d5b56727b746f5d

InvMixColumns(8dcab9dc035006bc8f57161e00cafd8d) =
5be3eb11928b5eaeeec9cc3bc55f5777

InvShiftRows(7b5b54657374566563746f725d53475d) =
5d7456657b536f65735b47726374545d

InvSubBytes(5d7456657b536f65735b47726374545d) =
8dcab9dc035006bc8f57161e00cafd8d

RotWord(3c4fcf09) = 093c4fcf   SubWord(73744765) = 8f92a04d

Fig. 1. The AES transformations expressed in "Little Endian" notation, as used in Intel's architecture

2.2   The Six AES Instructions

Intel's architecture offers six instructions to support AES (see Fig. 2). AESENC, AESENCLAST, support encryption. AESDEC and AESDECLAST are building blocks suitable for decryption using the Equivalent Inverse Cipher (see FIPS197 for definition). Each instruction has a register-memory and a register-register variant. AESIMC and AESKEYGENASSIST support the Key Expansion. AES-IMC facilitates the conversion of the encryption round keys to a form suitable for the Equivalent Inverse Cipher. AESKEYGENASSIST uses an immediate byte as part of the input (used as RCON).

3   Basic Usage of the AES Instructions

This section illustrates the basic usage of the AES instructions, using AES-128 (ECB mode) as an example. The general paradigm is that for AESENC, AESENCLAST, AESDEC, AESDECLAST, the inputs xmm1 and xmm2 are interpreted as xmm1 = State and xmm2 = Round Key. For AESIMC, the input xmm2 is interpreted as xmm2 = Round Key. Fig. 3 illustrates encryption/decryption flows. For AESKEYGENASSIST, the input should be interpreted as an intermediate step in the Key Expansion procedure, where the immediate byte is the value of RCON. An example for AES-128 Key Expansion is illustrated in Fig. 4 (Key Expansion for AES-192 and AES-256 is provided in the Appendix).

4   Some Design Considerations That Led to the Selection of the AES Architecture

Introducing a new instruction to Intel's processors implies long-term legacy commitment. Additionally, silicon area is a precious "real-estate". This mandates a great deal of care in the definitions and cost-performance-flexibility tradeoffs.

AESENC xmm1, xmm2/m128	AESENCLAST xmm1, xmm2/m128
Tmp := xmm1	Tmp := xmm1
RoundKey :=xmm2/m128	RoundKey := xmm2/m128
Tmp := ShiftRows (Tmp)	Tmp := ShiftRows (Tmp)
Tmp := SubBytes (Tmp)	Tmp := SubBytes (Tmp)
Tmp := MixColumns (Tmp)	
xmm1:= Tmp xor RoundKey	xmm1:= Tmp xor RoundKey
AESDEC xmm1, xmm2/m128	**AESDECLAST xmm1, xmm2/m128**
Tmp:=xmm1	Tmp:= xmm1
RoundKey := xmm2/m128	RoundKey := xmm2/m128
Tmp := InvShiftRows (Tmp)	Tmp := InvShiftRows (Tmp)
Tmp := InvSubBytes (Tmp)	Tmp := InvSubBytes (Tmp)
Tmp := InvMixColumns (Tmp)	
xmm1:= Tmp xor RoundKey	xmm1:= Tmp xor RoundKey

AESKEYGENASSIST xmm1, xmm2/m128, imm8
Tmp := xmm2/m128
RCON[31–8] := 0; RCON[7–0] := imm8;
X3[31–0] := Tmp[127–96]; X2[31–0] := Tmp[95–64];
X1[31–0] := Tmp[63–32]; X0[31–0] := Tmp[31–0];
xmm1 := [RotWord (SubWord (X3)) XOR RCON, SubWord (X3),
Rotword (SubWord (X1)) XOR RCON, SubWord (X1)]

AESIMC xmm1, xmm2/m128
RoundKey := xmm2/m128;
xmm1 := InvMixColumns (RoundKey)

Examples:	
xmm1 =	7b5b54657374566563746f725d53475d
xmm2 =	48692853686179295b477565726f6e5d
AESENC result:	a8311c2f9fdba3c58b104b58ded7e595
AESENCLAST result:	c7fb881e938c5964177ec42553fdc611
AESDEC result:	138ac342faea2787b58eb95eb730392a
AESDECLAST result:	c5a391ef6b317f95d410637b72a593d0
xmm2 =	7b5b54657374566563746f725d53475d
AESIMC result:	627a6f6644b109c82b18330a81c3b3e5
xmm2 =	3c4fcf098815f7aba6d2ae2816157e2b; imm8 = 1
AESKEYGENASSIST result:	01eb848beb848a013424b5e524b5e434

Fig. 2. Functional descriptions (architectural behavior) and examples of the AES instructions (note that ShiftRows and SubBytes, InvShiftRows and InvSubBytes commute)

Obviously, the AES architecture must offer an adequate solution for the short term requirements, but as importantly, it should have the ability to accommodate long range requirements that may emerge in the future. Therefore, the AES architecture needs to address the following considerations: a) Flexibility, b) Performance, c) Performance scalability, d) Security. We explain how these properties are achieved by the AES architecture.

4.1   Design for Software Flexibility

Software flexibility implies that the architecture should be able to support all of the current usage models for AES. Indeed, it is easy to realize that this the case with the new AES instructions: They are the building blocks that can support all the AES variants defined by FIPS197, uses of AES in cipher modes such as CBC or CTR, data authentication using CBC-MACs such as CMAC, random number generation using algorithms such as CTR-DRBG, and authenticated encryption using modes such as GCM. As an example, Fig. 5 shows encryption in CBC mode.

AES-128 encryption	Decryption Round Keys	AES-128 decryption
pxor xmm1, xmm2		pxor xmm1, xmm12;
AESENC xmm1, xmm3	AESIMC xmm3, xmm3	AESDEC xmm1, xmm11
AESENC xmm1, xmm4	AESIMC xmm4, xmm4	AESDEC xmm1, xmm10
AESENC xmm1, xmm5	AESIMC xmm5, xmm5	AESDEC xmm1, xmm9
AESENC xmm1, xmm6	AESIMC xmm6, xmm6	AESDEC xmm1, xmm8
AESENC xmm1, xmm7	AESIMC xmm7, xmm7	AESDEC xmm1, xmm7
AESENC xmm1, xmm8	AESIMC xmm8, xmm8	AESDEC xmm1, xmm6
AESENC xmm1, xmm9	AESIMC xmm9, xmm9	AESDEC xmm1, xmm5
AESENC xmm1, xmm10	AESIMC xmm10, xmm10	AESDEC xmm1, xmm4
AESENC xmm1, xmm11	AESIMC xmm11, xmm11	AESDEC xmm1, xmm3
AESENCLAST xmm1, xmm12		AESDECLAST xmm1, xmm2

Fig. 3. Left panel: AES-128 encryption. Register xmm1 holds the data to encrypt, xmm2 is the whitening key, and xmm3–xmm12 hold Round Keys 1–10. The AES flow starts with a whitening step (XOR with xmm2). Rounds 1–9 are implemented using AESENC, and round 10 is implemented using AESENCLAST. Middle panel: AESIMC is used for transforming the round keys for decryption using the Equivalent Inverse Cipher. Right panel: AES-128 decryption. Register xmm1 holds the data to decrypt. Registers xmm12-xmm2 hold the decryption round keys and the whitening key.

```
movdqu xmm1, XMMWORD PTR Key
movdqu XMMWORD PTR Key_Sched, xmm1        key_expansion_128:
mov rcx, OFFSET Key_Schedule+16
                                          pshufd xmm2, xmm2, 0xff
AESKEYGENASSIST xmm2, xmm1, 0x1           vpslldq xmm3, xmm1, 0x4
call key_expansion_128                    pxor xmm1, xmm3
AESKEYGENASSIST xmm2, xmm1, 0x2           vpslldq xmm3, xmm1, 0x4
call key_expansion_128                    pxor xmm1, xmm3
AESKEYGENASSIST xmm2, xmm1, 0x4           vpslldq xmm3, xmm1, 0x4
call key_expansion_128                    pxor xmm1, xmm3
AESKEYGENASSIST xmm2, xmm1, 0x8           pxor xmm1, xmm2
call key_expansion_128                    movdqu XMMWORD PTR [rcx], xmm1
AESKEYGENASSIST xmm2, xmm1, 0x10          add rcx, 0x10
call key_expansion_128                    ret
AESKEYGENASSIST xmm2, xmm1, 0x20
call key_expansion_128
AESKEYGENASSIST xmm2, xmm1, 0x40
call key_expansion_128
AESKEYGENASSIST xmm2, xmm1, 0x80
call key_expansion_128
AESKEYGENASSIST xmm2, xmm1, 0x1b
call key_expansion_128
AESKEYGENASSIST xmm2, xmm1, 0x36
call key_expansion_128
```

Fig. 4. AES-128 Key Expansion example (the cipher key is stored in the array "Key" and the generated key expansion is stored in the array "Key_Sched". (see comments in the Appendix)

Software has the flexibility to pre-expand the keys and re-use them (which is the typical usage model in bulk encryption) or to expand them on-the-fly. In addition, when compared with existing software implementations, one can realize that the AES instructions can help reduce the associated code size. We also point out here that the AES round instructions remain as useful as they are now, even if future analysis would change the standard to perform more rounds during

```
void AES_128_CBC_Encrypt (...) {
    int i, j, k;
    __m128i tmp, feedback;
    __m128i RKEY [11];
    for (k=0; k<11; k++) {
        RKEY [k] = _mm_load_si128 ( (__m128i*)&Key_Schedule [4*k]);
    }
    feedback = _mm_load_si128 ( (__m128i*)&IV [0]);
    for(i=0; i < NBLOCKS; i++) {
        tmp = _mm_load_si128 ( (__m128i*)&PLAINTEXT[i*4]);
        tmp = _mm_xor_si128 (tmp,feedback);
        tmp = _mm_xor_si128(tmp, RKEY[0]);
        for(j=1; j < 10; j++) {
            tmp = _mm_aesenc_si128 (tmp, RKEY [j]);
        }
        tmp = _mm_aesenclast_si128 (tmp, RKEY [10]);
        feedback = tmp;
        _mm_store_si128 ((__m128i*)&CIPHERTEXT[4*i], tmp);
    }
}
```

Fig. 5. Encryption in CBC mode. A C code snippet, using compiler intrinsics, illustrates a function that encrypts NBLOCKS data blocks.

encryption/decryption. Furthermore, as long as the Key Expansion procedure is not fundamentally changed, AESKEYGENASSIST (taking any Round Constant as an input byte) could be used for generating additional round keys.

4.2   Design for Performance

Performance is a main motivation for introducing the AES instructions. To this end, the architecture takes advantage of the 128-bit data-path available in the Intel's modern processors (compare with the 32-bit instructions proposed in [14], in a different setup, that does not have such a wide data-path).

The AES architecture is optimized for the common usage model for the PC platform where the round keys are generated once, stored in registers or in the cache memory, and then used for multiple data blocks. To this end, the hardware support for the key expansion is decoupled from the more performance-critical encryption/decryption acceleration. The four AES rounds instructions encapsulate the maximal sequence of transformations which is possible without having micro-architectural branches. To illustrate, consider a possible alternative instruction such as AESROUND xmm1, xmm2, imm8, where the immediate byte is a control that selects encryption/decryption and round/last round. Such architecture would require the implementation to have micro-branching which could incur some performance loss. To avoid this, four separate instructions are dedicated to each of the four "flavors" of the AES rounds.

4.3   Design for Performance Scalability

Performance scalability is also achieved by encapsulating the "maximal" possible flow in the performance-critical instructions, thus leaving room for micro architectural cost-performance tradeoffs. To illustrate this flexibility, consider the

AESENC instruction that performs tha sequence of transformation: ShiftRows; SubBytes; MixColumns; AddRoundKey (=XOR). These could be implemented by one piece of dedicated hardware, or by means of hardware elements that process the data in small granularity combined with some micro-instruction flows. Thus, it is possible to choose the cost-performance balance across processors and processors generations, according the performance requirements.

To show the benefit of bundling the maximal flow in one instruction, consider the following alternative of having two separate instructions, SUBBYTES xmm1, xmm2, and MIXCOL xmm1, xmm2. With these, the AES encryption round could be performed by the sequence PSHUFB (for ShiftRows), SUBBYTES, MIXCOL, PXOR. However, such an architectural approach limits the highest possible performance of the instruction.

4.4   Design for Security

We briefly explain here how side channel attacks can compromise the security of AES software implementations, and how the new architecture mitigates this problem.

Processor cache is a special type of memory that allows faster access compared to accessing main memory. The processor stores recently read memory areas in cache, with the speculative anticipation that these areas would be re-accessed "soon". In each memory access, the processor first checks if the data is in the cache (enjoying fast access) and if not, it reads from main memory (or lower level caches), and stores it in the cache for future usage. To place new data in the cache, the processor needs to evict less recent data.

Currently, many common efficient software implementations of AES use lookup tables (e.g., Gladman's code [4], OpenSSL [12], and Lipmaa [6,7]). The entries in the table(s), which are read during encryption, depend implicitly on the secret round key and on the processed data. A "spy process", which runs at the same privilege level, can exploit this fact: it runs in parallel to some AES code, fills the cache lines with its own data, and reads them again after the table was accessed by the AES code. Depending on the reading latency that the spy experiences (for its own data, as measured by using the RDTSC instruction), it can discover if the corresponding cache line was evicted or not, and therefore deduce which part of the table was accessed by the AES code. Repeatedly collected, and combined with the appropriate analysis, this information could eventually leak out the secret key (see e.g., [13,3]).

These side channel threat can be avoided by writing the AES software in a way that the memory access patterns hide the key dependence (e.g., [13,3]). However, these mitigation techniques may involve a significant performance penalty. There are also software implementations of AES that do not use table lookup at all (e.g., Matsui [8,9], Bernstein and Schwabe [2]).

The AES instructions are designed to mitigate all of the known timing and cache side channel leakage of sensitive data. Their latency is data-independent, and since all the computations are performed internally by the hardware, no lookup tables are required. Therefore, if the AES instructions are used properly, the AES encryption/decryption, as well as the key expansion, would have data-independent timing and would involve only data-independent memory access. Consequently, the AES instructions allow for easily writing high performance AES software which is, at the same time, protected against the currently known side channel threats.

5   Performance Optimizations for Parallel Modes of Operation

Significant performance optimization for encryption/decryption using the AES instructions can be achieved by re-ordering the code. This helps taking better advantage of parallelism in parallel modes of operation such as ECB, CTR, and CBC-Decrypt (with the CBC-Encrypt serial mode being the exception). This section explains how it can be done.

The hardware that supports the four AES round instructions is pipelined. This allows independent AES instructions to be dispatched theoretically every 1–2 CPU clock cycle, if data can be provided sufficiently fast. As a result, the AES throughput can be significantly enhanced for parallel modes of operation, if the "order of the loop" is reversed: instead of completing the encryption of one data block and then continuing to the subsequent block, it is preferable to write software sequences that compute one AES round on multiple blocks, using one round key, and only then continue to computing the subsequent round on

```
; load Round key
mov rdx, OFFSET keyex_addr          add rdx, 0x10
movdqu xmm0, XMMWORD PTR [rdx]      movdqu xmm0, XMMWORD PTR [rdx]

pxor xmm1, xmm0                     aesenclast xmm1, xmm0
pxor xmm2, xmm0                     aesenclast xmm2, xmm0
pxor xmm3, xmm0                     aesenclast xmm3, xmm0
pxor xmm4, xmm0                     aesenclast xmm4, xmm0
pxor xmm5, xmm0
pxor xmm6, xmm0                     aesenclast xmm5, xmm0
pxor xmm7, xmm0                     aesenclast xmm6, xmm0
pxor xmm8, xmm0                     aesenclast xmm7, xmm0
                                    aesenclast xmm8, xmm0
mov ecx, 9

main_loop:                         ; storing the encrypted blocks
; load Round key
add rdx, 0x10                       movdqu XMMWORD PTR [dest], xmm1
movdqu xmm0, XMMWORD PTR [rdx]      movdqu XMMWORD PTR [dest+0x10], xmm2
                                   movdqu XMMWORD PTR [dest+0x20], xmm3
aesenc xmm1, xmm0                   movdqu XMMWORD PTR [dest+0x30], xmm4
aesenc xmm2, xmm0                   movdqu XMMWORD PTR [dest+0x40], xmm5
aesenc xmm3, xmm0                   movdqu XMMWORD PTR [dest+0x50], xmm6
aesenc xmm4, xmm0                   movdqu XMMWORD PTR [dest+0x60], xmm7
aesenc xmm5, xmm0                   movdqu XMMWORD PTR [dest+0x70], xmm8
aesenc xmm6, xmm0
aesenc xmm7, xmm0
aesenc xmm8, xmm0

loop main_loop
```

Fig. 6. Encrypting multiple data blocks in parallel (ECB mode)

for multiple blocks (using another round key). For such optimization, one needs to choose the number of blocks that will be processed in parallel. The optimal parallelization parameter value depends on the scenario, for example on how many registers are available, and how many data blocks are to be (typically) processed. In general, it is useful to process 4–8 blocks in parallel, in order to achieve high throughput. We provide here two examples: Figure 6 outlines assembler code for encrypting 8 blocks in parallel, in ECB mode, and Figure 7 gives a C code snippet for decrypting 4 blocks in parallel in CBC mode.

```
void AES_128_CBC_Decrypt_C_4_blocks (...) {
    __m128i RKEY_DECRYPT [11];
    __m128i tmp1, tmp2, tmp3, tmp4, feedback;
    __m128i z1, z2, z3, z4;
    int j, k;
    for (k=0; k<11; k++) {
        RKEY_DECRYPT [10-k] =
        _mm_load_si128 ( (__m128i*)&Key_Schedule_Decrypt [4*k]);
    }
    feedback = _mm_load_si128 ( (__m128i*)&IV [0]);

    z1 = _mm_load_si128 ( (__m128i*)&CIPHERTEXT[0]);
    z2 = _mm_load_si128 ( (__m128i*)&CIPHERTEXT[4]);
    z3 = _mm_load_si128 ( (__m128i*)&CIPHERTEXT[8]);
    z4 = _mm_load_si128 ( (__m128i*)&CIPHERTEXT[12]);

    tmp1 = _mm_xor_si128(z1,RKEY_DECRYPT[0]);
    tmp2 = _mm_xor_si128(z2,RKEY_DECRYPT[0]);
    tmp3 = _mm_xor_si128(z3,RKEY_DECRYPT[0]);
    tmp4 = _mm_xor_si128(z4,RKEY_DECRYPT[0]);

    for(j=1; j <10; j++) {
        tmp1 = _mm_aesdec_si128 (tmp1, RKEY_DECRYPT [j]);
        tmp2 = _mm_aesdec_si128 (tmp2, RKEY_DECRYPT [j]);
        tmp3 = _mm_aesdec_si128 (tmp3, RKEY_DECRYPT [j]);
        tmp4 = _mm_aesdec_si128 (tmp4, RKEY_DECRYPT [j]);
    }
    tmp1 = _mm_aesdeclast_si128 (tmp1, RKEY_DECRYPT [10]);
    tmp2 = _mm_aesdeclast_si128 (tmp2, RKEY_DECRYPT [10]);
    tmp3 = _mm_aesdeclast_si128 (tmp3, RKEY_DECRYPT [10]);
    tmp4 = _mm_aesdeclast_si128 (tmp4, RKEY_DECRYPT [10]);

    tmp4 = _mm_xor_si128(tmp4,z3);
    tmp3 = _mm_xor_si128(tmp3,z2);
    tmp2 = _mm_xor_si128(tmp2,z1);
    tmp1 = _mm_xor_si128(tmp1,feedback);

    _mm_store_si128 ((__m128i*)&DECRYPTED_TEXT[0], tmp1);
    _mm_store_si128 ((__m128i*)&DECRYPTED_TEXT[4], tmp2);
    _mm_store_si128 ((__m128i*)&DECRYPTED_TEXT[8], tmp3);
    _mm_store_si128 ((__m128i*)&DECRYPTED_TEXT[12], tmp4);
}
```

Fig. 7. Decrypting 4 blocks in parallel, in CBC mode (C code using compiler intrinsics)

5.1 Parallelizing CBC Encryption for Performance

CBC encryption is a serial mode of operation, because encrypting a block requires the encryption result of the previous block. Therefore, CBC encryption does not allow for hiding the latency of the AES instructions by operating on

```
void AES_128_CBC_Encrypt_Parallel_4_Blocks (...) {

    int i, j, k;
    __m128i tmp, feedback, feedback1, feedback2, feedback3, feedback4;
    __m128i tmp1, tmp2, tmp3, tmp4;
    __m128i RKEY [11];

    for (k=0; k<11; k++) {
        RKEY [k] = _mm_load_si128 ( (__m128i*)&Key_Schedule [4*k]);
    }

    feedback1 = _mm_load_si128 ( (__m128i*)&IV1 [0]);
    feedback2 = _mm_load_si128 ( (__m128i*)&IV2 [0]);
    feedback3 = _mm_load_si128 ( (__m128i*)&IV3 [0]);
    feedback4 = _mm_load_si128 ( (__m128i*)&IV4 [0]);

    for(i=0; i < NBLOCKS; i++) {
        tmp1 = _mm_load_si128 ( (__m128i*)&PLAINTEXT1[i*4]);
        tmp2 = _mm_load_si128 ( (__m128i*)&PLAINTEXT2[i*4]);
        tmp3 = _mm_load_si128 ( (__m128i*)&PLAINTEXT3[i*4]);
        tmp4 = _mm_load_si128 ( (__m128i*)&PLAINTEXT4[i*4]);

        tmp1 = _mm_xor_si128 (tmp1, feedback1);
        tmp2 = _mm_xor_si128 (tmp2, feedback2);
        tmp3 = _mm_xor_si128 (tmp3, feedback3);
        tmp4 = _mm_xor_si128 (tmp4, feedback4);

        tmp1 = _mm_xor_si128(tmp1,RKEY[0]);
        tmp2 = _mm_xor_si128(tmp2,RKEY[0]);
        tmp3 = _mm_xor_si128(tmp3,RKEY[0]);
        tmp4 = _mm_xor_si128(tmp4,RKEY[0]);

        for(j=1; j <10; j++) {
            tmp1 = _mm_aesenc_si128 (tmp1, RKEY [j]);
            tmp2 = _mm_aesenc_si128 (tmp2, RKEY [j]);
            tmp3 = _mm_aesenc_si128 (tmp3, RKEY [j]);
            tmp4 = _mm_aesenc_si128 (tmp4, RKEY [j]);
        }
        tmp1 = _mm_aesenclast_si128 (tmp1, RKEY [10]);
        tmp2 = _mm_aesenclast_si128 (tmp2, RKEY [10]);
        tmp3 = _mm_aesenclast_si128 (tmp3, RKEY [10]);
        tmp4 = _mm_aesenclast_si128 (tmp4, RKEY [10]);

        feedback1 = tmp1;
        feedback2 = tmp2;
        feedback3 = tmp3;
        feedback4 = tmp4;

        _mm_store_si128 ((__m128i*)&CIPHERTEXT1[4*i], tmp1);
        _mm_store_si128 ((__m128i*)&CIPHERTEXT2[4*i], tmp2);
        _mm_store_si128 ((__m128i*)&CIPHERTEXT3[4*i], tmp3);
        _mm_store_si128 ((__m128i*)&CIPHERTEXT4[4*i], tmp4);
    }
}
```

Fig. 8. CBC encryption for 4 blocks in parallel (C code using compiler intrinsics)

independent blocks as shown above. However, in some cases it is possible to parallelize CBC encryption if the application needs to operate on multiple independent data streams. One possible example can be disk encryption applications where disk sectors are encrypted independently (not necessarily with the same key). If the software can encrypt multiple sectors in parallel, the application can enjoy the speedup of a parallel mode. Figure 8 gives a C code snippet for encrypting 4 blocks in parallel, in CBC mode (in this example, using the same key and different IV's).

6 More on Software Flexibility and Surprising Usage Models

6.1 Supporting RIJNDAEL with Block Size Larger Than 128 Bits

Although the main usage model for the AES instructions is AES, which operates on 128-bit blocks, they can also be used for processing the general RIJNDAEL cipher that supports any block size which is a multiple of 32 bits, from 128 to 256 bits. Figure 9 gives an example for computing a RIJNDAEL-256 round, using the new AES instructions.

6.2 Isolating the AES Transformations

Cipher designers may wish to build new cryptographic algorithms using components of AES. Such algorithms could benefit from the performance and side channel protection of the AES instructions if they are designed to use the AES transformations. In particular, the AES transformations can be a useful building block for hash functions. For example, the MixColumns transformation provides rapid diffusion and the AES S-box is a good nonlinear mixer. Manipulations on large block sizes could be useful for constructing hash functions, with a long digest size. This concept is already being used in quite a few of the new Secure Hash Function algorithms that have been recently submitted to the NIST cryptographic hash Algorithm Competition (some of the examples from the First Round Candidates list include LANE, SHAMATA, SHAvite-3, ECHO, GrØstl, Lesamnta (512-bit), and Vortex). Some algorithms use the whole AES round as a building block, some only one AES transformations, and some use variants of these transformations.

```
VPBLENDVB xmm3, xmm2, xmm1, xmm5
VPBLENDVB xmm4, xmm1, xmm2, xmm5
PSHUFB xmm3, xmm8
PSHUFB xmm4, xmm8
AESENC xmm3, xmm6
AESENC xmm4, xmm7
```

Fig. 9. Using the AES instructions for computing a RIJNDAEL round with a 256-bits block size. Register xmm1 holds the "left" half of RIJNDAEL input state (columns 0–3), xmm2 hold the right half of state (columns 4–7), xmm6 and xmm7 hold the left half and right half of RIJNDAEL round key, respectively. The output state is written into registers xmm1 (left half) and xmm2 (right half). Register xmm8 holds a mask (0x03020d0c0f0e0908b0a050407060100) used for the shuffling step which is necessary to account for the difference in ShiftRows offsets between the 256 (1,3,4) and 128-bit (1,2,3) versions of RIJNDAEL. Register xmm5 holds a mask for VPBLENDVB, selecting bytes 1–3, 6–7, 10–11, and 15 of the RIJNDAEL state from the first source operand, and all other bytes from the second source operand.

Therefore, it is important to note that although the AES instructions perform bundled sequences of AES transformations, each one of these transformations can be isolated by a proper combination of these instructions, and the use of the byte shuffling (PSHUFB instruction). This is shown in Figure 10.

6.3   Using the AES Instructions for RAID-6

We show here a surprising usage for the AES instructions for a non cryptographic application.

A Redundant Array of Independent Disks (RAID) combines a multiple physical hard disk drives into a logical drive for purposes of reliability, capacity, or performance. A level 6 RAID (RAID-6) system provides a high level of redundancy allowing recovery from two disk failures. Two syndromes (P and Q) are generated for the data and stored on hard disk drives in the RAID system. The P syndrome is generated by computing parity information for the data in a strip. The generation of the Q syndrome requires Finite Field multiplications in $GF(2^8)$ defined by the reduction polynomial $x^8 + x^4 + x^3 + x + 1$ (same as the one used for AES). Recovering data and/or P and/or Q syndromes requires both $GF(2^8)$ multiplications and inversions. In a RAID array with n data disks $D_0, D_1, D_2, \ldots, D_{n-1}$ (for $n \le 255$) P and Q are defined by: $P = D_0 + D_1 + D_2 + \ldots + D_{n-1}$, and $Q = g^0 \cdot D_0 + g^1 \cdot D_1 + g_2 \cdot D_2 + \ldots + g^{n-1} \cdot D_{n-1}$, where $g = \{02\}$ is a generator of $GF(2^8)$, and $+$ and $\cdot$ denote the operations in this field. The computational bottleneck associated with the RAID-6 system is the cost of computing Q. The performance of the generation of the Q syndrome may be improved by expressing Q in its Horner representation $Q = ((\ldots D_{n-1} \ldots) \cdot g + D_2) \cdot g + D_1) \cdot g + D_0$. The difficulty in the related software implementation stems from the fact that traditional processors have poor performance with Finite Fields computations. See [1] for a detailed overview.

We now note that the MixColumns transformation is a matrix multiplication in $GF(2^8)$, therefore useful for computing the Q syndrome. In order to use

```
Isolating ShiftRows
        PSHUFB xmm0, 0x0b06010c07020d08030e09040f0a0500
Isolating InvShiftRows
        PSHUFB xmm0, 0x0306090c0f0205080b0e0104070a0d00
Isolating MixColumns
        AESDECLAST xmm0, 0x00000000000000000000000000000000
        AESENC xmm0, 0x00000000000000000000000000000000
Isolating InvMixColumns
        AESENCLAST xmm0, 0x00000000000000000000000000000000
        AESDEC xmm0, 0x00000000000000000000000000000000
Isolating SubBytes
        PSHUFB xmm0, 0x0306090c0f0205080b0e0104070a0d00
        AESENCLAST xmm0, 0x00000000000000000000000000000000
Isolating InvSubBytes
        PSHUFB xmm0, 0x0b06010c07020d08030e09040f0a0500
        AESDECLAST xmm0, 0x00000000000000000000000000000000
```

Fig. 10. Isolating the AES transformations using combinations of AES instructions

```
__declspec (align(16)) unsigned int zero_ [4] =
        {0x0, 0x0, 0x0, 0x0};
__declspec (align(16)) unsigned int mask1 [4] =
        {0xff02ff00,0xff06ff04,0xff0aff08, 0xff0eff0c};
__declspec (align(16)) unsigned int mask2 [4] =
        {0x03ff01ff,0x07ff05ff,0x0bff09ff, 0x0fff0dff};
__declspec (align(16)) unsigned int mask3 [4] =
        {0x01000302,0x05040706,0x09080b0a, 0x0d0c0f0e};

void RAID6_1_block_in_parallel (...) {
    int ind1;
    __m128i MASK1, MASK2, MASK3, ZERO;
    __m128i XMM0, XMM1, XMM2;

    MASK3 = _mm_load_si128 ((__m128i*)&mask3[0]);
    MASK2 = _mm_load_si128 ((__m128i*)&mask2[0]);
    MASK1 = _mm_load_si128 ((__m128i*)&mask1[0]);
    ZERO = _mm_load_si128 ((__m128i*)&zero_[0]);

    for (ind1=0; ind1 < NBLOCKS; ind1++) {
        XMM0 = _mm_load_si128 ((__m128i*)&DATA[4*ind1]);
        XMM1 = _mm_shuffle_epi8(XMM0, MASK1);
        XMM1 = _mm_aesdeclast_si128 (XMM1, ZERO);
        XMM2 = _mm_shuffle_epi8(XMM0, MASK2);
        XMM0 = _mm_shuffle_epi8(XMM0, MASK3);
        XMM1 = _mm_aesenc_si128(XMM1, ZERO);
        XMM2 = _mm_aesdeclast_si128 (XMM2, ZERO);
        XMM1 = _mm_shuffle_epi8(XMM1, MASK1);
        XMM2 = _mm_aesenc_si128(XMM2, ZERO);
        XMM2 = _mm_shuffle_epi8(XMM2, MASK2);
        XMM2 = _mm_xor_si128(XMM2, XMM1);
        XMM0 = _mm_xor_si128(XMM0, XMM2);

        _mm_store_si128 ((__m128i*)&RES[4*ind1], XMM0);
    }
}
```

Fig. 11. Using the AES instructions for RAID-6: multiplying 16 bytes by {02}

the AES instructions, the MixColumns transformation needs to be isolated, as explained above. This transformation operates separately on the 4 columns of the state. If a column (32 bits) is denoted by the four bytes $[d, c, b, a]$, then the output $[d', c', b', a']$ of MixColumns is $a' = (\{02\} \cdot a) + (\{03\} \cdot b) + c + d$; $b' = a + (\{02\} \cdot b) + (\{03\} \cdot c) + d$; $c' = a + b + (\{02\} \cdot c) + (\{03\} \cdot d)$; $d' = (\{03\} \cdot a) + b + c + (\{02\} \cdot d)$ denoted in shorthand by $[3a + b + c + 2d, a + b + 2c + 3d, a + 2b + 3c + d, 2a + 3b + c + d]$. If the bytes b, d (odd positions) are set to 0, then the result of MixColumns becomes $[3a+c, a+2c, a+3c, 2a+c]$, and with the PSHUFB instruction odd position bytes can be zeroed to yield $[0, a+2c, 0, 2a+c]$. If this result is XOR-ed with $[0, a, 0, c]$ (a shuffled version of the input), the final result is $[0, 2c, 0, 2a]$, that is, two of the 4 bytes of the column were multiplied by {02}. Similar operations can be applied to the even-positioned bytes of the state. Figure 11 shows a code snippet that uses the AES instructions for RAID-6 (here, for clarity and brevity the code operates on a single block at a time. Operating on multiple blocks in parallel, improves the performance as explained above).

7   Conclusion

This paper provided some details and insights on Intel's new AES instructions which are expected to be widely used for security and privacy, by a wide range of applications and operating systems.

The AES instructions provide a substantial performance speedup to bulk data encryption and decryption. Exact performance measurements will be made available as soon as processors with these instructions are released. However, we can indicate that when using parallelizable modes of operation (e.g., CBC decryption, CTR, and CTR-derived modes GCM, XTS), the performance speedup could exceed an order of magnitude over the current performance of software-only AES implementations. In scenarios where pipelined operation is impossible, for example in CBC encryption, operating on a single buffer, the performance speedup would still be significant, around 2–3 times over software implementation. Note that AES implementations using the new instructions are inherently protected against the software side channel attacks associated with AES implementations based on table-lookup.

The paper showed some of the advantages of the AES instructions and how they can be used flexibly and efficiently.

An important observation that we pointed out was that due to the out-of-order execution capabilities of modern processors, hardware pipelining, and software techniques, parallel modes of operation can achieve a much higher throughput than serial modes. This is one point to consider when selecting modes of operation in future cryptosystems. For example, AES-GCM may become a favorable mode for achieving secrecy and authentication. In this context, we also mention that, together with the AES instructions, another instruction for computing carry-less (polynomial) multiplications (called PCLMULDQ) is released. This could give further speedup to AES-GCM (see [5]).

Acknowledgements. Many people contributed to the concepts, the studies, the definition of the architecture, and to the micro-architectural implementation. The list of contributors includes: Roee Bar, Frank Berry, Mayank Bomb, Brent Boswell, Ernie Brickell, Yuval Bustan, Mark Buxton, Srinivas Chennupaty, Tiran Cohen, Martin Dixon, Jack Doweck, Vivek Echambadi, Wajdi Feghali, Shay Fux, Vinodh Gopal, Eugene Gorkov, Amit Gradstein, Mostafa Hagog, Israel Hayun, Michael Kounavis, Ram Krishnamurthy, Sanu Mathew, Henry Ou, Efi Rosenfeld, Zeev Sperber, Kirk Yap.

References

1. Anvin, H.P.: The mathematics of RAID-6,
   http://www.kernel.org/pub/linux/kernel/people/hpa/raid6.pdf
2. Bernstein, D.J., Schwabe, P.: New AES Software Speed Records. In: Chowdhury, D.R., Rijmen, V., Das, A. (eds.) INDOCRYPT 2008. LNCS, vol. 5365, pp. 322–336. Springer, Heidelberg (2008)

3. Brickell, E., Graunke, G., Neve, M., Seifert, J.P.: Software mitigations to hedge AES against cache based software side channel vulnerabilties, IACR ePrint Archive, Report 2006/052 (2006), http://eprint.iacr.org/2006/052
4. Gladman, B.: Implementations of AES (Rijndael) in C/C++ and assembler, http://www.gladman.me.uk/cryptography_technology/rijndael
5. Gueron, S., Kounavis, M.E.: Carry-Less Multiplication and Its Usage for Computing the GCM Mode, http://softwarecommunity.intel.com/isn/downloads/intelavx/Carry-Less-Multiplication-and-The-1.GCM-Mode_WP%20.pdf
6. Lipmaa, H.: Fast Software Implementations of SC 2000. In: Chan, A.H., Gligor, V.D. (eds.) ISC 2002. LNCS, vol. 2433, pp. 63–74. Springer, Heidelberg (2002)
7. Lipmaa, H.: AES / Rijndael: speed, http://research.cyber.ee/~lipmaa/research/aes/rijndael.html
8. Matsui, M.: How far can we go on the x64 processors? In: Robshaw, M.J.B. (ed.) FSE 2006. LNCS, vol. 4047, pp. 341–358. Springer, Heidelberg (2006)
9. Matsui, M., Fukuda, S.: How to Maximize Software Performance of Symmetric Primitives on Pentium III and 4 Processors. In: Gilbert, H., Handschuh, H. (eds.) FSE 2005. LNCS, vol. 3557, pp. 398–412. Springer, Heidelberg (2005)
10. Microsoft, BitLocker, http://www.bitlocker.com
11. National Institute of Standards and Technology (NIST), FIPS-197: Advanced Encryption Standard (November 2001), http://www.itl.nist.gov/fipspubs/
12. OpenSSL: the open-source toolkit for SSL/TLS, http://www.openssl.org
13. Osvik, D.A., Shamir, A., Tromer, E.: Cache Attacks and Countermeasures: The Case of AES. In: Pointcheval, D. (ed.) CT-RSA 2006. LNCS, vol. 3860, pp. 1–20. Springer, Heidelberg (2006)
14. Tillich, S., Großschädl, J.: Instruction Set Extensions for Efficient AES Implementation on 32-bit Processors. In: Goubin, L., Matsui, M. (eds.) CHES 2006. LNCS, vol. 4249, pp. 270–284. Springer, Heidelberg (2006)

A   Code Sequences for AES-192 and AES-256 Key Expansion

```
movdqu xmm1, XMMWORD PTR Key          key_expansion_192:
movq xmm3, QWORD PTR Key_              pshufd xmm2, xmm2, 0x55
movdqu XMMWORD PTR Key_Sched, xmm1    vpslldq xmm4, xmm1, 0x4
movq QWORD PTR[Key_Sched+0x10], xmm3  pxor xmm1, xmm4
mov ecx, OFFSET Key_Sched+24          pslldq xmm4, 0x4

AESKEYGENASSIST xmm2, xmm3, 0x1       pxor xmm1, xmm4
call key_expansion_192                pslldq xmm4, 0x4
AESKEYGENASSIST xmm2, xmm3, 0x2
call key_expansion_192                pxor xmm1, xmm4
AESKEYGENASSIST xmm2, xmm3, 0x4       pxor xmm1, xmm2
call key_expansion_192                pshufd xmm2, xmm1, 0xff
AESKEYGENASSIST xmm2, xmm3, 0x8       vpslldq xmm4, xmm3, 0x4
call key_expansion_192
AESKEYGENASSIST xmm2, xmm3, 0x10      pxor xmm3, xmm4
call key_expansion_192                pxor xmm3, xmm2
AESKEYGENASSIST xmm2, xmm3, 0x20      movdqu XMMWORD PTR [rcx], xmm1
call key_expansion_192                add rcx, 0x10
AESKEYGENASSIST xmm2, xmm3, 0x40      movdqu XMMWORD PTR [rcx], xmm3
call key_expansion_192                add rcx, 0x8
AESKEYGENASSIST xmm2, xmm3, 0x80      ret
call key_expansion_192
jmp END;                              END:

movdqu xmm1, XMMWORD PTR Key          key_expansion_256:
movdqu xmm3, XMMWORD PTR Key_         pshufd xmm2, xmm2, 0xff
movdqu XMMWORD PTR Key_Sched, xmm1    vpslldq xmm4, xmm1, 0x4
movdqu XMMWORD PTR[Key_Sched+0x10], xmm3  pxor xmm1, xmm4
mov rcx, OFFSET Key_Sched+0x20        pslldq xmm4, 0x4
                                      pxor xmm1, xmm4
AESKEYGENASSIST xmm2, xmm3, 0x1       pslldq xmm4, 0x4
call key_expansion_256                pxor xmm1, xmm4
AESKEYGENASSIST xmm2, xmm3, 0x2       pxor xmm1, xmm2
call key_expansion_256                movdqu XMMWORD PTR [rcx], xmm1
AESKEYGENASSIST xmm2, xmm3, 0x4       add rcx, 0x10
call key_expansion_256                cmp rcx,OFFSET Key_Schedule+0xf0
AESKEYGENASSIST xmm2, xmm3, 0x8       jz ReachedLastKey
call key_expansion_256                AESKEYGENASSIST xmm4, xmm1, 0
AESKEYGENASSIST xmm2, xmm3, 0x10      pshufd xmm2, xmm4, 0xaa
call key_expansion_256                vpslldq xmm4, xmm3, 0x4
AESKEYGENASSIST xmm2, xmm3, 0x20      pxor xmm3, xmm4
call key_expansion_256                pslldq xmm4, 0x4
AESKEYGENASSIST xmm2, xmm3, 0x40      pxor xmm3, xmm4
call key_expansion_256                pslldq xmm4, 0x4
jmp END;                              pxor xmm3, xmm4
                                      pxor xmm3, xmm2
                                      movdqu XMMWORD PTR [rcx], xmm3
                                      add rcx, 0x10
                                      ReachedLastKey:
                                      ret
                                      END:
```

Fig. 12. AES-192 and AES-256 key expansion

Remark: There are several ways for expanding the key, using AESKEY-GENASSIST. These given examples use new Intel AVX instructions (http://software.intel.com/sites/avx/) with a nondestructive source. For example, instead of (A) movdqu xmm3, xmm1; pslldq xmm3, 0x4 we use (B)vpslldq xmm3, xmm1, 0x4. AVX extensions will be introduced only in the 2010 processors, and therefore option (B) would not be valid in the 2009 processors that require the form (A). The changes from form (B) to form (A) are straightforward.

Blockcipher-Based Hashing Revisited

Martijn Stam*

LACAL, EPFL, Switzerland
martijn.stam@epfl.ch

Abstract. We revisit the rate-1 blockcipher based hash functions as first studied by Preneel, Govaerts and Vandewalle (Crypto'93) and later extensively analysed by Black, Rogaway and Shrimpton (Crypto'02). We analyse a further generalization where any pre- and postprocessing is considered. This leads to a clearer understanding of the current classification of rate-1 blockcipher based schemes as introduced by Preneel et al. and refined by Black et al. In addition, we also gain insight in chopped, overloaded and supercharged compression functions. In the latter category we propose two compression functions based on a single call to a blockcipher whose collision resistance exceeds the birthday bound on the cipher's blocklength.

1 Introduction

One of the oldest ideas to create a hash function is to base it on a blockcipher (e.g.,[11, 12, 14, 15]). Preneel et al. [15] studied the general construction $H(M, V) = E(K, X) \oplus U$ where $K, X, U \in \{0, M, V, M \oplus V\}$ (or affine offsets thereof). They concluded that of the $4^3 = 64$ possibilities all but 12 allow collision attacks on the compression function with a complexity beating the birthday bound of $2^{n/2}$. Later Black et al. [5] showed that in the ideal cipher model these 12 compression functions are indeed collision resistant up to the birthday bound, an additional 8 constructions were shown secure when properly iterated. Duo and Li [7] later gave an alternative proof resulting in improved bounds. Unfortunately neither of these articles provides a deeper understanding of what makes these 12 respectively 8 schemes special to make them secure as compression function respectively as iterated hash function: what do they have in common that sets them apart from the other 44 schemes?

We isolate the properties that make Duo and Li's proof go through in the ideal cipher model for the collision resistance of rate-1 blockcipher based compression functions and their iterated hash functions. This sheds new light on what it is that provides the provable security for these schemes; indeed the classification by Black et al. can be *derived* from it. Central to our result is a more general type of compression function, consisting of the following three simple steps (see Figure 1):

---

* This work was partially funded by the European Commission through the ICT programme under Contract ICT-2007-216646 ECRYPT II.

O. Dunkelman (Ed.): FSE 2009, LNCS 5665, pp. 67–83, 2009.

1. Prepare key and plaintext: $(K, X) \leftarrow C^{\mathrm{PRE}}(M, V)$;
2. Make the call: $Y \leftarrow E(K, X)$;
3. Output the digest: $W \leftarrow C^{\mathrm{POST}}(M, V, Y)$.

Here E is a blockcipher (where key size $k = |K|$ and blocksize $n = |X| = |Y|$ may differ) and C^{PRE} and C^{POST} can be arbitrary functions given their respective domain and codomain. To avoid complications we will initially assume that input and output sizes of the compression function match those of the blockcipher, that is $m := |M| = k$ and $s := |V| = |W| = n$.

Similar to prior art we consider two types of schemes. Type-I schemes give rise to collision resistant compression functions whereas Type-II schemes give rise to compression functions that will turn into collision resistant hash functions when (Merkle-Damgård) iterated. Each type is defined by a set of three conditions on C^{PRE} and C^{POST}. Both types share the first two conditions and only differ in the third. The first condition is bijectivity of C^{PRE}, ensuring that each query to E (or its inverse) can only be used to evaluate the compression function for a single input. The second condition is that for all M, V the postprocessing $C^{\mathrm{POST}}(M, V, \cdot)$ is bijective. This causes optimal transfer of unpredictabilility of encryption answers to the output W. For Type-I schemes, the third condition is similar in nature to the second, making sure that the unpredictability of decryption answers carries over to the digest W as well. Formally, for all K, Y the modified postprocessing $C^{\mathrm{POST}}(C^{-\mathrm{PRE}}(K, \cdot), Y)$ should be bijective. For Type-II schemes, the third condition captures that for each decryption answer the corresponding input chaining variable V is highly unpredictable. Formally, for all K, the function $C^{-\mathrm{PRE}}(K, \cdot)$ restricted to its second output V is bijective.

We provide a proof in the ideal cipher model that the probability of finding a collision in the compression function (for Type-I) respectively in the iterated hash function (for Type-II) is upper bounded by $\frac{1}{2}q(q-1)/(2^n - q)$, where q is the number of queries allowed to the adversary and n is the block size. For Type-I schemes (everywhere) preimage resistance is upper bounded by $q/(2^n - q)$. We also investigate the ramifications of our general classification for the classical PGV schemes. We conclude that the Type-I schemes are exactly those 12 identified before by Preneel et al. and later Black et al. Our Type-II schemes include the 8 schemes identified as Type-II by Black et al., plus an additional 8 schemes that were already known to be Type-I.

The benefits of our generalized framework become even clearer when analysing three more complex scenarios, when the restrictions on the parameters n, k, s, and m are being relaxed. Here we achieve the following results:

Chopped Compression Functions. This corresponds to having an output size s of the compression function smaller than the blocksize n of the underlying blockcipher. A possible example is chopped Davies-Meyer; we show that, as one might expect, it is optimally collision resistant and preimage resistant. Note that chopping the output after each encryption frees up $n - s$ bits extra for message bits if we want to maintain $n + k = s + m$. In particular one can achieve compression even for fixed-key ($k = 0$) blockciphers.

Overloaded Compression Functions. Here one tries to cram the compression function by having more input to the compression function than the blockcipher can handle, i.e., $s + m > n + k$. Examples are the sponge construction [3, 4] or the (related) compression function of Cubehash [2]. Our bound on collision resistance of the compression function is worse than if we would chop the chaining variable (to make space for the message), which is partially due to an overly loose bound.

Supercharged Compression Functions. The exact opposite of the previous two cases, since here one attempts to boost collision resistance beyond the birthday bound on the blocksize by setting $s > n$. We present a general framework for the collision resistance of single call compression functions in the ideal cipher model. In particular, we give a variant of Stam's construction [18], collision resistant in the ideal cipher model (against adaptive adversaries). We also give a rate-1/2 compression function with collision resistance up to $2^{3n/4}$ queries based on a blockcipher with $k = n$ bit keys.

2   Background

For a positive integer n, we write $\{0,1\}^n$ for the set of all bitstrings of length n. When X and Y are strings we write $X \| Y$ to mean their concatenation and $X \oplus Y$ to mean their bitwise exclusive-or (xor).

For positive integers k and n, we let $\mathrm{Block}(k, n)$ denote the set of all blockciphers with k-bit key and operating on n-bit blocks. Given that $E(K, \cdot)$ is a permutation for all $K \in \{0,1\}^k$, we write $D(K, \cdot)$ for its inverse.

Unless otherwise specified, all finite sets are equipped with a uniform distribution for random sampling. We use the convention to write oracles that are provided to an algorithm as superscripts.

2.1   Compression Functions and Hash Functions

A *compression function* is a mapping H from $\{0,1\}^m \times \{0,1\}^s$ to $\{0,1\}^s$ for some $m, s > 0$. A *blockcipher-based* compression function is a mapping $H : \{0,1\}^m \times \{0,1\}^s \to \{0,1\}^s$ given by a program that, given (M, V), computes $H^E(M, V)$ via access to an oracle $E : \{0,1\}^k \times \{0,1\}^n \to \{0,1\}^n$ modeling an (ideal) blockcipher with k-bit key and operating on n-bit blocks. A *single-call* blockcipher-based compression function calls its encryption oracle only once. Compression of a message block then proceeds as follows: Given an s-bit state V and m-bit message M, compute output $W = H^E(M, V)$ by

1. Compute $(K, X) \leftarrow C^{\mathrm{PRE}}(M, V)$.
2. Set $Y \leftarrow E(K, X)$.
3. Output $W \leftarrow C^{\mathrm{POST}}(M, V, Y)$.

as illustrated by Figure 1. We will refer to $C^{\mathrm{PRE}} : \{0,1\}^m \times \{0,1\}^s \to \{0,1\}^k \times \{0,1\}^n$ as preprocessing and to $C^{\mathrm{POST}} : \{0,1\}^m \times \{0,1\}^s \times \{0,1\}^n \to \{0,1\}^s$ as postprocessing.

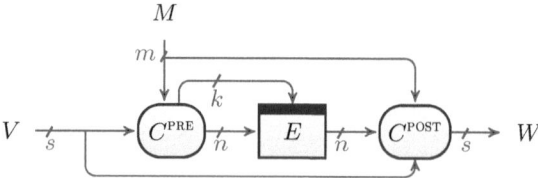

Fig. 1. General form of a $m + s$-to-s bit compression function based on a single call to the underlying blockcipher with k-bit key operating on n-bit block

Since a blockcipher is easy to invert (given its key), an adversary trying to find for instance collisions will also have access to D. To deal with inverse queries in our security analysis, we introduce the modified postprocessing $C^{\mathrm{AUX}}(K, X, Y) = C^{\mathrm{POST}}(C^{-\mathrm{PRE}}(K, X), Y)$. In general, this is a function mapping triplets of strings to subsets of strings, since the result of $C^{-\mathrm{PRE}}$ can have varying cardinality. For simplicity, when C^{PRE} is bijective, we understand C^{AUX} to have $\{0,1\}^n$ as its codomain.

A *hash function* is a mapping $\mathcal{H}$ from $\{0,1\}^*$ (the set of arbitrary length bitstrings) to $\{0,1\}^s$ for some $s > 0$. A compression function can be made into a hash function by iterating it. We briefly recall the standard Merkle-Damgård iteration [6, 13], where we assume that there is already some injective padding from $\{0,1\}^* \to (\{0,1\}^m)^* \backslash \emptyset$ in place (note that we disallow the empty message $\mathbf{M} = \emptyset$ as output of the injective padding). Given an initial vector $V_0 \in \{0,1\}^s$ define $\mathcal{H}^H : (\{0,1\}^m)^* \to \{0,1\}^s$ as follows for $\mathbf{M} = (M_1, \ldots, M_\ell)$ with $\ell > 0$:

1. Set $V_i \leftarrow H^E(M_i, V_{i-1})$ for $i = 1, \ldots, \ell$.
2. Output $\mathcal{H}^H(\mathbf{M}) = V_\ell$.

(Bearing this iteration in mind, given a compression function $H : \{0,1\}^m \times \{0,1\}^s \to \{0,1\}^s$ we will refer to the $\{0,1\}^m$ part of the input as 'message' and the $\{0,1\}^s$ part as the state or chaining variable.)

Collision Resistance. A *collision-finding adversary* is an algorithm with access to one or more oracles, whose goal it is to find collisions in some specified compression or hash function. It is standard practice to consider information-theoretic adversaries only. Currently this seems to provide the only handle to get any provable results. Information-theoretic adversaries are computationally unbounded and their complexity is measured only by the number of queries made to their oracles. Without loss of generality, such adversaries are assumed not to repeat queries to oracles nor to query an oracle outside of its specified domain. We also assume that the adversary, before outputting a message, makes all calls necessary to evaluate the compressing function on that message. This does not decrease the advantage of the adversary, though it does increase its query complexity.

Despite the concept of initial vector being somewhat alien to a compression function on its own, it turns out helpful to consider a preimage to the initial vector a collision [5].

Definition 1. *Let $n, k, m, s > 0$ be integer parameters. Let $H \colon \{0,1\}^m \times \{0,1\}^s \to \{0,1\}^s$ be a compression function taking oracle $E \in \mathrm{Block}(k,n)$. The collision-finding advantage of adversary $\mathcal{A}$ is defined to be*

$$\mathrm{Adv}_H^{\mathrm{coll}}(\mathcal{A}) = \max_{V_0 \in \{0,1\}^s} \Pr\left[E \xleftarrow{\$} \mathrm{Block}(k,n), ((M,V),(M',V')) \leftarrow \mathcal{A}^{E,D}(V_0) \colon \right.$$

$$\left. (M,V) \neq (M',V') \text{ and } H^E(M,V) \in \{V_0, H^E(M',V')\} \right] .$$

Define $\mathrm{Adv}_H^{\mathrm{coll}}(q)$ as the maximum advantage over all adversaries making at most q queries in total.

The quantity $\mathrm{Adv}_{\mathcal{H}}^{\mathrm{coll}}(q)$ denoting collision for the iterated hash function $\mathcal{H}^H$ is defined similarly: in this case the advantage of $\mathcal{A}$ is the maximum success probability taken over the choice of possible initial values V_0, which is input to $\mathcal{A}$. It is well known that the iterated hash function $\mathcal{H}$ is at least as secure as the compression function H it is based upon, as far as collision resistance is concerned [5, Lemma 1].

Theorem 2. *Let H be a blockcipher based compression function and let $\mathcal{H}$ be the iterated hash function based on H. Then*

$$\mathrm{Adv}_{\mathcal{H}}^{\mathrm{coll}}(q) \leq \mathrm{Adv}_H^{\mathrm{coll}}(q) .$$

Preimage Resistance. A *preimage-finding adversary* is an algorithm with access to one or more oracles, whose goal it is to find preimages in some specified compression function. There exist several definitions depending on the distribution of the element of which a preimage needs to be found. We opt for everywhere preimage resistance [16], which intuitively states that all points are hard to invert.

Definition 3. *Let $n, k, m, s > 0$ be integer parameters. Let $H \colon \{0,1\}^m \times \{0,1\}^s \to \{0,1\}^s$ be a compression function taking oracle $E \in \mathrm{Block}(k,n)$. The everywhere preimage-finding advantage of adversary $\mathcal{A}$ is defined to be*

$$\mathrm{Adv}_H^{\mathrm{epre}}(\mathcal{A}) = \max_{W \in \{0,1\}^s} \Pr_E \left(M',V') \leftarrow \mathcal{A}^{E,D}(W) \colon W = H^E(M',V') \right) .$$

Define $\mathrm{Adv}_H^{\mathrm{epre}}(q)$ as the maximum advantage over all adversaries making at most q queries in total.

The quantity $\mathrm{Adv}_{\mathcal{H}}^{\mathrm{epre}}(q)$ denoting preimage resistance for the iterated hash function $\mathcal{H}^H$ is defined similarly (in this case the advantage of $\mathcal{A}$ is the maximum success probability taken over the choice of possible initial values V_0, which is input to $\mathcal{A}$). Everywhere preimage resistance is preserved in the (MD-)iteration [1], so we get:

Theorem 4. *Let H be a blockcipher based compression function and let $\mathcal{H}$ be the iterated hash function based on H. Then*

$$\mathrm{Adv}_{\mathcal{H}}^{\mathrm{epre}}(q) \leq \mathrm{Adv}_H^{\mathrm{epre}}(q) \leq \mathrm{Adv}_H^{\mathrm{coll}}(q) .$$

3   Classical Rate-1 Blockcipher Based Compression Functions

In this section we will deal with classical rate-1 blockcipher based compression functions, where the state size s equals the block length n of the blockcipher and the message size m matches the keysize k of the blockcipher. This includes the famous PGV hash functions [15].

Following in the footsteps of Black et al. [5], we consider Type-I and Type-II compression functions. The former give optimal collision and preimage resistance in the compression function. The second type gives optimal collision resistance in the iteration; its preimage resistance can only be proved up to the birthday bound. One of the important differences with prior art is that we specify in very broad terms the requirements on C^{PRE} and C^{POST}. Essentially our primary concern here is for the proof to go through. In Section 3.3 we will discuss what our classification of Type-I and Type-II implies for the PGV hash functions.

The proof for Type-I schemes is fairly standard and straightforward. However, for the Type-II schemes we deviate from the one by Black et al. [5]. In particular, their proof is based upon colouring a directed graph where the vertices represent queries with all possible answers and arcs are drawn according to whether the input to one query is consistent with the output of the former, given the compression function under consideration. This leads to unwieldy graphs with a complicated notion of what consitutes a collision.

This counterintuitive use of graphs was fixed by Duo and Li [7] (as well as by Lucks [10]), who consider a directed graph where vertices correspond to chaining values and edges are drawn (or coloured) whenever a query has been made that would allow to move from one chaining value to the next. Moreover, for the actual bounding of collision resistance Duo and Li dispense with the direction of the arcs (that thus become edges). Although this seemingly aids the adversary (certain patterns in the graph will be deemed a success even when the underlying event on the hash function is not), this simplification leads to a tighter bound for the Type-II schemes, mainly because there is no longer any need to distinguish between several cases (whose success probability are subsequently added). Our proof (of Theorem 9) closely follows that of Duo and Li.

Note that even for Type-I schemes our bound appears a bit tighter than the one by Black et al., which is due to their simplification based on the the inequality $2(2^n - q) > 2^n$, at least for $q < 2^{n-1}$ (and for larger q most of the bounds become vacuous anyway). We believe the choice between tightness and simplicity in this case is one mainly of taste; we have opted for the former.

3.1   Type-I: Collision Resistant Compression Functions

Definition 5. *A single call blockcipher based compression function H^E is called rate-1 Type-I iff $n = s, k = m$ and the following three hold:*

1. *The preprocessing C^{PRE} is bijective.*
2. *For all M, V the postprocessing $C^{\mathrm{POST}}(M, V, \cdot)$ is bijective.*
3. *For all K, Y the modified postprocessing $C^{\mathrm{AUX}}(K, \cdot, Y)$ is bijective.*

Theorem 6. *Let H^E be a rate-1 Type-I compression function (based on a block-cipher with block size n). Then the advantage of an adversary in finding a collision in H^E after q queries can be upper bounded by*

$$\mathsf{Adv}_H^{\mathrm{coll}}(q) \le \frac{1}{2}q(q+1)/(2^n - q) \ .$$

Proof. Let $V_0 \in \{0,1\}^n$ be given. A collision consists of two pairs (M,V) and (M',V') satisfying $H^E(M,V) = \{V_0, H^E(M',V')\}$ yet $(M,V) \neq (M',V')$. We will maintain a list of triples (M,V,W) such that $W = H^E(M,V)$ and the adversary has made the relevant queries to E and/or D. The list is initialized with $(-,-,V_0)$. Since we require the adversary to have made all relevant queries when outputting a collision, we can upper bound the success probability of the adversary by bounding the probability of a collision occuring in this list. We show that any query, be it forward or inverse, will add at most one triple (M,V,W) to this list of computable compression functions, moreover the value W is almost completely out of the adversary's control.

Consider a forward query (K,X). By bijectivity of C^{PRE}, there is a unique pair (M,V) corresponding to this query. Thus, each forward query will add one triple (M,V,W) to the adversary's list of computable values. Since $C^{\mathrm{POST}}(M,V,\cdot)$ is bijective for all M,V, the distribution of compression function output W is closely related to that of blockcipher output Y, which is close to being uniform. More precisely, suppose that so far t queries to E (and D) have been made involving key K, resulting in t plaintext-ciphertext pairs (X_i,Y_i) with $Y_i = E(K,X_i)$ for $i = 1,\ldots,t$. The answer to a fresh query to $E(K,\cdot)$ will therefore be $Y^* \neq Y_i, i = 1,\ldots,t$. Moreover, each of the $2^n - t$ answers is equally likely if E is an ideal cipher. Each possible answer Y^* will combine under C^{POST} with the pair (M,V) consistent with the (K,X) query being made, leading to a possible compression function outcome W^*. Because C^{POST} is bijective when (M,V) are fixed, distinct Y^* lead to distinct W^*, so there are $2^n - t$ possible outcomes W^*, all equally likely.

Similarly, consider an inverse query (K,Y). This yields a unique X and hence by bijectivity of C^{PRE}, there is a unique pair (M,V) corresponding to this query once answered. Thus, each inverse query will add one triple (M,V,W) to the adversary's list of computable values. This time bijectivity of $C^{\mathrm{AUX}}(K,\cdot,Y)$ implies that the distribution of W is closely related to the (almost uniform) output distribution of D. Indeed, suppose that so far t queries to E have been made involving key K, resulting in t plaintext-ciphertext pairs (X_i,Y_i) with $Y_i = E(K,X_i)$ for $i = 1,\ldots,t$. The answer to a fresh query to $D(K,\cdot)$ will therefore be $X^* \neq X_i, i = 1,\ldots,t$. Moreover, each of the $2^n - t$ answers is equally likely if E is an ideal cipher. Each possible answer X^* will combine under $C^{-\mathrm{PRE}}$ and C^{POST} with K and Y to a triple (M,V,W). Because for all K and Y the mapping from X to W is bijective (by assumption on C^{AUX}), distinct X^* lead to distinct W^*, so there are $2^n - t$ possible outcomes W^*, all equally likely.

As a result, after $i - 1$ queries the list of computable values contains i triples (M,V,W). The i'th query will add one triple with W uniform over a set of size at least $2^n - i + 1$. Thus the probability that the i'th query causes a collision with any of these triples is at most $i/(2^n - i + 1)$. Using a union bound, the probability of a collision after q queries can then be upper bounded by $\sum_{i=1}^{q} i/(2^n - i + 1) \leq \frac{1}{2}q(q+1)/(2^n - q)$.   □

Theorem 7. *Let H^E be a rate-1 Type-I compression function (based on a a blockcipher with block size n). Then the advantage of an adversary in finding a preimage in H^E after q queries can be upper bounded by*

$$\mathsf{Adv}_H^{\mathsf{epre}}(q) \leq q/(2^n - q) .$$

Proof. Let $\mathcal{A}$ be an adversary that tries to find a preimage for its input σ. Assume that $\mathcal{A}$ asks its oracles E and D a total of q queries.

We recall the proof of Theorem 6, where we show that after $i - 1$ queries (to E or D) the list of computable values $W = H^E(M, V)$ contains $i - 1$ triples (M, V, W). The i'th query will add one triple with W uniform over a set of size at least $2^n - i + 1$. Thus the probability that the i'th query hits σ is at most $1/(2^n - i + 1)$. Using a union bound, the probability of finding a preimage for σ after q queries can then be upper bounded by $\sum_{i=1}^{q} 1/(2^n - i + 1) \leq q/(2^n - q)$.   □

3.2   Type-II: Collision Resistance in the Iteration

Definition 8. *A single call blockcipher based compression function H^E is called rate-1 Type-II iff $n = s, k = m$, and the following three hold:*

1. *The preprocessing C^{PRE} is bijective.*
2. *For all M, V the postprocessing $C^{\mathrm{POST}}(M, V, \cdot)$ is bijective.*
3. *For all K, $C^{-\mathrm{PRE}}(K, \cdot)$ restricted to V, its second output, is bijective.*

Theorem 9. *Let H^E be a rate-1 Type-II compression function. If E is an ideal cipher with block size n, then the advantage of an adversary in finding a collision in the iterated hash function $\mathcal{H}^H$ after q queries is upper bounded by*

$$\mathsf{Adv}_{\mathcal{H}}^{\mathsf{coll}}(q) \leq \frac{1}{2}q(q+1)/(2^n - q) .$$

Proof. Let $V_0 \in \{0, 1\}^n$ be $\mathcal{H}$'s initial vector.

We define an undirected graph $G = (V_G, E_G)$ with vertex set $V_G = \{0, 1\}^n$—corresponding to all 2^n possible chaining values—and initially an empty edge set $E_G = \emptyset$. We will dynamically add edges based on the queries to E and D. In particular, we add an edge (V, W), labelled by M, if we know a message M such that $W = H^E(M, V)$ (or $V = H^E(M, V)$) and the relevant query to either E or D has been made. We claim that to find a collision would require constructing a ρ-shape containing the initial vector V_0. Suppose that $\mathcal{H}(\mathbf{M}) = \mathcal{H}(\mathbf{M}')$ with $\mathbf{M} \neq \mathbf{M}'$. Write $\mathbf{M} = (M_1, \ldots, M_\ell)$ and $\mathbf{M}' = (M'_1, \ldots, M'_{\ell'})$ and correspondingly $V_0, \ldots, V_\ell$ respectively $V'_0, \ldots, V'_{\ell'}$ for the chaining values of the iterated hash. Note that $V_0 = V'_0$ and $V_\ell = V'_{\ell'}$. Assume $\ell \leq \ell'$. Because $\mathbf{M} \neq \mathbf{M}'$, there exists a t such that $M_i = M'_i$ for all $0 \leq i < t$ but $M_t \neq M'_t$ (or possibly

$\ell < t \le \ell'$). As a result, the paths $(V_0, \ldots, V_t)$ and $(V_0', \ldots, V_t')$ are identical, but the edges (V_t, V_{t+1}) and (V_t', V_{t+1}') are distinct, even when V_{t+1}' happens to equal V_{t+1} (in particular, the edges are labelled differently). Since $V_\ell = V_{\ell'}'$ at some point the paths need to come together again, completing the ρ-shape. Note that due to our use of an undirected graph not every ρ-shape will lead to a collision though.

Since we are dynamically adding edges to the graph, components in the graph will also grow dynamically. Let T be the set of all nodes that are in a component containing a cycle or the initial vector V_0. The first claim is that after i queries, the set T has cardinality at most $i+1$. Indeed, the component containing V_0 has at most $i'+1$ nodes when i' edges are used. A cyclic component based on i' edges has at most i' nodes. Thus the initial vector component is the only component in T that causes the number of nodes larger than the number of edges, by at most one. Bijectivity of C^{PRE} implies that a query (either forward or inverse) will add at most one edge to the graph, so after i queries, there are at most i edges in the entire graph and at most $i+1$ nodes in T.

The second claim is that to complete a ρ-shape, either a cycle has to be completed within the V_0-component, or the V_0-component needs to be connected with a cycle. Either way, an edge has to be found of which both nodes are already part of T. The probability that on the i'th query a collision is found by a forward query is at most $i/(2^n - i)$: bijectivity of $C^{\text{POST}}(M, V, \cdot)$ ensures that W is uniformly distributed over a set of size at least $2^n - i$, so hitting a set of size i occurs at most with said probability. Similarly, for an inverse query the probability of finding a collision on the i'th query using an inverse query is at most $i/(2^n - i)$: this time bijectivity of $C^{\text{AUX}}(K, \cdot, Y)$ ensures that V is uniformly distributed over a set of size at least $2^n - i$.

We can now wrap up and conclude that the probability of finding a collision on the i'th query is at most $i/(2^n - i)$ and the probability after q queries is at most $\sum_{i=1}^{q} i/(2^n - i) \le \frac{1}{2}q(q+1)/(2^n - q)$.   □

3.3   Implications to the PGV Schemes

In this section we investigate how the 64 PGV schemes [15] fit in the general Type-I and Type-II framework. Recall that for the PGV-style schemes the blockcipher has key size equal to the block length; the compression function will look like $H^E(M, V) = E(K, X) \oplus U$ where $K, X, U \in \{C, M, V, M \oplus V\}$ and C is some fixed, publicly known bitstring. These restrictions can also be expressed in terms of C^{PRE} and C^{POST}. Our results are in line with the classification of Black et al. [5] and the tighter bounds by Duo and Li [7].

Let us first set up some notation. As is customary [8] for schemes with linear processing C^{PRE} and C^{POST}, we will represent the linear PGV schemes using matrices. We will use $\mathbb{Z}_2^2$ to express the way K, X, and U are functions of M and V: a vector $\mathbf{X} \in \mathbb{Z}_2^2$ corresponds to $X = \mathbf{X} \cdot \binom{M}{V}$, making a distinction between the linear map $\mathbf{X} \in \mathbb{Z}_2^2$ and the value $X \in \{0,1\}^n$. We will also write $\mathbf{X} = (X_M, X_V)$. We can safely ignore any affine part, so $\mathbf{U} = (00)$ can be thought of to correspond to the aforementioned $U \leftarrow C$. (This is without loss of

Table 1. The 20 Secure PGV-style schemes, writing $E_K(X)$ for $E(K, X)$ and W for $M \oplus V$. Superscripted are the $\imath$-indices from [5, Fig. 1 and 2].

$\binom{\mathbf{k}}{\mathbf{x}} \backslash \mathbf{s}$	(00)	(01)	(10)	(11)
$\begin{pmatrix} 0 & 1 \\ 1 & 0 \end{pmatrix}$	insecure	insecure	$E_V(M) \oplus M^1$	$E_V(M) \oplus W^3$
$\begin{pmatrix} 0 & 1 \\ 1 & 1 \end{pmatrix}$	insecure	insecure	$E_V(W) \oplus M^4$	$E_V(W) \oplus W^2$
$\begin{pmatrix} 1 & 0 \\ 0 & 1 \end{pmatrix}$	$E_M(V)^{15}$	$E_M(V) \oplus V^5$	$E_M(V) \oplus M^{17}$	$E_M(V) \oplus W^7$
$\begin{pmatrix} 1 & 0 \\ 1 & 1 \end{pmatrix}$	$E_M(W)^{19}$	$E_M(W) \oplus V^8$	$E_M(W) \oplus M^{20}$	$E_M(W) \oplus W^6$
$\begin{pmatrix} 1 & 1 \\ 0 & 1 \end{pmatrix}$	$E_W(V)^{16}$	$E_W(V) \oplus V^{10}$	$E_W(V) \oplus M^{12}$	$E_W(V) \oplus W^{18}$
$\begin{pmatrix} 1 & 1 \\ 1 & 0 \end{pmatrix}$	$E_W(M)^{13}$	$E_W(M) \oplus V^{11}$	$E_W(M) \oplus M^9$	$E_W(M) \oplus W^{14}$

generality, since translation by a constant will not affect bijectivity in either of the criteria used in Definitions 5 and 8.) Since there are 4 elements in $\mathbb{Z}_2^2$ and we have to pick 3 ($\mathbf{K}$,$\mathbf{X}$, and $\mathbf{U}$), there are 64 constructions to consider in total, corresponding to the 64 PGV schemes.

We are now ready to see what the requirements from Definitions 5 and 8 mean in terms of the vectors $\mathbf{K}$, $\mathbf{X}$ and $\mathbf{U}$ and hence for the classification and security of the PGV schemes. The 20 interesting schemes are listed in Table 1, where we have also included the $\imath$-indices assigned to these schemes by Black et al. [5]. When we write $H_\imath$ resp. $\mathcal{H}_\imath$ for $\imath \in \{1, \ldots, 20\}$ we refer to this enumeration. Proofs are to be found in the full version [19].

Lemma 10. *A PGV scheme is Type-I iff $\binom{\mathbf{K}}{\mathbf{X}}$ and $\binom{\mathbf{K}}{\mathbf{U}}$ are both invertible matrices. In particular, $H_{1..12}$ are Type-I schemes.*

The requirements for the Type-II schemes turn out surprisingly simple: indeed apart from the preprocessing having full rank, the only requirement is that the key depends on the message. Consequently we end up with 16 Type-II schemes as opposed to only 8 given by Black et al. The 'additional' 8 schemes we identify are also Type-I, which explains why previously they were not classified as Type-II. Our results therefore suggest a subdivision of the PGV Type-I schemes, namely those that are also Type-II (being those with a key depending on the message) and those that are just Type-I (those whose key equals the chaining variable). The same subdivision was made by Duo and Li [7] in the context of second preimage resistance.

Lemma 11. *A PGV scheme is Type-II iff $\binom{\mathbf{K}}{\mathbf{X}}$ is an invertible matrix with $K_M = 1$. In particular, $\mathcal{H}_{5..20}$ are Type-II schemes.*

Combining Lemmas 10 and 11 with Theorems 6, 7, and 9 then yields Corollary 12 below. For completeness [5, 15], it is known that the given upper bounds on the advantages are tight up to a small constant factor. Moreover, for $\mathcal{H}_{13..20}$ preimage resistance is worse than desired, namely $\mathsf{Adv}_{\mathcal{H}}^{\mathsf{epre}}(q) = \Theta(q^2/2^n)$ (due to a meet-in-the-middle attack). The remaining 44 PGV schemes do not offer any collision resistance in the iteration.

Corollary 12. *(Security of the PGV schemes) For $H_{1..12}$ it holds that $\mathsf{Adv}_H^{\mathsf{coll}}(q)$ $\leq \frac{1}{2}q(q+1)/(2^n - q)$, and $\mathsf{Adv}_H^{\mathsf{epre}}(q) \leq q/(2^n - q)$; for $\mathcal{H}_{13..20}$ it holds that $\mathsf{Adv}_{\mathcal{H}}^{\mathsf{coll}}(q) \leq \frac{1}{2}q(q+1)/(2^n - q)$.*

4   Generalized Single Call Compression Functions

In the previous section we discussed the standard (single call) case where the input and output sizes of the compression function neatly matched those of the underlying blockcipher, in particular $m = k$ and $s = n$. In this section we let go of these restrictions and consider three more general scenarios.

First we will consider what could be called chopping the output of the compression (or really the scenario where $s < n$). For instance, the Davies-Meyer construction is optimally collision and preimage resistant, but what happens if you chop the output: is the security still optimal given the new output length (it is). A welcome benefit of chopping the output is that it frees up bits for the message. More precisely, if $s < n$ then we can have a larger m while maintaining $m + s = n + k$. In particular, compression becomes feasible even for fixed permutations (corresponding to $k = 0$). In view of the recent availability of huge size permutations constructions with $s < n$ gain traction; an example is Grindahl[9]. We will refer to this scenario as compression in the postprocessing, the corresponding H^E's are called chopped compression functions.

Similarly, one might also try to improve efficiency by squeezing in more bits of input in the compression function than can be input to the primitive (this corresponds to $m + s > n + k$). We call this compression in the preprocessing and speak of overloaded compression functions. Like the previous scenario, this opens up the possibility of achieving compression based on a single fixed permutation. We suggest a general Type-I compression function and give a bound on its collision resistance and preimage resistance. Security in the iteration is more complicated here: we discuss related work and point out some challenging open problems.

Finally we deal with the problem of getting security beyond the block length of the blockcipher, that is $s > n$. Here we say that expansion in the postprocessing gives rise to supercharged compression functions. Promising results were previously given by Lucks [10] in the iteration and Stam [18] for a compression function. We develop a general theory and give two concrete examples based on the latter work.

(Any missing proofs, as well as an expanded treatment of supercharged compression functions, can be found in the full version [19].)

4.1 Chopping: Compression in the Postprocessing

Let us consider an $m + s$-to-s bit compression function based on a single call to a blockcipher with key size k and block size n. In this section we will assume that $m + s = n + k$ and $s < n$. What can we say of the collision and preimage resistance of the compression function resp. iterated hash function, under which conditions will we achieve optimal security?

If we go through the criteria from the previous section, it is clear we can no longer satisfy them all. More to the point, whereas the first condition (bijectivity of the preprocessing) still applies, the postprocessing now becomes a mapping from n to s bits, which cannot be bijective since $s < n$. The natural generalization is to replace being a bijection with being balanced: all elements in the codomain should have the same number of preimages, namely 2^{n-s}. It turns out that this fairly simple modification works quite well. Again we have two types: the first one giving optimal collision and preimage resistance for the compression function; the second one giving optimal collision resistance in the iteration only (and guaranteed preimage resistance only up to the collision resistance).

Definition 13. *A single call blockcipher based compression function H^E is called chopped single call Type-I iff $s < n, m + k = n + s$, and the following three hold:*

1. *The preprocessing C^{PRE} is bijective.*
2. *For all M,V the postprocessing $C^{\text{POST}}(M, V, \cdot)$ is balanced.*
3. *For all K,Y the modified postprocessing $C^{\text{AUX}}(K, \cdot, Y)$ is balanced.*

Definition 14. *A single call blockcipher based compression function H^E is called chopped single call Type-II iff $s < n, m + k = n + s$ and the following three hold:*

1. *The preprocessing C^{PRE} is bijective.*
2. *For all M,V the postprocessing $C^{\text{POST}}(M, V, \cdot)$ is balanced.*
3. *For all K the inverse preprocessing $C^{-\text{PRE}}(K, \cdot)$ when restricted to its V output is balanced.*

Theorem 15. *Let H^E be a chopped single call Type-I compression function. Then the advantage of an adversary in finding a collision, resp. a preimage in H^E after q queries can be upper bounded by*

$$\text{Adv}_H^{\text{coll}}(q) \leq q(q+1)/2^s, \qquad \text{Adv}_H^{\text{epre}}(q) \leq q/2^{s-1} .$$

Theorem 16. *Let H^E be a chopped single call Type-II compression function. Then the advantage of an adversary in finding a collision in the iterated hash function $\mathcal{H}^H$ after q queries is upper bounded by*

$$\text{Adv}_{\mathcal{H}}^{\text{coll}}(q) \leq q(q+1)/2^s .$$

4.2   Overloading: Compression in the Preprocessing

Another way to improve efficiency it to keep $s = n$, but allow $m > k$. In this case bijectivity of the preprocessing can no longer be satisfied, which has ramifications throughout.

Firstly, for a given pair (K, X) it is now the case that $C^{-\mathrm{PRE}}$ yields a set of 2^{m-k} pairs (M, V). Consequently, the modified postprocessing $C^{\mathrm{AUX}}(K, \cdot, Y)$ becomes a function from n-bits to subsets of size (up to) 2^{m-k} of $\{0,1\}^n$. Our requirement on this new type of C^{AUX} is a natural generalization of balancedness.

Secondly, although the condition that $C^{\mathrm{POST}}(M, V, \cdot)$ is bijective is still well-defined, it is no longer sufficient. For instance, if $C^{\mathrm{PRE}}(M, V) = C^{\mathrm{PRE}}(M', V')$ for certain values of $(M, V) \neq (M', V')$ and the bijections $C^{\mathrm{POST}}(M, V, \cdot)$ and $C^{\mathrm{POST}}(M', V', \cdot)$ are identical, then collisions can very easily be found. To avoid this problem we explicitly rule out collisions in the output whenever (M, V) and (M', V') already collide during preprocessing (in C^{PRE}).

Definition 17. *A single call blockcipher based compression function H^E is called overloaded single call Type-I iff $s = n, m \geq k$, and the following four hold:*

1. *The preprocessing C^{PRE} is balanced.*
2. *For all $(M, V) \neq (M', V')$ with $C^{\mathrm{PRE}}(M, V) = C^{\mathrm{PRE}}(M', V')$ and all Y it holds that $C^{\mathrm{POST}}(M, V, Y) \neq C^{\mathrm{POST}}(M', V', Y)$.*
3. *For all M, V the postprocessing $C^{\mathrm{POST}}(M, V, \cdot)$ is bijective.*
4. *For all K, Y the modified postprocessing $C^{\mathrm{AUX}}(K, \cdot, Y)$ is balanced in the sense that for all V the number of X such that $V \in C^{\mathrm{AUX}}(K, X, Y)$ equals 2^{m-k}.*

Theorem 18. *Let H^E be an overloaded single call Type-I compression function. Then the advantage of an adversary in finding a collision, resp. a preimage in H^E after q queries can be upper bounded by*

$$\mathsf{Adv}_H^{\mathsf{coll}}(q) \leq q(q+1)/2^{2k+n-2m}, \qquad \mathsf{Adv}_H^{\mathsf{epre}}(q) \leq q/2^{n+k-m-1} \ .$$

Theorem 18 can be reinterpreted by saying that to find collisions roughly $2^{n/2+k-m}$ queries are required; to find preimages roughly 2^{n+k-m} queries should suffice. It is interesting to compare the collision resistance thus achieved with recently conjectured optimal bounds [17, 18]. A straightforward generalization of Rogaway and Steinberger's result [17] suggests the best we can achieve is collision resistance up to $2^{n/2+k-m}$ queries, neatly corresponding to our construction. However, Stam [18] conjectures collision resistance is feasible up to $2^{(n+k-m)/2}$ queries, based on an ideal state size s of $n + k - m$ bits. Using this state size actually brings us back exactly to compression in the postprocessing as discussed in the previous section: by reducing s we can increase m while maintaining $n + k = m + s$ and Theorem 15 essentially guarantees collision resistance up to $2^{(n+k-m)/2}$ queries. So here is another scenario where reducing the state size mysteriously seems to boost collision resistance.

But all is not as it seems. An example overloaded single call Type-I compression function is Davies-Meyer with the $m - k$ superfluous message bits xored directly into the output. It is not hard to show that in this case the collision finding advantage is much smaller than Theorem 18 makes believe:

$$\mathsf{Adv}_H^{\mathsf{coll}}(q) \leq q(q+1)/2^{k+n-m} \ .$$

Iterated Case. For rate-1 and chopped compression functions, looking at the iteration gave rise to a second class of schemes that had the same collision resistance in the iteration as the main schemes, but inferior preimage resistance. For overloaded compression functions, we do not give a classificiation of Type-II schemes (also in light of our Type-I bounds' lack of tightness). However, we do point out that some non-trivial results in this setting were previously achieved for sponge functions [4], whose collision resistance (in the iteration) holds roughly up to $2^{(n-m)/2}$ queries ($k = 0$). This matches the collision resistant compression function of the previous paragraph.

However, recent developments indicate that iteration might boost collision resistance even further. In particular, the sponge construction has rate $\alpha = m/(n-m)$ achieving collision resistance up to roughly $2^{n(1-\alpha)/2}$ queries. Rogaway and Steinberger [17] have shown that for any rate-α construction after $1.9n2^{n(1-\alpha)}$ queries collisions are guaranteed. This still leaves a considerable gap.

4.3   Supercharging: Expansion in the Postprocessing

Whereas for chopped and overloaded compression functions we sacrificed security for the sake of efficiency, in this section we will attempt the exact opposite: sacrificing efficiency for the sake of security. We do this by extending the state size, so $s > n$. Not to complicate things further, we will assume that $m+s = n+k$ (and let C^{PRE} be bijective). For any fixed pair (M, V) we have that C^{POST} maps $\{0,1\}^n$ to $\{0,1\}^s$. Since $n < s$ this cannot be a bijection, but at best an injection (similar for C^{AUX}). If all these injections have exactly the same range, we are not using our codomain of 2^s values to the full; indeed we might have well been padding the state with a constant. This leads us to the following formalization.

Definition 19. *A single call blockcipher based compression function H^E is called supercharged single call Type-I with overlap γ iff $s \geq n, m + s = n + k$ and the following three hold:*

1. *The preprocessing C^{PRE} is bijective.*
2. *For all M, V the postprocessing $C^{\mathrm{POST}}(M, V, \cdot)$ is injective, with effective range $R_{\mathrm{POST},(M,V)}$.*
3. *For all K, Y the modified postprocessing $C^{\mathrm{AUX}}(K, \cdot, Y)$ is injective, with effective range $R_{\mathrm{AUX},(K,Y)}$.*

Where the overlap γ is defined as:

$$\gamma = \max\left\{|R_Z \cap R_{Z'}| : Z, Z' \in \{\mathrm{POST}, \mathrm{AUX}\} \times \{0,1\}^{k+n}, Z \neq Z'\right\} \ .$$

Theorem 20. *Let H^E be a supercharged single call Type-I compression function with overlap γ. Then the advantage of an adversary in finding a collision after $q \leq 2^{n-1}$ queries can be upper bounded by*

$$\mathsf{Adv}_H^{\mathsf{coll}}(q) \leq q\kappa/2^{n-1} + 2^{m+s+1} \left(\frac{e\gamma q}{(\kappa - 1)2^{n-1}} \right)^{\kappa-1}$$

for arbitrary positive integer $\kappa > q\gamma/2^{n-1}$.

Corollary 21. *Let H^E be a supercharged single call Type-I compression function with overlap γ. Then for $q < 2^{n-1}/\gamma^{\frac{1}{2}}$ the probability of finding a collision can be upper bounded by*

$$\mathsf{Adv}_H^{\mathsf{coll}}(q) \leq 2 \max(2e\gamma^{\frac{1}{2}}, m + n + s + 2)q/2^n \ .$$

In practice this means that we get good security up to q of order $2^n/\gamma^{\frac{1}{2}}$. Stam [18] suggests that finding collisions can be expected after $2^{(n+k-m)/2}$ queries. Since $n + k = m + s$ this neatly corresponds to $2^{s/2}$, in other words optimal collision resistant compression functions of this type might actually exist. Note that the rate is lower than before, arguably m/n. As we show in Lemma 22, the best we can hope for is γ of order 2^{2n-s}, giving collision resistance up to $2^{s/2}$ queries. Whether for all relevant settings of n, s, k, and m there exists a postprocessing C^{POST} with overlap γ close to 2^{2n-s} is an open problem. Below we give two examples where it does though, based on an earlier construction [18].

Lemma 22. *Let H^E be a supercharged single call Type-I compression function then overlap*

$$\gamma \geq \frac{2(2^{2n+m} - 2^n)}{2^{s+m} - 1} \quad (\approx 2^{2n-s+1}) \ .$$

Example I: A Double-Length Construction. We recall the construction [18] for a double length compression function based on a single ideal $3n$-to-n compression function F. Split the $2n$-bit state V in two equally sized parts V_1 and V_2. Then given an n-bit message block M, compression proceeds as follows:

1. Compute $Y \leftarrow F(M, V_1, V_2)$.
2. Output $(W_1, W_2) \leftarrow (Y, V_2 Y^2 + V_1 Y + M)$.

where the polynomial evaluation is over $\mathbb{F}_{2^n}$. Originally only a proof of collision resistance against non-adaptive adversaries was given, based on random functions instead of random permutations (so in particular an adversary would not have access to an inversion oracle). We would like to port the scheme to the ideal cipher model, based on a blockcipher with $k = 2n$.

1. Set $K \leftarrow (V_1, V_2)$ and $X \leftarrow M$.
2. Compute $Y \leftarrow E(K, X)$.
3. Compute $W_1 \leftarrow Y + M$ and $W_2 \leftarrow MW_1^2 + V_1 W_1 + V_2$; output (W_1, W_2).

Lemma 23. *For the compression function above, $\gamma = 3$.*

Proof. To determine the overlap γ it helps to first write down the effective ranges $R_{\text{POST},(M,V)}$ and $R_{\text{AUX},(K,Y)}$ explicitly. It is easy to see that

$$R_{\text{POST},(M,V_1,V_2)} = \{(W, MW^2 + V_1W + V_2)|W \in \{0,1\}^n\}$$

and with a little bit more effort, using that $M = Y + W$ and $(K_1, K_2) = (V_1, V_2)$,

$$R_{\text{AUX},(K_1,K_2,Y)} = \{(W, W^3 + YW^2 + K_1W + K_2)|W \in \{0,1\}^n\} \ .$$

As a result, for (W_1, W_2) to be in the intersection of R_Z and $R_{Z'}$, we require W_1 to be a root of the difference of the two polynomials that define W_2 for Z resp. Z'. It can be readily verified that $Z \neq Z'$ implies the relevant two polynomials are distinct as well, and the resulting difference is a non-zero polynomial of degree at most three. It will therefore have at most three roots over $\mathbb{F}_{2^n}$.   □

Corollary 24. *For the compression function above, for $q \leq 2^{n-\frac{3}{2}}$:*

$$\mathsf{Adv}_H^{\text{coll}}(q) \leq (n + \frac{1}{2})q/2^{n-3} \ .$$

Curiously, if we would change the computation of W_2 even slightly, for instance $W_2 \leftarrow V_2W_1^2 + V_1W_1 + M$, the impact on the overlap γ is dramatic. Suddenly $R_{\text{AUX},(K_1,K_2,Y)} = \{W, K_2W^2 + (K_1 + 1)W + Y|W \in \{0,1\}^n\}$ and consequently $R_{\text{AUX},(V_1+1,V_2,M)} = R_{\text{POST},(V_1,V_2,M)}$, so that $\gamma = 2^n$. As a result, Theorem 20 can only be used to guarantee collision resistance up to roughly $2^{n/2}$ queries.

We note that like the original [18], our double length construction has some obvious shortcomings (see the full version [19] for more details).

Example II: An Intermediate Construction. We conclude with a construction based on a $3n/2$ bit state (split into three parts of $n/2$ bits each), that compresses $n/2$ message bits.

1. $X \leftarrow (M, V_1), K \leftarrow (V_2, V_3)$;
2. $Y \leftarrow E(K, X)$;
3. $W_1 \leftarrow Y_1 + M, W_2 \leftarrow Y_2 + V_1$, and $W_3 \leftarrow MW_1^3 + V_1W_1^2 + V_2W_1 + V_3$.

Lemma 25. *For the compression function above, $\gamma = 2^{2+n/2}$.*

Corollary 26. *For the compression function above and all $q < 2^{3n/4-2}$*

$$\mathsf{Adv}_H^{\text{coll}}(q) \leq eq/2^{3n/4-3} \ .$$

Acknowledgements

The author would like to thank John Black, Phil Rogaway and Onur Özen for useful ideas on the presentation; Elena Andreeva, Lars Knudsen and the anonymous FSE'09 referees for pointing out certain problems with preimage resistance; and Tom Shrimpton for great advice for the duration of the project.

References

1. Andreeva, E., Neven, G., Preneel, B., Shrimpton, T.: Seven-property-preserving iterated hashing: Rox. In: Kurosawa, K. (ed.) ASIACRYPT 2007. LNCS, vol. 4833, pp. 130–146. Springer, Heidelberg (2007)
2. Bernstein, D.J.: Cubehash specification (2.b.1). Submission to NIST (2008)
3. Bertoni, G., Daemen, J., Peeters, M., Van Assche, G.: Sponge functions. In: Ecrypt Hash Workshop (2007)
4. Bertoni, G., Daemen, J., Peeters, M., Van Assche, G.: On the indifferentiability of the sponge construction. In: Smart, N.P. (ed.) EUROCRYPT 2008. LNCS, vol. 4965, pp. 181–197. Springer, Heidelberg (2008)
5. Black, J., Rogaway, P., Shrimpton, T.: Black-box analysis of the block-cipher-based hash-function constructions from PGV. In: Yung, M. (ed.) CRYPTO 2002. LNCS, vol. 2442, pp. 320–335. Springer, Heidelberg (2002)
6. Damgård, I.: A design principle for hash functions. In: Brassard, G. (ed.) CRYPTO 1989. LNCS, vol. 435, pp. 416–427. Springer, Heidelberg (1990)
7. Duo, L., Li, C.: Improved collision and preimage resistance bounds on PGV schemes. IACR ePrint Archive, Report 2006/462 (2006), http://eprint.iacr.org/2006/462
8. Hohl, W., Lai, X., Meier, T., Waldvogel, C.: Security of iterated hash functions based on block ciphers. In: Stinson, D.R. (ed.) CRYPTO 1993. LNCS, vol. 773, pp. 379–390. Springer, Heidelberg (1993)
9. Knudsen, L.R., Rechberger, C., Thomsen, S.S.: The Grindahl hash functions. In: Biryukov, A. (ed.) FSE 2007. LNCS, vol. 4593, pp. 39–57. Springer, Heidelberg (2007)
10. Lucks, S.: A collision-resistant rate-1 double-block-length hash function. In: Symmetric Cryptography. Dagstuhl Seminar Proceedings, IBFI, Number 07021 (2007)
11. Matyas, S., Meyer, C., Oseas, J.: Generating strong one-way functions with cryptographic algorithms. IBM Technical Disclosure Bulletin 27(10a), 5658–5659 (1985)
12. Menezes, A., van Oorschot, P., Vanstone, S.: CRC-Handbook of Applied Cryptography. CRC Press, Boca Raton (1996)
13. Merkle, R.C.: One way hash functions and DES. In: Brassard, G. (ed.) CRYPTO 1989. LNCS, vol. 435, pp. 428–446. Springer, Heidelberg (1990)
14. Miyaguchi, S., Iwata, M., Ohta, K.: New 128-bit hash function. In: Proceedings 4th International Joint Workshop on Computer Communications, pp. 279–288 (1989)
15. Preneel, B., Govaerts, R., Vandewalle, J.: Hash functions based on block ciphers: A synthetic approach. In: Stinson, D.R. (ed.) CRYPTO 1993. LNCS, vol. 773, pp. 368–378. Springer, Heidelberg (1993)
16. Rogaway, P., Shrimpton, T.: Cryptographic hash-function basics: Definitions, implications and separations for preimage resistance, second-preimage resistance, and collision resistance. In: Roy, B., Meier, W. (eds.) FSE 2004. LNCS, vol. 3017, pp. 371–388. Springer, Heidelberg (2004)
17. Rogaway, P., Steinberger, J.: Security/efficiency tradeoffs for permutation-based hashing. In: Smart, N.P. (ed.) EUROCRYPT 2008. LNCS, vol. 4965, pp. 220–236. Springer, Heidelberg (2008)
18. Stam, M.: Beyond uniformity: Better security/efficiency tradeoffs for compression functions. In: Wagner, D. (ed.) CRYPTO 2008. LNCS, vol. 5157, pp. 397–412. Springer, Heidelberg (2008)
19. Stam, M.: Blockcipher based hashing revisited. IACR ePrint Archive, Report 2008/071 (2008), http://eprint.iacr.org/2008/071

On the Security of Tandem-DM

Ewan Fleischmann, Michael Gorski, and Stefan Lucks

Bauhaus-University Weimar, Germany
{ewan.fleischmann,michael.gorski,stefan.lucks}@uni-weimar.de

Abstract. We provide the first proof of security for Tandem-DM, one of
the oldest and most well-known constructions for turning a block cipher
with n-bit block length and $2n$-bit key length into a $2n$-bit cryptographic
hash function. We prove, that when Tandem-DM is instantiated with
AES-256, block length 128 bits and key length 256 bits, any adversary
that asks less than $2^{120.4}$ queries cannot find a collision with success probability greater than $1/2$. We also prove a bound for preimage resistance
of Tandem-DM.

Interestingly, as there is only one practical construction known turning such an $(n, 2n)$ bit block cipher into a $2n$-bit compression function
that has provably birthday-type collision resistance (FSE'06, Hirose),
Tandem-DM is one out of two constructions that has this desirable
feature.

Keywords: Cryptographic hash function, block cipher based, proof of
security, double-block length, ideal cipher model, Tandem-DM.

1 Introduction

A cryptographic hash function is a function which maps an input of arbitrary
length to an output of fixed length. It should satisfy at least collision-, preimage-
and second-preimage resistance and is one of the most important primitives in
cryptography [23].

Block Cipher-Based Hash Functions. Since their initial design by Rivest, MD4-
family hash functions (*e.g.* MD4, MD5, RIPEMD, SHA-1, SHA2 [26, 27, 29, 30])
have dominated cryptographic practice. But in recent years, a sequence of attacks
on these type of functions [7, 10, 37, 38] has led to a generalized sense of concern
about the MD4-approach. The most natural place to look for an alternative is in
block cipher-based constructions, which in fact predate the MD4-approach [22].
Another reason for the resurgence of interest in block cipher-based hash functions
is due to the rise of size restricted devices such as RFID tags or smart cards: A
hardware designer has to implement only a block cipher in order to obtain an
encryption function as well as a hash function. But since the output length of
most practical encryption functions is far too short for a collision resistant hash
function, *e.g.* 128-bit for AES, one is mainly interested in sound design principles
for *double block length* (DBL) hash functions [2]. A DBL hash-function uses a
block cipher with n-bit output as the building block by which it maps possibly
long strings to $2n$-bit ones.

O. Dunkelman (Ed.): FSE 2009, LNCS 5665, pp. 84–103, 2009.

Our Contribution. Four 'classical' DBL hash functions are known: MDC-2, MDC-4, ABREAST-DM and TANDEM-DM [3, 4, 20]. At EUROCRYPT'07, Steinberger [35] proved the first security bound for the hash function MDC-2: assuming a hash output length of 256 bits, any adversary asking less than $2^{74.9}$ queries cannot find a collision with probability greater than $1/2$.

In this paper, we prove the first security bound for the compression function TANDEM-DM in terms of collision resistance and preimage resistance. We will give an upper bound for success if an adversary is trying to find a collision. By assuming a hash output length of 256 bits, any adversary asking less than $2^{120.4}$ queries cannot find a collision with probability greater than $1/2$. We will also prove an upper bound for success if an adversary is trying to find a preimage. This bound is rather weak as it essentially only states, that the success probability of an adversary asking strictly less than 2^n queries is asymptotically negligible.

Beyond providing such a proof of security for TANDEM-DM in the first place, our result even delivers one of the most secure rate $1/2$ DBL compression functions known. The first *practical* DBL compression function with rate $1/2$ (without bit-fixing and other artificial procedures like employing two different block ciphers) that has a birthday-type security guarantee was presented at FSE'06 by Hirose [13]. He essentially states (see Appendix B for more details) that no adversary asking less than $2^{124.55}$ queries, again for $2n = 256$, can find a collision with probability greater then $1/2$. These two compression functions (Hirose's FSE '06 proposal and TANDEM-DM) are the only rate $1/2$ practical compression functions that are known to have a birthday-type security guarantee.

Outline. The paper is organized as follows: Section 2 includes formal notations and definitions as well as a review of related work. In Section 3 we proof that an adversary asking less than $2^{120.4}$ oracle queries has negligible advantage in finding a collision for the TANDEM-DM compression function. A bound for preimage resistance of TANDEM-DM is given in Section 4. In Section 5 we discuss our results and conclude the paper.

2 Preliminaries

2.1 Iterated DBL Hash Function Based on Block Ciphers

Ideal Cipher Model. An (n, k)-bit block cipher is a keyed family of permutations consisting of two paired algorithms $E : \{0,1\}^n \times \{0,1\}^k \to \{0,1\}^n$ and $E^{-1} : \{0,1\}^n \times \{0,1\}^k \to \{0,1\}^n$ both accepting a key of size k bits and an input block of size n bits. For simplicity, we will call it an (n, k)-block cipher. Let $\mathrm{BC}(n, k)$ be the set of all (n, k)-block ciphers. Now, for any one fixed key $K \in \{0,1\}^k$, decryption $E_K^{-1} = E^{-1}(\cdot, K)$ is the inverse function of encryption $E_K = E(\cdot, K)$, so that $E_K^{-1}(E_K(x)) = x$ holds for any input $X \in \{0,1\}^n$.

The security of block cipher based hash functions is usually analyzed in the *ideal cipher model* [2, 9, 17]. In this model, the underlying primitive, the block cipher E, is modeled as a family of random permutations $\{E_k\}$ whereas the random permutations are chosen independently for each key K, i.e. formally E is selected randomly from $\mathrm{BC}(n, k)$.

DBL Compression Functions. Iterated DBL hash functions with two block cipher calls in their compression function are discussed in this article. A hash function $H : \{0,1\}^* \to \{0,1\}^{2n}$ can be built by iterating a compression function $F : \{0,1\}^{3n} \to \{0,1\}^{2n}$ as follows: Split the padded message M into n-bit blocks $M_1, \ldots, M_l$, fix (G_0, H_0), apply $(G_i, H_i) = F(G_{i-1}, H_{i-1}, M_i)$ for $i = 1, \ldots, l$ and finally set $H(M) := (G_l, H_l)$. Let the compression function F be such that

$$(G_i, H_i) = F(G_{i-1}, H_{i-1}, M_i),$$

where $G_{i-1}, H_{i-1}, G_i, H_i, M_i \in \{0,1\}^n$. We assume that the compression function F consists of F_T, the top row, and F_B, the bottom row. We explicitly allow the results of F_T to be fed into the calculation of F_B. Each of the component functions F_B and F_T performs exactly *one* call to the block cipher and can be defined as follows:

$$\begin{aligned} G_i &= \quad F_T(G_{i-1}, H_{i-1}, M_i) &= E(X_T, K_T) \oplus Z_T, \\ H_i &= F_B(G_i, G_{i-1}, H_{i-1}, M_i) &= E(X_B, K_B) \oplus Z_B, \end{aligned}$$

where X_T, K_T, Z_T are uniquely determined by G_{i-1}, H_{i-1}, M_i and X_B, K_B, Z_B are uniquely determined by $G_i, G_{i-1}, H_{i-1}, M_i$.

We define the rate r of a block cipher based compression/hash function F by

$$r = \frac{|M_i|}{(\text{number of block cipher calls in F}) \times n}$$

It is a measure of efficiency for such block cipher based constructions.

2.2   The Tandem-DM Compression Function

The TANDEM-DM compression function was proposed by Lai and Massey at EUROCRYPT'92 [20]. It uses two cascaded Davies-Meyer [2] schemes. The compression function is illustrated in Figure 1 and is formally given in Definition 1.

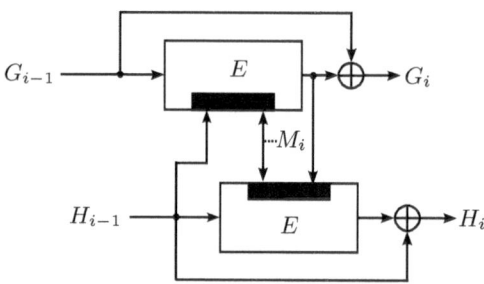

Fig. 1. The compression function TANDEM-DM F^{TDM} where E is an $(n, 2n)$ block cipher, the small rectangle inside the cipher rectangle indicates which input is used as key

Definition 1. *Let* $F^{TDM} : \{0,1\}^{2n} \times \{0,1\}^n \to \{0,1\}^{2n}$ *be a compression function such that* $(G_i, H_i) = F^{TDM}(G_{i-1}, H_{i-1}, M_i)$ *where* $G_i, H_i, M_i \in \{0,1\}^n$. F^{TDM} *is built upon an* $(n, 2n)$ *block cipher* E *as follows:*

$$W_i = E(G_{i-1}, H_{i-1}|M_i)$$
$$G_i = \mathrm{F}_T(G_{i-1}, H_{i-1}, M_i) = W_i \oplus G_{i-1}$$
$$H_i = \mathrm{F}_B(G_{i-1}, H_{i-1}, M_i) = E(H_{i-1}, M_i|W_i) \oplus H_{i-1}.$$

2.3   Related Work

Our work is largely motivated by Steinberger [35] in order to provide rigorous proofs for well-known block cipher based hash functions. As is reviewed in the following, there are many papers on hash functions composed of block ciphers.

Schemes with non-optimal or unknown collision resistance. The security of SBL hash functions against several generic attacks is discussed by Preneel *et al.* in [28]. They concluded that 12 out of 64 hash functions are secure against the attacks. However, formal proofs were first given by Black *et al.* [2] about 10 years later. Their most important result is that 20 hash functions – including the 12 mentioned above – are optimally collision resistant. Knudsen *et al.* [18] discussed the insecurity of DBL hash functions with rate 1 composed of (n, n)-block ciphers. Hohl *et al.* [14] analyzed the security of DBL compression functions with rate 1 and 1/2. Satoh *et al.* [33] and Hattoris *et al.* [11] discussed DBL hash functions with rate 1 composed of $(n, 2n)$ block ciphers. MDC-2 and MDC-4 [15, 1, 4] are (n, n)-block cipher based DBL hash functions with rates 1/2 and 1/4, respectively. Steinberger [35] proved that for MDC-2 instantiated with, *e.g.*, AES-128 no adversary asking less than $2^{74.9}$ can find a collision with probability greater than 1/2. Nandi *et al.* [25] proposed a construction with rate 2/3 but it is not optimally collision resistant. Furthermore, Knudsen and Muller [19] presented some attacks against it. At EUROCRYPT'08 and CRYPTO'08, Steinberger [31, 32] proved some security bounds for fixed-key (n, n)-block cipher based hash functions, *i.e.* permutation based hash functions, that all have small rates and low security guarantees. None of these schemes/techniques mentioned so far are known to have birthday-type collision resistance.

Schemes with Birthday-Type Collision Resistance. Merkle [24] presented three DBL hash functions composed of DES with rates of at most 0.276. They are optimally collision resistant in the ideal cipher model. Lucks [21] gave a rate 1 DBL construction with birthday-type collision resistance using a $(n, 2n)$ block cipher, but it involves some multiplications over $\mathbb{F}_{128}$. Hirose [12] presented a class of DBL hash functions with rate 1/2 which are composed of two different and independent $(n, 2n)$ block ciphers that have birthday-type collision resistance. At FSE'06, Hirose [13] presented a rate 1/2 and $(n, 2n)$ block cipher based DBL hash function that has birthday-type collision resistance. As he stated the proof only for the hash function, we have given the proof for his compression function in Appendix B.

3   Collision Resistance

In this section we will discuss the collision resistance of the compression function
TANDEM-DM.

3.1   Defining Security – Collision Resistance of a Compression Function (Pseudo Collisions)

Insecurity is quantified by the success probability of an optimal resource-bounded
adversary. The resource is the number of backward and forward queries to an
ideal cipher oracle E. For a set S, let $z \xleftarrow{R} S$ represent random sampling from
S under the uniform distribution. For a probabilistic algorithm $\mathcal{M}$, let $z \xleftarrow{R} \mathcal{M}$
mean that z is an output of $\mathcal{M}$ and its distribution is based on the random
choices of $\mathcal{M}$.

An adversary is a computationally unbounded but always-halting collision-
finding algorithm $\mathcal{A}$ with access to an oracle $E \in \mathrm{BC}(n, k)$. We can assume (by
standard arguments) that $\mathcal{A}$ is deterministic. The adversary may make a *forward*
query $(K, X)_{fwd}$ to discover the corresponding value $Y = E_K(X)$, or the adversary may make a *backward* query $(K, Y)_{bwd}$, so as to learn the corresponding
value $X = E_K^{-1}(Y)$ for which $E_K(X) = Y$. Either way the result of the query
is stored in a triple (X_i, K_i, Y_i) and the *query history*, denoted $\mathcal{Q}$, is the tuple
$(Q_1, \dots, Q_q)$ where $Q_i = (X_i, K_i, Y_i)$ is the result of the i-th query and q is
the total number of queries made by the adversary. The value $X_i \oplus Y_i$ is called
'XOR'-output of the query. Without loss of generality, it is assumed that $\mathcal{A}$ asks
at most only once on a triplet of a key K_i, a plaintext X_i and a ciphertext Y_i
obtained by a query and the corresponding reply.

The goal of the adversary is to output two different triplets, (G, H, M) and
(G', H', M'), such that $F(G, H, M) = F(G', H', M')$. We impose the reasonable
condition that the adversary must have made all queries necessary to compute
$F(G, H, M)$ and $F(G', H', M')$. We will in fact dispense the adversary from
having to output these two triplets, and simply determine whether the adversary
has been successful or not by examining its query history $\mathcal{Q}$. Formally, we say
that $\mathrm{COLL}(\mathcal{Q})$ holds if there is such a collision and $\mathcal{Q}$ contains all the queries
necessary to compute it.

Definition 2. (Collision resistance of a compression function) *Let F
be a block cipher based compression function, $F : \{0, 1\}^{3n} \to \{0, 1\}^{2n}$. Fix an
adversary $\mathcal{A}$. Then the advantage of $\mathcal{A}$ in finding collisions in F is the real
number*

$$\mathbf{Adv}_F^{\mathrm{COLL}}(\mathcal{A}) = \Pr[E \xleftarrow{R} \mathrm{BC}(n, k); ((G, H, M), (G', H', M')) \xleftarrow{R} \mathcal{A}^{E, E^{-1}} :$$
$$((G, H, M) \neq (G', H', M')) \wedge F(G, H, M) = F(G', H', M')].$$

For $q \geq 1$ we write

$$\mathbf{Adv}_F^{\mathrm{COLL}}(q) = \max_{\mathcal{A}}\{\mathbf{Adv}_F^{\mathrm{COLL}}(\mathcal{A})\},$$

*where the maximum is taken over all adversaries that ask at most q oracle
queries, i.e. E and E^{-1} queries.*

3.2   Security Results

Our discussion will result in a proof for the following upper bound:

Theorem 1. *Let $F := F^{TDM}$ be as in Definition 1 and n, q be natural numbers with $q < 2^n$. Let $N' = 2^n - q$ and let α be any positive number with $eq/N' \leq \alpha$ and $\tau = \alpha N'/q$ (and e^x being the exponential function). Then*

$$\mathbf{Adv}_F^{COLL}(q) \leq q 2^n e^{q\tau(1-\ln \tau)/N'} + 4q\alpha/N' + 6q/(N')^2 + 2q/(N')^3.$$

The proof is given on page 96 and is a simple corollary of the discussion and lemmas below. As this theorem is rather incomprehensible, we will investigate what this theorem means for AES-256. The bound obtained by this theorem depends on a parameter α. We do not require any specific value α as any α (meeting to the conditions mentioned in Theorem 1) leaves us with a correct bound. For Theorem 1 to give a *good* bound one must choose a suitable value for the parameter α. Choosing large values of α reduces the value of the first term but increases the value of the second term. There seems to be no good closed form for α as these will change with every q. The meaning of α will be explained in the proof. We will optimize the parameter α numerically as given in the following corollary.

Corollary 1. *For the compression function* TANDEM-DM, *instantiated e.g. with AES-256[1], any adversary asking less than $2^{120.4}$ (backward or forward) oracle queries cannot find a collision with probability greater than 1/2. In this case, $\alpha = 24.0$.*

3.3   Proof of Theorem 1

Analysis Overview. We will analyze if the queries made by the adversary contain the means for constructing a collision of the compression function F^{TDM}. Effectively we look to see whether there exist four queries that form a collision (see Figure 2).

To upper bound the probability of the adversary obtaining queries than can be used to construct a collision, we upper bound the probability of the adversary making a query that can be used as the final query to complete such a collision. Namely for each i, $1 \leq i \leq q$, we upper bound the probability that the answer to the adversary's i-th query $(K_i, X_i)_{fwd}$ or $(K_i, Y_i)_{bwd}$ will allow the adversary to use the i-th query to complete the collision. In the latter case, we say that the i-th query is 'successful' and we give the attack to the adversary.

As the probability depends naturally on the first $i - 1$ queries, we need to make sure that the adversary hasn't already been too lucky with these (or else the probability of the i-th query being successful would be hard to upper bound). Concretely, being lucky means, that there exists a large subset of the first $i - 1$ queries that all have the same XOR output (see below for a formal definition).

---

[1] Formally, we model the AES-256 block cipher as an ideal block cipher.

Our upper bound thus breaks down into two pieces: an upper bound for the probability of the adversary getting lucky in one defined specific way and the probability of the adversary ever making a successful i-th query, conditioned on the fact that the adversary has not yet become lucky by its $(i-1)$-th query.

Analysis Details. Fix numbers n, q and an adversary $\mathcal{A}$ asking q queries to its oracle. We say COLL^{TDM} if the adversary wins. Note that *winning* does not necessarily mean finding a collision as will be explained in the following. We upper bound $\Pr[\text{COLL}^{TDM}(\mathcal{Q})]$ by exhibiting predicates $\text{LUCKY}(\mathcal{Q})$, $\text{WIN1}(\mathcal{Q})$, $\text{WIN2}(\mathcal{Q})$ and $\text{WIN3}(\mathcal{Q})$ such that $\text{COLL}^{TDM}(\mathcal{Q}) \Rightarrow \text{LUCKY}(\mathcal{Q}) \vee \text{WIN1}(\mathcal{Q}) \vee \text{WIN2}(\mathcal{Q}) \vee \text{WIN3}(\mathcal{Q})$ and then by upper bounding separately the probabilities $\Pr[\text{LUCKY}(\mathcal{Q})]$, $\Pr[\text{WIN1}(\mathcal{Q})]$, $\Pr[\text{WIN2}(\mathcal{Q})]$ and $\Pr[\text{WIN3}(\mathcal{Q})]$. Then, obviously, $\Pr[\text{COLL}(\mathcal{Q})] \leq \Pr[\text{LUCKY}(\mathcal{Q})] + \Pr[\text{WIN1}(\mathcal{Q})] + \Pr[\text{WIN2}(\mathcal{Q})] + \Pr[\text{WIN3}(\mathcal{Q})]$. The event $\text{LUCKY}(\mathcal{Q})$ happens if the adversary is lucky, whereas if the adversary is not lucky but makes a successful i-th query then one of the other predicates hold.

To state the predicates, we need one additional definition. Let $a(\mathcal{Q})$ be a function defined on query sequences of length q as follows:

$$a(\mathcal{Q}) = \max_{Z \in \{0,1\}^n} |\{i : X_i \oplus Y_i = Z\}|$$

is the maximum size of a set of queries in $\mathcal{Q}$ whose XOR outputs are all the same. The event $\text{LUCKY}(\mathcal{Q})$ is now defined by

$$\text{LUCKY}(\mathcal{Q}) = a(\mathcal{Q}) > \alpha,$$

where α is the constant from Theorem 1 (it is chosen depending on n and q by a numerical optimization process). Thus as α is chosen larger $\Pr[\text{LUCKY}(\mathcal{Q})]$ diminishes. The other events, $\text{WIN1}(\mathcal{Q})$, $\text{WIN2}(\mathcal{Q})$ and $\text{WIN3}(\mathcal{Q})$ are different in nature from the event $\text{LUCKY}(\mathcal{Q})$. Simply put, they consider mutually exclusive configurations on how to find a collision for TANDEM-DM (see Figure 2 for an overview).

Notation. As in Figure 2, the four queries that can be used to form a collision will be labeled as TL for the query (X_i, K_i, Y_i) that is used for the position *top left*, BL for *bottom left*, TR for *top right* and BR for *bottom right*. Given

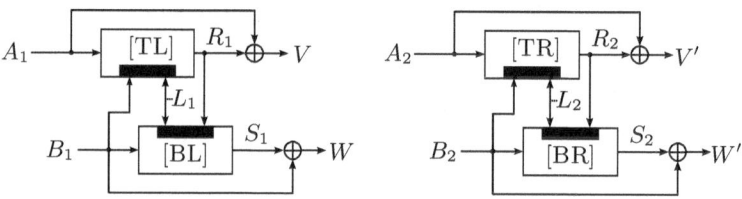

Fig. 2. Generic configuration for a collision, *i.e.* $V = V'$ and $W = W'$, for the TANDEM-DM compression function

$K \in \{0,1\}^n$, we will denote the first a bits as $K^{(0...a-1)} \in \{0,1\}^a$ and the leftover bits of K as $K^{(a...n)} \in \{0,1\}^{n-a}$. Furthermore, we say that two queries, e.g. BL and BR, are equal, i.e. BL=BR, if and only if there exists an i such that $BL = (X_i, K_i, Y_i) \in \mathcal{Q}$ and $BR = (X_i, K_i, Y_i)$.

We will call the configuration necessary for, e.g., predicate WIN1$a(\mathcal{Q})$ simply $1a$. Now, take for example just this configuration of predicate WIN1$a(\mathcal{Q})$ (i.e. all four queries are different and a collision is found; this case will be defined formally in Definition 3). We say, that the four queries $Q_i, Q_j, Q_k, Q_l \in \mathcal{Q}$ fit configuration $1a$ if and only if

$$(i \neq j) \wedge (i \neq k) \wedge (i \neq l) \wedge (j \neq k) \wedge (j \neq l) \wedge (k \neq l) \wedge$$
$$(X_i \oplus Y_i = X_k \oplus Y_k) \wedge (X_j \oplus Y_j = X_l \oplus Y_l) \wedge$$
$$(K_i = X_j | K_j^{(0...n/2-1)}) \wedge (K_j = K_i^{(n/2...n-1)} | Y_i) \wedge$$
$$(K_k = X_l | K_k^{(0...n/2-1)}) \wedge (K_l = K_k^{(n/2...n-1)} | Y_k).$$

We say, that FIT1$a(\mathcal{Q})$ holds if there exist $i, j, k, l \in \{1, 2, \ldots, q\}$ such that queries Q_i, Q_j, Q_k, Q_l fit configuration $1a$. The other predicates, namely FIT1$b(\mathcal{Q})$, FIT1$c(\mathcal{Q})$, FIT1$d(\mathcal{Q})$, FIT2$a(\mathcal{Q})$, ..., FIT2$d(\mathcal{Q})$, FIT3$a(\mathcal{Q})$, ..., FIT3$d(\mathcal{Q})$, whose configurations are given in Definition 3, are likewise defined. We also let

$$\text{FIT}j(\mathcal{Q}) := \text{FIT}ja(\mathcal{Q}) \vee \ldots \vee \text{FIT}jd(\mathcal{Q}) \quad \text{for} \quad j = 1, 2, 3.$$

Definition 3. Fit1$(\mathcal{Q})$: *The last query is used only once in position TL. Note that this is equal to the case where the last query is used only once in position TR.*

    Fit1$a(\mathcal{Q})$ *all queries used in the collision are pairwise different,*
    Fit1$b(\mathcal{Q})$ *BL = TR and BR is different to TL, BL, TR,*
    Fit1$c(\mathcal{Q})$ *BL = BR and TR is different to TL, BL, BR,*
    Fit1$d(\mathcal{Q})$ *TR = BR and BL is different to TL, TR, BR.*

Fit2$(\mathcal{Q})$: *The last query is used only once in position BL. Note that this is equal to the case where the last query is used only once in position BR.*

    Fit2$a(\mathcal{Q})$ *all queries used in the collision are pairwise different,*
    Fit2$b(\mathcal{Q})$ *TL = TR and BR is different to TL, BL, TR,*
    Fit2$c(\mathcal{Q})$ *TL = BR and TR is different to TL, BL, BR,*
    Fit2$d(\mathcal{Q})$ *TR = BR and TL is different to BL, TR, BR.*

Fit3$(\mathcal{Q})$: *The last query is used twice in a collision.*

    Fit3$a(\mathcal{Q})$ *last query used in TL, BL (TL = BL) and TR $\neq$ BR,*
    Fit3$b(\mathcal{Q})$ *last query used in TL, BL (TL = BL) and TR = BR,*
    Fit3$c(\mathcal{Q})$ *last query used in TL, BR (TL = BR) and BL $\neq$ TR,*
    Fit3$d(\mathcal{Q})$ *last query used in TL, BR (TL = BR) and BL = TR.*

In Lemma 1 we will show that these configurations cover all possible cases of a collision. We now define the following predicates:

$$\text{WIN1}(\mathcal{Q}) = \neg \text{LUCKY}(\mathcal{Q}) \wedge \text{FIT1}(\mathcal{Q}),$$
$$\text{WIN2}(\mathcal{Q}) = \neg(\text{LUCKY}(\mathcal{Q}) \vee \text{FIT1}(\mathcal{Q})) \wedge \text{FIT2}(\mathcal{Q}),$$
$$\text{WIN3}(\mathcal{Q}) = \neg(\text{LUCKY}(\mathcal{Q}) \vee \text{FIT1}(\mathcal{Q}) \vee \text{FIT2}(\mathcal{Q})) \wedge \text{FIT3}(\mathcal{Q}).$$

Thus $\text{WIN3}(\mathcal{Q})$, for example, is the predicate which is true if and only if $a(\mathcal{Q}) \leq \alpha$ (i.e. $\neg \text{LUCKY}(\mathcal{Q})$) and $\mathcal{Q}$ contains queries that fit configurations $3a, 3b, 3c$ or $3d$ but $\mathcal{Q}$ does *not* contain queries fitting configurations $1a, \ldots, 1d, 2a, \ldots 2d$. We now show, that $\text{COLL}^{TDM}(\mathcal{Q}) \implies \text{LUCKY}(\mathcal{Q}) \vee \text{WIN1}(\mathcal{Q}) \vee \text{WIN2}(\mathcal{Q}) \vee \text{WIN3}(\mathcal{Q})$.

Lemma 1

$$\text{COLL}^{TDM}(\mathcal{Q}) \implies \text{LUCKY}(\mathcal{Q}) \vee \text{WIN1}(\mathcal{Q}) \vee \text{WIN2}(\mathcal{Q}) \vee \text{WIN3}(\mathcal{Q}).$$

Proof. If the adversary is *not* lucky, *i.e.* $\neg \text{LUCKY}(\mathcal{Q})$, then

$$\text{FIT1}a(\mathcal{Q}) \vee \ldots \vee \text{FIT3}d(\mathcal{Q}) \implies \text{WIN1}a(\mathcal{Q}) \vee \ldots \text{WIN3}d(\mathcal{Q})$$

holds. So it is sufficient to show that $\text{COLL}^{TDM}(\mathcal{Q}) \implies \text{FIT1}a(\mathcal{Q}) \vee \ldots \vee \text{FIT3}d(\mathcal{Q})$. Now, say $\text{COLL}^{TDM}(\mathcal{Q})$ and $\neg \text{LUCKY}(\mathcal{Q})$. Then a collision can be constructed from the queries $\mathcal{Q}$. That is, our query history $\mathcal{Q}$ contains queries Q_i, Q_j, Q_k, Q_l (see Figure 2) such that we have a collision, i.e. $V = V'$ and $W = W'$ and $\text{TL} \neq \text{TR}$. Note that the last condition suffices to ensure a real collision (a collision from two different inputs).

First assume that the last query is used once in the collision. If it is used in position TL, then we have to consider the queries BL, TR and BR. If these three queries are all different (and as the last query is only used *once*), then $\text{FIT1}a(\mathcal{Q})$. If $\text{BL} = \text{TR}$ and BR is different, then $\text{FIT1}b(\mathcal{Q})$. If $\text{BL} = \text{BR}$ and TR is different, then $\text{FIT1}c(\mathcal{Q})$. If $\text{TR} = \text{BR}$ and BL is different, then $\text{FIT1}d(\mathcal{Q})$. If $\text{BL} = \text{TR} = \text{BR}$, then we have $\text{BL} = \text{BR}$ and $\text{TL} = \text{TR}$ and this would not result in a collision since the inputs to the two compression functions would be the same. As no cases are left, we are done (for the case that the last query is used only in position TL).

If the last query is used once in the collision and is used in position BL, then we have to consider the queries TL, TR and BR. If these three queries are all different (and as the last query is only used *once*), then $\text{FIT2}a(\mathcal{Q})$. If $\text{TL} = \text{TR}$ and BR is different, then $\text{FIT2}b(\mathcal{Q})$. If $\text{TL} = \text{BR}$ and TR is different, then $\text{FIT2}c(\mathcal{Q})$. If $\text{TR} = \text{BR}$ and TL is different, then $\text{FIT2}d(\mathcal{Q})$. If $\text{TL} = \text{TR} = \text{BR}$, it follows $\text{TL} = \text{TR}$ and $\text{BL} = \text{BR}$ and this would not result in a collision since the inputs to the two compression functions would be the same. As no cases are left, we are done.

We now analyze the case when the last query is used twice in the collision. First, assume that the query is used for the positions TL and BL ($\text{TL} = \text{BL}$). If $\text{TR} \neq \text{BR}$, then $\text{FIT3}a(\mathcal{Q})$, if $\text{TR} = \text{BR}$, then $\text{FIT3}b(\mathcal{Q})$. Now assume that the query is employed for the pair TL and BR ($\text{TL} = \text{BR}$). Note, that this case is equal to the case where the query is employed for BL and TR. If $BL \neq TR$, then $\text{FIT}3c(\mathcal{Q})$, if $BL = TR$, then $\text{FIT}3d(\mathcal{Q})$. The other cases, *i.e.* the last query is employed either for $TL = TR$ or $BL = BR$, do not lead to a real collision as this would imply the same compression function input. As no cases are left, we are done.

If the last query is used more than twice for the collision we do not get a real collision as this case would imply either $TL = TR$ or $BL = BR$ and we have the same input, again, for both compression functions.   ∎

The next step is to upper bound the probability of the predicates $\text{LUCKY}(\mathcal{Q})$, $\text{WIN}1(\mathcal{Q})$, $\text{WIN}2(\mathcal{Q})$ and $\text{WIN}3(\mathcal{Q})$.

Lemma 2. *Let α be as in Theorem 1. If $\alpha > e$ and $\tau = N'\alpha/q$, then*

$$\Pr[\text{LUCKY}(\mathcal{Q})] \leq q2^n e^{\tau\nu(1-\ln\tau)}.$$

The proof is quite technical and is given in Appendix A.

Lemma 3. $\Pr[\text{WIN}1(\mathcal{Q})] \leq q\alpha/N' + 2q/(N')^2 + q/(N')^3.$

Proof. As $\text{WIN}1(\mathcal{Q}) = \neg\text{LUCKY}(\mathcal{Q}) \wedge \text{FIT}1(\mathcal{Q})$, we will upper bound the probabilities of $\text{FIT}1a(\mathcal{Q})$, $\text{FIT}1b(\mathcal{Q})$, $\text{FIT}1c(\mathcal{Q})$ and $\text{FIT}1d(\mathcal{Q})$ separately in order to get an upper bound for $\Pr[\text{FIT}1(\mathcal{Q})] \leq \text{FIT}1a(\mathcal{Q}) + \ldots + \text{FIT}1d(\mathcal{Q})$. We will use the notations given in Figure 3.

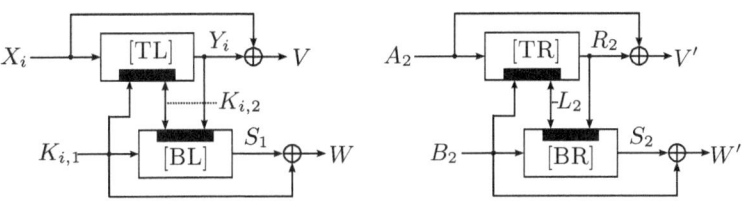

Fig. 3. Notations used for $\text{WIN}1(\mathcal{Q})$

Let $\mathcal{Q}_i$ denote the first i queries made by the adversary. The term 'last query' means the latest query made by the adversary (we examine the adversary's queries $(K_i, X_i)_{fwd}$ or $(K_i, Y_i)_{bwd}$ one at a time as they come in). The last query is always given index i. We say the last query is successful if the output X_i or Y_i for the last query is such that $a(\mathcal{Q}_i) < \alpha$ and such that the adversary can use the query (X_i, K_i, Y_i) to fit the configuration given in Figure 3 using only queries in $\mathcal{Q}_i$ (in particular, the last query *must* be used once in the fitting for that query to count as successful). The goal is thus to upper bound the adversary's chance of ever making a successful last query. The basic setup for upper bounding the probability of success in a given case is to upper bound the maximum number of different outputs Y_i or X_i (depending on whether the last query is a forward or a backward query) that would allow the query (X_i, K_i, Y_i) to be used to fit the configuration, and then divide this number by $N' = 2^n - q$ (since either Y_i or X_i, depending, is chosen randomly among a set of at least $2^n - q$ different values). The ratio is then multiplied by q, since the adversary makes q queries in all, each of which could become a successful last query.

(i) $\textsc{Fit}1a(\mathcal{Q})$: The last query, wlog. $(X_i, K_{i,1}|K_{i,2}, Y_i)$, is used in position TL. We do not care whether the last query was a forward or backward query since the analysis below is the same. All queries are, as claimed, pairwise different. We give the adversary for free the answer to the forward query BL, $(K_{i,1}, K_{i,2}|Y_i, S_1)$. Then we have $V = Y_i \oplus X_i$ and $W = S_1 \oplus K_{i,1}$. This pair of queries is successful if the adversary's query history $\mathcal{Q}_{i-1}$ contains a pair $(A_2, B_2|L_2, R_2), (B_2, L_2|R_2, S_2)$ such that $V = X_i \oplus Y_i = R_2 \oplus A_2 = V'$ and $W = S_1 \oplus K_{i,1} = S_2 \oplus B_2 = W'$. There are at most α queries in $\mathcal{Q}_{i-1}$ that can possibly be used for query in TR that all lead to a collision in the top row, i.e. $V = V'$. Therefore we have at most α possibilities for the query in BR since the query in TR uniquely determines the query BR. Thus, the last query has a chance of $\leq \alpha/N'$ of succeeding. So the total chance of making a successful query of this type is $\leq q\alpha/N'$.

(ii) $\textsc{Fit}1b(\mathcal{Q})$: Again, the last query, wlog. $(X_i, K_{i,1}|K_{i,2}, Y_i)$, is used in position TL. We give the adversary for free the answer to the forward query BL, $(K_{i,1}, K_{i,2}|Y_i, S_1)$. By our claim, as BL=TR, we have $A_2 = K_{i,1}, B_2 = K_{i,2}, L_2 = Y_i$ and $R_2 = S_1$. It follows that for any given query i for TL, we have at most one query for TR to form a collision $V = V'$ (as the query TL uniquely determines the query BL and the queries BL and TR are equal) and therefore have at most one query BR in our query history to form a collision $W = W'$. The last query has a chance of $\leq 1/(N' \cdot N')$ of succeeding and so the total chance of making a successful query in the attack is $\leq q/(N')^2$.

(iii) $\textsc{Fit}1c(\mathcal{Q})$: As this analysis is essentially the same as for $\textsc{Fit}1b(\mathcal{Q})$ we conclude with a total chance of success for this type of query of $\leq q/(N')^2$.

(iv) $\textsc{Fit}1d(\mathcal{Q})$: Again, the last query, wlog. $(X_i, K_{i,1}|K_{i,2}, Y_i)$, is used in position TL. We give the adversary for free the answer to the forward query BL, $(K_{i,1}, K_{i,2}|Y_i, S_1)$. Note, that this query is trivially different from the query in TL as we assume that the last query is only used once in this configuration (the case in which the two queries, TL and BL, are equal is discussed in the analysis of $\textsc{Win}3(\mathcal{Q})$). We have $V = Y_i \oplus X_i$ and $W = S_1 \oplus K_{i,1}$. As by our claim, we assume TR = BR. The pair of queries for TL and BL is successful if the adversary's query history $\mathcal{Q}_{i-1}$ contains a query $(A_2, B_2|L_2, R_2)$ such that $V = R_2 \oplus A_2 = V'$ and $W = R_2 \oplus A_2 = W'$, i.e. $V = W = V' = W'$. Moreover, it follows from $B_2 = R_2 = L_2$ that $V = W = V' = W' = 0$. As at least three of them are chosen randomly by the initial query input (wlog. V, W, V'), the query has a chance of success in the i-th query $\leq 1/(N' \cdot N' \cdot N')$ and therefore a total chance of success $\leq q/(N')^3$.

The claim follows by adding up the individual results. ∎

Lemma 4. $\Pr[\text{WIN2}(\mathcal{Q})] \leq q\alpha/N' + 2q/(N')^2 + q/(N')^3$.

As the proof and the result is (in principle) identical to the proof of $\Pr[\text{WIN1}(\mathcal{Q})]$ we omitted the details of the proof.

Lemma 5. $\Pr[\text{WIN3}(\mathcal{Q})] \leq 2q\alpha/N' + 2q/(N')^2$.

Proof. The same notations and preliminaries as in the proof of Lemma 3 are used.

(i) WIN3$a(\mathcal{Q})$: The last query, wlog. $(X_i, K_{i,1}|K_{i,2}, Y_i)$ is used in positions TL and BL. We do not care whether the last query is a forward or backward query since the analysis is the same. It follows, that $X_i = K_{i,1} = K_{i,2} = Y_i$ and therefore $V = X_i \oplus Y_i = W = 0$. We assume that the adversary is successful concerning these restraints, i.e. has found a query TL that can also be used for BL such as $X_i = Y_i = K_{i,1} = K_{i,2}$. (Note, that this condition is quite hard.) We do have at most α queries in $\mathcal{Q}_{i-1}$ that can possibly be used for a query in TR and that lead to a collision in the top row, i.e. $0 = V = V'$. For every such query TR we have at most one corresponding query in $\mathcal{Q}_{i-1}$ that can be used in position BR. So the last query has a chance of $\leq \alpha/N'$ of succeeding and so the total chance of making a successful query of this type during the attack is $\leq q\alpha/N'$.

(ii) WIN3$b(\mathcal{Q})$: The last query, wlog. $(X_i, K_{i,1}|K_{i,2}, Y_i)$ is used in positions TL and BL. We do not care whether the last query is a forward or backward query since the analysis is the same. It follows, that $X_i = K_{i,1} = K_{i,2} = Y_i$ and therefore $V = X_i \oplus Y_i = W = 0$. We assume again that the adversary is successful concerning these restraints, i.e. has found a query TL that can also be used for BL. We do have at most α queries in $\mathcal{Q}_{i-1}$ that can possibly be used for a query in TR and that lead to a collision in the top row, i.e. $0 = V = V'$. We assume that we can use any such query equally as the corresponding query for BR. In reality, this gives the adversary with high probability more power than he will have. Thus, the last query has a chance of $\leq \alpha/N'$ of succeeding and so the total chance of making a successful query of this type during the attack is $\leq q\alpha/N'$. As discussed above, this upper bound is likely to be generous.

(iii) WIN3$c(\mathcal{Q})$: The last query, wlog. $(X_i, K_{i,1}|K_{i,2}, Y_i)$ is used in positions TL and BR. Note, that this situation is equal to the last query being used in position BL and TR. We do not care whether the last query is a forward or backward query. We give the adversary for free the answer to the forward query BL, $(K_{i,1}, K_{i,2}|Y_i, S_1)$. We also give the adversary for free the answer to the backward query TR, $(A_2, X_i|K_{i,1}, K_{i,2})$. The probability for the i-th query to be successful is equal to $\Pr[V = V'] \cdot \Pr[W = W']$, and as W and V' are guaranteed to be chosen independently and randomly the chance of success is $\leq 1/(N')^2$. The total chance of success is therefore $\leq q/(N')^2$.

(iv) WIN3$d(\mathcal{Q})$: The last query, wlog. $(X_i, K_{i,1}|K_{i,2}, Y_i)$ is used in positions TL and BR. Note, that this situation is equal to the last query being used in position BL and TR. We do not care whether the last query is a forward or backward query. We give the adversary for free the answer to the forward query BL, $(K_{i,1}, K_{i,2} | Y_i, S_1)$. (This query is also used for position TR and it follows (by comparing the input values of query BL that is used for TR with them of BR) $K_{i,2} | Y_i = X_i | K_{i,1}$ and $S_1 = K_{i,2}$. Comparing the outputs we get a collision in the top row of the compression functions $\Pr[V = V'] = \Pr[E_{K_{i,1} | K_{i,2}}(X_i) \oplus X_i = E_{K_{i,2} | Y_i}(K_{i,1}) \oplus K_{i,1}]$, where $Y_i = E_{K_{i,1} | K_{i,2}}$, with probability $\leq 1/N'$. This is, because the input values $X_i, K_{i,1}, K_{i,2}$ have to be in such a way that the two inputs to the E oracle are different (if they are not, we would have no colliding inputs for the two compression functions). For the bottom row of the compression function we get, similarly, a collision with probability $\leq 1/N'$. So the total chance for succeeding is in this case $\leq q/(N')^2$ as we have again at most q queries by the adversary.   ∎

We now give the proof for Theorem 1.

Proof. (of Theorem 1)
The proof follows directly with Lemma 1, 2, 3, 4 and Lemma 5.   ∎

4   Preimage Resistance

Although, the main focus is on collision resistance, we are also interested in the difficulty of inverting the compression function of TANDEM-DM. Generally speaking, second-preimage resistance is a stronger security requirement than preimage resistance. A preimage may have some information of another preimage which produces the same output. However, in the ideal cipher model, for the compression function TANDEM-DM, a second-preimage has no information useful to find another preimage. Thus, only preimage resistance is analyzed. Note, that there have be various results that discuss attacks on iterated hash functions in terms of pre- and second-preimage, *e.g.* long-message second-preimage attacks [6, 16], in such a way that the preimage-resistance level cannot easily be transferred to an iterated hash function built on it.

The adversary's goal is to output a preimage (G, H, M) for a given σ, where σ is taken randomly from the output domain, such as $F(G, H, M) = \sigma$. As in the proof of Theorem 1 we will again dispense the adversary from having to output such a preimage. We will determine whether the adversary has been successful or not by examining its query history $\mathcal{Q}$. We say, that PREIMG($\mathcal{Q}$) holds if there is such a preimage and $\mathcal{Q}$ contains all the queries necessary to compute it.

Definition 4. (Inverting random points) *Let F be a block cipher based compression function, $F : \{0, 1\}^{3n} \rightarrow \{0, 1\}^{2n}$. Fix an adversary $\mathcal{A}$ that has access to oracles E, E^{-1}. Then the advantage of $\mathcal{A}$ of inverting F is the real number*

$$\mathbf{Adv}_F^{\text{INV}}(\mathcal{A}) = \Pr[E \xleftarrow{R} \text{BC}(n, k); \sigma \xleftarrow{R} \{0, 1\}^{2n} :$$
$$(G, H, M) \xleftarrow{R} \mathcal{A}^{E, E^{-1}}(\sigma) : F(G, H, M) = \sigma].$$

Again, for $q \geq 1$, we write

$$\mathbf{Adv}_F^{\mathrm{INV}}(q) = \max_{\mathcal{A}}\{\mathbf{Adv}_F^{\mathrm{INV}}(\mathcal{A})\}$$

where the maximum is taken over all adversaries that ask at most q oracle queries. Note that there has been a discussion on formalizations of preimage resistance. For details we refer to [2, Section 2, Appendix B].

4.1 Preimage Security

The preimage resistance of the compression function TANDEM-DM is given in the following Theorem.

Theorem 2. *Let $F := F^{TDM}$ be as in Definition 1. For every $N' = 2^n - q$ and $q > 1$*

$$\mathbf{Adv}_F^{\mathrm{INV}}(q) \le 2q/(N')^2.$$

Proof. Fix $\sigma = (\sigma_1, \sigma_2) \in \{0,1\}^{2n}$ where $\sigma_1, \sigma_2 \in \{0,1\}^n$ and an adversary $\mathcal{A}$ asking q queries to its oracles. We upper bound the probability that $\mathcal{A}$ finds a preimage for a given σ by examining the oracle queries as they come in and upper bound the probability that the last query can be used to create a preimage, i.e. we upper bound $\Pr[\mathrm{PREIMG}(\mathcal{Q})]$. Let $\mathcal{Q}_i$ denote the first i queries made by the adversary. The term 'last query' means the latest query made by the adversary since we examine again the adversary's queries $(K_i, X_i)_{fwd}$ or $(K_i, X_i)_{bwd}$ one at a time as they come in. The last query is always given index i.

Case 1: The last query (X_i, K_i, Y_i) is used in the top row. Either X_i or Y_i was randomly assigned by the oracle from a set of at least the size N'. The query is successful *in the top row* if $X_i \oplus Y_i = \sigma_1$ and thus has a chance of success of $\le 1/N'$. In $\mathcal{Q}_i$ there is at most one query Q_j that matches for the bottom row. If there is no such query in $\mathcal{Q}_i$ we give this query Q_j the adversary for free. This 'bottom' query is successful if $X_j \oplus Y_j = \sigma_2$ and therefore has a chance of success of $\le 1/N'$. So the total chance of success after q queries is $\le q/(N')^2$.

Case 2: The last query (X_i, K_i, Y_i) is used in the bottom row. The analysis is essentially the same as in Case 1. The total chance of success is $\le q/(N')^2$, too.

As any query can either be used in the top or in the bottom row, the claim follows.

5 Discussion and Conclusion

In this paper, we have investigated the security of TANDEM-DM, a long outstanding DBL compression function based on an $(n, 2n)$ block cipher. In the ideal cipher model, we showed that this construction has birthday-type collision resistance. As there are some generous margins in the proof it is likely, that

TANDEM-DM is even more secure. Our bound for preimage resistance is far from optimal, but we have not found an attack that would classify this bound as tight.

Somewhat surprisingly, there seems to be only one practical rate 1/2 DBL compression function that also has a birthday-type security guarantee. It was presented at FSE'06 by Hirose [13]. Taking into account that it was presented about 15 years after TANDEM-DM, it is clear that there needs still to be a lot of research done in the field of block cipher based hash functions, *e.g.* there are still security proofs missing for the aforementioned ABREAST-DM and MDC-4 compression or hash functions.

Acknowledgments

The authors wish to thank the anonymous reviewers for helpful comments.

References

1. ANSI. ANSI X9.31:1998: Digital Signatures Using Reversible Public Key Cryptography for the Financial Services Industry (rDSA). American National Standards Institute (1998)
2. Black, J., Rogaway, P., Shrimpton, T.: Black-Box Analysis of the Block-Cipher-Based Hash-Function Constructions from PGV. In: Yung, M. (ed.) CRYPTO 2002. LNCS, vol. 2442, pp. 320–335. Springer, Heidelberg (2002)
3. Meyer, C., Matyas, S.: Secure program load with manipulation detection code (1988)
4. Coppersmith, D., Pilpel, S., Meyer, C.H., Matyas, S.M., Hyden, M.M., Oseas, J., Brachtl, B., Schilling, M.: Data authentication using modification dectection codes based on a public one way encryption function. U.S. Patent No. 4, 908, 861, March 13 (1990)
5. Cramer, R. (ed.): EUROCRYPT 2005. LNCS, vol. 3494. Springer, Heidelberg (2005)
6. Dean, R.D.: Formal aspects of mobile code security. PhD thesis, Princeton, NJ, USA (1999); Adviser-Andrew Appel
7. den Boer, B., Bosselaers, A.: Collisions for the Compression Function of MD5. In: Helleseth, T. (ed.) EUROCRYPT 1993. LNCS, vol. 765, pp. 293–304. Springer, Heidelberg (1993)
8. Freitag, E., Busam, R.: Complex Analysis, 1st edn. Springer, Heidelberg (2005)
9. Even, S., Mansour, Y.: A Construction of a Cipher From a Single Pseudorandom Permutation. In: Imai, H., Rivest, R.L., Matsumoto, T. (eds.) ASIACRYPT 1991. LNCS, vol. 739, pp. 210–224. Springer, Heidelberg (1991)
10. Dobbertin, H.: The status of MD5 after a recent attack (1996)
11. Hattori, M., Hirose, S., Yoshida, S.: Analysis of Double Block Length Hash Functions. In: Paterson, K.G. (ed.) Cryptography and Coding 2003. LNCS, vol. 2898, pp. 290–302. Springer, Heidelberg (2003)
12. Hirose, S.: Provably Secure Double-Block-Length Hash Functions in a Black-Box Model. In: Park, C., Chee, S. (eds.) ICISC 2004. LNCS, vol. 3506, pp. 330–342. Springer, Heidelberg (2004)

13. Hirose, S.: Some Plausible Constructions of Double-Block-Length Hash Functions. In: Robshaw, M.J.B. (ed.) FSE 2006. LNCS, vol. 4047, pp. 210–225. Springer, Heidelberg (2006)
14. Hohl, W., Lai, X., Meier, T., Waldvogel, C.: Security of Iterated Hash Functions Based on Block Ciphers. In: Stinson [36], pp. 379–390
15. ISO/IEC. ISO DIS 10118-2: Information technology - Security techniques - Hash-functions, Part 2: Hash-functions using an n-bit block cipher algorithm. First released in 1992 (2000)
16. Kelsey, J., Schneier, B.: Second Preimages on n-Bit Hash Functions for Much Less than 2^n Work. In: Cramer [5], pp. 474–490
17. Kilian, J., Rogaway, P.: How to Protect DES Against Exhaustive Key Search. In: Koblitz, N. (ed.) CRYPTO 1996. LNCS, vol. 1109, pp. 252–267. Springer, Heidelberg (1996)
18. Knudsen, L.R., Lai, X., Preneel, B.: Attacks on Fast Double Block Length Hash Functions. J. Cryptology 11(1), 59–72 (1998)
19. Knudsen, L.R., Muller, F.: Some Attacks Against a Double Length Hash Proposal. In: Roy, B.K. (ed.) ASIACRYPT 2005. LNCS, vol. 3788, pp. 462–473. Springer, Heidelberg (2005)
20. Lai, X., Massey, J.L.: Hash Function Based on Block Ciphers. In: Rueppel, R.A. (ed.) EUROCRYPT 1992. LNCS, vol. 658, pp. 55–70. Springer, Heidelberg (1992)
21. Lucks, S.: A Collision-Resistant Rate-1 Double-Block-Length Hash Function. In: Biham, E., Handschuh, H., Lucks, S., Rijmen, V. (eds.) Symmetric Cryptography, Internationales Begegnungs- und Forschungszentrum fuer Informatik (IBFI), Schloss Dagstuhl, Germany. Dagstuhl Seminar Proceedings, vol. 07021 (2007)
22. Rabin, M.: Digitalized Signatures. In: DeMillo, R., Dobkin, D., Jones, A., Lipton, R. (eds.) Foundations of Secure Computation, pp. 155–168. Academic Press, London (1978)
23. Menezes, A., van Oorschot, P.C., Vanstone, S.A.: Handbook of Applied Cryptography. CRC Press, Boca Raton (1996)
24. Merkle, R.C.: One Way Hash Functions and DES. In: Brassard, G. (ed.) CRYPTO 1989. LNCS, vol. 435, pp. 428–446. Springer, Heidelberg (1989)
25. Nandi, M., Lee, W.I., Sakurai, K., Lee, S.-J.: Security Analysis of a 2/3-Rate Double Length Compression Function in the Black-Box Model. In: Gilbert, H., Handschuh, H. (eds.) FSE 2005. LNCS, vol. 3557, pp. 243–254. Springer, Heidelberg (2005)
26. NIST National Institute of Standards and Technology. FIPS 180-1: Secure Hash Standard (April 1995), http://csrc.nist.gov
27. NIST National Institute of Standards and Technology. FIPS 180-2: Secure Hash Standard (April 1995), http://csrc.nist.gov
28. Preneel, B., Govaerts, R., Vandewalle, J.: Hash Functions Based on Block Ciphers: A Synthetic Approach. In: Stinson [36], pp. 368–378
29. Rivest, R.L.: RFC 1321: The MD5 Message-Digest Algorithm. Internet Activities Board (April 1992)
30. Rivest, R.L.: The MD4 Message Digest Algorithm. In: Menezes, A., Vanstone, S.A. (eds.) CRYPTO 1990. LNCS, vol. 537, pp. 303–311. Springer, Heidelberg (1990)
31. Rogaway, P., Steinberger, J.P.: Constructing Cryptographic Hash Functions from Fixed-Key Blockciphers. In: Wagner, D. (ed.) CRYPTO 2008. LNCS, vol. 5157, pp. 433–450. Springer, Heidelberg (2008)
32. Rogaway, P., Steinberger, J.P.: Security/Efficiency Tradeoffs for Permutation-Based Hashing. In: Smart, N.P. (ed.) EUROCRYPT 2008. LNCS, vol. 4965, pp. 220–236. Springer, Heidelberg (2008)

33. Satoh, Haga, Kurosawa: Towards Secure and Fast Hash Functions. TIEICE: IEICE Transactions on Communications/Electronics/Information and Systems (1999)
34. Steinberger, J.P.: The collision intractability of mdc-2 in the ideal cipher model. IACR ePrint Archive, Report 2006/294 (2006), http://eprint.iacr.org/2006/294
35. Steinberger, J.P.: The Collision Intractability of MDC-2 in the Ideal-Cipher Model. In: Naor, M. (ed.) EUROCRYPT 2007. LNCS, vol. 4515, pp. 34–51. Springer, Heidelberg (2007)
36. Stinson, D.R. (ed.): CRYPTO 1993. LNCS, vol. 773. Springer, Heidelberg (1994)
37. Wang, X., Lai, X., Feng, D., Chen, H., Yu, X.: Cryptanalysis of the Hash Functions MD4 and RIPEMD. In: Cramer [5], pp. 1–18
38. Wang, X., Yin, Y.L., Yu, H.: Finding Collisions in the Full SHA-1. In: Shoup, V. (ed.) CRYPTO 2005. LNCS, vol. 3621, pp. 17–36. Springer, Heidelberg (2005)

A   Proof of Lemma 2

Note that this proof is essentially due to Steinberger [34]. We can rephrase the problem of upper bounding $\Pr[\textsc{Lucky}(\mathcal{Q})] = \Pr[a(\mathcal{Q}) > \alpha]$ as a balls-in-bins question. Let $N = 2^n$ be the number of bins and q be the number of balls to be thrown. The i-th ball falls into the j-th bin if the XOR output of the i-th query is equal to the XOR output of the j-th query, i.e. $X_i \oplus Y_i = X_j \oplus Y_j$. In the following we will upper bound the probability that some bin contains more than α balls. As the balls are thrown independent of each other, the i-th ball always has probability $\leq p = 1/(2^n - q)$ of falling in the j-th bin. This is because the XOR output of the i-th query is chosen uniformly at random from a set of size at least $2^n - q$. If we let $B(k)$ be the probability of having exactly k balls in a particular bin, say bin 1, then

$$B(k) \leq p^k \binom{q}{k}.$$

Let $\nu = qp$, where ν is an upper bound for the expected number of balls in any bin. By Stirlings approximation [8] (and e^x being the exponential function)

$$n! \leq \sqrt{2\pi n} \cdot \left(\frac{n}{e}\right)^n \cdot e^{1/(12n)}$$

we can upper bound $B(k)$ as follows:

$$
\begin{aligned}
B(k) &\leq p^k \frac{q!}{k!(q-k)!} \\
&\leq \frac{p^k}{\sqrt{2\pi}} \sqrt{\frac{q}{k(q-k)}} \cdot \frac{q^q}{k^k(q-k)^{1-k}} \cdot \frac{e^k \cdot e^{q-k}}{e^q} \cdot e^{\frac{1}{12}(q-k-(q-k))} \\
&\leq k^{-k}\nu^k \left(\frac{q}{q-k}\right) \\
&\leq \nu^k \cdot k^{-k} \cdot e^k.
\end{aligned}
$$

Since $\alpha = \tau\nu$ we get

$$B(\alpha) \leq \frac{\nu^{\tau\nu}e^{\tau\nu}}{(\tau\nu)^{\tau\nu}} = \frac{e^{\tau\nu}}{\tau^{\tau\nu}} = e^{\tau\nu(1-\ln\tau)}.$$

As $B(\alpha)$ is a decreasing function of α if $(1 - \ln\tau) < 0$, it follows that $B(\alpha)$ is a decreasing function if $\alpha > e$. And so we have

$$\Pr[a(\mathcal{Q}) > \alpha] \leq 2^n \cdot \sum_{j=\alpha}^{q} B(j)$$

$$\leq q2^n B(\alpha) \leq q2^n e^{\tau\nu(1-\ln\tau)}.$$

This proves our claim. ∎

B   Security of the FSE'06 Proposal by Hirose for a DBL Compression Function

At FSE'06, Hirose [13] proposed a DBL compression function (Definition 5 and Figure 4). He proved that when his compression function F^{Hirose} is employed in an iterated hash function H, then no adversary asking less than $2^{125.7}$ queries can have more than a chance of 0.5 in finding a collision for $n = 128$. As he has not stated a security result for the compression function we do here for comparison with TANDEM-DM.

B.1   Compression Function

Definition 5. *Let* $F^{Hirose} : \{0,1\}^{2n} \times \{0,1\}^n \to \{0,1\}^{2n}$ *be a compression function such that* $(G_i, H_i) = F^{Hirose}(G_{i-1}, H_{i-1}, M_i)$ *where* $G_i, H_i, M_i \in \{0,1\}^n$. F^{Hirose} *is built upon a* $(n, 2n)$ *block cipher* E *as follows:*

$$G_i = F_T(G_{i-1}, H_{i-1}, M_i) = E(G_{i-1}, H_{i-1}|M_i) \oplus G_{i-1}$$
$$H_i = F_B(G_{i-1}, H_{i-1}, M_i) = E(G_{i-1} \oplus C, H_{i-1}|M_i) \oplus G_{i-1} \oplus C,$$

where $'|'$ *represents concatenation and* $c \in \{0,1\}^n - \{0^n\}$ *is a constant.*

A visualization of this compression function is given in Figure 4.

B.2   Collision Resistance of the Compression Function

As the security proof of Hirose [13, Theorem 4] only states a collision resistance bound for a hash function built using F^{Hirose}, we will give a bound for the compression function itself. In particular, we will show:

Theorem 3. *Let* $F := F^{Hirose}$ *be a compression function as in Defintion 5. Then,*

$$\mathbf{Adv}_F^{\mathrm{COLL}}(q) \leq \frac{2q^2}{(2^n - 2q)^2} + \frac{2q}{2^n - 2q}.$$

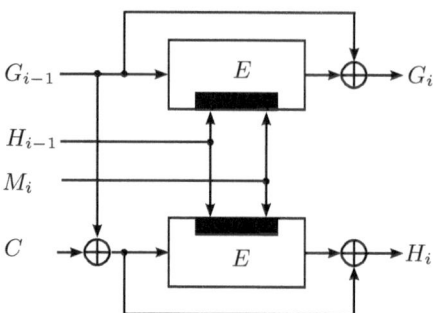

Fig. 4. The compression function F^{Hirose}, E is an $(n, 2n)$ block cipher, the small rectangle inside the cipher rectangle indicates the position of the key

In numerical terms, it means that no adversary performing less than $2^{124.55}$ oracle calls can have more than an even chance, *i.e.* 0.5, in finding a collision.

Due to the special structure of the compression function, the following definition is useful for the proof.

Definition 6. *A pair of distinct inputs* $(G_{i-1}, H_{i-1}, M_i), (G'_{i-1}, H'_{i-1}, M'_i)$ *to* F^{Hirose} *is called a* matching *pair if* $(G'_{i-1}, H'_{i-1}, M'_i) = (G_{i-1}, H_{i-1}, M_i \oplus C.$ *Otherwise they are called a* non-matching *pair.*

Note, that the proof is essentially due to Hirose [13], but as he stated it only for the hash function and not for the compression function itself. We will give a proof here for the compression function.

Proof. Let $\mathcal{A}$ be an adversary that asks q queries to oracles E, E^{-1}. Since

$$G_i = E(G_{i-1}, H_{i-1}|M_i) \oplus G_{i-1}$$

depends both on the plaintext and the ciphertext of E and one of them is fixed by a query and the other is determined by the answer, it follows that G_i is determined randomly. We give the adversary for free the answer to the query for H_i. Let $(X_i, K_{i,1}|K_{i,2}, Y_i)$ and $(X_i \oplus C, K_{i,1}|K_{i,2}, Z_i)$ be the triplets of E obtained by the i-th pair of queries and the corresponding answers.

For any $2 \leq i \leq q$, let C_i be the event that a colliding pair of non-matching inputs is found for F with the i-th pair of queries. Namely, it is the event that, for some $i' < i$

$$F(X_i, K_{1,i}, K_{2,i}) = F(X_{i'}, K_{1,i'}, K_{2,i'}) \text{ or } F(X_{i'} \oplus C, K_{1,i'}, K_{2,i'})$$

or

$$F(X_i \oplus C, K_{1,i}, K_{2,i}) = F(X_{i'}, K_{1,i'}, K_{2,i'}) \text{ or } F(X_{i'} \oplus C, K_{1,i'}, K_{2,i'})$$

which is equivalent to

$$(Y_i \oplus X_i, Z_i \oplus X_i \oplus C) = (Y_{i'} \oplus X_{i'}, Z_{i'} \oplus X_{i'} \oplus C)$$
$$\text{or } (Z_{i'} \oplus X_{i'} \oplus C, Y_{i'} \oplus X_{i'}).$$

It follows, that

$$\Pr[C_i] \leq \frac{2(i-1)}{(2^n - (2i-2))(2^n - (2i-1))} \leq \frac{2q}{(2^n - 2q)^2}.$$

Let C be the event that a colliding pair of non-matching inputs is found for F^{Hirose} with q (pairs) of queries. Then,

$$\Pr[C] \leq \sum_{i=2}^{q} \Pr[C_i] \leq \sum_{i=2}^{q} \frac{2q}{(2^n - 2q)^2} \leq \frac{2q^2}{(2^n - 2q)^2}.$$

Now, let $\hat{C}_i$ be the event that a colliding pair of matching inputs is found for F. It follows, that

$$\Pr[\hat{C}_i] \leq \frac{2}{(2^n - 2q)}.$$

Let $\hat{C}$ be the event that a colliding pair of matching inputs is found for F^{Hirose} with q (pairs) of queries. Then,

$$\Pr[\hat{C}] \leq \sum_{i=2}^{q} \Pr[\hat{C}_i] \leq \frac{2q}{2^n - 2q}.$$

Since $\mathbf{Adv}_F^{\mathrm{COLL}}(q) = \Pr[C \vee \hat{C}] \leq \Pr[C] + \Pr[\hat{C}]$, the claim follows.

Indifferentiability of Permutation-Based Compression Functions and Tree-Based Modes of Operation, with Applications to MD6

Yevgeniy Dodis[1], Leonid Reyzin[2], Ronald L. Rivest[3], and Emily Shen[3]

[1] New York University
dodis@cs.nyu.edu
[2] Boston University
reyzin@cs.bu.edu
[3] Massachusetts Institute of Technology
{rivest,eshen}@csail.mit.edu

Abstract. MD6 [17] is one of the earliest announced SHA-3 candidates, presented by Rivest at CRYPTO'08 [16]. Since then, MD6 has received a fair share of attention and has resisted several initial cryptanalytic attempts [1,11].

Given the interest in MD6, it is important to formally verify the soundness of its design from a theoretical standpoint. In this paper, we do so in two ways: once for the MD6 compression function and once for the MD6 mode of operation. Both proofs are based on the indifferentiability framework of Maurer et al. [13] (also see [9]).

The first proof demonstrates that the "prepend/map/chop" manner in which the MD6 compression function is constructed yields a compression function that is indifferentiable from a fixed-input-length (FIL), fixed-output-length random oracle.

The second proof demonstrates that the tree-based manner in which the MD6 mode of operation is defined yields a hash function that is indifferentiable from a variable-input-length (VIL), fixed-output-length random oracle.

Both proofs are rather general and apply not only to MD6 but also to other sufficiently similar hash functions.

These results may be interpreted as saying that the MD6 design has no structural flaws that make its input/output behavior clearly distinguishable from that of a VIL random oracle, even for an adversary who has access to inner components of the hash function. It follows that, under plausible assumptions about those inner components, the MD6 hash function may be safely plugged into any application proven secure assuming a monolithic VIL random oracle.

1 Introduction

In light of recent devastating attacks on existing hash functions, such as MD4 [18,22], MD5 [20], SHA-0 [21,7], and SHA-1 [19], NIST recently announced a competition for a new hash function standard, to be called SHA-3 [14]. NIST

O. Dunkelman (Ed.): FSE 2009, LNCS 5665, pp. 104–121, 2009.

received 64 submissions, one of which is MD6. The current status of the SHA-3 competition can be found on the NIST web site[1].

Crutchfield [10] showed that MD6 has many attractive properties required of a good hash function, such as preservation of collision-resistance, unpredictability, preimage-resistance, and pseudorandomness. As observed by Coron et al. [9], however, the above "traditional" properties of hash functions are often insufficient for many applications, which require that the hash function behaves "like a random oracle." Moreover, the initial NIST announcement for the SHA-3 competition [15] states that the candidate submissions will be judged in part by

"The extent to which the algorithm output is indistinguishable from a random oracle."

Thus, it is important to show that the design of the hash function is "consistent" with the proofs of security in the random oracle (RO) model [3]. Such a notion of consistency with the random oracle model was recently defined by Coron et al. [9]; it is called *indifferentiability from a random oracle* and is based on the general indifferentiability framework of Maurer et al. [13].

Given the importance of the random oracle model in the design of practical cryptographic schemes and the increased popularity of the indifferentiability framework in the analysis of hash functions [9,2,8,12,4], we suggest that it is critical that the winner of the SHA-3 competition satisfy such an "indifferentiability from a random oracle" property.

The main result of this paper is a formal proof that the design of MD6, both at the compression function level and at the mode of operation level, provides indifferentiability from a random oracle. Thus, the MD6 mode of operation and compression function have no structural flaws that would allow them to be distinguished from (VIL or FIL, respectively) random oracles. It follows, from results due to Maurer et al. [13], that, given reasonable assumptions about the permutation inside the compression function, the MD6 hash function may be safely plugged into any higher-level application whose security is proven assuming the hash function is a VIL random oracle.

These results generalize to other hash functions built on permutation-based compression functions and tree-based modes of operations, if they are sufficiently similar to MD6 in structure to meet the conditions of our proofs. We note that similar results on tree-based hashing have been obtained in independent work by Bertoni et al. [6].

To explain our results more precisely, we briefly recall the indifferentiability framework and the high level design of the MD6 hash function.

1.1   Indifferentiability

The notion of indifferentiability was first introduced by Maurer et al. in [13]. Informally, it gives sufficient conditions under which a primitive F can be "safely replaced" by some construction C^G (using an ideal primitive G). The formal definition is as follows.

---

[1] http://csrc.nist.gov/groups/ST/hash/sha-3/index.html

Definition 1. *A Turing machine C with oracle access to an ideal primitive G is (t, q_F, q_S, ϵ)-indifferentiable from an ideal primitive F if there exists a simulator S such that, for any distinguisher D, it holds that:*

$$\left| \Pr \left[D^{C,G} = 1 \right] - \Pr \left[D^{F,S} = 1 \right] \right| < \epsilon$$

The simulator S has oracle access to F (but does not see the queries of the distinguisher D to F) and runs in time at most t. The distinguisher makes at most q_F queries to C or F and at most q_S queries to G or S.

Indifferentiability is a powerful notion; Maurer et al. [13] show that if C^G is indifferentiable from F, then F may be replaced by C^G in any cryptosystem, and the resulting cryptosystem is at least as secure in the G model as in the F model.

In this paper, F will always be a random oracle — either fixed-input-length, fixed-output-length or variable-input-length, fixed-output-length. Thus, we will be showing that a certain construction C is indifferentiable from a random oracle, meaning that any cryptosystem proven secure in the RO model will still be secure in the G model, when the hash function is implemented using C^G.

1.2   The MD6 Hash Function: High-Level View

The MD6 function consists of the following two high-level steps. First, there is a *compression function* which operates on fixed-length inputs. Second, there is a *mode of operation*, which uses the compression functions as a black-box and to evaluate the hash function on arbitrary-length inputs. Naturally, our indifferentiability results will consist of two parts as well: (a) that the compression function, under some natural assumptions, is indifferentiable from a fixed-input-length RO, and (b) that, assuming the compression function is a fixed-input-length RO, the MD6 mode of operation yields a hash function that is indifferentiable from a variable-input-length RO.

The Compression Function. The MD6 compression function f maps an input N of length $n = 89$ words (consisting of 25 words of auxiliary input followed by 64 words of data) to an output of $c = 16$ words. Thus, the compression function reduces the length of its data input by a factor of 4. The compression function f is computed as a series of operations, which we view as the application of a random permutation π over the set of 89-word strings, followed by a truncation operation which returns the last 16 words of $\pi(N)$.

In each call to f in an MD6 computation, the first 15 words of auxiliary input are a constant Q (a representation of the fractional part of $\sqrt{6}$). Therefore, in our analysis, we consider the "reduced" compression function f_Q, where $f_Q(x) = f(Q\|x)$. For the full specification of the MD6 compression function, we refer the reader to [17].

In Section 3, we prove that f_Q is indifferentiable from a random oracle F, assuming that the main operation of the compression function is the application of a fixed public random permutation π.

The Mode of Operation. The standard MD6 mode of operation is a hierarchical, tree-based construction to allow for parallelism. However, for devices with limited storage, the MD6 mode of operation can be iterative. There is an optional level parameter L which allows a smooth transition between the fully hierarchial mode of operation and the fully iterative mode of operation. At each node in the tree, the input to the compression function includes auxiliary information. The auxiliary information includes a unique identifier U for each node (consisting of a tree level and index). In addition, there is a bit z which equals 1 in the input to the root node (the final compression call) and 0 in the input to all other nodes. For the full specification of the MD6 mode of operation, we refer the reader to [17].

In Section 4, we prove that the MD6 mode of operation is indifferentiable from a random oracle F when the compression function f_Q is modeled as a random oracle. In fact, our proof is quite general and applies essentially to any tree-like construction, as long as the final computation node has a distinguishable input structure and compression function inputs are uniquely parsable into blocks that are either raw message bits, metadata, or outputs of the compression function on "child" nodes.

2   Notation

We first introduce some notation.

Let $\mathbf{W} = \{0,1\}^w$ denote the set of all $w = 64$-bit words. Let $\chi_a(X)$ denote a function that returns the last a bits of X.

3   Indifferentiability from Random Oracle of MD6 Compression Function

In this section, we will prove the indifferentiability from a random oracle of the MD6 compression function construction, under certain assumptions. The compression function construction involves three steps:

- *Prepending* a constant value Q to the compression function input.
- *Mapping* the result by applying a fixed (pseudo)-random permutation π to it, and
- *Chopping* (removing) bits off the front of the result, so that what remains has the desired length as a compression function output.

Our proof applies in general to compression functions constructed in this manner. Our presentation of the proof will use notation shared with the formal specification of MD6, for convenience.

We view the compression function as based on a fixed public random permutation $\pi(\cdot)$, i.e., $f(N) = \chi_{cw}(\pi(N))$. (Recall that $\chi_{cw}(\cdot)$ returns the last cw bits of its input.) Since the permutation is public, an adversary can compute both π and π^{-1} easily. Therefore, we need to consider an adversarial model where the adversary has these powers.

Note that in this model the adversary can both invert f and find collisions for f easily, if we do not do something additional. (This is because the adversary can take the c-word output C, prepend $n - c$ words of random junk, then apply π^{-1} to get a valid pre-image for C. He can do this twice, with different junk values, to get a collision.) However, MD6 does have an important additional feature: a valid compression function input must begin with a fixed constant Q. We now proceed to show that this yields a compression function that behaves like a random oracle when π is a random permutation.

Recall that $\mathbf{W} = \{0,1\}^w$ denotes the set of all w-bit words and that f takes n-word inputs. We let $f_Q(x) = f(Q||x)$ denotes the "reduced" compression function that takes $(n - q)$-word inputs, prepends the fixed prefix Q, and runs f. To make it explicit that in this section we are modeling f and f_Q in terms of a random permutation π on $\mathbf{W}^n$, we will write

$$f_Q^\pi(x) = \chi_{cw}(\pi(Q||x)), \tag{1}$$

where $\chi_{cw}(y)$ returns the last cw bits of y, and where x is in $\mathbf{W}^{n-q}$.

Let the ideal functionality be represented by $F : \mathbf{W}^{n-q} \rightarrow \mathbf{W}^c$, a random oracle with same signature as f_Q^π. We will show that f_Q^π is indifferentiable from F, as stated below.

Theorem 1. *If π is a random permutation and Q is arbitrary, the reduced MD6 compression function f_Q^π defined by equation (1) is (t, q_F, q_S, ϵ)-indifferentiable from a random oracle F, for any number of queries q_F and q_S, for distinguishing advantage*

$$\epsilon = \frac{(q_S + q_F)^2}{2^{nw}} + \frac{q_S}{2^{qw}} + \frac{q_S q_F}{2^{(n-c)w}}, \tag{2}$$

and for running time of the simulator $t = O(q_S n w)$.

Proof. We use the approach of Coron et al. [9], who showed that the indifferentiability framework can be successfully applied to the analysis of hash functions built from simpler primitives (such as block ciphers or compression functions). We note that related results have been obtained by Bertoni et al. [5] in their proof of indifferentiability of "sponge functions."

In our case, because π is a permutation, the oracle G contains both π and π^{-1}, and we need to simulate them both. Slightly abusing notation, we will write S for the simulator of π and S^{-1} for the simulator of π^{-1}. Thus, we need to construct simulator programs S and S^{-1} for π and π^{-1} such that no distinguisher D can distinguish (except with negligible probability) between the following two scenarios:

(A) The distinguisher has oracle access to f_Q^π, to π, and to π^{-1}.
(B) The distinguisher has oracle access to F, S, and S^{-1}.

We define the simulators S, S^{-1} for π, π^{-1} as follows:

1. S and S^{-1} always act consistently with each other and with previous calls, if possible. If not possible (i.e., there are multiple answers for a given query), they abort.

2. To evaluate $S(X)$ where $X = Q||x$, compute $y = F(x)$, then return $R||y$ where R is chosen randomly from in $\mathbf{W}^{n-c}$.
3. To evaluate $S(X)$ where X does not start with Q, return a value R chosen randomly from $\mathbf{W}^n$.
4. To evaluate $S^{-1}(Y)$: return a random N in $\mathbf{W}^n$ which does not start with Q (i.e., from $\mathbf{W}^n \backslash (Q||\mathbf{W}^{n-q})$).

The running time of the simulators is at most $t = O(q_S n w)$. Next, we argue the indifferentiability of our construction. To this end, consider any distinguisher D making at most q_S to queries to S/π and S^{-1}/π^{-1} and at most q_F queries to F/f_Q^π. To analyze the advantage of this distinguisher, we consider several games $G_0, G_1, \ldots, G_7$. For each game G_i below, let $p_i = \Pr(D$ outputs 1 in $G_i)$. Intuitively, G_0 will be the "real" game, G_7 will be the "ideal" game, and the intermediate game will slowly transform these games into each other.

Game G_0. This is the interaction of D with f_Q^π, π, π^{-1}.

Game G_1. The game is identical to G_0 except the permutation π is chosen in a "lazy" manner. Namely, we introduce a *controller C_π* which maintains a table T_π consisting of all currently defined values (X, Y) such that $\pi(X) = Y$. Initially, this table is empty. Then, whenever a value $\pi(X)$ or $\pi^{-1}(Y)$ is needed, C_π first checks in T_π whether the corresponding value is already defined. If yes, it supplies it consistently. Else, it chooses the corresponding value at random subject to the "permutation constraint". Namely, if $T_\pi = \{(X_i, Y_i)\}$, then $\pi(X)$ is drawn uniformly from $\mathbf{W}^n \backslash \{Y_i\}$ and $\pi^{-1}(Y)$ is drawn uniformly from $\mathbf{W}^n \backslash \{X_i\}$. It is clear that G_1 is simply a syntactic rewriting of G_0. Thus, $p_1 = p_0$.

Game G_2. This game is identical to G_1 except the controller C_π does not make an effort to respect the permutation constraint above. Instead, it simply chooses undefined values $\pi(X)$ and $\pi^{-1}(Y)$ completely at random from $\mathbf{W}^n$, but explicitly aborts the game in case the permutation constraint is not satisfied. It is clear that $|p_2 - p_1|$ is at most the probability of such an abort, which, in turn, is at most $(q_S + q_F)^2 / 2^{nw}$.

Game G_3. This game is identical to G_2 except the controller C_π does not choose values starting with Q when answering the new inverse queries $\pi^{-1}(Y)$. Namely, instead of choosing such queries at random from $\mathbf{W}^n$, it chooses them at random from $\mathbf{W}^n \backslash (Q||\mathbf{W}^{n-q})$. It is easy to see that $|p_3 - p_2|$ is at most the probability that C_π would choose an inverse starting with Q in the game G_2, which is at most $q_S / 2^{qw}$.

Game G_4. This game is identical to G_3 except we modify the controller C_π as follows. Notice that there are three possible ways in which C_π would add an extra entry to the table T_π:

1. D makes a query $\pi(X)$ to π, in which case a new value (X, Y) might be added (for random Y). We call such additions *forward*.

2. D makes a query $\pi^{-1}(Y)$ to π^{-1}, in which case a new value (X, Y) is added (for random X not starting with Q). We call such additions *backward*.
3. D makes a query $f_Q^\pi(x) = \chi_{cw}(\pi(Q||x))$, in which case C_π needs to evaluate $\pi(Q||x)$ and add a value $(Q||x, Y)$ (for random Y). We call such additions *forced*.

We start by making a syntactic change. When a forced addition $(Q||x, Y)$ to T_π is made, C_π will mark it with a special symbol and will call this entry *marked*. C_π will keep it marked until D asks the usual forward query to $\pi(Q||x)$, in which case the entry will become *unmarked*, just like all the regular forward and backward additions to T_π. With this syntactic addition, we can now make a key semantic change in the behavior of the controller C_π.

– In game G_3, when a backward query $\pi^{-1}(Y)$ is made, C_π scans the entire table T_π to see if an entry of the form (X, Y) is present. In the new game G_4, C_π will only scan the *unmarked* entries in T_π, completely ignoring the currently marked entries.

We can see that the only way the distinguisher D will notice a difference between G_3 and G_4 is if D can produce a backward query $\pi^{-1}(Y)$ such that the current table T_π contains a *marked* entry of the form $(Q||x, Y)$. Let us call this event E, and let us upper-bound the probability of E. For each forced addition $(Q||x, Y)$, the value Y is chosen at random from $\mathbf{W}^n$, and the distinguisher D only learns the "chopped" value $y = \chi_{cw}(Y)$. In other words, D does not see $(n-c)w$ completely random bits of Y. Thus, for any particular forced addition, the probability that D ever "guesses" these missing bits is $2^{-(n-c)w}$. Since D gets at most q_S attempts, and there are at most q_F forced values to guess, we get that $\Pr(E) \le q_S q_F / 2^{(n-c)w}$. Thus, $|p_4 - p_3| \le \frac{q_S q_F}{2^{(n-c)w}}$.

Game G_5. We introduce a new controller C_F, which is simply imitating a random function $F : \mathbf{W}^{n-q} \to \mathbf{W}^c$. Namely, C_F keeps a table T_F, initially empty. When a query x is made, C_F checks if there is an entry (x, y) in T_F. If so, it outputs y. Else, it picks y at random from $\mathbf{W}^c$, adds (x, y) to T_F, and outputs y. Now, we modify the behaviors of the controller C_π for π/π^{-1} from the game G_4 as follows. In game G_4, when a new forward query $(Q||x)$ was made to π, or a new query x was made to f_Q^π, C_π chose a random Y from $\mathbf{W}^n$ and set $\pi(Q||x) = Y$. In game G_5, in either one of these cases, C_π will send a query x to the controller C_F, get the answer y, and then set $Y = R||y$, where R is chosen at random from $\mathbf{W}^{n-c}$.

We notice that the game G_5 is simply a syntactic rewriting of the game G_4, since choosing a random value in $\mathbf{W}^n$ is equivalent to concatenating two random values in $\mathbf{W}^{n-c}$ and $\mathbf{W}^c$. Thus, $p_5 = p_4$.

Game G_6. Before describing this game, we make the following observations about the game G_5. First, we claim that all the entries of the form $(Q||x, Y)$ in T_π, whether marked or unmarked, have come from the explicit interaction with the controller C_F. Indeed, because in game G_3 we restricted C_π to never answer a backward query so that the answer starts with Q, all such entries in T have come either from a forward query $\pi(Q||x)$, or the f_Q^π-query $f_Q^\pi(x)$. In either case, in game G_5 the controller C_π "consulted" C_F before making the answer. In fact, we can say more about the f_Q^π-query $f_Q^\pi(x)$. The answer to this query was simply the value y which C_F returned to C_π on input x. Moreover, because of the rules introduced in game G_4, C_π immediately marked the entry $(Q||x, R||y)$ which it added to T_π, and completely ignored this entry when answering the future backward queries to π^{-1} (until a query $\pi(Q||x)$ to π was made).

Thus, we will make the following change in the new game G_6. When D asks a new query $f_Q^\pi(x)$, the value x no longer goes to C_π (which would then attempt to define $\pi(Q||x)$ by consulting C_F). Instead, this query goes directly to C_F, and D is given the answer y. In particular, C_π will no longer need to mark any of the entries in T_π, since all the f_Q^π queries are now handled directly by C_F. More precisely, C_π will only "directly" define the forward queries $\pi(X)$ and the backward queries $\pi^{-1}(Y)$ (in the same way it did in Game G_5), but no longer define $\pi(Q||x)$ as a result of D's call to $f_Q^\pi(x)$.

We claim that game G_6 is, once again, only a syntactic rewriting of game G_5. Indeed, the only change between the two games is that, in game G_5, T_π will contain some marked entries $(Q||x, R||y)$, which will be ignored anyway in answering all the inverse queries, while in Game G_6 such entries will be simply absent. There is only one very minor subtlety. In Game G_5, if D first asks $f_Q^\pi(x)$, and later asks $\pi(Q||x)$, the latter answer $R||y$ will already be stored in T_π at the time of the first question $f_Q^\pi(x)$. However, it will be marked and ignored until the second question $\pi(Q||x)$ is made. In contrast, in Game G_6 this answer will only be stored in T_π after the second question. However, since in both cases C_π would answer by choosing a random R and concatenating it with C_F's answer y to x, this minor difference results in the same view for D. To sum up, $p_6 = p_5$.

Game G_7. This is our "ideal" game where D interacts with S/S^{-1} and a true random oracle F. We claim this interaction is identical to the one in Game G_6. Indeed, C_F is simply a "lazy" evaluation of the random oracle F. Also, after all our changes, the controller C_π in Game G_6 is *precisely* equivalent to our simulators S and S^{-1}. Thus, $p_7 = p_6$.

Collecting all the pieces together, we get that the advantage of D in distinguishing Game G_0 and Game G_7 is at most the claimed value

$$\epsilon \leq \frac{(q_S + q_F)^2}{2^{nw}} + \frac{q_S}{2^{qw}} + \frac{q_S q_F}{2^{(n-c)w}}$$

$\square$

(In practice, there are other inputs to consider, such as the key input K, the unique ID U, and the control word V. The above proof applies as given, assuming that these inputs are available for the distinguisher to control. This is the correct assumption to make from the viewpoint of the MD6 mode of operation or other applications using the MD6 compression function.)

Remark 1. We remark that our indifferentiability proof for the function $f_Q^\pi(x) = \chi_{cw}(\pi(Q||x))$ trivially generalizes to any compression function $f^\pi(x)$ of the form $f^\pi(x) = h_{cw}(\pi(g_{qw}(x)))$, where:

- $h_{cw} : \mathbf{W}^n \to \mathbf{W}^c$ is any *regular*[2] function. For the case of MD6, we use the "chop function" $h_{cw} = \chi_{cw}$.
- $g_{qw} : \mathbf{W}^{n-q} \to \mathbf{W}^n$ is any *injective* function which is (a) efficiently *invertible* and (b) efficiently *verifiable* (i.e., one can determine whether or not a point $y \in \mathbf{W}^n$ belongs to the range of g_{qw}). For the case of MD6, we use the "prepend function" $g_{qw}(x) = (Q||x)$, for some constant $Q \in \mathbf{W}^q$.

4   Indifferentiability of Tree-Based Modes of Operation

In this section, we prove that any tree-based mode of operation, with certain properties (defined below), is indifferentiable from a random oracle when the compression function is a random oracle. (In fact, our result applies to modes of operation that can be described as straight-line programs, which are more general than trees.) We then derive the indifferentiability of the MD6 mode of operation as a consequence of this result.

Consider a compression function $\phi : \mathbf{W}^n \to \mathbf{W}^c$. Let $\mu^\phi : \{0,1\}^* \to \{0,1\}^d$ denote the mode of operation μ applied to ϕ. We will prove that, if ϕ is a random oracle and μ satisfies the required properties, then μ^ϕ is indifferentiable from a random oracle.

4.1   Required Properties of Mode of Operation

To prove indifferentiability of a mode of operation μ, we will require μ to have the following properties. These are properties of the mode μ itself, independent of the particular compression function ϕ.

Unique Parsing. Every compression function input $x \in \mathbf{W}^\eta$ that occurs in the computation of $\mu^\phi(M)$ for some $M \in \{0,1\}^*$ must be efficiently and uniquely parsable into a sequence of blocks (not necessarily of the same size), each of which is a compression function output, raw message bits (from the message being hashed), or metadata. Note that there may be parsable inputs that do not actually occur in the computation of $\mu^\phi(M)$ for any M.

Parent Predicate. Given the unique parsing property, we define a predicate $parent(x, y, i)$ which, given oracle access to a compression function ϕ, takes two inputs $x, y \in \mathbf{W}^\eta$, and an index i. It outputs true iff $\phi(y)$ is equal to the ith compression function output in the parsing of x. We say that x is a parent of y, or equivalently, y is a child of x if $parent(x, y, i)$ is true for some i. We also say that y is the ith child of x if $parent(x, y, i)$ is true. Note that actual parent-child pairs occurring during the execution of μ may satisfy

---

[2] A function is regular if every value in the range has an equal number of preimages in the domain.

additional conditions: for example, x and y may contain consecutive level numbers. Our definition of *parent* does not verify these conditions, and thus may create a parent relationship where none should exist. These conditions (among others) will be verified by a function ρ defined below.

Leaf Predicate. Given the unique parsing property, we also define a predicate *leaf*(x) on compression function inputs x which returns true iff the parsing of x contains only message bits and metadata but no compression function outputs. We say that x is a leaf if *leaf*(x) is true.

Root Predicate. There must be a well-defined, efficiently testable predicate *root*(x) such that: for any $M \in \{0,1\}^*$, for every non-final compression function call x in the computation of $\mu^\phi(M)$, *root*(x) is false, and for the final compression function call y, *root*(y) is true. For strings x that are neither non-final nor final compression function calls in the computation of $\mu^\phi(M)$ for any M, *root*(x) can be either true or false.

Note that this condition implies that the set of strings $x \in \mathbf{W}^\eta$ that are the final compression call for some M must be disjoint from the set of strings $y \in \mathbf{W}^\eta$ that are a non-final compression call for some M.

Straight-Line Program Structure. The mode of operation μ must be a straight-line program in the following sense. It carries out a sequence of calls to the compression function, where input to call number i in the sequence is computed from the message itself, metadata, and outputs of some j calls numbered $i_1, i_2, \ldots, i_j < i$. This sequence (that is, the exact dependence of ith call on the previous calls and the message) must be deterministically computable from M alone, regardless of ϕ. For every call in the sequence except the last one, its output value must be used to compute some other input value. Moreover, for any ϕ, the output values of calls $i_1, i_2, \ldots, i_j$ must occur in the parsing of the input to the ith call. The last call x in that sequence must have *root*$(x) = $ true; for all the others, the root predicate must be false. Denote by $\Sigma(M)$ the set of all calls to ϕ (input and output pairs) during the computation of $\mu^\phi(M)$.

Final Output Processing. It must be the case that $\mu^\phi(M) = \zeta(\phi(x))$ where x is the final compression input, where $\zeta : \mathbf{W}^c \to \{0,1\}^d$ is an efficiently computable, regular function. The set of all preimages $\zeta^{-1}(h)$ of a value h must be efficiently sampleable given h.

Message Reconstruction. There must be an efficiently computable function ρ that takes a set Π of compression function calls and returns a message M if $\Pi = \Sigma(M)$, and $\bot$ otherwise. Because μ is deterministic, it follows that if $\rho(\Pi_1) = M_1$ and $\rho(\Pi_2) = M_2$ and $\Pi_1 \neq \Pi_2$, then $M_1 \neq M_2$.

We let $\kappa(\ell)$ denote an upper bound on the running time of ρ on an input set Π containing at most ℓ compression function calls. We assume that the other efficiently computable operations defined above (compression function input parsing, computing $\zeta(C)$, sampling from $\zeta^{-1}(h)$, and evaluating *root*(x)) run in constant time.

4.2 The Simulator

For the proof of indifferentiability of μ, we will define a polynomial-time simulator S_0 for the compression function ϕ. S_0 works as follows.

S_0 maintains a set T, initially empty, of pairs $(x, C) \in \mathbf{W}^\eta \times \mathbf{W}^c$ such that it
has responded with C to a query x.

Upon receiving a compression function query x^*, S_0 searches its set T for a
pair (x^*, C^*). If it finds such a pair, S_0 returns C^*.

Otherwise, S_0 evaluates $root(x^*)$. If false, S_0 chooses a fresh random string
$C^* \in \mathbf{W}^c$, inserts (x^*, C^*) into T, and returns C^*.

If true, S_0 executes the following "reconstruction procedure" to determine
whether x^* is the final compression function call in a computation of $\mu^\phi(M)$,
all of whose non-final compression function calls have already been seen by S_0.
In the reconstruction procedure, S_0 will build a set Π, initially empty, of pairs
(x, C) of compression function inputs and outputs.

1. Parse x^* into an ordered sequence of message bits, metadata, and compression function outputs. Let j be the number of compression function outputs.
2. For $i = 1, \ldots, j$:
   For each pair (x, C) in T, evaluate $parent(x^*, x, i)$. If no pair (x, C) satisfies
   $parent(x^*, x, i)$, or if multiple pairs (x, C) satisfy $parent(x^*, x, i)$, quit the
   reconstruction procedure; choose a fresh random string $C^* \in \mathbf{W}^c$, insert
   (x^*, C^*) into T, and return C^*. If a unique child pair (x, C) is found, add
   (x, C) to Π. If $leaf(x)$ is false, execute steps 1 and 2 for x.

If S_0 completes the reconstruction procedure (without returning a fresh random
string), S_0 now calls $\rho(\Pi)$. If ρ returns $\bot$, S_0 chooses a fresh random string
$C^* \in \mathbf{W}^c$, inserts (x^*, C^*) into T, and returns C^*. If ρ returns a message M^*,
S_0 calls F on M^* and samples a string $C^* \in \mathbf{W}^c$ randomly from $\zeta^{-1}(F(M^*))$.
S_0 inserts (x^*, C^*) into T and returns C^*.

Running Time. Let q_t be the total number of compression function queries to the
simulator. To answer each compression function query, the S_0 takes $O(q_t^2 + \kappa(q_t))$
time. Therefore, the total running time of the simulator is $O(q_t^3 + q_t \cdot \kappa(q_t))$.

Correctness. In our proof, we will use the following correctness properties of the
simulator.

Property 1: Suppose that there are no collisions on the output of S_0 (equivalently, there are no Type 1 events, which we define later). If all of the non-final
compression calls in the computation of $\mu^\phi(M^*)$ have been made to S_0, then
on the final compression call x^*, S_0 will reconstruct M^* and consult F on M^*.
This property can be easily proven by induction, starting at the final query
and working backward; the straight-line program property ensures that all the
compression function calls in $\Sigma(M^*)$ will be found, and the assumption of no
collisions ensures that no extra calls will be found.

Property 2: Suppose S_0 reconstructs M in response to a query at time τ. Then
all of the compression calls in the computation of the $\mu^\phi(M)$ have been made
to S_0 by time τ (indeed, otherwise $\rho(\Pi)$ would not return M, by definition).

4.3   Games

To prove the indifferentiability of the mode of operation μ, we consider a distinguisher D making compression function queries and mode of operation queries. We define a sequence of games, G_0 through G_4. Game G_0 is the "ideal" game, where D has oracle access to the random oracle F and the simulator S_0. Game G_4 is the "real" game, where D has oracle access to μ^ϕ and the random oracle ϕ.

For each game G_i, let p_i denote the probability that D outputs 1 in game G_i. We will argue that the view of the distinguisher cannot differ between consecutive games with more than negligible probability.

The games are defined as follows.

Game G_0. In Game G_0, the distinguisher D interacts with the random oracle F and the polynomial-time simulator S_0 defined above.

Game G_1. In Game G_1, we modify the simulator. At the start of the game, the new simulator S_1 uses its random coins to specify a random oracle $\mathcal{O}_S$: $\mathbf{W}^\eta \to \mathbf{W}^c$. The subsequent behavior of S_1 is identical to that of S_0 except, whereas S_0 generates its random bits in a lazy manner, S_1 gets its random bits from $\mathcal{O}_S$. Specifically, wherever S_0 answers a query x^* with a fresh random string $C^* \in \mathbf{W}^c$, S_1 answers with $\mathcal{O}_S(x^*)$. Similarly, wherever S_0 answers a query x^* by sampling randomly from $\zeta^{-1}(F(M^*))$, S_1 samples randomly from $\zeta^{-1}(F(M^*))$ using $\mathcal{O}_S(x^*)$ as its source of random bits.

The view of the distinguisher is the same in Game G_1 and Game G_0, so we have $p_1 = p_0$.

Game G_2. In Game G_2, we introduce a relay algorithm R_0 between D and F. The relay algorithm R_0 has oracle access to F and simply relays D's mode of operation queries to F and relays F's responses back to D.

The view of the distinguisher is the same in Game G_2 and Game G_1, so we have $p_2 = p_1$.

Game G_3. In Game G_3, we modify the relay algorithm. Instead of querying F, the new relay algorithm R_1 computes the mode of operation on its input, querying S_1 on each compression function call.

Game G_4. This is the final game. In Game G_4, we modify the simulator so that it no longer consults F. The new simulator S_2 always responds to a new query x^* with $\mathcal{O}_S(x^*)$. Thus, in this game D interacts with the mode of operation $\mu^{\mathcal{O}_S}$ and the random oracle $\mathcal{O}_S$.

Bad Events. In order to argue that the view of D cannot differ between Game G_2 and Game G_3 with more than negligible probability, we first define three types of "bad events" that can occur in Game G_2 or Game G_3.

- Type 1: S_1 inserts a pair (x_2, C) into T when there is already a pair (x_1, C) such that $x_1 \neq x_2$.

- Type 2: S_1 inserts a pair (x_2, C_2) into T when there is already a pair (x_1, C_1) in T such that x_1 is a parent of x_2.
- Type 3: D makes a query x_2 to S_1 such that x_2 is a parent of an x_1 which was previously queried by R but never directly by D.

We now prove that if none of these bad events occur in Game G_3, then the view of D is identical in G_2 and G_3.

Lemma 1. *For fixed coins of F, S_1, and D, if no bad events occur in Game G_3, then the view of D is identical in Game G_2 and Game G_3.*

Proof. We fix the random coins of F, S_1, and D and assume no bad events occur. We show by induction that D's observable values (the responses of the relay algorithm and the simulator) are identical in Games G_2 and G_3.

Suppose that the observable values have been identical in Games G_2 and G_3 so far. Consider D's next query. It is either a mode of operation query or a compression function query.

Mode of Operation Query. Consider a query M^* to μ^ϕ. In Game G_2, R_0 always returns $F(M^*)$. In Game G_3, R_1 will return $F(M^*)$ if the response of S_1 on the final compression call x^* is sampled from $\zeta^{-1}(F(M^*))$. There are two cases two consider.

1. x^* is a new query to S_1. Since R_1 has made all of the non-final calls for M^* before x^*, assuming there are no Type 1 events, by Correctness Property 1 of the simulator, S_1 will reconstruct M^* and return a string sampled from $\zeta^{-1}(F(M^*))$.
2. x^* has already been queried to S_1 before. Suppose it was first queried at time τ. Consider x^*'s children in the current computation $\mu^\phi(x)$. All of these children must have also been seen before time τ. Otherwise, a Type 2 event occurred in Game G_3 during one of the calls by R_1 after time τ. By induction, all of R_1's calls in the computation of $\mu^\phi(x)$ must have been seen before time τ. Therefore, assuming there are no Type 1 events, by Correctness Property 1 of the simulator, S_1 reconstructed M^* and returned a string sampled from $\zeta^{-1}(F(M^\prime))$ at time τ.

Therefore, given that the observables have been the same so far, if the next query is a mode of operation query, the next observable will be the same in games G_2 and G_3.

Compression Function Query. Consider a query x^* to ϕ.

We first make the following observation. Let T_2 be the set of queries S_1 has seen so far in Game G_2. Let T_3 be the set of queries S_1 has seen so far in Game G_3. Assuming the observables in the two games have been identical so far, T_2 must be a subset of T_3. This is because T_2 contains only the compression function queries made by D and T_3 contains these queries along with the queries made by R_1 in computing responses to D's mode of operation queries.

Now, suppose that x^* has already previously been queried to S_1 in Game G_2. Then S_1's response will be the same in both games, since by assumption it was the same in both games the first time x^* was queried.

So suppose that x^* is a new query in Game G_2. If in Game G_2, S_1 reconstructs an M^* and returns an element of $\zeta^{-1}(F(M^*))$ sampled using $\mathcal{O}_S(x^*)$ for its random bits, then by Correctness Property 2 of the simulator, all of M^*'s queries are in T_2, and therefore all of M^*'s queries are in T_3. Therefore, assuming no Type 1 events, by Correctness Property 1, S_1 must also reconstruct M^* in Game G_3 and return an element of $\zeta^{-1}(F(M^*))$ sampled using $\mathcal{O}_S(x^*)$ for its random bits.

If in Game G_2, S_1 instead returns $\mathcal{O}_S(x^*)$, then S_1's answer in Game G_3 is the same unless in Game G_3 S_1 reconstructs an M^*. If S_1 reconstructs an M^* in Game G_3 but not in Game G_2, then at least one of the queries used in the reconstruction must have come from R_1. But D made a query for the final compression function input x^*. Consider all of x^*'s children in G_3. All of these queries must have been asked by D at some time. Otherwise, a Type 3 event occurred in game G_3. By induction, all of the queries for M^* must have been asked by D before x^*, but then (assuming no Type 1 events) S_1 would have reconstructed M^* in Game G_2, contradicting the assumption that S_1 returns $\mathcal{O}_S(x^*)$ in Game G_2.

Therefore, given that the observables have been the same so far, if the next query is a mode of operation query, the next observable will be the same in games G_2 and G_3.

Therefore, conditioned on there being no occurrences of bad events in G_3, the view of D is identical in games G_2 and G_3.

We now bound the probability of bad events in Game G_3.

Lemma 2. *Suppose D makes q_S compression function queries and generates q_t compression function calls from its mode of operation queries and its compression function queries. Let $\Pr[\text{Bad}]$ denote the probability that a bad event occurs in Game G_3. $\Pr[\text{Bad}] \leq (2\eta/c + 1)q_t^2/2^{cw}$.*

Proof. We first make the observation that on a new query x, S_1 always responds with a fresh random string in $\mathbf{W}^c$. This is because S_1 responds with either (a) $\mathcal{O}_S(x) \in \mathbf{W}^c$ or (b) a random sample from $\zeta^{-1}(F(M))$ using $\mathcal{O}_S(x)$ as its source of randomness. In case (a), S_1's response is a fresh random string in $\mathbf{W}^c$. In case (b), S_1's is a fresh random string in $\mathbf{W}^c$ as long as S_1 has not queried F on M before. Since a message M has a unique final compression call and x is a new query, S_1 cannot have queried F on M before.

We now consider the three types of bad events.

Type 1 event: A Type 1 event corresponds to a collision between two random c-word strings among at most q_t compression function queries. We can bound the probability using the birthday bound over q_t random c-word strings: $\Pr[\text{Type 1}] \leq q_t^2/2^{cw+1}$.

Type 2 event: A Type 2 event occurs when S_1's random c-word response C_2 to a query x_2 equals one of the compression function outputs in the parsing of a previous query x_1. There at most η/c compression function outputs in the parsing of any query. So, there are at most $q_t \cdot \eta/c$ compression function outputs for S_1 to "guess" and S_1 has at most q_t guesses. Therefore, $\Pr[\text{Type 2}] \leq (\eta/c \cdot q_t^2)/2^{cw}$.

Type 3 event: A Type 3 event occurs when D makes a query which is a parent of a query that R made but D did not. The only output that D sees from R's queries is $\zeta(f(x))$ for the final compression call x, but it is not possible to make a query which is a parent of a final compression call. The probability of D "guessing" a non-final compression function output that it has not seen is $1/2^{cw}$. There are at most $(q_t - q_S)$ outputs to guess and D has at most $q_S \cdot \eta/c$ guesses. Therefore, $\Pr[\text{Type 3}] \leq (\eta/c \cdot q_S(q_t - q_S))/2^{cw}$.

Summing together, we get $\Pr[\text{Bad}] \leq q_t^2/2^{cw+1} + (\eta/c \cdot q_t^2)/2^{cw} + (\eta/c \cdot q_S(q_t - q_S))/2^{cw} \leq (2\eta/c + 1)q_t^2/2^{cw}$.

4.4   Indifferentiability Theorem

We now state our theorem of indifferentiability of μ.

Theorem 2. *If $\phi : \mathbf{W}^\eta \to \mathbf{W}^c$ is a random oracle, then μ^ϕ is (t, q_F, q_S, ϵ)-indifferentiable from a random oracle $F : \{0,1\}^* \to \{0,1\}^d$, with $\epsilon = (2\eta/c + 1)q_t^2/2^{cw}$ and $t = O(q_t^3 + q_t \cdot \kappa(q_t))$, for any q_F and q_S such that the total number of compression function calls from the mode of operation queries and the compression function queries of the distinguisher is at most q_t.*

Proof. Consider a distinguisher D that makes q_S compression function queries and generates q_t compression function calls from its mode of operation queries and its compression function queries.

It can easily be seen that the view of D is the same in Games G_0, G_1, and G_2, so $p_2 = p_1 = p_0$.

Combining Lemma 1 and Lemma 2, we have that $|p_3 - p_2| \leq (2\eta/c+1)q_t^2/2^{cw}$.

It is straightforward to see that $p_4 = p_3$. To see this, consider a query x^* (from either D or R_1) to S_2. If S_2 has seen the query x^* before, S_2 repeats the answer it gave the first time it was queried on x^*. If x^* is a new query, then in G_4, S_2 responds with the fresh random string $\mathcal{O}_S(x^*)$. As argued previously in the proof of Lemma 2, S_1 always responds to a new query x^* with a fresh random string in $\mathbf{W}^c$. Thus, the view of D is unchanged from G_3 to G_4, and $p_4 = p_3$.

Summing over all the games, the total advantage of D in distinguishing between G_0 and G_4 is at most the claimed $(2\eta/c + 1)q_t^2/2^{cw}$.

This completes the proof of the theorem.

It follows from our indifferentiability theorem that the MD6 mode of operation $\mathcal{M}^{f_Q}$ (specified in [17]) is indifferentiable from a random oracle when f_Q is a random oracle. This result is stated below.

Corollary 1. *If $f_Q : \mathbf{W}^{n-q} \to \mathbf{W}^c$ is a random oracle, the MD6 mode of operation $\mathcal{M}^{f_Q}$ is (t, q_F, q_S, ϵ)-indifferentiable from a random oracle $F : \{0,1\}^* \to \{0,1\}^d$, with $\epsilon = 9q_t^2/2^{cw}$ and $t = O(q_t^3)$, for any q_F and q_S such that the total number of compression function calls from the mode of operation queries and the compression function queries of the distinguisher is at most q_t.*

Proof. It can easily be seen that the MD6 mode of operation $\mathcal{M}$ satisfies the required properties. In particular, for MD6 ζ is the function χ_d which returns the last d bits of its input. The $root(x)$ predicate corresponds to testing whether the z bit of x is 1. It is straightforward to define a message reconstruction algorithm ρ for MD6 that runs in time $\kappa(q_t) = O(q_t)$.

Remark 2. The distinguishing probability ϵ and simulator running time t stated above follow directly from Theorem 2.

However, for MD6 specifically, we can easily get a tighter bound on ϵ because each compression function input has a unique node identifier, consisting of a tree level and index. Therefore, for a given compression function input y, in any other compression function input x, there is at most one (not $\eta/c = 4$) compression function output that could cause y to be a child of x. Thus, in bounding the probability of bad events for MD6, we can replace all instances of η/c with 1 and get $\epsilon = \Pr[\text{Bad}] \leq 2q_t^2/2^{cw}$.

Similarly, the presence of unique node identifiers in MD6 gives us a tighter bound on the running time t of the simulator. By maintaining its previously seen queries in sorted order (sorted by node identifier), the simulator can run in time $O(q_t^2)$ instead of $O(q_t^3)$.

5   Conclusion

We have shown that the (reduced) compression function $\chi_{cw}(\pi(Q\|x))$ of MD6 is indifferentiable from a fixed-input-length random oracle when π is a random permutation. We have also shown that any tree-based mode of operation with certain properties is indifferentiable from a variable-input-length random oracle when applied to a compression function that is a fixed-input-length random oracle. As a consequence of this result, the MD6 mode of operation is indifferentiable from a random oracle. Combined, these results imply that the design of MD6 has no stuctural weaknesses, such as "extension attacks", and that MD6 can be plugged into any application proven secure assuming a variable-input-length random oracle to obtain a scheme secure in a (fixed-length) random permutation model.

References

1. Aumasson, J.-P., Meier, W.: Nonrandomness observed on a reduced version of the compression function with 18 rounds in about 2^{17} operations
2. Bellare, M., Ristenpart, T.: Multi-property-preserving hash domain extension and the EMD transform. In: Lai, X., Chen, K. (eds.) ASIACRYPT 2006. LNCS, vol. 4284, pp. 299–314. Springer, Heidelberg (2006)
3. Bellare, M., Rogaway, P.: Random oracles are practical: A paradigm for designing efficient protocols. In: ACM Conference on Computer and Communications Security, pp. 62–73 (1993)
4. Bertoni, G., Daemen, J., Peeters, M., Van Assche, G.: Sponge functions (May 2007), http://www.csrc.nist.gov/pki/HashWorkshop/PublicComments/2007May.html

5. Bertoni, G., Daemen, J., Peeters, M., Van Assche, G.: On the indifferentiability of the sponge construction. In: Smart, N.P. (ed.) EUROCRYPT 2008. LNCS, vol. 4965, pp. 181–197. Springer, Heidelberg (2008)

6. Bertoni, G., Daemen, J., Peeters, M., Van Assche, G.: Sufficient conditions for sound tree hashing modes. In: Handschuh, H., Lucks, S., Preneel, B., Rogaway, P. (eds.) Symmetric Cryptography. Dagstuhl Seminar Proceedings (2009), http://www.dagstuhl.de/Materials/index.en.phtml?09031

7. Chabaud, F., Joux, A.: Differential collisions of SHA-0. In: Krawczyk, H. (ed.) CRYPTO 1998. LNCS, vol. 1462, pp. 56–71. Springer, Heidelberg (1998)

8. Chang, D., Lee, S., Nandi, M., Yung, M.: Indifferentiable security analysis of popular hash functions with prefix-free padding. In: Lai, X., Chen, K. (eds.) ASIACRYPT 2006. LNCS, vol. 4284, pp. 283–298. Springer, Heidelberg (2006)

9. Coron, J.-S., Dodis, Y., Malinaud, C., Puniya, P.: Merkle-Damgård revisited: How to construct a hash function. In: Shoup, V. (ed.) CRYPTO 2005. LNCS, vol. 3621, pp. 430–448. Springer, Heidelberg (2005)

10. Crutchfield, C.Y.: Security proofs for the MD6 hash function mode of operation. Master's thesis, MIT EECS Department (2008), http://groups.csail.mit.edu/cis/theses/crutchfield-masters-thesis.pdf

11. Dinur, I., Shamir, A.: Cube attack on a reduced version of the compression function with 15 rounds

12. Dodis, Y., Pietrzak, K., Puniya, P.: A new mode of operation for block ciphers and length-preserving MACs. In: Smart, N.P. (ed.) EUROCRYPT 2008. LNCS, vol. 4965, pp. 198–219. Springer, Heidelberg (2008)

13. Maurer, U., Renner, R., Holenstein, C.: Indifferentiability, impossibility results on reductions, and applications to the random oracle methodology. In: Naor, M. (ed.) TCC 2004. LNCS, vol. 2951, pp. 21–39. Springer, Heidelberg (2004)

14. National Institute of Standards and Technology. Announcing request for candidate algorithm nominations for a new cryptographic hash algorithm (SHA-3) family. Federal Register Notices, vol. 72(212), pp. 62212–62220 (November 2, 2007), http://csrc.nist.gov/groups/ST/hash/documents/FR_Notice_Nov07.pdf

15. National Institute of Standards and Technology. Announcing the development of new hash algorithm(s) for the revision of Federal Information Processing Standard (FIPS) 1802, secure hash standard. Federal Register Notices, vol. 72(14), pp. 2861–2863 (January 23, 2007), http://csrc.nist.gov/groups/ST/hash/documents/FR_Notice_Jan07.pdf

16. Rivest, R.L.: Slides from invited talk at Crypto 2008 (2008), http://group.csail.mit.edu/cis/md6/Rivest-TheMD6HashFunction.ppt

17. Rivest, R.L., Agre, B., Bailey, D.V., Crutchfield, C., Dodis, Y., Fleming, K.E., Khan, A., Krishnamurthy, J., Lin, Y., Reyzin, L., Shen, E., Sukha, J., Sutherland, D., Tromer, E., Yin, Y.L.: The MD6 hash function: A proposal to NIST for SHA-3 (2008), http://groups.csail.mit.edu/cis/md6/submitted-2008-10-27/Supporting_Documentation/md6_report.pdf

18. Wang, X., Lai, X., Feng, D., Chen, H., Yu, X.: Cryptanalysis of the hash functions MD4 and RIPEMD. In: Cramer, R. (ed.) EUROCRYPT 2005. LNCS, vol. 3494, pp. 1–18. Springer, Heidelberg (2005)

19. Wang, X., Yin, Y.L., Yu, H.: Finding collisions in the full SHA-1. In: Shoup, V. (ed.) CRYPTO 2005. LNCS, vol. 3621, pp. 17–36. Springer, Heidelberg (2005)

20. Wang, X., Yu, H.: How to Break MD5 and Other Hash Functions. In: Cramer, R. (ed.) EUROCRYPT 2005. LNCS, vol. 3494, pp. 19–35. Springer, Heidelberg (2005)
21. Wang, X., Yu, H., Yin, Y.L.: Efficient Collision Search Attacks on SHA-0. In: Shoup, V. (ed.) CRYPTO 2005. LNCS, vol. 3621, pp. 1–16. Springer, Heidelberg (2005)
22. Yu, H., Wang, X.: Multicollision attack on the compression functions of MD4 and 3-pass HAVAL. IACR ePrint Archive, Report 2007/085 (2007), http://eprint.iacr.org/2007/085

Cryptanalysis of RadioGatún

Thomas Fuhr[1] and Thomas Peyrin[2]

[1] DCSSI Labs
thomas.fuhr@sgdn.gouv.fr
[2] Ingenico
thomas.peyrin@ingenico.com

Abstract. In this paper we study the security of the RadioGatún family of hash functions, and more precisely the collision resistance of this proposal. We show that it is possible to find differential paths with acceptable probability of success. Then, by using the freedom degrees available from the incoming message words, we provide a significant improvement over the best previously known cryptanalysis. As a proof of concept, we provide a colliding pair of messages for RadioGatún with 2-bit words. We finally argue that, under some light assumption, our technique is very likely to provide the first collision attack on RadioGatún.

Keywords: hash functions, RadioGatún, cryptanalysis.

1 Introduction

A cryptographic hash function is a very important tool in cryptography, used in many applications such as digital signatures, authentication schemes or message integrity. Informally, a cryptographic hash function H is a function from $\{0,1\}^*$, the set of all finite length bit strings, to $\{0,1\}^n$ where n is the fixed size of the hash value. Moreover, a cryptographic hash function must satisfy the properties of preimage resistance, 2nd-preimage resistance and collision resistance [27]:

- **collision resistance**: finding a pair $x \neq x' \in \{0,1\}^*$ such that $H(x) = H(x')$ should require $2^{n/2}$ hash computations.
- **2nd preimage resistance**: for a given $x \in \{0,1\}^*$, finding a $x' \neq x$ such that $H(x) = H(x')$ should require 2^n hash computations.
- **preimage resistance**: for a given $y \in \{0,1\}^n$, finding a $x \in \{0,1\}^*$ such that $H(x) = y$ should require 2^n hash computations.

Generally, hash functions are built upon a *compression function* and a *domain extension algorithm*. A compression function h, usually built from scratch, should have the same security requirements as a hash function but takes fixed length inputs instead. Wang *et al.* [32, 33, 34, 35] recently showed that most standardized compression functions (e.g. MD5 or SHA-1) are not collision resistant. Then, a domain extension method allows the hash function to handle arbitrary length

O. Dunkelman (Ed.): FSE 2009, LNCS 5665, pp. 122–138, 2009.

inputs by defining an (often iterative) algorithm using the compression function as a black box. The pioneering work of Merkle and Damgård [15, 28] provided to designers an easy way in order to turn collision resistant compression functions onto collision resistant hash functions. Even if preserving collision resistance, it has been recently shown that this iterative process presents flaws [16, 19, 20, 21] and new algorithms [1, 2, 7, 25, 26] with better security properties have been proposed.

Most hash functions instantiating the Merkle-Damgård construction use a block-cipher based compression function. Some more recent hash proposals are based on construction principles which are closely related to stream ciphers. For example we can cite Grindahl [24] or RadioGatún [4]. The underlying idea of *stream-oriented* functions is to first absorb m-bit message blocks into a big internal state of size $c + m$ using a simple round function, and then squeeze the hash output words out. As the internal state is larger than the output of the hash function, the cryptanalytic techniques against the iterative constructions can not be transposed to the case of stream-oriented functions. In 2007, Bertoni *et al.* published a new hash construction mode, namely the *sponge functions* [6]. At Eurocrypt 2008, the same authors [5] published a proof of security for their construction : when assuming that the internal function F is a random permutation or a random transformation, then the sponge construction is indifferentiable from a random oracle up to $2^{c/2}$ operations.

However, even though the same authors designed RadioGatún and defined the sponge construction, RadioGatún does not completely fulfill the sponge definition. For evident performance reasons, the internal function F of RadioGatún is not a very strong permutation and this might lead to correlations between some input and output words. This threat is avoided by applying blank rounds (rounds without message incorporation) just after adding the last padded message word. More recently, some NIST SHA-3 candidates are using permutation-based modes as well, for example SHABAL [10], or sponge functions, for example Keccak [3].

Regarding the Grindahl family of hash functions, apart from potential slide attacks [18], it has been shown [23, 29] that it can not be considered as collision resistant. However, RadioGatún remains yet unharmed by the preliminary cryptanalysis [22]. The designers of RadioGatún claimed that for an instance manipulating w-bit words, one can output as much as $19 \times w$ bits and get a collision resistant hash function. That is, no collision attack should exist which requires less than $2^{9.5 \times w}$ hash computations. The designers also stated [4] that the best collision attack they could find (apart from generic birthday paradox ones) requires $2^{46 \times w}$ hash computations. A first cryptanalysis result by Bouillaguet and Fouque [8] using algebraic technique showed that one can find collisions for RadioGatún with $2^{24,5 \times w}$ hash computations. Finally, Khovratovich [22] described an attack using $2^{18 \times w}$ hash computations and memory, that can find collisions with the restriction that the IV must chosen by the attacker (semi-free-start collisions).

Our Contributions. In this paper, we provide an improved cryptanalysis of RadioGatún regarding collision search. Namely, using an improved computer-aided backtracking search and symmetric differences, we provide a technique that can find a collision with $2^{11 \times w}$ hash computations and negligible memory. As a proof of concept, we also present a colliding pair of messages for the case $w = 2$. Finally, we argue that this technique has a good chance to lead to the first collision attack on RadioGatún (the computation cost for setting up a complete collision attack is below the ideal bound claimed by the designers, but still unreachable for nowadays computers).

Outline. The paper is organized as follows. First, in Section 2, we describe the hash function proposal RadioGatún. Then, in Section 3, we introduce the concepts of *symmetric differences* and *control words*, that will be our two mains tools in order to cryptanalyze the scheme. In Section 4, we explain our differential path generation phase and in Section 5 we present our overall collision attack. Finally, we draw the conclusion in last section.

2 Description of RadioGatún

RadioGatún is a hash function using the design approach and correcting the problems of Panama [14], StepRightUp [13] or Subterranean [11, 13].

RadioGatún maintains an internal state of 58 words of w bits each, divided in two parts and simply initialized by imposing the zero value to all the words. The first part of the state, the *mill*, is composed of 19 words and the second part, the *belt*, can be represented by a matrix of 3 rows and 13 columns of words. We denote by M_i^k the i-th word of the mill state before application of the k-th iteration (with $0 \leq i \leq 18$) and $B_{i,j}^k$ represents the word located at column i and row j of the belt state before application of iteration k (with $0 \leq i \leq 12$ and $0 \leq j \leq 2$).

The message to hash is first padded and then divided into blocks of 3 words of w bits each that will update the internal state iteratively. We denote by m_i^k the i-th word of the message block m^k (with $0 \leq i \leq 2$). Namely, for iteration k, the message block m^k is firstly incorporated into the internal state and then a permutation P is applied on it. The incorporation process at iteration k is defined by :

$$B_{0,0}^k = B_{0,0}^k \oplus m_0^k \quad B_{0,1}^k = B_{0,1}^k \oplus m_1^k \quad B_{0,2}^k = B_{0,2}^k \oplus m_2^k$$
$$M_{16}^k = M_{16}^k \oplus m_0^k \quad M_{17}^k = M_{17}^k \oplus m_1^k \quad M_{18}^k = M_{18}^k \oplus m_2^k$$

where $\oplus$ denotes the bitwise *exclusive or* operation.

After having processed all the message blocks, the internal state is finally updated with N_{br} blank rounds (simply the application of the permutation P, without incorporating any message block). Eventually, the hash output value is generated by successively applying P and then outputting M_1^k and M_2^k as many time as required by the hash output size.

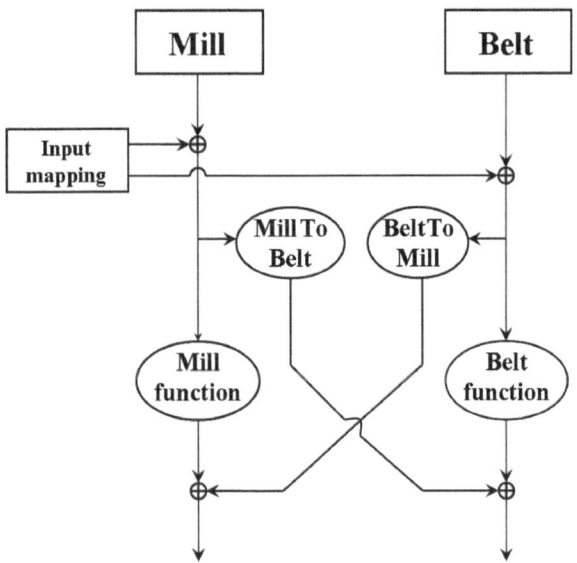

Fig. 1. The permutation P in RadioGatún

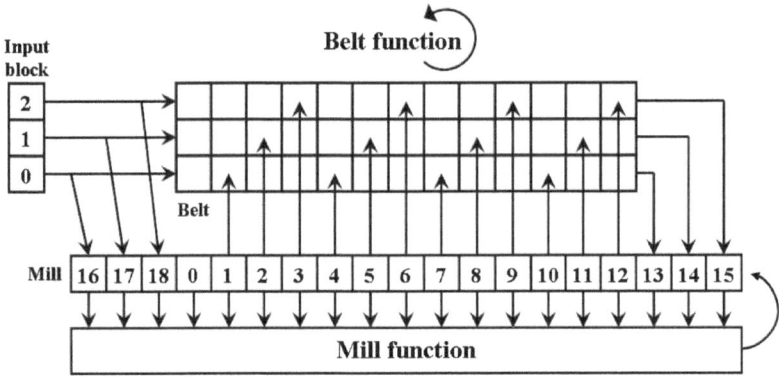

Fig. 2. The permutation P in RadioGatún

The permutation P can be divided into four parts. First, the *Belt* function is applied, then the *MillToBelt* function, the *Mill* function and eventually the *BeltToMill* function. This is depicted in Figures 1 and 2.

The *Belt* function simply consists of a row-wise rotation of the belt part of the state. That is, for $0 \leq i \leq 12$ and $0 \leq j \leq 2$:

$$B'_{i,j} = B_{i+1 \bmod 13, j}.$$

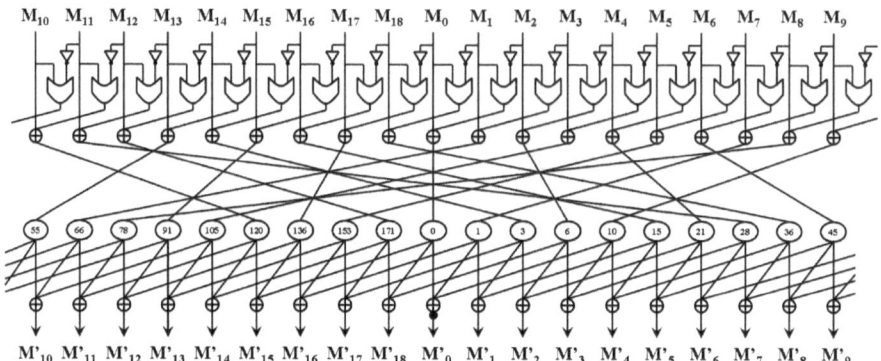

Fig. 3. The *Mill* function in `RadioGatún`

The *MillToBelt* function allows the mill part of the state to influence the belt one. For $0 \le i \le 11$, we have:

$$B'_{i+1,i \bmod 3} = B_{i+1,i \bmod 3} \oplus M_{i+1}.$$

The *Mill* function is the most complex phase of the permutation P and it updates the mill part of the state (see Figure 3). In the following, all indexes should be taken modulo 19. First, a nonlinear transformation is applied on all the words. For $0 \le i \le 18$:

$$M'_i = M_i \oplus \overline{M_{i+1} \wedge M_{i+2}}$$

where $\overline{X}$ denotes the bitwise negation of X and $\wedge$ represents the bitwise *and* operation. Then, a diffusion phase inside the words is used. For $0 \le i \le 18$:

$$M'_i = M_{7 \times i} \ggg (i \times (i+1)/2)$$

where $X \ggg (y)$ denotes the rotation of X on the right over y positions. Then, a diffusion phase among all the words is applied. For $0 \le i \le 18$:

$$M'_i = M_i \oplus M_{i+1} \oplus M_{i+4}.$$

Finally, an asymmetry is created by simply setting $M_0 = M_0 \oplus 1$.

The *BeltToMill* function allows the belt part of the state to influence the mill one. For $0 \le i \le 2$, we have :

$$M'_{i+13} = M_{i+13} \oplus B_{12,i}.$$

The `RadioGatún` security claims. Although `RadioGatún` has some common features with the sponge functions, the security proof of the sponge construction does not apply for this proposal. In their original paper [4], the authors claim that `RadioGatún` can output as much as 19 words and remain a secure hash function. Thus, it should not be possible for an attacker to find a collision attack running in less than $2^{9,5 \times w}$ hash computations.

3 Symmetric Differences and Control Words

3.1 Symmetric Differences

The first cryptanalysis tool we will use are symmetric differences. This technique has first been described in [30]. It was mentioned as a potential threat for RadioGatún in [4]. More precisely, a symmetric difference is an intra-word *exclusive or* difference that is part of a stable subspace of all the possible differences on a w-bit word. For example, in the following we will use the two difference values 0^w and 1^w (where the exponentiation by x denotes the concatenation of x identical strings), namely either a zero difference or either a difference on every bit of the word.

Considering those symmetric differences will allow us to simplify the overall scheme. Regarding the intra-word rotations during the *Mill* function, a 0^w or a 1^w difference will obviously remain unmodified. Moreover, the result of an *exclusive or* operation between two symmetric differences will naturally be a symmetric difference itself:

$$0^w \oplus 0^w = 0^w \quad 0^w \oplus 1^w = 1^w \quad 1^w \oplus 0^w = 1^w \quad 1^w \oplus 1^w = 0^w$$

The nonlinear part of the *Mill* function is more tricky. We can write:

$$\overline{\overline{a} \wedge b} = a \vee \overline{b}.$$

The output of this transformation will remain a symmetric difference with a certain probability of success, given in Table 1.

Table 1. Differential transitions for symmetric differences during the nonlinear part of the *Mill* function of RadioGatún. Δ_a and Δ_b denote the difference applied on a and b respectively, and $\Delta_{a \vee \overline{b}}$ the difference expected on the output of $a \vee \overline{b}$. The last column gives the corresponding conditions on the values of a and b in order to validate the differential transition. By $a = b$ (respectively $a \neq b$) we mean that all the bits of a and b are equal (respectively different), i.e. $a \oplus b = 0^w$ (respectively $a \oplus b = 1^w$).

Δ_a	Δ_b	$\Delta_{a \vee \overline{b}}$	Probability	Condition
0^w	0^w	0^w	1	
0^w	1^w	0^w	2^{-w}	$a = 1^w$
0^w	1^w	1^w	2^{-w}	$a = 0^w$
1^w	0^w	0^w	2^{-w}	$b = 0^w$
1^w	0^w	1^w	2^{-w}	$b = 1^w$
1^w	1^w	0^w	2^{-w}	$a = b$
1^w	1^w	1^w	2^{-w}	$a \neq b$

Due to the use of symmetric differences, the scheme to analyze can now be simplified : we can concentrate our efforts on a $w = 1$ version of RadioGatún, for which the intra-word rotations can be discarded. However, when building a differential path, for each differential transition during the nonlinear part of the *Mill* function, we will have to take the corresponding probability from Table 1 in account[1]. Note that this probability will be the only source of uncertainty in the differential paths we will consider (all the differential transitions through exclusive or operation always happen with probability equal to 1) and the product of all probabilities will be the core of the final complexity of the attack.

Also, one can check that the conditions on the *Mill* function input words are not necessarily independent. One may have to control differential transitions for nonlinear subfonctions located on adjacent positions (for example the first subfunction, involving M_0 and M_1, and the second, involving M_1 and M_2). This has two effects : potential incompatibility or condition compression (concerning M_1 in our example). In the first case, two conditions are located on the same input word and are contradicting (for example, one would have both $M_1 = 0^w$ and $M_1 = 1^w$). Thus, the differential path would be impossible to verify and, obviously, one has to avoid this scenario. For the second case, two conditions apply on the same input word but are not contradicting. Here, there is a chance that those conditions are redundant and we only have to account one time for a probability 2^{-w}. Finally, note that all those aspects have to be handled during the differential path establishment and not during the search for a valid pair of messages.

3.2 Control Words

When trying to find a collision attack for a hash function, two major tools are used : the differential path and the freedom degrees. In the next section, we will describe how to find good differential paths using symmetric differences. If a given path has probability of success equal to P, the complexity of a naive attack would be $1/P$ operations : if one chooses randomly and non-adaptively $1/P$ random message input pairs that are coherent with the differential constraints, there is a rather good chance that a one of them will follow the differential path entirely. However, for the same differential path, the complexity of the attack can be significantly decreased if the attacker chooses its inputs in a clever and adaptive manner.

In the case of RadioGatún, 3 w-bit message words are incorporated into the internal state at each round. Those words will naturally diffuse into the whole internal state, but not immediately. Thus, it is interesting to study how this diffusion behaves. Since the events we want to control through the differential path are the transitions of the nonlinear part of the *Mill* function (which depend on the input words of the *Mill* function), we will only study the diffusion regarding the input words of the *Mill* function.

Table 2 gives the dependencies between the message words incorporated at an iteration k, and the 19 input words of the *Mill* function at iteration k, $k+1$ and $k+2$. One can argue that a modification of a message block does not necessarily impacts the input word marked by a tick in Table 2 because the nonlinear

---

[1] In a dual view, all the conditions derived from Table 1 must be fulfilled.

Table 2. Dependencies between the message words incorporated at an iteration k, and the 19 input words of the *Mill* function of RadioGatún at iteration k, $k+1$ and $k+2$. The first table (respectively second and third) gives the dependencies regarding the message block m_0^k (respectively m_1^k and m_2^k). The columns represent the input words of the *Mill* function considered and a tick denotes that a dependency exists between the corresponding input word and message block.

iteration	M_0	M_1	M_2	M_3	M_4	M_5	M_6	M_7	M_8	M_9	M_{10}	M_{11}	M_{12}	M_{13}	M_{14}	M_{15}	M_{16}	M_{17}	M_{18}
k																	✓		
k+1		✓	✓		✓	✓				✓			✓	✓				✓	
k+2	✓	✓	✓	✓	✓	✓	✓	✓	✓	✓	✓	✓	✓	✓	✓	✓	✓	✓	✓

iteration	M_0	M_1	M_2	M_3	M_4	M_5	M_6	M_7	M_8	M_9	M_{10}	M_{11}	M_{12}	M_{13}	M_{14}	M_{15}	M_{16}	M_{17}	M_{18}
k																	✓		
k+1		✓			✓	✓				✓			✓	✓		✓	✓		
k+2	✓	✓	✓	✓	✓	✓	✓	✓	✓	✓	✓	✓	✓	✓	✓	✓	✓	✓	✓

iteration	M_0	M_1	M_2	M_3	M_4	M_5	M_6	M_7	M_8	M_9	M_{10}	M_{11}	M_{12}	M_{13}	M_{14}	M_{15}	M_{16}	M_{17}	M_{18}
k																			✓
k+1		✓			✓	✓		✓	✓					✓			✓	✓	
k+2	✓	✓	✓	✓	✓	✓	✓	✓	✓	✓	✓	✓	✓	✓	✓	✓	✓	✓	✓

function can sometimes "absorb" the diffusion of the modification. However, we emphasize that even if we depict here a behavior on average for the sake of clarity, all those details are taken in account thanks to our computer-aided use of the control words.

4   An Improved Backtracking Search

Our aim is to find internal collisions, i.e. collisions on the whole internal state before application of the blank rounds.

In order to build a good differential path using symmetric differences, we will use a computer-aided meet-in-the-middle approach, similar to the technique in [29]. More precisely, we will build our differential path DP by connecting together separate paths DP_f and DP_b. We emphasize that, in this section, we only want to build the differential path and not to look for a colliding pair of messages. DP_f will be built in the forward direction starting from an internal state containing no difference (modeling the fact that we have no difference after the initialization of the hash function), while DP_b will be built in the backward direction of the hash computation starting from an internal state containing no difference (modeling the fact that we want a collision at the end of the path).

Starting from an internal state with no difference, for each round the algorithm will go through all the possible difference incorporations of the message input (remember that we always use symmetric differences, thus we only have $2^3 = 8$ different cases to study) and all the possible symmetric differences transitions during the *Mill* function according to Table 1 (the differential transitions through exclusive or operations are fully deterministic). The algorithm can be

compared to a search tree in which the depth represents the number of rounds of
RadioGatún considered and each node is a reachable differential internal state.

4.1   Entropy

An exhaustive search in this tree would obviously imply making useless computations (some parts of the tree provide too costly differential paths anyway). To
avoid this, we always compute an estimation of the cost of finding a message
pair fulfilling the differential paths during the building phase of the tree, from
an initial state to the current leaf in the forward direction, and from the current
leaf to colliding states in the backward direction.

A first idea would be to compute the current cost of DP_f and DP_b during
the *meet-in-the-middle* phase. But, as mentioned in Section 3, some words of the
mill only depend on the inserted message block after 1 or 2 rounds. Therefore,
some conditions on the mill value have to be checked 2 rounds earlier, and some
degrees of freedom may have to be used to fulfill conditions two rounds later.
As DP_f and DP_b are computed round per round, it is difficult to compute
their complexity during the search phase, while having an efficient early-abort
algorithm.

Therefore, we use an *ad hoc* parameter, denoted H^k and defined as follows.
If c^k is the total number of conditions on the mill input words at round k (from
Table 1), we have for a path of length n:

$$\begin{cases} H^k = \max(H^{k+1} + c^k - 3, 0), \ \forall k < n \\ H^n = 0 \end{cases}$$

The idea is to evaluate the number of message pairs required at step k in order
to get $2^{w \times H^{k+1}}$ message pairs at step $k + 1$ of the exhaustive search phase. To
achieve this, one needs to fulfill $c^k \times w$ bit conditions on the mill input values,
with $3 \times w$ degrees of freedom. Therefore, the values of H^k can be viewed as the
relative entropies on the successive values of the internal state during the hash
computation.

The final collision search complexity would be $2^{w \times H_{max}}$, where H_{max} is the
maximum value of H^i along the path, if the adversary could choose 3 words of
his choice at each step, and if each output word of the *Mill* function depended
on all the input words. In the case of RadioGatún, the computation cost is more
complex to evaluate, and this is described in Section 5. The maximum entropy
can be linked to the *backtracking cost* C_b, as defined in [4]. One has the relation
$C_b = H_{max}+3$. The difference between these two notions is that the backtracking
cost takes in account the randomization of the input message pairs, which has a
cost 2^{3w}.

4.2   Differential Path Search Algorithm

The path search algorithm works as follows. Keep in mind that the values of
the entropy along the path are relative values - any constant value can therefore

be added or subtracted to all the H_i. A zero entropy at step i means that one expects $2^0 = 1$ message pair to follow the path until step i. To evaluate a path, we then set the minimal value of the entropy along the path to zero, the cost being the maximal value of the entropy. Therefore we first compute candidates for DP_f with a modified breadth-first search algorithm, eliminating those for which the maximum entropy exceeds the minimum entropy by more than $8 \times w$ (because we want to remain much lower than the $9, 5 \times w$ bound from the birthday paradox). The algorithm differs from a traditional breadth-first search as **we do not store all the nodes, but only those with an acceptable entropy** : to increase the probability of linking it to DP_b, one only stores the nodes whose entropy is at least $(H_{max} - 4) \times w$. We also store the state value of the previous node with entropy at least $(H_{max} - 4) \times w$, to enable an efficient backtracking process once the path is found.

We then compute DP_b, using a depth-first search among the backwards transitions of the *Mill* function, starting from colliding states. We set the initial entropy to $H^n = 0$, and we do not search the states for which $H > 8$ (same reason as for DP_f : we want to remain much lower than the bound from the birthday paradox). For each node having an entropy at most 4, we try to link it with a candidate for DP_f.

4.3   Complexity of the Path Search Phase

The total amount of possible values for a symmetric differential on the whole state is $2^{13 \times 3 + 19} = 2^{58}$. We use the fact that for RadioGatún, the insertion of $M \oplus M'$ can be seen as the successive insertions of M and M' without applying the round function. Therefore, we can consider setting the words $16, 17, 18$ of the stored mill to 0 by a message insertion before storing it in the forward phase, and doing the same in the backward phase before comparing it to forward values. Therefore, the space on which the meet-in-the-middle algorithm has to find a collision has approximately 2^{55} elements. We chose to store 2^{27} values of DP_f, and thus we have to compare approximately 2^{28} values for DP_b.

5   The Collision Attack

In this section, we depict the final collision attack, and compute its complexity. Once a differential path is settled, the derived collision attack is classic : we will use the control words to increase as much as possible the probability of success of the differential path.

5.1   Description

The input for this attack is a differential path, with a set of sufficient conditions on the values of the mill to ensure that a pair of messages follow the path. The adversary searches the colliding pairs in a tree, in which the nodes are messages following a prefix of the differential path. The leaves are messages following the whole differential path. Thanks to an early-abort approach, the adversary eliminates candidates as soon as they differ from the differential path. Nodes are associated with message pairs, or equivalently by the first message of a message pair – the second message is specified by the differential trail. Therefore, they will be denoted by the message they stand for. The sons of node M are then messages $M||b$, where b is a given message block, and the hash computation of $M||b$ fulfills all the conditions.

The adversary then uses a depth-first approach to find at least one node at depth n, where n is the length of the differential path. It is based on the trail backtracking technique, described in [4, 29]. To decrease the complexity of the algorithm, we check the conditions on the words of the mill as soon as they cannot be modified anymore by a message word inserted later.

From Table 2, we know that the k-th included message block impacts some words of the mill before the k-th iteration of the *Mill* function, some other words before the $k + 1$-th iteration, and the rest of the mill words before the $k + 2$-th iteration. We recall that m^k is the k-th inserted block, and we now set that M_j^k is the value of the j-th mill word after the k-th message insertion. Let also $\hat{M}_j^k$ be the value of the j-th word of the mill after the k-th nonlinear function computation.

After inserting m^k, one can then compute $M_{16}^k, M_{17}^k, M_{18}^k$, but also M_j^{k+1} for $j = \{1, 2, 4, 5, 7, 8, 9, 12, 13, 15\}$, and M_j^{k+2} for $j = \{0, 3, 6, 10, 11, 14\}$.

Some other conditions imply differences or non-differences between state words, $M_j^k \oplus M_{j+1}^k$. When writing these variables as functions of the input message words at step k and $k - 1$, and of the state variables before message insertion $k - 1$, one can notice the following : before the k-th message insertion, one can compute $M_j^k \oplus M_{j+1}^k$, for $j = \{15, 16, 17, 18\}$, $M_j^{k+2} \oplus M_{j+1}^{k+2}$ for $j = \{7, 10\}$, and $M_j^{k+1} \oplus M_{j+1}^{k+1}$ for all other possible values of j. Therefore, the adversary has to check conditions on three consecutive values of the mill on message insertion number k.

The most naive way to do it would consist in choosing m^k at random and hoping the conditions are verified, but one can use the following facts to decrease the number of messages to check:

- The conditions on words M_{16}^k, M_{17}^k and M_{18}^k as well as these on the values $M_{15}^k \oplus M_{16}^k$, $M_{16}^k \oplus M_{17}^k$, $M_{17}^k \oplus M_{18}^k$ and $M_{18}^k \oplus M_0^k$ at step k can be fulfilled by xor-ing the adequate message values at message insertion k.
- Using the linearity of all operations except the first one, the adversary can rewrite the values M_j^{k+1} as a linear combination of variables $\hat{M}_j^k$, with $j = \{0, \ldots, 18\}$. Words $\hat{M}_0^k$ to $\hat{M}_{13}^k$ do not depend on the last inserted message value, therefore can be computed before the message insertion.
- A system of equations in variables $\hat{M}_{14}^k, \ldots, \hat{M}_{18}^k$ remains. These equations are derived from conditions on round $k + 1$, by reversing the linear part of the *Mill* function. More precisely, these equations define the possible values of these variables, or of the xor of two of these variables, one of them being rotated.

The computation of the sons of a node at depth k work as follows:

1. The adversary checks the consistency of the equations on $\hat{M}_{14}^k, \ldots, \hat{M}_{18}^k$. If these equations are not consistent, the adversary does not search the node. The probability that this system is consistent depends on dimension of the Kernel of the system and can be computed *a priori*.
2. The adversary exhausts the possible joint values of $\hat{M}_{14}^k, \ldots, \hat{M}_{18}^k, M_{16}^k, M_{17}^k$ and M_{18}^k, considering all the conditions on these variables, which can be expressed bitwise (as the nonlinear part of the *Mill* function also works bitwise). The cost of this phase is then linear in w. The mean number of sons depends on the number of conditions.
3. For each remaining message block, the adversary checks all the other linear conditions on $\hat{M}_{14}^k, \ldots, \hat{M}_{18}^k$ and the conditions on the mill values 2 rounds later.

5.2   Computation of the Cost

We will now explain how to compute the complexity of the collision search algorithm. The most expensive operation is the search of the sons of nodes. The total complexity of a given depth level k is the product of the number of nodes that have to be explored at depth k by the average cost of the search of these nodes. These parameters are exponential in w, therefore the total cost of the search can be approximated by the search of the most expensive nodes.

To compute the search cost, we assume that for all considered messages, the words of the resulting states for which no condition is imposed are independent and identically distributed. This is true at depth 0, provided the attacker initializes the search phase with a long random message prefix. The identical distribution of the variables can be checked recursively, their independence is an hypothesis for the attack to work. This assumption is well-known in the field of hash function cryptanalysis for computing the cost associated to a differential path (see e.g. [29]).

Let A^k be the number of nodes that have to be reached at depth k, and C^k the average cost of searching one of these nodes. Let P^k be the probability that a random son of a node at depth k follows the differential path, and Q^k the probability that a given node at depth k has at least one valid son. At depth k, the average number of explored nodes is related to the average number of explored nodes at depth $k + 1$. When only a few nodes are needed, the average case is not sufficient, and one has to evaluate the cost of finding at least one valid node of depth $k + 1$.

One has the following relations, for $k \in \{0, \ldots, n - 1\}$:

$$\begin{cases} A^k = \max(\dfrac{A^{k+1}}{2^{3w} P^k}, \dfrac{1}{Q^k}) \\ A^n = 1 \end{cases}$$

Let K^k be the dimension of the Kernel of the linear system that has to be solved at depth k, and $\hat{P}^k$ the probability that the bitwise system of equations on the values of the mill before and after the nonlinear function has solutions. $\hat{P}^k$ can be computed exhaustively *a priori* for each value of k. A random node at depth k has at least one valid son if the two following conditions happen :

- The bitwise conditions at depth k and $k+1$ can be fulfilled,
- The remaining freedom degrees can be used to fulfill all the remaining conditions.

The first item takes in account the fact that some conditions might not depend on all the freedom degrees. Therefore, we have :

$$Q^k = \min(2^{-K^k}\hat{P}^k, 2^{3w-N^k_{COND}}),$$

where N^k_{COND} is the total number of conditions that has to be checked on the k-th message insertion. We also have $P^k = 2^{-N^k_{COND}}$, because each condition is supposed to be fulfilled with probability half in the average case, which is true provided the free words - *i.e.* without conditions fixing their values, or linking it to another word - are *i.i.d.* .

Searching a node works as follows : one solves the bitwise system of equations on the values of $M_{16}, M_{17}, M_{18}, \hat{M}_{14}, \ldots, \hat{M}_{18}$. The set of message blocks that fulfill this equations system then has to be searched exhaustively to fulfill the other conditions, and to generate nodes at depth $k+1$. C^k is then the cost of this exhaustive search, and can be computed as the average number of message blocks that fulfill the system of equations. Therefore, we have $C^k = 2^{3w}\hat{P}^k$.

For each node at depth k, the attacker can first check the consistency of the conditions on the mill words at steps k and $k+1$, which allows him not to search inconsistent nodes. Therefore, we have the following overall complexity:

$$T = O(\max_k(\frac{C^k A^k}{2^{K^k}}))$$

The best path we found has complexity about $2^{11 \times w}$, which is above the security claimed by the designers of RadioGatún[4], it is given in Appendix. As a proof of concept, we also provide in Appendix an example of a colliding pair of messages following our differential path for RadioGatún with $w = 2$. One can check that the observed complexity confirms the estimated one.

5.3   Breaking the Birthday Bound

Finding a final collision attack for RadioGatún with a computation complexity of 2^{11w} required us to own a computer with a big amount of RAM for a few hours of computation. Yet, the memory and computation cost of the differential path search phase is determined by the H_{max} chosen by the attacker. We conducted tests that tend to show that the search tree is big enough in order to find a collision attack with an overall complexity lower than the birthday bound claimed by the designers[2]. **The problem here is that the memory**

---

[2] Note also that the size of the search tree can be increased by considering more complex symmetric differences, such as 0^w, 1^w, $01^{w/2}$ and $10^{w/2}$.

and computation cost of the differential path search will be too big for nowadays computers, but much lower than the birthday bound. This explains why we are now incapable of providing a fully described collision attack for RadioGatún. However, we conjecture that applying our techniques with more memory and computation resources naturally leads to a collision attack for RadioGatún, breaking the ideal birthday bound.

6 Conclusion

In this paper, we presented an improved cryptanalysis of RadioGatún regarding collision search. Our attack can find collisions with a computation cost of about 2^{11w} and negligible memory, which is by far the best known attack on this proposal.

We also gave arguments that shows that RadioGatún might not be a collision resistant hash function. We conjecture that applying our differential path search technique with more constraints will lead to collision attacks on RadioGatún.

Acknowledgments

The authors would like to thank Guido Bertoni, Joan Daemen, Michal Peeters annd Gilles Van Assche for their comments on sponge functions.

References

1. Andreeva, E., Neven, G., Preneel, B., Shrimpton, T.: Seven-Property-Preserving Iterated Hashing: ROX. In: Kurosawa, K. (ed.) ASIACRYPT 2007. LNCS, vol. 4833, pp. 130–146. Springer, Heidelberg (2007)
2. Bellare, M., Ristenpart, T.: Multi-Property-Preserving Hash Domain Extension and the EMD Transform. In: Lai, X., Chen, K. (eds.) ASIACRYPT 2006. LNCS, vol. 4284, pp. 299–314. Springer, Heidelberg (2006)
3. Bertoni, G., Daemen, J., Peeters, M., Van Assche, G.: Keccak specifications. Submission to NIST (2008)
4. Bertoni, G., Daemen, J., Peeters, M., Van Assche, G.: Radiogatun, a belt-and-mill hash function. Presented at Second Cryptographic Hash Workshop, Santa Barbara, August 24-25 (2006), http://radiogatun.noekeon.org/
5. Bertoni, G., Daemen, J., Peeters, M., Van Assche, G.: On the Indifferentiability of the Sponge Construction. In: Smart, N.P. (ed.) EUROCRYPT 2008. LNCS, vol. 4965, pp. 181–197. Springer, Heidelberg (2008)
6. Bertoni, G., Daemen, J., Peeters, M., Van Assche, G.: Sponge Functions. Presented at ECRYPT Hash Workshop (2007)
7. Biham, E., Dunkelman, O.: A framework for iterative hash functions: Haifa. In: Second NIST Cryptographic Hash Workshop (2006)
8. Bouillaguet, C., Fouque, P.-A.: Analysis of radiogatun using algebraic techniques. In: Keliher, L., Avanzi, R., Sica, F. (eds.) SAC 2008. LNCS. Springer, Heidelberg (2008)

9. Brassard, G. (ed.): CRYPTO 1989. LNCS, vol. 435. Springer, Heidelberg (1990)
10. Bresson, E., Canteaut, A., Chevallier-Mames, B., Clavier, C., Fuhr, T., Gouget, A., Icart, T., Misarsky, J.-F., Naya-Plasencia, M., Paillier, P., Pornin, T., Reinhard, J.-R., Thuillet, C., Videau, M.: Shabal – a submission to advanced hash standard. Submission to NIST (2008)
11. Claesen, L.J.M., Daemen, J., Genoe, M., Peeters, G.: Subterranean: A 600 mbit/sec cryptographic vlsi chip. In: ICCD, pp. 610–613 (1993)
12. Cramer, R. (ed.): EUROCRYPT 2005. LNCS, vol. 3494. Springer, Heidelberg (2005)
13. Daemen, J.: Cipher and hash function design strategies based on linear and differential cryptanalysis. PhD thesis, Katholieke Universiteit Leuven (1995)
14. Daemen, J., Clapp, C.S.K.: Fast hashing and stream encryption with panama. In: Vaudenay, S. (ed.) FSE 1998. LNCS, vol. 1372, pp. 60–74. Springer, Heidelberg (1998)
15. Damgård, I.: A Design Principle for Hash Functions. In: Brassard [9], pp. 416–427
16. Dean, R.D.: Formal aspects of mobile code security. PhD thesis. Princeton University, Princeton (1999)
17. Fuhr, T., Peyrin, T.: Cryptanalysis of Radiogatún (2008)
18. Gorski, M., Lucks, S., Peyrin, T.: Slide attacks on hash functions. In: Pieprzyk, J. (ed.) ASIACRYPT 2008. LNCS, vol. 5350, pp. 143–160. Springer, Heidelberg (2008)
19. Joux, A.: Multicollisions in Iterated Hash Functions. Application to Cascaded Constructions. In: Franklin, M.K. (ed.) CRYPTO 2004. LNCS, vol. 3152, pp. 306–316. Springer, Heidelberg (2004)
20. Kelsey, J., Kohno, T.: Herding Hash Functions and the Nostradamus Attack. In: Vaudenay, S. (ed.) EUROCRYPT 2006. LNCS, vol. 4004, pp. 183–200. Springer, Heidelberg (2006)
21. Kelsey, J., Schneier, B.: Second Preimages on n-Bit Hash Functions for Much Less than 2^n Work. In: Cramer [12], pp. 474–490
22. Khovratovich, D.: Two attacks on radiogatun. In: Chowdhury, D.R., Rijmen, V., Das, A. (eds.) INDOCRYPT 2008. LNCS, vol. 5365, pp. 53–66. Springer, Heidelberg (2008)
23. Khovratovich, D.: Cryptanalysis of hash functions with structures. Presented at ECRYPT Hash Workshop (2008)
24. Knudsen, L.R., Rechberger, C., Thomsen, S.S.: The Grindahl Hash Functions. In: Biryukov, A. (ed.) FSE 2007. LNCS, vol. 4593, pp. 39–57. Springer, Heidelberg (2007)
25. Lucks, S.: A Failure-Friendly Design Principle for Hash Functions. In: Roy, B.K. (ed.) ASIACRYPT 2005. LNCS, vol. 3788, pp. 474–494. Springer, Heidelberg (2005)
26. Maurer, U.M., Tessaro, S.: Domain Extension of Public Random Functions: Beyond the Birthday Barrier. In: Menezes, A. (ed.) CRYPTO 2007. LNCS, vol. 4622, pp. 187–204. Springer, Heidelberg (2007)
27. Menezes, A.J., Vanstone, S.A., Van Oorschot, P.C.: Handbook of applied cryptography. CRC Press, Inc., Boca Raton (1996)
28. Merkle, R.C.: One Way Hash Functions and DES. In: Brassard [9], pp. 428–446
29. Peyrin, T.: Cryptanalysis of Grindahl. In: Kurosawa, K. (ed.) ASIACRYPT 2007. LNCS, vol. 4833, pp. 551–567. Springer, Heidelberg (2007)
30. Rijmen, V., Van Rompay, B., Preneel, B., Vandewalle, J.: Producing collisions for panama. In: Matsui, M. (ed.) FSE 2001. LNCS, vol. 2355, pp. 37–51. Springer, Heidelberg (2001)

31. Shoup, V. (ed.): CRYPTO 2005. LNCS, vol. 3621. Springer, Heidelberg (2005)

32. Wang, X., Lai, X., Feng, D., Chen, H., Yu, X.: Cryptanalysis of the hash functions md4 and ripemd. In: Cramer [12], pp. 1–18

33. Wang, X., Yin, Y.L., Yu, H.: Finding collisions in the full sha-1. In: Shoup [31], pp. 17–36

34. Wang, X., Yu, H.: How to break md5 and other hash functions. In: Cramer [12], pp. 19–35

35. Wang, X., Yu, H., Yin, Y.L.: Efficient collision search attacks on sha-0. In: Shoup [31], pp. 1–16

Appendix A: Collision for RadioGatún[2]

To generate a collision for RadioGatún[2], we use a 143-block differential path of cost 2^{11w}.

We give here a collision for the 2-bit version of RadioGatún. One can easily check that it follows the differential path given above. We write the message words using values between 0 and 3, which stand for the possible values of 2-bit words. The differential path, and some statistics about the collision search, can be found in the longer version of this paper [17].

To ensure that one has enough starting points, we used a 5-block common prefix.

The two colliding messages are :

$M_0 =$ 330 000 000 000 000 113 311 012 012 112 300 202

      020 302 233 030 030 000 223 222 220 111 000 010

      031 001 033 020 000 000 222 103 110 312 231 321

      102 012 322 023 323 232 001 023 032 220 130 103

      203 003 200 232 023 011 222 222 133 110 211 031

      232 122 033 122 021 202 302 003 120 003 300 203

      133 021 302 311 101 031 200 003 013 231 032 312

      002 202 131 331 122 201 333 301 032 230 031 220

      012 130 312 100 020 322 222 220 201 012 000 201

      200 010 230 130 310 330 201 103 130 210 102 001

      200 321 112 110 232 223 010 301 213 000 133 123

      323 222 331 132 103 021 012 330 201 100 203 321

      013 332 020 000

$M_1 =$ 330 000 000 000 000 113 311 312 022 122 030 202
020 332 103 303 303 003 113 222 120 121 030 020
031 001 303 313 000 330 222 103 110 312 202 321
201 011 022 010 313 202 031 023 032 120 130 103
200 303 233 232 013 321 111 211 203 123 121 031
132 112 300 122 011 202 032 003 210 300 300 100
203 311 302 012 101 002 100 303 013 231 302 322
032 131 102 001 211 232 300 301 302 230 301 120
011 103 022 200 013 022 212 113 131 311 003 131
200 010 230 200 020 000 231 103 100 113 132 031
233 321 112 220 232 220 010 332 223 300 100 123
013 122 302 131 200 311 012 300 202 230 133 321
013 331 023 003

The common value of the internal state is then :

$$\texttt{belt}[0] = (0, 0, 2, 1, 2, 0, 3, 0, 2, 1, 1, 1, 3),$$
$$\texttt{belt}[1] = (3, 1, 0, 2, 3, 2, 2, 3, 1, 2, 3, 0, 2),$$
$$\texttt{belt}[2] = (2, 3, 3, 2, 2, 2, 1, 1, 1, 3, 2, 0, 3),$$
$$\texttt{mill} = (2, 0, 2, 2, 1, 0, 1, 0, 3, 1, 3, 3, 2, 2, 3, 3, 0, 3, 3)$$

Preimage Attacks on Reduced Tiger and SHA-2

Takanori Isobe and Kyoji Shibutani

Sony Corporation
1-7-1 Konan, Minato-ku, Tokyo 108-0075, Japan
{Takanori.Isobe,Kyoji.Shibutani}@jp.sony.com

Abstract. This paper shows new preimage attacks on reduced Tiger and SHA-2. Indesteege and Preneel presented a preimage attack on Tiger reduced to 13 rounds (out of 24) with a complexity of $2^{128.5}$. Our new preimage attack finds a one-block preimage of Tiger reduced to 16 rounds with a complexity of 2^{161}. The proposed attack is based on meet-in-the-middle attacks. It seems difficult to find "independent words" of Tiger at first glance, since its key schedule function is much more complicated than that of MD4 or MD5. However, we developed techniques to find independent words efficiently by controlling its internal variables. Surprisingly, the similar techniques can be applied to SHA-2 including both SHA-256 and SHA-512. We present a one-block preimage attack on SHA-256 and SHA-512 reduced to 24 (out of 64 and 80) steps with a complexity of 2^{240} and 2^{480}, respectively. To the best of our knowledge, our attack is the best known preimage attack on reduced-round Tiger and our preimage attack on reduced-step SHA-512 is the first result. Furthermore, our preimage attacks can also be extended to second preimage attacks directly, because our attacks can obtain random preimages from an arbitrary IV and an arbitrary target.

Keywords: hash function, preimage attack, second preimage attack, meet-in-the-middle, Tiger, SHA-256, SHA-512.

1 Introduction

Cryptographic hash functions play an important role in the modern cryptology. Many cryptographic protocols require a secure hash function which holds several security properties such as classical ones: collision resistance, preimage resistance and second preimage resistance. However, a lot of hash functions have been broken by collision attacks including the attacks on MD4 [3], MD5 [11] and SHA-1 [12]. These hash functions are considered to be broken in theory, but in practice many applications still use these hash functions because they do not require collision resistance. However, (second) preimage attacks are critical for many applications including integrity checks and encrypted password systems. Thus analyzing the security of the hash function with respect to (second) preimage resistance is important, even if the hash function is already broken by a collision attack. However, the preimage resistance of hash functions has not been studied well.

O. Dunkelman (Ed.): FSE 2009, LNCS 5665, pp. 139–155, 2009.

Table 1. Summary of our results

Target	Attack (first or second preimage)	Attacked steps (rounds)	Complexity
Tiger (full 24 rounds)	first [4]	13	$2^{128.5}$
	first (**this paper**)	16	2^{161}
	second [4]	13	$2^{127.5}$
	second (**this paper**)	16	2^{160}
SHA-256 (full 64 steps)	first [10]	36	2^{249}
	first (**this paper**)	24	2^{240}
	second (**this paper**)	24	2^{240}
SHA-512 (full 80 steps)	first (**this paper**)	24	2^{480}
	second (**this paper**)	24	2^{480}

Tiger is a dedicated hash function producing a 192-bit hash value designed by Anderson and Biham in 1996 [2]. As a cryptanalysis of Tiger, at FSE 2006, Kelsey and Lucks proposed a collision attack on 17-round Tiger with a complexity of 2^{49} [5], where full-version Tiger has 24 rounds. They also proposed a pseudo-near collision attack on 20-round Tiger with a complexity of 2^{48}. This attack was improved by Mendel et al. at INDOCRYPT 2006 [8]. They proposed a collision attack on 19-round Tiger with a complexity of 2^{62}, and a pseudo-near collision attack on 22-round Tiger with a complexity of 2^{44}. Later, they proposed a pseudo-near-collision attack of full-round (24-round) Tiger with a complexity of 2^{44}, and a pseudo-collision (free-start-collision) attack on 23-round Tiger [9]. The above results are collision attacks and there is few evaluations of preimage resistance of Tiger. Indesteege and Preneel presented preimage attacks on reduced-round Tiger [4]. Their attack found a preimage of Tiger reduced to 13 rounds with a complexity of $2^{128.5}$.

In this paper, we introduce a preimage attack on reduced-round Tiger. The proposed attack is based on meet-in-the-middle attacks [1]. In this attack, we need to find independent words ("neutral words") in the first place. However, the techniques used for finding independent words of MD4 or MD5 cannot be applied to Tiger directly, since its key schedule function is much more complicated than that of MD4 or MD5. To overcome this problem, we developed new techniques to find independent words of Tiger efficiently by adjusting the internal variables. As a result, the proposed attack finds a preimage of Tiger reduced to 16 (out of 24) rounds with a complexity of about 2^{161}. Surprisingly, our new approach can be applied to SHA-2 including both SHA-256 and SHA-512. We present a preimage attack on SHA-256 and SHA-512 reduced to 24 (out of 64 and 80) steps with a complexity of about 2^{240} and 2^{480}, respectively. As far as we know, our attack is the best known preimage attack on reduced-round Tiger and our preimage attack on reduced-step SHA-512 is the first result. Furthermore, we show that our preimage attacks can also be extended to second preimage attacks directly and all of our attacks can obtain one-block preimages, because our preimage attacks can obtain random preimages from an arbitrary IV and an arbitrary target. These results are summarized in Table 1.

This paper is organized as follows. Brief descriptions of Tiger, SHA-2 and the meet-in-the-middle approach are given in Section 2. A preimage attack on reduced-round Tiger and its extensions are shown in Section 3. In Section 4, we present a preimage attack on reduced-step SHA-2. Finally, we present conclusions in Section 5.

2 Preliminaries

2.1 Description of Tiger

Tiger is an iterated hash function that compresses an arbitrary length message into a 192-bit hash value. An input message value is divided into 512-bit message blocks $(M^{(0)}, M^{(1)}, ..., M^{(t-1)})$ by the padding process as well as the MD family. The compression function of Tiger shown in Fig. 1 generates a 192-bit output chaining value $H^{(i+1)}$ from a 512-bit message block $M^{(i)}$ and a 192-bit input chaining value $H^{(i)}$ where chaining values consist of three 64-bit variables, $A_j^{(i)}$, $B_j^{(i)}$ and $C_j^{(i)}$. The initial chaining value $H^{(0)} = (A_0^{(0)}, B_0^{(0)}, C_0^{(0)})$ is as follows:

$$A_0^{(0)} = \texttt{0x0123456789ABCDEF},$$
$$B_0^{(0)} = \texttt{0xFEDCBA9876543210},$$
$$C_0^{(0)} = \texttt{0xF096A5B4C3B2E187}.$$

In the compression function, a 512-bit message block $M^{(i)}$ is divided into eight 64-bit words $(X_0, X_1, ..., X_7)$. The compression function consists of three pass functions and between each of them there is a key schedule function. Since each pass function has eight round functions, the compression function consists of 24 round functions. The pass function is used for updating chaining values, and the key schedule function is used for updating message values. After the third pass function, the following feedforward process is executed to give outputs of the compression function with input chaining values and outputs of the third pass function,

$$A'_{24} = A_0 \oplus A_{24}, \ B'_{24} = B_0 - B_{24}, \ C'_{24} = C_0 + C_{24},$$

where A_i, B_i and C_i denote the i-th round chaining values, respectively, and A'_{24}, B'_{24} and C'_{24} are outputs of the compression function.

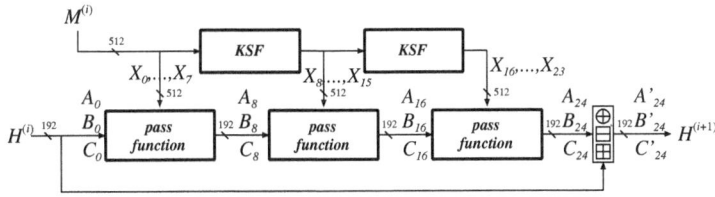

Fig. 1. Compression function f of Tiger

In each round of the pass function, chaining values A_i, B_i and C_i are updated by a message word X_i as follows:

$$B_{i+1} = C_i \oplus X_i, \tag{1}$$
$$C_{i+1} = A_i - even(B_{i+1}), \tag{2}$$
$$A_{i+1} = (B_i + odd(B_{i+1})) \times mul, \tag{3}$$

where mul is the constant value $\in \{5, 7, 9\}$ which is different in each pass function. The nonlinear functions $even$ and odd are expressed as follows:

$$even(W) = T_1[w_0] \oplus T_2[w_2] \oplus T_3[w_4] \oplus T_4[w_6], \tag{4}$$
$$odd(W) = T_4[w_1] \oplus T_3[w_3] \oplus T_2[w_5] \oplus T_1[w_7], \tag{5}$$

where 64-bit value W is split into eight bytes $\{w_7, w_6, ..., w_0\}$ with w_7 is the most significant byte and $T_1, ..., T_4$ are the S-boxes: $\{0,1\}^8 \rightarrow \{0,1\}^{64}$. Figure 2 shows the round function of Tiger.

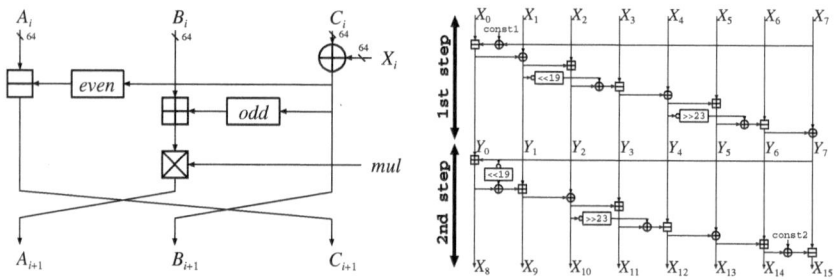

Fig. 2. Tiger round function   **Fig. 3.** Key schedule function

The key schedule function (KSF) updates message values. In the first pass function, eight message words $X_0, ..., X_7$, which are identical to input message blocks of the compression function, are used for updating chaining values. Remaining two pass functions use sixteen message words which are generated by applying KSF:

$$(X_8, ..., X_{15}) = KSF(X_0, ..., X_7), \tag{6}$$
$$(X_{16}, ..., X_{23}) = KSF(X_8, ..., X_{15}). \tag{7}$$

The function KSF which updates the inputs $X_0, ..., X_7$ in two steps, is shown in Table 2. The first step shown in the left table generates internal variables $Y_0, ..., Y_7$ from inputs $X_0, ..., X_7$, and the second step shown in the right table calculates outputs $X_8, ..., X_{15}$ from internal variables $Y_0, .., Y_7$, where const1 is 0xA5A5A5A5A5A5A5A5 and const2 is 0x0123456789ABCDEF. By using the same function, $X_{16}, ..., X_{23}$ are also derived from $X_8, ..., X_{15}$. Figure 3 shows the key schedule function of Tiger.

Table 2. Algorithm of the key schedule function KSF

$Y_0 = X_0 - (X_7 \oplus \texttt{const1}),$ (8)	$X_8 = Y_0 + Y_7,$ (16)
$Y_1 = X_1 \oplus Y_0,$ (9)	$X_9 = Y_1 - (X_8 \oplus (\overline{Y_7} \ll 19)),$ (17)
$Y_2 = X_2 + Y_1,$ (10)	$X_{10} = Y_2 \oplus X_9,$ (18)
$Y_3 = X_3 - (Y_2 \oplus (\overline{Y_1} \ll 19)),$ (11)	$X_{11} = Y_3 + X_{10},$ (19)
$Y_4 = X_4 \oplus Y_3,$ (12)	$X_{12} = Y_4 - (X_{11} \oplus (\overline{X_{10}} \gg 23)),$ (20)
$Y_5 = X_5 + Y_4,$ (13)	$X_{13} = Y_5 \oplus X_{12},$ (21)
$Y_6 = X_6 - (Y_5 \oplus (\overline{Y_4} \gg 23)),$ (14)	$X_{14} = Y_6 + X_{13},$ (22)
$Y_7 = X_7 \oplus Y_6.$ (15)	$X_{15} = Y_7 - (X_{14} \oplus \texttt{const2}).$ (23)

2.2 Description of SHA-256

We only show the structure of SHA-256, since SHA-512 is structurally very similar to SHA-256 except for the number of steps, word size and rotation values. The compression function of SHA-256 consists of a message expansion function and a state update function. The message expansion function expands 512-bit message block into 64 32-bit message words $W_0, ..., W_{63}$ as follows:

$$W_i = \begin{cases} M_i & (0 \leq i < 16), \\ \sigma_1(W_{i-2}) + W_{i-7} + \sigma_0(W_{i-15}) + W_{i-16} & (16 \leq i < 64), \end{cases}$$

where the functions $\sigma_0(X)$ and $\sigma_1(X)$ are given by

$$\sigma_0(X) = (X \ggg 7) \oplus (X \ggg 18) \oplus (X \gg 3),$$
$$\sigma_1(X) = (X \ggg 17) \oplus (X \ggg 19) \oplus (X \gg 10).$$

The state update function updates eight 32-bit chaining values, $A, B, ..., G, H$ in 64 steps as follows:

$$T_1 = H_i + \Sigma_1(E_i) + Ch(E_i, F_i, G_i) + K_i + W_i, \tag{24}$$
$$T_2 = \Sigma_0(A_i) + Maj(A_i, B_i, C_i), \tag{25}$$
$$A_{i+1} = T_1 + T_2, \tag{26}$$
$$B_{i+1} = A_i, \tag{27}$$
$$C_{i+1} = B_i, \tag{28}$$
$$D_{i+1} = C_i, \tag{29}$$
$$E_{i+1} = D_i + T_1, \tag{30}$$
$$F_{i+1} = E_i, \tag{31}$$
$$G_{i+1} = F_i, \tag{32}$$
$$H_{i+1} = G_i, \tag{33}$$

where K_i is a step constant and the function Ch, Maj, Σ_0 and Σ_1 are given as follows:

$$Ch(X, Y, Z) = XY \oplus \overline{X}Z,$$
$$Maj(X, Y, Z) = XY \oplus YZ \oplus XZ,$$
$$\Sigma_0(X) = (X \ggg 2) \oplus (X \ggg 13) \oplus (X \ggg 22),$$
$$\Sigma_1(X) = (X \ggg 6) \oplus (X \ggg 11) \oplus (X \ggg 25).$$

After 64 step, a feedfoward process is executed with initial state variable by using word-wise addition modulo 2^{32}.

2.3   Meet-in-the-Middle Approach for Preimage Attack

We assume that a compression function F consists of a key scheduling function (KSF) and a round/step function as shown in Fig. 4. The function F has two inputs, an n-bit chaining variable H and an m-bit message M, and outputs an n-bit chaining variable G. The function KSF expands the message M, and provides them into the round/step function.

We consider a problem that given H and G, find a message M satisfying $G = F(H, M)$. This problem corresponds to the preimage attack on the compression function with a fixed input chaining variable. In this model, a feedforward function does not affect the attack complexity, since the targets H and G are arbitrary values. If we obtain a preimage from arbitrary values of H and G, we can also compute a preimage from H and $H \oplus G$ instead of G.

In the meet-in-the-middle preimage attack, we first divide the round function into two parts: the forward process (FP) and the backward process (BP) so that each process can compute an ℓ-bit meet point S independently. We also need independent words X and Y in KSF to compute S independently. The meet point S can be determined from FP and BP independently such that $S = FP(H, X)$ and $S = BP(G, Y)$.

If there are such two processes FP and BP, and independent words X and Y, we can obtain a message M satisfying S with a complexity of $2^{\ell/2}$ F evaluations, assuming that FP and BP are random ones, and the computation cost of BP is almost same as that of inverting function of BP. Since remaining internal state value

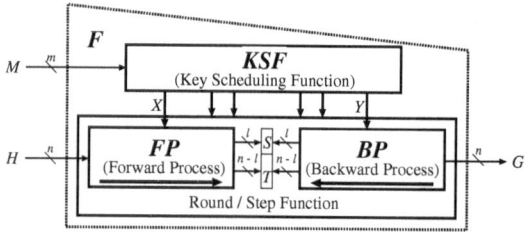

Fig. 4. Meet-in-the-middle approach

T is $(n - \ell)$ bits, the desired M can be obtained with a complexity of $2^{n-\ell/2}(= 2^{n-\ell+\ell/2})$. Therefore, if FP and BP up to the meet point S can be calculated independently, a preimage attack can succeed with a complexity of $2^{n-\ell/2}$. This type of preimage attacks on MD4 and MD5 was presented by Aoki and Sasaki [1].

In general, it is difficult to find such independent words in a complicated KSF. We developed new techniques to construct independent transforms in KSF by controlling internal variabes to obtain independent words.

3   Preimage Attack on Reduced-Round Tiger

In this section, we propose a preimage attack on 16-round Tiger with a complexity of 2^{161}. This variant shown in Fig. 5 consists of two pass functions and one key schedule function. First, we show properties of Tiger which are used for applying the meet-in-the-middle attack. Next, we show how to apply the meet-in-the-middle attack to Tiger, and then introduce the algorithm of our attack. Finally, we evaluate the required complexity and memory of our attack.

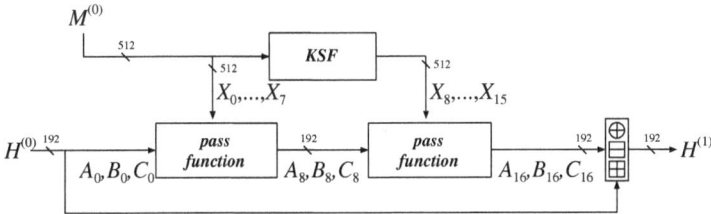

Fig. 5. Reduced-round Tiger (2-pass = 16-round)

3.1   Properties of Tiger

We show five properties of Tiger, which enable us to apply the meet-in-the-middle attack.

Property 1: *The pass function is easily invertible.*

Property 1 can be obtained from the design of the round function. From Eq. (1) to Eq. (3), A_i, B_i, and C_i can be determined from A_{i+1}, B_{i+1}, C_{i+1} and X_i. The computation cost is almost same as the cost of calculating A_{i+1}, B_{i+1} and C_{i+1} from A_i, B_i, C_i and X_i. Since the round function is invertible, we can construct the inverse pass function.

Property 2: *In the inverse pass function, the particular message words are independent of particular state value.*

The detail of the Property 2 is that once X_i, A_{i+3} B_{i+3} and C_{i+3} are fixed, then C_i, B_{i+1}, A_{i+2} and B_{i+2} can be determined from Eq. (1) to Eq. (3) independently of X_{i+1} and X_{i+2}. Thus the property 2 implies that X_{i+1} and X_{i+2} are independent of C_i in the inverse pass function.

Property 3: *In the round function, C_{i+1} is independent of odd bytes of X_i.*

The property 3 can be obtained from the property of the non-linear function *even*.

Property 4: *The key schedule function KSF is easily invertible.*

The property 4 implies that we can build the inverse key schedule function KSF^{-1}. Moreover, the computation cost of KSF^{-1} is almost the same as that of KSF.

Property 5: *In the inverse key schedule function KSF^{-1}, if input values are chosen appropriately, there are two independent transforms.*

The property 5 is one of the most important properties for our attack. In the next section, we show this in detail.

3.2 How to Obtain Two Independent Transforms in the KSF^{-1}

Since any input word of KSF^{-1} affects all output words of KSF^{-1}, it appears that there is no independent transform in the KSF^{-1} at first glance.

However, we analyzed the relation among the inputs and the outputs of KSF^{-1} deeply, and then found a technique to construct two independent transforms in the KSF^{-1} by choosing inputs carefully and controlling internal variables. Specifically, we can show that a change of input word X_8 only affects output words X_0, X_1, X_2 and X_3, and also modifications of X_{13}, X_{14} and X_{15} only affect X_5 and X_6 if these input words are chosen properly. We present the relation among inputs, outputs and internal variables of KSF^{-1} and then show how to build independent transforms in the KSF^{-1}.

As shown in Fig. 6, changes of inputs X_{13}, X_{14} and X_{15} only propagate internal variables Y_0, Y_1, Y_5, Y_6 and Y_7. If internal variables Y_6 and Y_7 are fixed

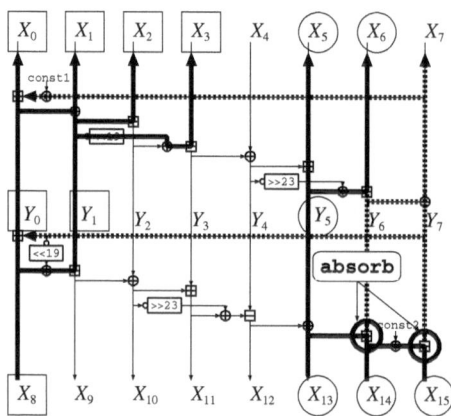

Fig. 6. Relation among inputs and outputs of KSF^{-1}

even when X_{13}, X_{14} and X_{15} are changed, it can be considered that an internal variable Y_0, Y_1 and an output X_7 are independent of changes of X_{13}, X_{14} and X_{15}. From Eq. (22) and (23), Y_6 and Y_7 can be fixed to arbitrary values by choosing X_{13}, X_{14} and X_{15} satisfying the following formulae:

$$X_{14} = Y_6 + X_{13}, \tag{34}$$

$$X_{15} = Y_7 - (X_{14} \oplus \texttt{const2}). \tag{35}$$

Therefore modifications of inputs X_{13}, X_{14} and X_{15} only propagate X_5 and X_6 by selecting these input values appropriately. In addition, a modification of X_8 only affects X_0, ..., X_3.

As a result, we obtain two independent transforms in KSF^{-1} by choosing X_{13}, X_{14} and X_{15} properly, since in this case a change of X_8 only affects X_0, ..., X_3, and changes of X_{13}, X_{14} and X_{15} only propagate X_5 and X_6.

3.3 Applying Meet-in-the-Middle Attack to Reduced-Round Tiger

We show the method for applying the meet-in-the-middle attack to Tiger by using above five properties. We define the meet point as 64-bit C_6, the process 1 as rounds 1 to 6, and the process 2 as rounds 7 to 16.

In the process 2, intermediate values A_9, B_9 and C_9 can be calculated from A_{16}, B_{16}, C_{16} and message words X_9 to X_{15}, since Tiger without the feedforward function is easily invertible. From the property 2, C_6 can be determined from A_8, B_8 and X_6. It is also observed that A_8 and B_8 are independent of X_8, because these values are calculated from A_9, B_9 and C_9. From the property 5, X_8 does not affect X_6. Therefore, C_6, the output of the process 2, can be determined from X_6, X_9 to X_{15}, A_{16}, B_{16} and C_{16}.

In the process 1, the output C_6 can be calculated from X_0 to X_5, A_0, B_0 and C_0. If some changes of the message words used in each process do not affect the message words used in the other process, C_6 can be determined independently in each process.

The message words X_0 to X_4 are independent of changes of X_6 and X_{13} to X_{15}, if X_9 to X_{12} are fixed and X_{13} to X_{15} are calculated as illustrated in the section 3.2. Although changes of X_{13}, X_{14} and X_{15} propagate X_5, from the property 3, C_6 in the process 1 is not affected by changes of odd bytes of X_5. Therefore, if even bytes of X_5 are fixed, C_6 in the process 1 can be determined independently from a change of X_5.

We show that the even bytes of X_5 can be fixed by choosing X_{11}, X_{12} and X_{13} properly. From Eq. (21), Y_5 is identical to X_{13} when X_{12} equals zero, and from Eq. (13), X_5 is identical to Y_5 when Y_4 equals zero. Thus X_5 is identical to X_{13} when both X_{12} and Y_4 are zero. Consequently, if the even bytes of X_{13} are fixed, and X_{12} and Y_4 equal zero, the even bytes of X_5 can be fixed. Y_4 can be fixed to zero by choosing X_{11} as $X_{11} \leftarrow \overline{X_{10}} \gg 23$. Therefore, if the following conditions are satisfied, C_6 in the process 1 can be independent of changes of X_{13}, X_{14} and X_{15}.

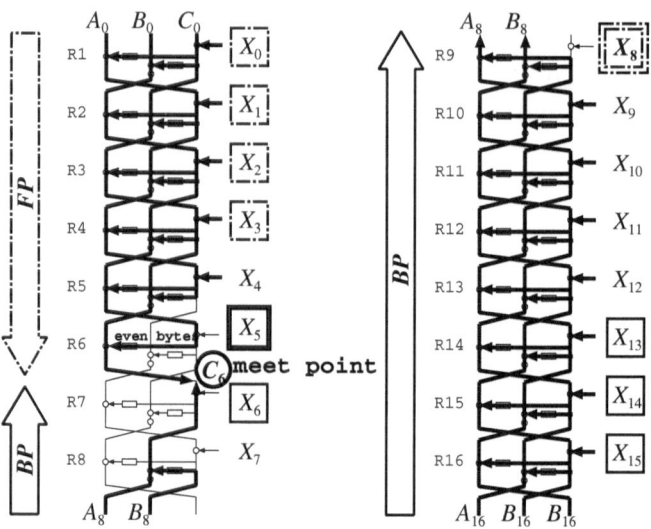

Fig. 7. Meet-in-the-middle attack on 16-round Tiger

- X_9 and X_{10} are fixed arbitrarily,
- $X_{11} = \overline{X_{10}} \ggg 23$, $X_{12} = 0$,
- X_{13}, X_{14} and X_{15} are chosen properly.

By choosing inputs of the inverse pass function satisfying the above conditions, we can execute the process 1 and the process 2 independently. Specifically, if only X_{13}, X_{14} and X_{15} are treated as variables in the process 2, then the process 2 can be executed independently from the process 1. Similarly, if only X_8 is treated as a variable in the process 1, then the process 1 is independent of the process 2, as long as X_8 to X_{15} satisfy the above conditions. These results are shown in Fig. 7.

3.4 (Second) Preimage Attack on 16-Round Tiger Compression Function

We present the whole algorithm of the (second) preimage attack on the compression function of Tiger reduced to 16 rounds. The attack consists of three phases: preparation, first and second phase.

The preparation phase sets $X_i (i \in \{4, 7, 9, 10, 11, 12\})$, $Y_i (i \in \{2, 3, 4, 6, 7\})$ and even bytes of X_{13} as follows:

Preparation

1: Let A'_{16}, B'_{16} and C'_{16} be given targets. Choose A_0, B_0 and C_0 arbitrarily, and set A_{16}, B_{16} and C_{16} as follows:

$$A_{16} \leftarrow A_0 \oplus A'_{16}, \quad B_{16} \leftarrow B_0 - B'_{16}, \quad C_{16} \leftarrow C'_{16} - C_0.$$

2: Choose X_9, X_{10}, Y_6, Y_7 and even bytes of X_{13} arbitrarily, set X_{12} and Y_4 to zero, and set X_7, X_{11}, Y_2, Y_3 and X_4 as follows:

$$X_7 \leftarrow Y_6 \oplus Y_7, \ X_{11} \leftarrow \overline{X_{10}} \ggg 23, \ Y_2 \leftarrow X_9 \oplus X_{10}, \ Y_3 \leftarrow X_{11} - X_{10}, \ X_4 \leftarrow Y_3.$$

The first phase makes a table of $(C_6, \text{odd bytes of } X_{13})$ pairs in the process 2 as follows:

First Phase

1: Choose odd bytes of X_{13} randomly.
2: Set X_5, X_6, X_{14} and X_{15} as follows:

$$X_5 \leftarrow X_{13}, X_6 \leftarrow Y_6 + X_{13}, X_{14} \leftarrow Y_6 + X_{13}, X_{15} \leftarrow Y_7 - ((Y_6 + X_{13}) \oplus \text{const2}).$$

3: Compute C_6 from $A_{16}, B_{16}, C_{16}, X_6$ and X_9 to X_{15}.
4: Place a pair $(C_6, \text{odd bytes of } X_{13})$ into a table.
5: If all 2^{32} possibilities of odd bytes of X_{13} have been checked, terminate this phase. Otherwise, set another value, which has not been set yet, to odd bytes of X_{13} and return to the step 2.

The second phase finds the desired message values X_0 to X_{15} in the process 1 by using the table as follows:

Second Phase

1: Choose X_8 randomly.
2: Set Y_0, Y_1, X_0, X_1, X_2 and X_3 as follows:

$$Y_0 \leftarrow X_8 - X_7,$$
$$Y_1 \leftarrow X_9 + (X_8 \oplus (\overline{Y_7} \lll 19)),$$
$$X_0 \leftarrow Y_0 + (X_7 \oplus \text{const1}),$$
$$X_1 \leftarrow Y_0 \oplus Y_1,$$
$$X_2 \leftarrow Y_2 - Y_1,$$
$$X_3 \leftarrow Y_3 + (Y_2 \oplus (\overline{Y_1} \lll 19)).$$

3: Compute C_6 from X_0 to X_4, even bytes of X_5, A_0, B_0 and C_0.
4: Check whether this C_6 is in the table generated in the first phase. If C_6 is in the table, the corresponding X_0 to X_7 are a preimage for the compression function of the target $A'_{16}, B'_{16}, C'_{16}$ and successfully terminates the attack. Otherwise, set another value, which has not been set yet, to X_8 and return to the step 2.

By repeating the second phase about 2^{32} times for different choices of X_8, we expect to obtain a matched C_6. The complexity of the above algorithm is $2^{32} (= 2^{32} \cdot \frac{6}{16} + 2^{32} \cdot \frac{10}{16})$ compression function evaluations, and success probability is about 2^{-128}. By executing the above algorithm 2^{128} times with different fixed values, we can obtain a preimage of the compression function. In the preparation phase, A_0, B_0, C_0, X_9, X_{10}, Y_6, Y_7 and even bytes of X_{13} can be chosen arbitrarily. In other words, this attack can use these values as free words. These free words are enough for searching 2^{128} space. Accordingly, the complexity of the preimage attack on the compression function is $2^{160} (= 2^{32} \cdot 2^{128})$. Also, this algorithm requires 2^{32} 96-bit or $2^{35.6}$ bytes memory.

3.5   One-Block (Second) Preimage Attack on 16-Round Tiger

The preimage attack on the compression function can be extended to the one-block preimage attack on 16-round Tiger hash function. For extending the attack, A_0, B_0, C_0 are fixed to the IV words, the padding word X_7 is fixed to 447 encoded in 64-bit string, and the remaining 224 bits are used as free bits in the preparation phase. Although our attack cannot deal with another padding word X_6, the attack still works when the least significant bit of X_6 equals one.

Hence, the success probability of the attack on the hash function is half of that of the attack on the compression function. The total complexity of the one-block preimage attack on 16-round Tiger hash function is 2^{161} compression function computations.

This preimage attack can also be extended to the one-block second preimage attack directly. Our second preimage attack obtains a one-block preimage with the complexity of 2^{161}. Moreover, the complexity of our second preimage attack can be reduced by using the technique given in [4]. In this case, the second preimage attack obtains the preimage which consists of at least two message blocks with a complexity of 2^{160}.

4   Preimage Attack on Reduced-Round SHA-2

We apply our techniques to SHA-2 including both SHA-256 and SHA-512 in straightforward and present a preimage attack on SHA-2 reduced to 24 (out of 64 and 80, respectively) steps. We first check the properties of SHA-2, then introduce the algorithm of the preimage attack on 24-step SHA-2.

4.1   Properties of 24-Step SHA-2

We first check whether SHA-2 has similar properties of Tiger. The pass function of Tiger corresponds to the 16-step state update function of SHA-2, and the key schedule function of Tiger corresponds to the 16-step message expansion function of SHA-2. Since the state update function and the message expansion function of SHA-2 are easily invertible, the compression function of SHA-2 without the feedforward function is also invertible.

In the inverse state update function, $A_{18}, B_{18}, ..., H_{18}$ are determined from $A_{24}, B_{24}, ..., H_{24}$ and W_{18} to W_{23}, and A_{11} only depends on $A_{18}, ..., H_{18}$. Thus A_{11} is independent of W_{11} to W_{17} when $A_{18}, ..., H_{18}$ and W_{18} to W_{23} are fixed. It corresponds to the property 2 of Tiger.

Then we check whether there are independent transforms in the inverse message expansion function of SHA-2. It corresponds to the property 5 of Tiger. For

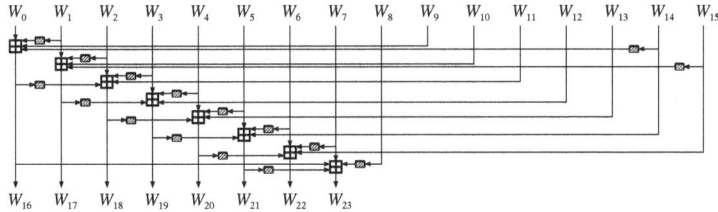

Fig. 8. Message expansion function of 24-step SHA-2

the 24-step SHA-2, 16 message words W_0 to W_{15} used in the first 16 steps are identical to input message blocks of the compression function, and 8 message words W_{16} to W_{23} used in the remaining eight steps are derived from W_0 to W_{15} by the message expansion function shown in Fig. 8. Table 3 shows the relation among message words in the message expansion function. For example, W_{16} is determined from W_{14}, W_9, W_1 and W_0. By using these relation and techniques introduced in previous sections, we can configure two independent transforms in the message expansion function of SHA-2.

We show that, in the inverse message expansion function of 24-step SHA-2, i) a change of W_{17} only affects W_0, W_1, W_3 and W_{11}, and ii) W_{19}, W_{21} and W_{23} only affect W_{12} by using the message modification techniques. In Tab. 3, asterisked values are variables of i), and underlined values are variables of ii).

First, we consider the influence of W_{23}. Though W_{23} affects W_7, W_8, W_{16} and W_{21}, this influence can be absorbed by modifying $W_{21} \rightarrow W_{19} \rightarrow W_{12}$. Consequently, we obtain a result that W_{19}, W_{21} and W_{23} only affect W_{12} by choosing these values properly, since W_{12} does not affect any other values in the inverse message expansion function.

Similarly, we consider the influence of W_{17} in the inverse message expansion function. W_{17} affects W_1, W_2, W_{10} and W_{15}. This influence can be absorbed by modifying $W_1 \rightarrow W_0$. W_{17} is also used for generating W_{19}. In order to cancel this influence, $W_3 \rightarrow W_{11}$ are also modified. As a result, we obtain a result that W_{17} only affects W_0, W_1, W_3 and W_{11} by choosing these values appropriately.

Table 3. Relation among message values W_{16} to W_{23}

computed value	values for computing
W_{16}	W_{14}, W_9, W_1*, W_0*
$W_{17}*$	$W_{15}, W_{10}, W_2, W_1*$
W_{18}	$W_{16}, W_{11}*, W_3*, W_2$
$\underline{W_{19}}$	$W_{17}*, \underline{W_{12}}, W_4, W_3*$
W_{20}	W_{18}, W_{13}, W_5, W_4
$\underline{W_{21}}$	$\underline{W_{19}}, W_{14}, W_6, W_5$
W_{22}	W_{20}, W_{15}, W_7, W_6
$\underline{W_{23}}$	$\underline{W_{21}}, W_{16}, W_8, W_7$

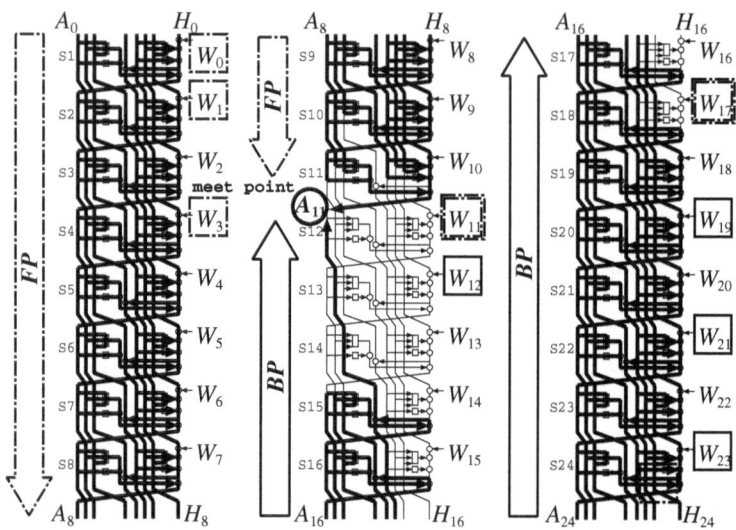

Fig. 9. Meet-in-the-middle attack on 24-step SHA-2

4.2 (Second) Preimage Attack on 24-Step SHA-256 Compression Function

As shown in Fig. 9, we define the meet point as 32-bit A_{11}, the process 1 as steps 1 to 11, and the process 2 as steps 12 to 24. In the process 1, A_{11} can be derived from $A_0, ..., H_0$ and W_0 to W_{10}. Similarly, in the process 2, A_{11} can be determined from $A_{24}, ..., H_{24}$ and W_{18} to W_{23}. Since the process 1 and process 2 are independent of each other for A_{11} by using the above properties of SHA-2, we apply the meet-in-the-middle attack to SHA-2 as follows:

Preparation

1: Let $A'_{24}, ..., H'_{24}$ be given targets. Choose $A_0, ..., H_0$ arbitrarily, and compute $A_{24}, ..., H_{24}$ by the feedforward function.

2: Choose 32-bit value CON and $W_i (i \in \{2, 4, 5, 6, 7, 8, 9, 10, 13, 14, 15, 16, 18\})$ arbitrarily, and then calculate W_{20} and W_{22}.

First Phase

1: Choose W_{23} randomly.

2: Determine W_{21}, W_{19} and W_{12} as follows[1]:

$$W_{21} \leftarrow \sigma_1^{-1}(W_{23} - W_{16} - \sigma_0(W_8) - W_7),$$
$$W_{19} \leftarrow \sigma_1^{-1}(W_{21} - W_{14} - \sigma_0(W_6) - W_5),$$
$$W_{12} \leftarrow W_{19} - \text{CON}.$$

---

[1] The method how to calculate σ_1^{-1} is illustrated in the appendix.

3: Compute A_{11} from $A_{24}, ..., H_{24}$ and W_{18} to W_{23}.

4: Place a pair (A_{11}, W_{23}) into a table.

5: If 2^{16} pairs of (A_{11}, W_{23}) have been listed in the table, terminate this algorithm. Otherwise, set another value, which has not been set yet, to W_{23} and return to the step 2.

Second Phase

1: Choose W_{17} randomly.

2: Determine W_0, W_1, W_3 and W_{11} as follows:

$$W_1 \leftarrow W_{17} - \sigma_1(W_{15}) - W_{10} - \sigma_0(W_2),$$
$$W_0 \leftarrow W_{16} - \sigma_1(W_{14}) - W_9 - \sigma_0(W_1),$$
$$W_3 \leftarrow \mathtt{CON} - \sigma_1(W_{17}) - \sigma_0(W_4),$$
$$W_{11} \leftarrow W_{18} - \sigma_1(W_{16}) - \sigma_0(W_3) - W_2.$$

3: Compute A_{11} from $A_0, ..., H_0$ and W_0 to W_{10}.

4: Check whether this A_{11} is in the table generated in the first phase. If A_{11} is in the table, the corresponding W_0 to W_{23} is a preimage of the compression function of the target $A'_{24}, ..., H'_{24}$ and successfully terminates the attack. Otherwise, set another value, which has not been set yet, to W_{17} and return to the step 2.

By repeating the second phase about 2^{16} times for different W_{17}, we expect to obtain a matched A_{11}. The complexity of the preimage attack on the compression function is $2^{240} (= 2^{256-32/2})$ compression function evaluations. The required memory is 2^{16} 64-bit or 2^{19} bytes. In this attack, the words $A_0, ..., H_0$, $\mathtt{CON}$ and $W_i (i \in \{2, 4, 5, 6, 7, 8, 9, 10, 13, 14, 15, 16, 18\})$ can be used as free words. The total free words are 22 words or 704 bits.

4.3   One-Block (Second) Preimage Attack on 24-Step SHA-2 Hash Function

The preimage attack on the compression function can be extended to the (second) preimage attack on the hash function directly, since our preimage attack can obtain random preimages from an arbitrary IV and an arbitrary target, and can deal with the padding words W_{14} and W_{15}. Thus the complexities of the preimage attack and the second preimage attack on 24-step SHA-256 are 2^{240}. Furthermore, this attack can also be extended to the (second) preimage attack on 24-step SHA-512. The complexities of the (second) preimage attack on 24-step SHA-512 are $2^{480} (= 2^{512-64/2})$.

5   Conclusion

In this paper, we have shown preimage attacks on reduced-round Tiger, reduced-step SHA-256 and reduced-step SHA-512. The proposed attacks are based on meet-in-the-middle attack. We developed new techniques to find "independent words" of the compression functions. In the attack on reduced-round Tiger, we found the "independent transforms" in the message schedule function by adjusting the internal variables, then we presented there are independent words in the compression function of Tiger. In the attack on reduced-round SHA-2, we found the "independent transforms" in the message expansion function by modifying the messages, then we showed that there are independent words in the compression function of SHA-2.

Our preimage attack can find a preimage of 16-step Tiger, 24-step SHA-256 and 24-step SHA-512 with a complexity of 2^{161}, 2^{240} and 2^{480}, respectively. These preimage attacks can be extended to second preimage attacks with the almost same complexities. Moreover, our (second) preimage attacks can find a one-block preimage, since it can obtain random preimages from an arbitrary IV an arbitrary target, and can also deal with the padding words.

Acknowledgments. The authors thank to the anonymous referees for their valuable comments.

References

1. Aoki, K., Sasaki, Y.: Preimage attacks on one-block MD4, 63-step MD5 and more. In: Avanzi, R., Keliher, L., Sica, F. (eds.) SAC 2008. LNCS, pp. 82–98. Springer, Heidelberg (2008)
2. Anderson, R., Biham, E.: Tiger: A fast new hash function. In: Gollmann, D. (ed.) FSE 1996. LNCS, vol. 1039, pp. 89–97. Springer, Heidelberg (1996)
3. Dobbertin, H.: Cryptanalysis of MD4. In: Gollmann, D. (ed.) FSE 1996. LNCS, vol. 1039, pp. 53–69. Springer, Heidelberg (1996)
4. Indesteege, S., Preneel, B.: Preimages for reduced-round Tiger. In: Lucks, S., Sadeghi, A.-R., Wolf, C. (eds.) WEWoRC 2007. LNCS, vol. 4945, pp. 90–99. Springer, Heidelberg (2008)
5. Kelsey, J., Lucks, S.: Collisions and near-collisions for reduced-round Tiger. In: Robshaw, M.J.B. (ed.) FSE 2006. LNCS, vol. 4047, pp. 111–125. Springer, Heidelberg (2006)
6. Leurent, G.: MD4 is not one-way. In: Nyberg, K. (ed.) FSE 2008. LNCS, vol. 5086, pp. 412–428. Springer, Heidelberg (2008)
7. Mendel, F., Pramstaller, N., Rechberger, C.: A (second) preimage attack on the GOST hash function. In: Nyberg, K. (ed.) FSE 2008. LNCS, vol. 5086, pp. 224–234. Springer, Heidelberg (2008)
8. Mendel, F., Preneel, B., Rijmen, V., Yoshida, H., Watanabe, D.: Update on Tiger. In: Barua, R., Lange, T. (eds.) INDOCRYPT 2006. LNCS, vol. 4329, pp. 63–79. Springer, Heidelberg (2006)
9. Mendel, F., Rijmen, V.: Cryptanalysis of the Tiger hash function. In: Kurosawa, K. (ed.) ASIACRYPT 2007. LNCS, vol. 4833, pp. 536–550. Springer, Heidelberg (2007)
10. Sasaki, Y., Aoki, K.: Preimage Attacks on MD, HAVAL, SHA, and Others. In: Rump session at CRYPTO 2008 (2008)
11. Wang, X., Yu, H.: How to break MD5 and other hash functions. In: Cramer, R. (ed.) EUROCRYPT 2005. LNCS, vol. 3494, pp. 19–35. Springer, Heidelberg (2005)
12. Wang, X., Yin, Y.L., Yu, H.: Finding collisions in the full SHA-1. In: Shoup, V. (ed.) CRYPTO 2005. LNCS, vol. 3621, pp. 17–36. Springer, Heidelberg (2005)

Appendix A

Here, we show how to calculate the inverse function σ_1^{-1}. Let $(x_{31}, ..., x_0)$ and $(y_{31}, ..., y_0)$ be outputs and inputs of σ_1^{-1} respectively, where $x_i, y_i \in \{0, 1\}$, and x_{31} and y_{31} are the most significant bit. The inverse function σ_1^{-1} is calculated as follows:

$$(x_{31}, x_{30}, ..., x_0)^t = M_{\sigma_1^{-1}} \cdot (y_{31}, y_{30}, ..., y_0)^t,$$

where

$$M_{\sigma_1^{-1}} = \begin{pmatrix}
1\,0\,0\,1\,0\,1\,1\,1\,0\,0\,1\,0\,0\,0\,1\,0\,1\,1\,1\,0\,1\,0\,1\,0\,0\,0\,0\,1\,1\,0\,0\,0 \\
0\,1\,0\,0\,1\,0\,1\,1\,1\,0\,0\,1\,0\,0\,0\,1\,0\,1\,1\,1\,0\,1\,0\,1\,0\,0\,0\,0\,1\,1\,0\,0 \\
0\,0\,1\,0\,0\,1\,0\,1\,1\,1\,0\,0\,1\,0\,0\,0\,1\,0\,1\,1\,1\,0\,1\,0\,1\,0\,0\,0\,0\,1\,1\,0 \\
1\,0\,0\,0\,0\,1\,0\,1\,1\,1\,0\,0\,0\,1\,1\,0\,1\,0\,1\,1\,0\,1\,1\,1\,0\,1\,0\,1\,1\,0\,1\,1 \\
0\,1\,1\,1\,0\,0\,0\,1\,0\,1\,0\,1\,1\,1\,0\,0\,0\,0\,0\,1\,0\,1\,1\,0\,0\,1\,1\,1\,0\,0\,1 \\
1\,0\,0\,1\,1\,1\,0\,0\,0\,0\,1\,1\,0\,0\,1\,1\,1\,0\,1\,1\,1\,1\,1\,1\,0\,0\,0\,1\,0\,0\,0\,0 \\
0\,1\,0\,0\,1\,1\,1\,0\,0\,0\,0\,1\,1\,0\,0\,1\,1\,1\,0\,1\,1\,1\,1\,1\,1\,0\,0\,0\,1\,0\,0\,0 \\
0\,0\,1\,0\,0\,1\,1\,1\,0\,0\,0\,0\,1\,1\,0\,0\,1\,1\,1\,0\,1\,1\,1\,1\,1\,1\,0\,0\,0\,1\,0\,0 \\
1\,0\,0\,0\,0\,1\,0\,0\,1\,0\,1\,0\,0\,1\,0\,0\,1\,0\,0\,1\,1\,1\,0\,1\,1\,1\,1\,1\,1\,0\,1\,0 \\
0\,1\,0\,0\,0\,0\,1\,0\,0\,1\,0\,1\,0\,0\,1\,0\,0\,1\,0\,0\,1\,1\,1\,0\,1\,1\,1\,1\,1\,1\,0\,1 \\
0\,0\,0\,1\,0\,0\,1\,0\,1\,0\,0\,1\,0\,1\,1\,0\,0\,1\,1\,1\,0\,1\,1\,1\,1\,1\,1\,0\,1\,0\,1\,0 \\
0\,0\,0\,0\,1\,0\,0\,1\,0\,1\,0\,0\,1\,0\,1\,1\,0\,0\,1\,1\,1\,0\,1\,1\,1\,1\,1\,1\,0\,1\,0\,1 \\
0\,0\,1\,1\,0\,1\,1\,1\,0\,0\,0\,1\,1\,0\,1\,0\,1\,1\,0\,0\,1\,1\,0\,1\,0\,1\,1\,0\,1\,1\,1\,0 \\
1\,0\,0\,0\,1\,1\,0\,0\,1\,0\,1\,0\,1\,1\,1\,1\,0\,0\,0\,1\,1\,0\,0\,1\,0\,1\,0\,1\,1\,1\,1 \\
0\,1\,1\,1\,0\,1\,0\,1\,1\,1\,1\,0\,1\,0\,0\,0\,1\,0\,0\,1\,0\,1\,1\,0\,1\,1\,0\,0\,0\,0\,1\,1 \\
1\,0\,0\,1\,1\,1\,1\,0\,0\,1\,1\,0\,1\,0\,0\,1\,1\,1\,1\,0\,0\,0\,1\,1\,1\,1\,0\,1\,1\,0\,1 \\
0\,1\,1\,1\,1\,1\,0\,0\,1\,0\,0\,0\,1\,0\,1\,1\,1\,0\,1\,0\,1\,0\,0\,0\,0\,1\,1\,0\,0\,0\,1\,0 \\
1\,0\,1\,0\,1\,0\,0\,1\,0\,1\,1\,0\,0\,1\,1\,1\,0\,0\,1\,1\,1\,1\,1\,0\,0\,0\,1\,0\,1\,0\,0\,1 \\
1\,1\,1\,1\,0\,0\,0\,0\,0\,0\,1\,0\,1\,1\,1\,0\,0\,0\,1\,0\,0\,1\,0\,1\,1\,0\,0\,1\,1\,0\,0\,0 \\
1\,1\,1\,0\,1\,1\,1\,1\,0\,0\,1\,1\,0\,1\,0\,1\,1\,1\,1\,1\,1\,0\,0\,0\,1\,1\,0\,1\,0\,1\,0\,0 \\
1\,1\,1\,0\,0\,0\,0\,0\,1\,0\,1\,1\,1\,0\,0\,0\,0\,0\,0\,1\,0\,1\,1\,0\,0\,1\,1\,1\,0\,0\,1\,0 \\
1\,1\,1\,0\,0\,1\,1\,1\,0\,1\,1\,1\,1\,1\,1\,0\,1\,1\,1\,0\,0\,0\,0\,1\,0\,0\,1\,0\,0\,0\,0\,1 \\
1\,1\,0\,1\,0\,1\,1\,1\,0\,0\,1\,0\,0\,0\,1\,0\,1\,1\,0\,0\,1\,0\,1\,0\,0\,0\,0\,1\,1\,1\,0\,0 \\
0\,1\,1\,0\,1\,0\,1\,1\,1\,0\,0\,1\,0\,0\,0\,1\,0\,1\,1\,0\,0\,1\,0\,1\,0\,0\,0\,0\,1\,1\,1\,0 \\
1\,0\,1\,0\,0\,0\,1\,0\,1\,1\,1\,0\,1\,0\,1\,0\,0\,1\,0\,1\,1\,0\,0\,0\,1\,0\,0\,1\,1\,1\,1\,1 \\
1\,1\,1\,1\,0\,1\,0\,1\,1\,1\,1\,0\,1\,0\,0\,0\,1\,0\,0\,1\,0\,1\,1\,0\,1\,1\,0\,0\,0\,0\,1\,1 \\
1\,1\,0\,1\,1\,1\,1\,0\,0\,1\,1\,0\,1\,0\,0\,1\,1\,1\,1\,1\,0\,0\,0\,1\,1\,1\,1\,0\,1\,1\,0\,1 \\
0\,1\,0\,1\,1\,1\,0\,0\,1\,0\,0\,0\,1\,0\,1\,1\,1\,0\,1\,0\,1\,0\,0\,0\,0\,1\,1\,0\,0\,0\,1\,0 \\
0\,0\,1\,0\,1\,1\,1\,0\,0\,1\,0\,0\,0\,1\,0\,1\,1\,1\,0\,1\,0\,1\,0\,0\,0\,0\,1\,1\,0\,0\,0\,1 \\
1\,0\,1\,1\,0\,0\,1\,1\,1\,0\,1\,1\,1\,1\,1\,0\,1\,0\,1\,0\,0\,0\,0\,1\,0\,0\,1\,0\,1\,0\,0 \\
1\,1\,0\,0\,1\,1\,1\,0\,1\,1\,1\,1\,1\,1\,0\,1\,0\,1\,0\,0\,0\,0\,1\,0\,0\,1\,0\,1\,0\,0\,1\,0 \\
0\,1\,1\,0\,0\,1\,1\,1\,0\,1\,1\,1\,1\,1\,1\,0\,1\,0\,1\,0\,0\,0\,0\,1\,0\,0\,1\,0\,1\,0\,0\,1
\end{pmatrix}.$$

Cryptanalysis of the LAKE Hash Family

Alex Biryukov[1], Praveen Gauravaram[3], Jian Guo[2], Dmitry Khovratovich[1],
San Ling[2], Krystian Matusiewicz[3], Ivica Nikolić[1], Josef Pieprzyk[4],
and Huaxiong Wang[2]

[1] University of Luxembourg, Luxembourg
{alex.biryukov,dmitry.khovratovich,ivica.nikolic}@uni.lu
[2] School of Physical and Mathematical Sciences,
Nanyang Technological University, Singapore
{guojian,lingsan,hxwang}@ntu.edu.sg
[3] Department of Mathematics,
Technical University of Denmark, Denmark
{P.Gauravaram,K.Matusiewicz}@mat.dtu.dk
[4] Centre for Advanced Computing - Algorithms and Cryptography,
Macquarie University, Australia
josef@ics.mq.edu.au

Abstract. We analyse the security of the cryptographic hash function
LAKE-256 proposed at FSE 2008 by Aumasson, Meier and Phan. By
exploiting non-injectivity of some of the building primitives of LAKE,
we show three different collision and near-collision attacks on the compression function. The first attack uses differences in the chaining values
and the block counter and finds collisions with complexity 2^{33}. The second attack utilizes differences in the chaining values and salt and yields
collisions with complexity 2^{42}. The final attack uses differences only in
the chaining values to yield near-collisions with complexity 2^{99}. All our
attacks are independent of the number of rounds in the compression function. We illustrate the first two attacks by showing examples of collisions
and near-collisions.

1 Introduction

The recent cryptanalytical results on the cryptographic hash functions following
the attacks on MD5 and SHA-1 by Wang et al. [17,16,15] have seriously undermined the confidence in many currently deployed hash functions. Around the
same time, new generic attacks such as multicollision attack [7], long message
second preimage attack [9] and herding attack [8], exposed some undesirable
properties and weaknesses in the Merkle-Damgård (MD) construction [12,5].
These developments have renewed the interest in the design of hash functions.
Subsequent announcement by NIST of the SHA-3 hash function competition,
aiming at augmenting the FIPS 180-2 [13] standard with a new cryptographic
hash function, has further stimulated the interest in the design and analysis of
hash functions.

O. Dunkelman (Ed.): FSE 2009, LNCS 5665, pp. 156–179, 2009.

The hash function family LAKE [1], presented at FSE 2008, is one of the new designs. It follows the design principles of the HAIFA framework [2,3] – a strengthened alternative to the MD construction.

As the additional inputs to the compression function, LAKE uses a random value (also called salt) and an index value, which counts the number of bits/blocks in the input message processed so far.

The first analysis of LAKE, presented by Mendel and Schläffer [11], has shown collisions for 4 out of 8 rounds. The complexity of their attack is 2^{109}. The main observation used in the attack is the non-injectivity of one of the internal functions. This property allows to introduce difference in the message words, which is canceled immediately, when the difference goes through the non-injective function.

Our contributions. Our attacks focus on finding collisions for the compression function of LAKE. Let $f(H, M, S, t)$ be a compression function of a HAIFA hash function using chaining values H, message block M, salt S and the block index t. We present the following three types of collision attacks. The first attack uses differences in the chaining values H and block index t, so we are looking for collisions of form $f(H, M, S, t) = f(H', M, S, t')$. We call it a (H, t)-type attack. The complexity of this attack is 2^{33} compression calls. The second attack deals with the differences injected in the chaining values and salt S, we call it a (H, S)-attack. We present how to find near-collisions of the compression function with the complexity 2^{30} of compression calls and extend it to full collisions with the complexity 2^{42}. The final attack, called a H-type attack, uses only differences in the chaining values and finds near-collisions for the compression function with the complexity 2^{99}. The success of our collision attacks relies on solving the systems of equations that originate from the differential conditions imposed by the attacks. We present some efficient methods to solve these systems of equations.

Our attacks demonstrate that increasing the number of rounds of LAKE does not increase its security as they all aim at canceling the differences within the first ProcessMessage function of the compression function.

2   Description of LAKE

In this section, we provide a brief description of the LAKE compression function, skipping details that are not relevant to our attacks. See [1] for details.

Basic functions – LAKE uses two functions f and g defined as follows

$$f(a, b, c, d) = (a + (b \vee C_0)) + ((c + (a \wedge C_1)) \ggg 7) +$$
$$((b + (c \oplus d)) \ggg 13) ,$$
$$g(a, b, c, d) = ((a + b) \ggg 1) \oplus (c + d) ,$$

where each variable is a 32-bit word and C_0, C_1 are constants.

The compression function of LAKE has three components: SaltState, ProcessMessage and FeedForward. The functionality of these components are

Input: $H = H_0 \| \ldots \| H_7$, $S = S_0 \| \ldots \| S_3$, $t = t_0 \| t_1$
Output: $F = F_0 \| \ldots \| F_{15}$
for $i = 0, \ldots, 7$ do
$\quad | \quad F_i = H_i;$
end
$F_8 = g(H_0, S_0 \oplus t_0, C_8, 0);$
$F_9 = g(H_1, S_1 \oplus t_1, C_9, 0);$
for $i = 10, \ldots, 15$ do
$\quad | \quad F_i = g(H_i, S_i, C_i, 0);$
end

Algorithm 1. LAKE's SaltState

Input: $F = F_0 \| \ldots \| F_{15}$, $M = M_0 \| \ldots \| M_{15}$, σ
Output: $W = W_0 \| \ldots \| W_{15}$
for $i = 0, \ldots, 15$ do
$\quad | \quad L_i = f(L_{i-1}, F_i, M_{\sigma(i)}, C_i);$
end
$W_0 = g(L_{15}, L_0, F_0, L_1);$
$L_0 = W_0;$
for $i = 1, \ldots, 15$ do
$\quad | \quad W_i = g(W_{i-1}, L_i, F_i, L_{i+1});$
end

Algorithm 2. LAKE's ProcessMessage

Input: $W = W_0 \| \ldots \| W_{15}$, $H = H_0 \| \ldots \| H_7$, $S = S_0 \| \ldots \| S_3$, $t = t_0 \| t_1$
Output: $H = H_0 \| \ldots \| H_7$
$H_0 = f(W_0, W_8, S_0 \oplus t_0, H_0);$
$H_1 = f(W_1, W_9, S_1 \oplus t_1, H_1);$
for $i = 2, \ldots, 7$ do
$\quad | \quad H_i = f(W_i, W_{i+8}, S_i, H_i);$
end

Algorithm 3. LAKE's FeedForward

Input: $H = H_0 \| \ldots \| H_7$, $M = M_0 \| \ldots \| M_{15}$, $S = S_0 \| \ldots \| S_3$, $t = t_0 \| t_1$
Output: $H = H_0 \| \ldots \| H_7$
$F = \mathsf{SaltState}(H, S, t);$
for $i = 0, \ldots, r - 1$ do
$\quad | \quad F = \mathsf{ProcessMessage}(F, M, \sigma_i);$
end
$H = \mathsf{FeedForward}(F, H, S, t);$

Algorithm 4. LAKE's CompressionFunction

described in Algorithms 1, 2 and 3, respectively. The whole compression function of LAKE is presented as Algorithm 4. Our attacks do not depend on the constants C_i for $i = 0, \ldots, 15$ and hence we do not provide their actual values here.

3 Properties and Observations

We first present some properties of the f function used in our analysis.

Observation 1. *Function $f(x, y, z, t)$ is non-injective with respect to the first three arguments x, y, z.*

For example, for x there exist two different values x and x' such that $f(x, y, z, t) = f(x', y, z, t)$ for some y, z, t. The same property holds for y and z. This observation was mentioned by Lucks at FSE'08. Mendel and Schläffer independently found and used this property to successfully attack four out of eight rounds of LAKE-256. Non-injectivity of the function f can be used to cancel a difference in one of the first three arguments of f, when the rest of the arguments are fixed.

The following observation of the rotation on the modular addition allows us to simplify the analysis of f.

Lemma 1 ([6]). $(a + b) \ggg k = (a \ggg k) + (b \ggg k) + \alpha - \beta \cdot 2^{n-k}$, *where* $\alpha = 1[a_k^R + b_k^R \geq 2^k]$ *and* $\beta = 1[a_k^L + b_k^L + \alpha \geq 2^{n-k}]$.

Using Lemma (1), the function f can be written as

$$f(a, b, c, d) = a + b \vee C_0 + (c \ggg 7) + ((a \wedge C_1) \ggg 7) + (b \ggg 13)$$
$$+ ((c \oplus d) \ggg 13) + \alpha_1 + \alpha_2 - \beta_1 \cdot 2^{25} - \beta_2 \cdot 2^{19}, \quad (1)$$

where

$$\alpha_1 = 1[c_7^L + (a \wedge C_1)_7^L \geq 2^7], \qquad \beta_1 = 1[c_7^R + (a \wedge C_1)_7^R + \alpha_1 \geq 2^{25}],$$
$$\alpha_2 = 1[b_{13}^L + (c \oplus d)_{13}^L \geq 2^{13}], \qquad \beta_2 = 1[b_{13}^R + (c \oplus d)_{13}^R + \alpha_2 \geq 2^{19}].$$

Note that α_2 and β_2 are independent of a. Consider now the difference of the outputs of f induced by the difference in the variable a, i.e.

$$\Delta f = f(a', b, c, d) - f(a, b, c, d)$$
$$= [a' + (a' \wedge C_1) + \alpha_1' - \beta_1' \cdot 2^{25}] - [a + (a \wedge C_1) + \alpha_1 - \beta_1 \cdot 2^{25}]$$
$$= a' + ((a' \wedge C_1) \ggg 7) - [a + ((a \wedge C_1) \ggg 7)] + (\alpha_1' - \alpha_1) - (\beta_1' - \beta_1) \cdot 2^{25}$$
$$= f_a(a') - f_a(a) + (\alpha_1' - \alpha_1) - (\beta_1' - \beta_1) \cdot 2^{25},$$

where $f_a(a) \overset{\text{def}}{=} a + ((a \wedge C_1) \ggg 7)$.

A detailed analysis (cf. Lemma 5) shows that given random a, a' and c, $P(\alpha_1 = \alpha_1', \beta_1 = \beta_1') = \frac{4}{9}$, so with the probability $\frac{4}{9}$, a collision for f_a is also a collision of f when the input difference is in a only. Let us call this a *carry effect*. However, if we have control over the variable c, we can adjust the values of $\alpha_1, \alpha_1', \beta_1, \beta_1'$ and always satisfy this condition. From here we can see that $(a + b) \ggg k$ is not a good mixing function when modular differences are concerned.

This reasoning can be repeated for differences in the variable b and similarly for differences in a pair of the variables c, d. It is easy to see that also for those cases, with a high probability, collisions in f happen when the following functions collide

$$f_b(b) \stackrel{\text{def}}{=} b \vee C_0 + (b \ggg 13) \ ,$$

$$f_{cd}(c, d) \stackrel{\text{def}}{=} (c \ggg 7) + ((c \oplus d) \ggg 13) \ .$$

So, when we follow differences in one or two variables only, we can consider those variables without the side effects from other variables. We summarize the above observations below.

Observation 2. *Collisions or output differences of f for input differences in one variable can be made independent from the values of other variables.*

We denote the set of solutions for f_a and f_b with respect to input pairs and modular differences as

$$S_{fa} \stackrel{\text{def}}{=} \{(x, x') | f_a(x) = f_a(x')\} \ , \qquad S_{fa}^A \stackrel{\text{def}}{=} \{x - x' | f_a(x) = f_a(x')\} \ ,$$

$$S_{fb} \stackrel{\text{def}}{=} \{(x, x') | f_b(x) = f_b(x')\} \ , \qquad S_{fb}^A \stackrel{\text{def}}{=} \{x - x' | f_b(x) = f_b(x')\} \ .$$

Choose the odd elements from S_{fb}^A and define them to be $S_{fb_{odd}}^A$. Note that we can easily precompute all the above solution sets using 2^{32} evaluations of the appropriate functions and 2^{32} words of memory (or some more computations with proportionally less memory).

4 (H, t)-Type Attack

First, let us try to attack only the middle part of the compression function, i.e. ProcessMessage function. It consists of 8 rounds (10 rounds for LAKE-512). In every round, first all of the 16 internal variables are updated by the function f, and then all of them are updated by the function g.

Our differential trail is as follows:

1. Introduce a carefully chosen difference in F_0.
2. After the first application of the function f from all L_i, only L_0 has a non-zero difference.
3. After the first application of the function g none of W_i have any difference.

Let us show that this differential is possible. First let us prove that Step 2 is achievable. Considering that $L_i = f(L_{i-1}, F_i, M_{\sigma(i)}, C_i)$, we get that in L_i a difference can be introduced only through L_{i-1} and F_i (message words do not have differences, C_i are simply constants). Note that in the first round $\sigma(i)$ is defined as the identity permutation hence we can write M_i instead of $M_{\sigma(i)}$.

For ΔL_0 we require a non-zero difference

$$\Delta L_0 = f(F_{15}, F_0', M_0, C_0) - f(F_{15}, F_0, M_0, C_0) \neq 0. \tag{2}$$

For ΔL_1 we require the zero difference

$$\Delta L_1 = f(L_0', F_1, M_1, C_1) - f(L_0, F_1, M_1, C_1) = 0. \tag{3}$$

From Observation 1, it follows that it is possible to get zero for ΔL_1. For all the other $\Delta L_i, i = 2..15$ we require the zero difference. This is trivially fulfilled because there are no inputs with difference. Now, let us consider Step 3. Note that $W_i = g(W_{i-1}, L_i, F_i, L_{i+1})$, so we can introduce a difference in W_i by any of W_{i-1}, L_i, F_i and L_{i+1}.

For ΔW_0, we require the zero difference, so we get

$$\Delta W_0 = g(L_{15}, L_0', F_0', L_1) - g(L_{15}, L_0, F_0, L_1) = 0. \tag{4}$$

Note that there are differences in two variables, L_0 and F_0, hence the above equation can be solved. For the indexes $i = 1, \ldots, 14$, we obtain

$$\Delta W_i = g(W_{i-1}, L_i, F_i, L_{i+1}) - g(W_{i-1}, L_i, F_i, L_{i+1}) = 0. \tag{5}$$

All the above equations hold as there are no differences in any of the arguments. For W_{15}, we have

$$\Delta W_{15} = g(W_{14}, L_{15}, F_{15}, W_0) - g(W_{14}, L_{15}, F_{15}, W_0) = 0.$$

Notice that the last argument is not L_0 but rather W_0 because there are no temporal variables that store the previous values of L_i (see ProcessMessage). This non-symmetry in the ProcessMessage, which updates L registers stops the flow of the difference from L_0 to W_{15}.

So, after only one round, we can obtain an internal state with all-zero differences in the variables. Then the following rounds can not introduce any difference because there are no differences in the internal state variables or in the message words. So, if we are able to solve the equations that we have got then *the attack is applicable to any number of rounds, i.e. increasing the number of rounds in the ProcessMessage function does not improve the security of LAKE.*

Let us take a closer look at our equations. Equation (2) can be written as

$$\Delta L_0 = f(F_{15}, F_0', M_0, C_0) - f(F_{15}, F_0, M_0, C_0) =$$
$$= (F_0' \vee C_0) - (F_0 \vee C_0) + [F_0' + (M_0 \oplus C_0)] \ggg 13 - [F_0 + (M_0 \oplus C_0)] \ggg 13.$$

Hereafter we will use that $(A + B) \ggg r = (A \ggg r) + (B \ggg r)$ with the probability $\frac{1}{4}$ (see [6]). The same holds when rotation to the left is used. Therefore, the above equation can be rewritten as

$$\Delta L_0 = (F_0' \vee C_0) - (F_0 \vee C_0) + F_0' \ggg 13 - F_0 \ggg 13. \tag{6}$$

Equation (3) can be written as

$$\Delta L_1 = f(L_0', F_1, M_1, C_1) - f(L_0, F_1, M_1, C_1) =$$
$$= L_0' - L_0 + [M_1 + (L_0' \wedge C_1)] \ggg 7 - [M_1 + (L_0 \wedge C_1)] \ggg 7 =$$
$$= L_0' - L_0 + (L_0' \wedge C_1) \ggg 7 - (L_0 \wedge C_1) \ggg 7 = 0.$$

Equation (4) can be written as

$$\Delta W_0 = g(L_{15}, L_0', F_0', L_1) - g(L_{15}, L_0, F_0, L_1) =$$
$$= [(L_{15} + L_0') \ggg 1] \oplus (F_0' + L_1) - [(L_{15} + L_0) \ggg 1] \oplus (F_0 + L_1) = 0.$$

Let us try to extend the collision attack on the ProcessMessage function to the full compression function. First, let us deal with the initialization (function SaltState).

From the initialization of LAKE, it can be seen that the variables H_0 through H_7 are copied into F_0 through F_7. The variable F_8 depends on H_0 and t_0. Similarly, F_9 depends on H_1 and t_1. The rest of the variables do not depend on either t_0 or t_1. Since we need a difference in F_0 (for the previous attack on ProcessMessage function), we will introduce difference in H_0. Further, we can follow our previous attack on the ProcessMessage block and get collisions after the ProcessMessage function. The only difficulty is how to deal with F_8 since it does depend on H_0, which now has a non-zero difference. As a way out, we use the block index t_0. By introducing a difference in t_0 we can cancel the difference from H_0 in F_8. So we get the following equation

$$\Delta F_8 = g(H_0', S_0 \oplus t_0', C_0, 0) - g(H_0, S_0 \oplus t_0, C_0, 0) =$$
$$= ((H_0' + (S_0 \oplus t_0')) \ggg 1 \oplus C_0) - ((H_0 + (S_0 \oplus t_0)) \ggg 1 \oplus C_0) = 0.$$

Let $\tilde{t}_0' = t_0' \oplus S_0$ and $\tilde{t}_0 = t_0 \oplus S_0$. Then, the above equation gets the following form

$$\Delta F_8 = H_0' - H_0 + \tilde{t}_0' - \tilde{t}_0 = 0.$$

Now, let us deal with the last building block of the compression function, the FeedForward function. Note that we have differences in H_0 and t_0 only. If we take a glance at the FeedForward procedure, we can see that H_0 and t_0 can be found in the same equation, and only there, which defines the new value for H_0. Since we require the zero difference in all of the output variables, we get the following equation

$$\Delta H_0 = f(F_0, F_8, H_0', S_0 \oplus t_0') - f(F_0, F_8, H_0, S_0 \oplus t_0) =$$
$$= \tilde{t}_0' \ggg 7 - \tilde{t}_0 \ggg 7 + (\tilde{t}_0' \oplus H_0') \ggg 13 - (\tilde{t}_0 \oplus H_0) \ggg 13 = 0.$$

This concludes our attack. We have shown that if we introduce a difference in the chaining value H_0 and the block index t_0 only, it is possible to reduce the problem of finding collisions for the compression function of LAKE to the problem of solving a system of equations.

4.1   Solving Equation Systems

To find a collision for the full compression function of LAKE, we have to solve the equations that were mentioned in the previous sections. As a result, we get the following system equations (note that $H_0 = F_0$)

$$L_0' - L_0 + (L_0' \wedge C_1) \ggg 7 - (L_0 \wedge C_1) \ggg 7 = 0; \tag{7}$$

$$L_0' - L_0 = (H_0' \vee C_0) - (H_0 \vee C_0) + H_0' \ggg 13 - H_0 \ggg 13; \tag{8}$$

$$[(L_{15} + L_0') \ggg 1] \oplus (H_0' + L_1) - [(L_{15} + L_0) \ggg 1] \oplus (H_0 + L_1) = 0; \tag{9}$$

$$H_0' - H_0 + \tilde{t}_0' - \tilde{t}_0 = 0; \tag{10}$$

$$\tilde{t}_0' \ggg 7 - \tilde{t}_0 \ggg 7 + (\tilde{t}_0' \oplus H_0') \ggg 13 - (\tilde{t}_0 \oplus H_0) \ggg 13 = 0. \tag{11}$$

Let us analyze Equation (7). By fixing $L_0' - L_0 = R$ and rotating to the left by 7 bits, this equation can be rewritten as

$$(X + A) \wedge C = X \wedge C + B, \tag{12}$$

where $X = L_0, A = R, B = (-R) \ll 7, C = C_1$. Now, let us analyze Equation (8). Again, let us fix $L_0' - L_0 = R$ and $H_0' - H_0 = D$. Then Equation(8) gets the following form

$$(X + A) \vee C = X \vee C + B, \tag{13}$$

where $X = H_0, A = D, B = R - (D \ggg 13), C = C_0$. In Equation (9), if we regroup the components, we obtain

$$[(L_{15} + L_0') \oplus (L_{15} + L_0)] \ggg 1 = (H_0' + L_1) \oplus (H_0 + L_1).$$

Then, the above equation is of the following form

$$((X + A) \oplus X) \ggg 1 = (Y + B) \oplus Y, \tag{14}$$

where $X = L_{15} + L_0, A = L_0' - L_0, Y = L_1 + H_0, B = H_0' - H_0$.

Now, let us analyze Equations (10) and (11). Let us fix $H_0' - H_0 = D$. Note that from Equation (10), we have $\tilde{t}_0' - \tilde{t}_0 = -D$. If we rotate everything by 13 bits to the left in Equation (11), we get

$$(-D) \ll 6 + (\tilde{t}_0' \oplus H_0') - (\tilde{t}_0 \oplus H_0) = 0; \tag{15}$$

$$\tilde{t}_0 = [(\tilde{t}_0' \oplus H_0') - D \ll 6] \oplus H_0. \tag{16}$$

If we substitute $\tilde{t}_0$ in Equation (10) by the above expression, then we have

$$D + \tilde{t}_0' - [(\tilde{t}_0' \oplus H_0') - D \ll 6] \oplus H_0 = 0; \tag{17}$$

$$\tilde{t}_0' = [(\tilde{t}_0' \oplus H_0') - D \ll 6] \oplus H_0 - D. \tag{18}$$

If we XOR the value of H_0' to the both sides, we get

$$\tilde{t}_0' \oplus H_0' = ([(\tilde{t}_0' \oplus H_0') - D \ll 6] \oplus H_0 - D) \oplus H_0'. \tag{19}$$

Let us denote $\tilde{t_0} \oplus H_0' = X$. Then we can write

$$X = [(X - D \ll 6) \oplus H_0 - D] \oplus H_0'; \tag{20}$$
$$X \oplus H_0' = (X - D \ll 6) \oplus H_0 - D. \tag{21}$$

Finally, we get an equation of the following form

$$(X \oplus K_1) + A = (X + B) \oplus K_2, \tag{22}$$

where $K_1 = H_0', A = R, B = -R \ll 6, K_2 = H_0$.

Lemma 2. *There exist efficient algorithms* **Al1, Al2, Al3, Al4** *for finding solutions for equations of type (12),(13),(14),(22).*

The description of these algorithms can be found in Appendix B.

Now, we can present our algorithm for finding solutions for the system of equations. With **Al1** we find a difference R (and values for L_0, L_0') such that Equation (7) holds. Actually, for the same difference R many distinct solutions (L_0, L_0') exist (experiments show that when Equation (7) is solvable, then there are around 2^5 solutions). Next, we pass as an input to **Al2** the difference R and we find a difference D (and values for H_0, H_0') such that Equation (8) holds. Again for a fixed R and D, many pairs (H_0, H_0') exist. We verified experimentally that for a random R and a "good" D, there are around 2^{10} solutions. Using Algorithm **Al3**, we check if we can find solutions for Equation (9), i.e. we try to find L_1 and L_{15}. Note that the input of **Al3** is the previously found sequence (L_0, L_0', H_0, H_0'). If **Al3** can not find a solution, then we get another pair (H_0, H_0') (or generate first a new difference D and then generate another 2^{10} pairs (H_0, H_0')). If **Al3** finds a solution to (9), then we use Algorithm **Al4** and try to find solutions for Equations (10) and (11), where the input to **Al4** is already found as the pair (H_0, H_0'). If **Al4** can not find a solution, then we can take a different pair (H_0, H_0') (or generate first a new difference D and then generate (H_0, H_0')) and then apply first **Al3** and then **Al4**.

4.2   Complexity of the Attack

Let us try to find the complexity of the algorithm. Note that when analyzing the initial equations, we have used the assumption that $(A + B) \gg r = (A \gg r) + (B \gg r)$, which holds with the probability $\frac{1}{4}$ (see [6]). In total, we used this assumption 5 times. In the equation for ΔF_0, we can control the exact value of M_1, so in total, we have used the assumption 4 times. Therefore, the probability that a solution of the system is a solution for the initial equations is 2^{-8}. This means that we have to generate 2^8 solutions for the system. Let us find the cost for a single solution.

The average complexity for both **Al1** and **Al2** is 2^1 steps. We confirmed experimentally that, for a random difference R, there exists a solution for Equation (7) with the probability 2^{-27}. So this takes $2^{27} \cdot 2^1 = 2^{28}$ steps using **Al1** and it finds 2^5 solutions for Equation (7). Similarly, for a random difference D, there is a solution for Equation (8) with the probability 2^{-27}. Therefore, this consumes $2^{27} \cdot 2^1 = 2^{28}$ steps and finds 2^{10} pairs (H_0, H_0) for Equation (8). The probability that a pair is a good pair for Equation (9) is 2^{-1} and that it is a good pair for Equations (10) and (11) is 2^{-12} (as explained in Appendix B). Thus, we need $2^1 \cdot 2^{12} = 2^{13}$ pairs, which we can be generated in $2^{28} \cdot 2^3 = 2^{31}$ steps. Since we need 2^8 solutions, the total complexity is 2^{39}. Note that this complexity estimate (a step) is measured by the number of calls to the algorithms that solve our specific equations. If we assume that a call to the algorithms is four times less efficient than the call to the functions f or g (which on average seems to be true), and consider the fact that the compression function makes a total of around 2^8 calls to the functions f or g, then we get that the total complexity of the collision search is around 2^{33} compression function calls.

Note that when a solution for the system exists, then this still does not mean that we have a collision. This is partially because we cannot control some of the values directly. Indeed, we can control directly only H_0, H_0', t_0, t_0'. The rest of the variables, i.e. L_0, L_0', L_1, L_{15}, we can control through the message words M_i or with the input variables H_i, where $i > 0$. Since we pass these values as arguments for the non-injective function f, we may experience situation when we cannot get the exact value that we need. Yet, with an overwhelming probability, we can find the exact values. Let us suppose that we have a solution $(H_0, H_0', L_0, L_0', L_1, L_{15}, t_0, t_0')$ for the system of equations. First, we find a message word M_0 such that $f(F_{15}, H_0, M_0, C_0) = L_0$. Notice that F_{15} can be previously fixed by choosing some value for H_7. Then, $f(F_{15}, H_0', M_0, C_0) = L_0'$. We choose M_1 such that $[M_1 + (L_0' \wedge C_1)] \ggg 7 - [M_1 + (L_0 \wedge C_1)] \ggg 7 = (L_0' \wedge C_1) \ggg 7 - (L_0 \wedge C_1) \ggg 7$. This way the probability that the previous identity holds becomes 1. Then we find H_1 such that $f(L_0, H_1, M_1, C_1) = L_1$. At last, we find M_{15} such that $f(L_{14}, F_{15}, M_{15}, C_{15}) = L_{15}$. If such M_{15} does not exist, then we can change the value of L_{14} by changing M_{14} and then try to find M_{15}.

5 (H, S)-Type Attack

The starting idea for this attack is to inject differences in the input chaining variable H and the salt S and then cancel them within the first iteration of ProcessMessage. Consequently, no difference appears throughout the compression function until the FeedForward step. If the differences in the chaining and salt variables are selected properly, we can hope they cancel each other, so we get no difference at the output of the compression function.

5.1 Finding High-Level Differentials

To find a suitable differential for the attack, an approach similar to the one employed to analyse FORK-256 [10, Section 6] can be used. We model each of the registers a, b, c, d, as a single binary value $\delta a, \delta b, \delta c, \delta d$ that denotes whether there is a difference in the register or not. Moreover, we assume that we are able to make any two differences cancel each other to obtain a model that can be expressed in terms of arithmetics over $\mathbb{F}_2$. We model the differential behavior of function g simply as $\delta g(\delta a, \delta b, \delta c, \delta d) = \delta a \oplus \delta b \oplus \delta c \oplus \delta d$, where $\delta a, \delta b, \delta c, \delta d \in \mathbb{F}_2$, as this description seems to be functionally closest to the original. For example, it is impossible to get collisions for g when only one variable has differences and such a model ensures that we always have two differences to cancel each other if we need no output difference of g. When deciding how to model $f(a, b, c, d)$, we have more options. First, note that when looking for collisions, there are no differences in message words and the last parameter of f is a constant, so we need to deal with differences in only two input variables a and b. Since we can find collisions for f when differences are only in a single variable (either a or b), we can model f not only as $\delta f(\delta a, \delta b) = \delta a \oplus \delta b$ but more generally as $\delta f(\delta a, \delta b) = \gamma_0(\delta a) \oplus \gamma_1(\delta b)$, where $\gamma_0, \gamma_1 \in \mathbb{F}_2$ are fixed parameters. Let us call the pair (γ_0, γ_1) a γ-configuration of δf and denote it by $\delta f_{[\gamma_0, \gamma_1]}$. As an example, $\delta f_{[1,0]}$ corresponds to $\delta f(\delta a, \delta b) = \delta a$, which means that whenever a difference appears in register b, we need to use the properties of f to find collisions in the coordinate b. For functions f appearing in FeedForward, we use the model $\delta f = \delta a \oplus \delta b \oplus \delta c \oplus \delta d$.

With these assumptions, it is easy to see that such a model of the whole compression function is linear over $\mathbb{F}_2$ and finding the set of input differences (in chaining variables $H_0, \ldots, H_7$ and salt registers $S_0, \ldots, S_3$) is just a matter of finding the kernel of a linear map. Since we want to find only simple differentials, we are interested in those that use as few registers as possible. To find them, we can think of all possible states of the linear model as a set of codewords of a linear code over $\mathbb{F}_2$. That way, finding differentials affecting only few registers corresponds to finding low-weight codewords. So instead of an enumeration of all 2^{12} possible states of of $H_0, \ldots, H_7, S_0, \ldots, S_3$ for each γ-configuration of f functions, this can be done more efficiently by using tools like MAGMA [4].

We implemented this method in MAGMA and performed such a search for all possible γ-configurations of the 16 functions f appearing in the first ProcessMessage. We used the following search criteria: (a) as few active f functions as possible; (b) as few active g functions as possible; (c) non-zero differences appear only in the first few steps using function g as it is harder to adjust the values for later steps due to lack of variables we control; (d) we prefer γ-configurations $[1, 0]$ and $[0, 1]$ over $[1, 1]$ because it seems easier to deal with differences in one register than in two registers simultaneously.

The optimal differential for this set of criteria contains differences in registers $H_0, H_1, H_4, H_5, S_0, S_1$ with the following γ-configurations of the first seven f functions in ProcessMessage: $[0, 1], [1, 1], [0, 1], [\cdot, \cdot], [0, 1], [1, 1], [0, 1]$ (Note a simpler configuration (H_0, H_4, S_0) is not possible here). Unfortunately, the system of constraints resulting from that differential has no solutions, so we introduced a small modification of it, adding differences in registers H_2, H_6, S_2, ref. Figure 1. After introducing these additional differences, we gain more freedom at the expense of dealing with more active functions and we can find solutions for

SALTSTATE
input: $H_0, \ldots, H_7, S_0, \ldots, S_3, t_0, t_1$

$\Delta F_0 \leftarrow \Delta H_0$
$\Delta F_1 \leftarrow \Delta H_1$
$\Delta F_2 \leftarrow \Delta H_2$
$F_3 \leftarrow H_3$
$\Delta F_4 \leftarrow \Delta H_4$
$\Delta F_5 \leftarrow \Delta H_5$
$\Delta F_6 \leftarrow \Delta H_6$
$F_7 \leftarrow H_7$
$F_8 \leftarrow g(\Delta H_0, \Delta S_0 \oplus t_0, C_8, 0)$ {s1}
$F_9 \leftarrow g(\Delta H_1, \Delta S_1 \oplus t_1, C_9, 0)$ {s2}
$F_{10} \leftarrow g(\Delta H_2, \Delta S_2, C_{10}, 0)$ {s3}
$F_{11} \leftarrow g(H_3, S_3, C_{11}, 0)$
$F_{12} \leftarrow g(\Delta H_4, \Delta S_0, C_{12}, 0)$ {s4}
$F_{13} \leftarrow g(\Delta H_5, \Delta S_1, C_{13}, 0)$ {s5}
$F_{14} \leftarrow g(\Delta H_6, \Delta S_2, C_{14}, 0)$ {s6}
$F_{15} \leftarrow g(H_7, S_3, C_{15}, 0)$
output: $F_0, \ldots, F_{15}$

FEEDFORWARD
input: $R_0, \ldots, R_{15}, H_0, \ldots, H_7,$
$\qquad\quad S_0, \ldots, S_3, t_0, t_1$
$H_0 \leftarrow f(R_0, R_8, \Delta S_0 \oplus t_0, \Delta H_0)$ {f1}

$H_1 \leftarrow f(R_1, R_9, \Delta S_1 \oplus t_1, \Delta H_1)$ {f2}

$H_2 \leftarrow f(R_2, R_{10}, \Delta S_2, \Delta H_2)$ {f3}
$H_3 \leftarrow f(R_3, R_{11}, S_3, H_3)$
$H_4 \leftarrow f(R_4, R_{12}, \Delta S_0, \Delta H_4)$ {f4}
$H_5 \leftarrow f(R_5, R_{13}, \Delta S_1, \Delta H_5)$ {f5}
$H_6 \leftarrow f(R_6, R_{14}, \Delta S_2, \Delta H_6)$ {f6}
$H_7 \leftarrow f(R_7, R_{15}, S_3, H_7)$
output: $H_0, \ldots, H_7$

PROCESSMESSAGE
input: $F_0, \ldots, F_{15}, M_0, \ldots, M_{15}, \sigma$
$L_0 \leftarrow f(F_{15}, \Delta F_0, M_{\sigma(0)}, C_0)$ {p1}
$\Delta L_1 \leftarrow f(L_0, \Delta F_1, M_{\sigma(1)}, C_1)$ {p2}
$\Delta L_2 \leftarrow f(\Delta L_1, \Delta F_2, M_{\sigma(2)}, C_2)$ {p3}
$L_3 \leftarrow f(\Delta L_2, F_3, M_{\sigma(3)}, C_3)$ {p4}
$L_4 \leftarrow f(L_3, \Delta F_4, M_{\sigma(4)}, C_4)$ {p5}
$\Delta L_5 \leftarrow f(L_4, \Delta F_5, M_{\sigma(5)}, C_5)$ {p6}
$\Delta L_6 \leftarrow f(\Delta L_5, \Delta F_6, M_{\sigma(6)}, C_6)$ {p7}
$L_7 \leftarrow f(\Delta L_6, F_7, M_{\sigma(7)}, C_7)$ {p8}
$L_8 \leftarrow f(L_7, F_8, M_{\sigma(8)}, C_8)$

$\vdots$

$L_{15} \leftarrow f(L_{14}, F_{15}, M_{\sigma(15)}, C_{15})$

$W_0 \leftarrow g(L_{15}, L_0, \Delta F_0, \Delta L_1)$ {p9}
$W_1 \leftarrow g(W_0, \Delta L_1, \Delta F_1, \Delta L_2)$ {p10}
$W_2 \leftarrow g(W_1, \Delta L_2, \Delta F_2, L_3)$ {p11}
$W_3 \leftarrow g(W_2, L_3, F_3, L_4)$
$W_4 \leftarrow g(W_3, L_4, \Delta F_4, \Delta L_5)$ {p12}
$W_5 \leftarrow g(W_4, \Delta L_5, \Delta F_5, \Delta L_6)$ {p13}
$W_6 \leftarrow g(W_5, \Delta L_6, \Delta F_6, L_7)$ {p14}
$W_7 \leftarrow g(W_6, L_7, F_7, L_8)$

$\vdots$

$W_{15} \leftarrow g(W_{14}, L_{15}, F_{15}, W_0)$
output: $W_0, \ldots, W_{15}$

Fig. 1. High-level differential used to look for (H, S)-type collisions

the system of constraints. The labels for all constraints are defined by Figure 1, we will refer to them throughout the text.

The process of finding the actual pair of inputs following the differential can be split into two phases. The first one is to solve the constraints from ProcessMessage to get the required Fs (same as Hs used in SaltState). Then, in the second phase, we look at the SaltState to find appropriate salts to have constraints in FeedForward satisfied. We can do this because the output from ProcessMessage has only a small effect on the solutions for FeedForward.

5.2 Solving the ProcessMessage

An important feature of our differentials in ProcessMessage is that it can be separated into two disjoint groups, i.e. $(F_0, F_1, F_2, L_1, L_2)$ and $(F_4, F_5, F_6, L_5, L_6)$. Differentials for these two groups have exactly the same structure. Thanks to that, if we can find values for the differences in the first group, we can reuse them for the second group by making corresponding registers in the second group equal to the ones from the first group. Following Observation 2 we can safely say that the second group also follows the differential path with a high probability. Algorithm 5 gives the details of solving the constrains in the first group of ProcessMessage.

1: Randomly pick $(L_2, L_2') \in S_{fa}$
2: **repeat**
3:    Randomly pick F_1, compute $F_1' = -1 - \Delta L_2 - F_1$
4: **until** $f_b(F_1) - f_b(F_1') \in S_{fb_{odd}}^A$
5: **repeat**
6:    Randomly pick L_1, F_2
7:    Compute $L_1' = f_b(F_1') - f_b(F_1) + L_1$
8:    Compute F_2' so that $f_b(F_2') = \Delta L_2 + f_a(L_1) - f_a(L_1') + f_b(F_2)$
9: **until** $p11$ is fulfilled
10: Pick $(F_0, F_0') \in S_{fb}$ so that $\Delta F_0 + \Delta L_1 = 0$

Algorithm 5. Find solutions for the first group of differences of ProcessMessage

Correctness. We show that after the execution of Algorithm 5, it indeed finds values conforming to the differential. In other words, we show that constraints $p1 - p4$ and $p9 - p11$ hold. Referring to Algorithm 5:

Line 1: (L_2, L_2') is chosen in such a way that $p4$ is satisfied.
Line 3: F_1' is computed in such a way that $(F_1 + L_2) \oplus (F_1' + L_2') = -1$
Line 4: $\Delta L_1 = \Delta f_b(F_1)$ is odd together with $(F_1 + L_2) \oplus (F_1' + L_2') = -1$. This implies that $p10$ could hold, which will be discussed later in Lemma 3. The fact that $\Delta L_1 \in S_{fb_{odd}}^A$ makes it possible that $p1$ and $p9$ hold.
Line 7: L_1' is computed in such a way that $p2$ holds.
Line 8: F_2' is computed in such a way that $p3$ holds.
Line 9: after exiting the loop $p11$ holds.
Line 10: (F_0, F_0') is chosen in such a way that $p1, p9$ hold.

Probability and Complexity Analysis. Let us consider the probability for exiting the loops in Algorithm 5. We require $f_a(F_1) - f_a(F_1') \in S_{fb_{odd}}^A$ and the constraint $p11$ to hold. The size of the set $S_{fb_{odd}}^A$ is around 2^{11}. By assuming that $f_a(F_1) - f_a(F_1')$ is random, the probability to have it in $S_{fb_{odd}}^A$ is 2^{-21}. This needs to be done only once. Now we show that the constraint $p11$ is satisfied with the probability 2^{-24}. We have sufficiently many choices, i.e. 2^{64}, for (L_1, F_2) to have p11 satisfied. The constraint p11 requires that $[(W_1 + L_2) \ggg 1] \oplus (F_2 + L_3) = [(W_1 + L_2')] \ggg 1] \oplus (F_2' + L_3)$, which is equivalent to $[(W_1 + L_2) \oplus (W_1 + L_2')] \ggg 1 = (F_2 + L_3) \oplus (F_2' + L_3)$, where $W_1, L_2, L_2', F_2, F_2'$ are given from previous steps. We have choices for L_3 by choosing an appropriate $M_{\sigma(3)}$. The problem could be rephrased as follows: *given random A and D, what is the probability to have at least one x such that $x \oplus (x + D) = A$?*

To answer this question, let us note first that $x \oplus y = (1, \ldots, 1)$ iff $x + y = -1$. This is clear as $y = \bar{x}$ and always $(x \oplus \bar{x}) + 1 = 0$. Now we can show the following result.

Lemma 3. *For any odd integer d, there exist exactly two x such that $x \oplus (x + d) = (1, \ldots, 1)$. They are given by $x = (-1 - d)/2$ and $x = (-1 - d)/2 + 2^{n-1}$.*

Proof. $x \oplus (x + d) = -1$ implies that $x + x + d = -1 + k2^n$ for an integer k, so $x = \frac{-1-d+k2^n}{2}$. Only when d is odd, $x = \frac{-1-d}{2} + k2^{n-1}$ an integer and a solution exists. As we are working in modulo 2^n, $k = 0, 1$ are the only solutions. $\square$

Following the lemma, given an odd ΔL_1 and $(F_1 + L_2) \oplus (F_1' + L_2') = -1$, we can always find two W_0 such that $(W_0 + L_1) \oplus (W_0 + L_1') = -1$, then p10 follows. Such W_0 could be found by choosing an appropriate L_{15}, which could be adjusted by choosing $M_{\sigma(15)}$ (if such $M_{\sigma(15)}$ does not exist, although the chance is low, we can adjust L_{14} by choosing $M_{\sigma(14)}$).

Coming back to the original question, consider A as "0"s and blocks of "1"s. Following the lemma above, for $A_i = 0$, we need $D_i = 0$ (except "0" as MSB followed by a "1"); for a block of "1"s, say $A_k = A_{k+1} = \cdots = A_{k+l} = 1$, the condition that needs to be imposed on D is $D_k = 1$. By counting the number of "0"s and the number of blocks of "1"s, we can get number of conditions needed. For an n-bit A, the number is $\frac{3n}{4}$ on average (cf. Appendix Lemma 4).

For LAKE-256, it is 24, so the probability for p11 to hold is 2^{-24}. We will need to find an appropriate L_3 so that p11 holds. Note that we have control over L_3 by choosing the appropriate $M_{\sigma(3)}$. For each differential path found, we need to find message words fulfilling the path. The probability to find a correct message is $1 - \frac{1}{e}$ for the first path by assuming f_c is random (because for a random function from n bits to n bits, the probability that a point from the range has a preimage is $1 - \frac{1}{e}$), and $\frac{4}{9}$ for the second path because of the carry effect. For example, given L_0, F_{15}, F_0, C_0, the probability to have $M_{\sigma(0)}$ so that $L_0 = f(F_{15}, F_0, M_{\sigma(0)}, C_0)$ is $1 - \frac{1}{e}$. The same $M_{\sigma(0)}$ satisfies $L_0' = f(F_{15}', F_0', M_{\sigma(0)}, C_0)$ (note for this case $F_{15}' = F_{15}$ and $L_0 = L_0'$) with the probability $\frac{4}{9}$. So for each message word, the probability for it to fulfill the differential path is 2^{-2}. We have such restrictions on $M_{\sigma(0)} - M_{\sigma(2)}, M_{\sigma(4)} - M_{\sigma(6)}$ (we don't have such restriction on $M_{\sigma(3)}$ and $M_{\sigma(7)}$ because we still have control over F_3 and F_7), so overall complexity for solving ProcessMessage is $5 \cdot 2^{36}$ in terms of calls to f_a or f_b. The compression function of LAKE-256 calls functions f and g 136 times each and f_a, f_b contain less than half of the operations used in f. So the complexity for this part of the attack is 2^{30} in terms of the number of calls to the compression function.

Solving the second group of ProcessMessage. After we are done with the first group, we can have the second group of differential path for free by assigning $F_{i+4} = F_i$, $F'_{i+4} = F'_i$ for $i = 0, 1, 2$ and $L_{i+4} = L_i$, $L'_{i+4} = L'_i$ for $i = 1, 2$. In this way, we can have the constrains $p5 - p8$ and $p12$ automatically satisfied. Similarly, for the constraints $p13$ and $p14$, we will need appropriate W_4 and L_7. We have control over W_4 by choosing F_3 and L_4 (note we need to keep L_3 stable to have $p11$ satisfied, this can be achieved by choosing appropriate $M_{\sigma(3)}$). We also have control over L_7 by choosing $M_{\sigma(7)}$. That way we can force the difference to vanish within the first ProcessMessage. Table 2 in Appendix shows an example of a set of solutions.

5.3   Near Collisions

In this section we explain how to get a near collision directly from collisions of ProcessMessage. Refer to SaltState and FeedForward in Fig. 1. Note that the function $g(a, b, c, d)$ with differences at positions (a, b) means $\Delta a + \Delta b = 0$, then constraints $(s1 - s6)$ in SaltState can be simplified to

$$s1 : \Delta H_0 + \Delta S_0 = 0; \tag{23}$$

$$s2 : \Delta H_1 + \Delta S_1 = 0; \tag{24}$$

$$s3 : \Delta H_2 + \Delta S_2 = 0. \tag{25}$$

Note that $H_{i+4} = H_i$, $H'_{i+4} = H'_i$ for $i = 0, 1, 2$ as required by ProcessMessage, Let $t_0 = t_1 = 0$, then conditions $s4 - s6$ follow $s1 - s3$. Conditions in FeedForward could be simplified to

$$f1 : f_{cd}(S_0, H_0) = f_{cd}(S'_0, H'_0), \tag{26}$$

$$f2 : f_{cd}(S_1, H_1) = f_{cd}(S'_1, H'_1), \tag{27}$$

$$f3 : f_{cd}(S_2, H_2) = f_{cd}(S'_2, H'_2) \tag{28}$$

and $f4 - f6$ follow $f1 - f3$. This set of constraints can be grouped into three independent sets (si, fi) for $i = 0, 1, 2$ each one of the same type, i.e. $\Delta H + \Delta S = 0$ and $f_{cd}(S, H) = f_{cd}(S', H')$.

To find near collisions, we proceed as follows. First we choose those S_i with $S'_i = S_i - \Delta H_i$ so that the Hamming weight of $f_{cd}(S'_i, H'_i) - f_{cd}(S_i, H_i)$ is small for $i = 0, 1, 2$. Thanks to that, only small differences are expected in the final output of the compression function, due to the fact that inputs from a, b of the function f have only carry effect to the final difference of f when inputs differ in c, d only. We choose values of S_i without going through the compression function, so the number of rounds of the compression function does not affect our algorithm. Further, the complexity for finding values of S_i is much smaller than that of ProcessMessage, so it does not increase the 2^{30} complexity. Experiments show that, based on the collision in ProcessMessage, we can have near collisions with a very little additional effort. Table 3 in Appendix shows a sample result with 16-bit of differences out of 256 bits of the output.

5.4   Extending the Attack to Full Collisions

It is clear that finding full collisions is equivalent to solving Equations (26)-(28). The complexity to solve a single equation is around 2^{12} (as done for solving Equations (10) and (11)). Looking at Algorithm 5, $(s1, f1)$ can be checked when F_1 and F_1' are chosen, so it does not affect the overall complexity. The pair $(s0, f0)$ can be checked immediately after (L_1, L_1') is given as show in Line 7 of Algorithm 5. Similarly, $(s2, f2)$ can be checked after (F_2, F_2') is chosen in Line 8. So the overall complexity for our algorithm to get a collision for the full compression function is 2^{54}.

5.5   Reducing the Complexity

In this subsection, we show a better way (rather than randomly) to choose (L_2, L_2') so that the probability for the constraint $p11$ to hold increases, which reduces the complexity for collision finding to 2^{42}.

Note the constraint $p11$ is as follows. Given W_1, L_2, L_2', what is the probability to have L_3 and (F_2, F_2') so that $((W_1 + L_2) \oplus (W_1 + L_2')) \ggg 1 = (F_2 + L_3) \oplus (F_2' + L_3)$. We calculate the probability by counting the number of 0s and block of 1s in $((W_1 + L_2) \oplus (W_1 + L_2')) \ggg 1$ (let's denote it as $\alpha = \#(((W_1 + L_2) \oplus (W_1 + L_2')) \ggg 1))$. Now we show that the number α can be reduced within the first loop of the algorithm, i.e. given only (L_2, L_2') and (F_1, F_1'), we are able to get α and hence, by repeating the loop sufficiently many times, we can reduce α to a number smaller than 24 (we don't fix it here, but will give it later).

Note that to find α, we still need W_1 besides (L_2, L_2'). Now we show W_1 can be computed from (L_2, L_2') and (F_1, F_1') only. $W_1 \stackrel{\text{def}}{=} ((W_0 + L_1) \ggg 1) \oplus (F_1 + L_2)$, where we restrict $(W_0 + L_1) \oplus (W_0 + L_1') = -1$. Denote $S = (W_0 + L_1)$, then the equation can be derived to $S \oplus (S + \Delta L_1) = -1$, where $\Delta L_1 \stackrel{\text{def}}{=} f_b(F_1') - f_b(F_1)$.

So let's make 2^y more effort in the first loop so that α is reduced by y. The probability for the first loop to exit becomes 2^{-33-y} and for the second loop, the probability becomes 2^{-60+y}. Choosing the optimal value $y = 13$ (y must be an integer), the probabilities are 2^{-46} and 2^{-47}, respectively. Hence this gives final complexity 2^{42} for collision searching.

6   (H)-Type Attack

Let us introduce difference only in the chaining value H_0. Hence, this difference after the SaltState procedure, will produce differences in F_0 and F_8. In the first application of the ProcessMessage procedure the following differential is used:

1. Let F_0 has some specially chosen difference. Also, F_8 has some difference that depends on the difference in F_0.
2. After the first application of the function f only $L_0, L_1, \ldots, L_8$ have non-zero differences
3. After the first application of the function g all W_i have zero differences

Again, we should prove that this differential is possible. Basically, we should check only for the updates with non-zero input differences and zero output difference (other updates hold trivially). Hence, we should prove that we can get the zero difference in L_9 and $W_i, i = 0, \ldots, 8$. Since f is non-injective, it is possible to get the zero difference in L_9. For $W_0, \ldots, W_8$ is also possible to get zero differences because their updating functions g always have at least two arguments with differences. Therefore, this differential is valid.

Now, let us write the system of equations that we require. Note that $L_i - L_i' = \delta_i, i = 0, \ldots, 8$. The system is as follows

$$f\,(F_{15}, L_0, M_0, C_0) = L_0, f(F_{15}, L_0', M_0, C_0) = L_0', \tag{29}$$
$$f\,(L_0, F_1, M_1, C_1) = L_1, f(L_0', F_1, M_1, C_1) = L_1', \tag{30}$$
$$f\,(L_{i-1}, F_i, M_i, C_i) = L_i, f(L_{i-1}', F_i, M_i, C_i) = L_i', i = 2, \ldots, 6, \tag{31}$$
$$f\,(L_7, L_8, M_8, C_9) = L_8, f(L_7, L_8', M_8, C_9) = L_8', \tag{32}$$
$$f\,(L_8, F_9, M_9, C_9) = L_9, f(L_8', F_9, M_9, C_9) = L_9, \tag{33}$$
$$g\,(L_{15}, L_0, F_0, L_1) = W_0, g(L_{15}, L_0', F_0', L_1') = W_0, \tag{34}$$
$$g\,(W_{i-1}, L_i, F_i, L_{i+1}) = W_i, g(W_{i-1}, L_i', F_i, L_{i+1}') = W_i, i = 1, \ldots, 7, \tag{35}$$
$$g\,(W_7, L_8, L_8, L_9) = g(W_7, L_8', L_8', L_9). \tag{36}$$

Let us focus on Equation (35). It can be rewritten as

$$(W_{i-1} + L_i) \ggg 1 \oplus (F_i + L_{i+1}) = (W_{i-1} + L_i') \ggg 1 \oplus (F_i + L_{i+1}') \quad (= W_i).$$

Similarly as in the previous attacks, we get the following equation

$$((X + A) \oplus X) \ggg 1 = (Y + B) \oplus Y, \tag{37}$$

where $X = W_{i-1} + L_i', A = L_i - L_i', Y = F_i + L_{i+1}', B = L_{i+1} - L_{i+1}'$. In **A13** of Appendix B, we have explained how to split this equation into two equations, $((X + A) \oplus X) = -1, (Y + B) \oplus Y = -1$, and solve them separately. The solution $X = \overline{A \gg 1}, Y = \overline{B \gg 1}$ exists when LSB of A and B are 1. Hence, for W_{i-1} and F_i we get

$$W_{i-1} = \overline{(L_i - L_i') \gg 1} - L_i' = \overline{\delta_i \gg 1} - L_i', \tag{38}$$
$$F_i = \overline{(L_{i+1} - L_{i+1}') \gg 1} - L_{i+1}' = \overline{\delta_{i+1} \gg 1} - L_{i+1}'. \tag{39}$$

If we put these values in the equation for W_i we obtain

$$W_i = (W_{i-1} + L_i') \ggg 1 \oplus (F_i + L_{i+1}') = \overline{\delta_i \gg 1} \ggg 1 \oplus \overline{\delta_{i+1} \gg 1}. \tag{40}$$

This means that we can split equations of the type (35) into two equations and solve them separately. Also, from (38) and (39) we get that $W_i = F_i$.

Now let us explain how to get two pairs that satisfy the whole differential. First, by choosing randomly $L_0, L_0', F_{15}, M_0, F_1$, and M_1, we produce a solution for Equations (29),(30), (34) and (35). Actually, we need to satisfy only Equation (35), i.e. $W_0 = \overline{(L_1 - L_1') \gg 1} - L_1' = \overline{\delta_1 \gg 1} - L_1'$, because the values of

$L_0^j, L_1^j, j = 1, 2$ can be any, and finding a solution for (34) is trivial. Then, by taking some M_2 and F_2 we produces $L_2^j = f(L_1^j, F_2, M_2, C_2), j = 1, 2$. Having the values of δ_1 and δ_2, we can find the new value of F_1

$$F_1 = W_1 = \overline{\delta_1 \gg 1} \ggg 1 \oplus \overline{\delta_2 \gg 1}.$$

Since we have changed the value of F_1, then the values of L_1 and L_1' might change. Therefore, we find another value of M_1 such that the old values of L_1, L_1' stay the same. Note, that is is not always possible. Yet, with the probability 2^{-2} this value can be found. As a result, we have fixed the values of M_1, F_1, L_2, and L_2'. Using the same technique, we can fix the values of $M_2, \ldots, M_6, F_2, F_6, L_3^j, L_7^j, j = 1, 2$ such that (35) would hold for $i = 2, \ldots, 6$. In short, the following is done. Let the values of W_{i-1}, M_i, F_i, L_i, and L_i' be fixed. First we generate any L_{i+1} and L_{i+1}'. Then we find the value of F_i from (39). Then, we change the value of M_i. This way, the values of L_i, L_i' stay the same, but now $W_{i+1}, L_i^j, M_i, F_i, L_{i+1}^j, j = 1, 2$ satisfy (35).

Now let us fix the right L_8, L_8' such that

$$f(L_8, F_9, M_9, C_9) = f(L_8', F_9, M_9, C_9). \tag{41}$$

We try different M_8, S_0 (notice that the values of F_8, F_8' depend on F_0, F_0', and S_0), and create different pairs (L_8, L_8'). If this pair satisfies (41) and (38) then we change M_7 and F_7 as described previously. Finally, we change M_9 and F_9 so that (36) will hold. First, we find the good value of L_9 from the equation $L_9 = \overline{\Delta_2 \gg 1} - L_8'$ and than change M_9 and F_9 to achieve this value. As a result, we have fixed all the values such that all equations hold.

After the ProcessMessage procedure, there are no differences in any of the state variables. The FeedForward procedure, which produces the new chaining value, depends on the initial chaining value, the internal state variables, the salt, and the block index. Since there is a difference only in the initial chaining value (only in H_0), it means that there has to be a difference in the new chaining variable H_0 (and only there). If we repeat the attack on ProcessMessage with different input difference Δ_1, we can produce a near collision with a low Hamming difference. If, in the truncated digest LAKE-224, the first 32 bits were truncated instead of the last 32 bits, we could find a real collisions for the compression function of LAKE-224.

Now, let us estimate the complexity of our attack. For finding good random $L_0, L_0', F_{15}, M_0, F_1$, and M_1 that satisfies the first set of equations we have to try 2^{32} different values. For successfully fixing the correct $F_i, M_i, i = 1, \ldots, 7$, we have to start with $(2^2)^7 = 2^{14}$ different δ_1. For finding a good pair (L_8, L_8') that satisfies (41) and (38) we have to try $2^{27} \cdot 2^{32} = 2^{59}$ different M_8, S_8. Hence, the total attack complexity is around 2^{105} computations. If we apply the same reasoning for computing the complexity in the number of compression function calls as it was done in the two previous attacks, we will get that the near collision algorithm requires around 2^{99} calls to the compression function of LAKE-256.

7   Conclusions

We presented three different collision attacks on the compression function of LAKE-256. All of them make use of some weaknesses of the functions used to build the compression function. The first two of them facilitate the additional variables of salt and block counter required by the HAIFA compression functions. Due to a weak mixing of those variables, we were able to better control diffusion of differences.

All our attacks cancel the injected differences within the first ProcessMessage and later only in the final FeedForward again and therefore are independent of the number of rounds.

The SHA-3 first round candidate BLAKE, a successor of LAKE, uses a different ProcessMessage function. Hence, our attacks do not apply to BLAKE. We believe that the efficient methods to solve the systems of equations and to find high level differentials presented in this paper may be useful to analyse other dedicated designs based on modular additions, rotations and XORs and constitute a nice illustration of how very small structural weaknesses can lead to attacks on complete designs.

Acknowledgments

This work is supported in part by Singapore National Research Foundation under Research Grant NRF-CRP2-2007-03, the Singapore Ministry of Education under Research Grant T206B2204, and the European Commission through the ICT programme under contract ICT-2007-216676 ECRYPT II. Praveen Gauravaram is supported by the Danish Research Council for Technology and Production Sciences grant number 274-08-0052. Dmitry Khovratovich is supported by PRP "Security & Trust" grand of the University of Luxembourg. Krystian Matusiewicz is supported by the Danish Research Council for Technology and Production Sciences grant number 274-07-0246. Ivica Nikolić is supported by the Fonds National de la Recherche Luxembourg grant TR-PHD-BFR07-031. Josef Pieprzyk is supported by Australian Research Council grants DP0663452 and DP0987734. We thank Wei Lei for helpful discussions and anonymous reviewers of ASIACRYPT 2008 and FSE 2009 for their useful comments.

References

1. Aumasson, J.-P., Meier, W., Phan, R.: The hash function family LAKE. In: Nyberg, K. (ed.) FSE 2008. LNCS, vol. 5086, pp. 36–53. Springer, Heidelberg (2008)
2. Biham, E., Dunkelman, O.: A framework for iterative hash functions – HAIFA. IACR ePrint Archive, Report 2007/278 (2007),
   http://eprint.iacr.org/2007/278
3. Biham, E., Dunkelman, O., Bouillaguet, C., Fouque, P.-A.: Re-visiting HAIFA and why you should visit too. In: ECRYPT workshop Hash functions in cryptology: theory and practice (June 2008),
   http://www.lorentzcenter.nl/lc/web/2008/309/presentations/
   Dunkelman.pdf (accessed on 11/23/2008)

4. Bosma, W., Cannon, J., Playoust, C.: The Magma algebra system I: The user language. Journal of Symbolic Computation 24(3-4), 235–265 (1997), http://magma.maths.usyd.edu.au/
5. Damgård, I.B.: A design principle for hash functions. In: Brassard, G. (ed.) CRYPTO 1989. LNCS, vol. 435, pp. 416–427. Springer, Heidelberg (1989)
6. Daum, M.: Cryptanalysis of Hash Functions of the MD4-Family. PhD thesis, Ruhr-Universität Bochum (May 2005)
7. Joux, A.: Multicollisions in iterated hash functions. Application to cascaded constructions. In: Franklin, M. (ed.) CRYPTO 2004. LNCS, vol. 3152, pp. 306–316. Springer, Heidelberg (2004)
8. Kelsey, J., Kohno, T.: Herding Hash Functions and the Nostradamus Attack. In: Vaudenay, S. (ed.) EUROCRYPT 2006. LNCS, vol. 4004, pp. 183–200. Springer, Heidelberg (2006)
9. Kelsey, J., Schneier, B.: Second preimages on n-bit hash functions for much less than 2^n work. In: Cramer, R. (ed.) EUROCRYPT 2005. LNCS, vol. 3494, pp. 474–490. Springer, Heidelberg (2005)
10. Matusiewicz, K., Peyrin, T., Billet, O., Contini, S., Pieprzyk, J.: Cryptanalysis of FORK-256. In: Biryukov, A. (ed.) FSE 2007. LNCS, vol. 4593, pp. 19–38. Springer, Heidelberg (2007)
11. Mendel, F., Schläffer, M.: Collisions for round-reduced LAKE. In: Mu, Y., Susilo, W., Seberry, J. (eds.) ACISP 2008. LNCS, vol. 5107, pp. 267–281. Springer, Heidelberg (2008)
12. Merkle, R.C.: One way hash functions and DES. In: Brassard, G. (ed.) CRYPTO 1989. LNCS, vol. 435, pp. 428–446. Springer, Heidelberg (1989)
13. National Institute of Standards and Technology. Secure hash standard (SHS). FIPS 180-2 (August 2002)
14. Paul, S., Preneel, B.: Solving systems of differential equations of addition. In: Boyd, C., Nieto, J.M.G. (eds.) ACISP 2005. LNCS, vol. 3574, pp. 75–88. Springer, Heidelberg (2005)
15. Wang, X., Yin, Y.L., Yu, H.: Collision search attacks on SHA-1 (Feburary 13, 2005), http://theory.csail.mit.edu/~yiqun/shanote.pdf
16. Wang, X., Yin, Y.L., Yu, H.: Finding collisions in the full SHA-1. In: Shoup, V. (ed.) CRYPTO 2005. LNCS, vol. 3621, pp. 17–36. Springer, Heidelberg (2005)
17. Wang, X., Yu, H.: How to break MD5 and other hash functions. In: Cramer, R. (ed.) EUROCRYPT 2005. LNCS, vol. 3494, pp. 19–35. Springer, Heidelberg (2005)

A   Collision Examples

Table 1. (H, t)-colliding pair for the compression function of LAKE

h_0	63809228 6cc286da 00000000 00000000 00000000 00000000 00000000 00000540
h'_0	ba3f5d77 6cc286da 00000000 00000000 00000000 00000000 00000000 00000540
M	55e07658 00000009 00000000 00000000 00000000 00000000 00000000 00000000
	00000000 00000000 00000000 00000000 00000000 00000000 00000002 5c41ab0e
F_0	0265e384 00000000
F_1	aba71835 00000000
S	00000000 00000000 00000000 00000000
H	79725351 e61a903f 730aace9 756be78a b679b09d de58951b f5162345 14113165

Table 2. Example of a pair of chaining values F, F' and a message block M that yield a collision in ProcessMessage

F	1E802CB8	799491C5	1FE58A14	07069BED	1E802CB8	799491C5	1FE58A14	74B26C5B
	00000000	00000000	00000000	00000000	00000000	00000000	00000000	00000000
F'	C0030007	B767CE5E	30485AE7	07069BED	C0030007	B767CE5E	30485AE7	74B26C5B
	00000000	00000000	00000000	00000000	00000000	00000000	00000000	00000000
M	683E64F1	9B0FC4D9	0E36999A	A9423F09	27C2895E	1B76972D	BEF24B1C	78F25F25
	00000000	00000000	00000000	00000000	00000000	00000000	657C34F5	3A992294
L	D0F3077A	31A06494	395A0001	10E105FC	82026885	31A06494	395A0001	10E105FC
	ECF7389A	2F4D466F	9FFC71E1	54BAFAE6	FCDDBCDB	E635FFB7	5D302719	CD102144
L'	D0F3077A	901D9145	95A99FDB	10E105FC	82026885	901D9145	95A99FDB	10E105FC
	ECF7389A	2F4D466F	9FFC71E1	54BAFAE6	FCDDBCDB	E635FFB7	5D302719	CD102144
$L^{\oplus}$	00000000	A1BDF5D1	ACF39FDA	00000000	00000000	A1BDF5D1	ACF39FDA	00000000
	00000000	00000000	00000000	00000000	00000000	00000000	00000000	00000000
W	1F210513	1A8E2515	1932829B	1C00C039	1F210513	1A8E2515	1932829B	F4A060BE
	5F868AC3	D8959978	E8F3FF4A	E20AC1C3	8941C0F8	EA8BC74E	6ECDD677	82CFFECE
W'	1F210513	1A8E2515	1932829B	1C00C039	1F210513	1A8E2515	1932829B	F4A060BE
	5F868AC3	D8959978	E8F3FF4A	E20AC1C3	8941C0F8	EA8BC74E	6ECDD677	82CFFECE

Table 3. Example of a pair of chaining values F, F', salts S, S' and a message block M that yield near collision in CompressionFunction with 16 bits differences out of 256 bits output. Hs are final output.

F	7B2000C4	23E79FBD	73D102C3	88E0E02B	7B2000C4	23E79FBD	73D102C3	00000000
F'	801FF801	18C0005E	846FD480	88E0E02B	801FF801	18C0005E	846FD480	00000000
S	00010081	23043423	03C5B03E	D44CFD2C				
S'	FB010944	2E2BD382	F326DE81	D44CFD2C				
M	00000012	64B31375	CFA0A77E	8F7BE61F	1E30C9D3	6A9FB0DA	290E506E	3AAE159C
	00000000	00000000	00000000	00000000	00000000	00000000	00000000	1B89AA75
H	261B50AA	3873E2BE	BDD7EC4D	7CE4BFF8	007BB4D4	869473FF	833D9EFA	9DABEDDA
H'	361150AA	387BE23E	FDD6E84D	7CE4BFF8	1071B4D4	869C737F	C33C9AFA	9DABEDDA
$H^{\oplus}$	100A0000	00080080	40010400	00000000	100A0000	00080080	40010400	00000000

B   Lemmas and Proofs

Lemma 4. *Given random x of length n, then the average number of "0"s and block of "1"s, excluding the case "0" as MSB followed by "1", is $\frac{3n}{4}$.*

Proof. Denote C_n as the sum of the counts for "0"s and blocks of "1"s for all x of length n, denote such x as x_n. Similarly we define P_n as the sum of the counts for all x of length n with MSB "0" (let's denote such x as x_n^0); and Q_n for the sum of the counts for all x of length n with MSB "1" (denote such x as x_n). It is clearly that

$$C_n = P_n + Q_n \tag{42}$$

Note that there are 2^{n-1} many x with length $n-1$, half of them with MSB "0", which contribute to P_{n-1} and the other half with MSB "1", which contribute to Q_{n-1}. Now we construct x_n of length n from x_{n-1} of length $n-1$ in the following way:

– Append "0" with each x_{n-1}, this "0" contribute to C_n once for each x_{n-1} and there are 2^{n-2} many such x_{n-1}.

- Append "1" with each x_{n-1}, this "1" does not contribute to C_n
- Append "0" with each x^0_{n-1}, this contributes 2^{n-2} to C_n
- Append "1" with each x^0_{n-1}, this contributes 2^{n-2} to C_n

So overall we have $C_n = P_{n-1} + P_{n-1} + 2^{n-2} + Q_{n-1} + 2^{n-2} + Q_{n-1} + 2^{n-2} = 3 \cdot 2^{n-2} + 2C_{n-1}$. Note $C_1 = 2$, solving the recursion, we get $C_n = \frac{3n+1}{4} \cdot 2^n$. Exclude the exceptional case, we have final result $\frac{3n}{4}$ on average.

Lemma 5. *Given random $a, a', x \in \mathbb{Z}_{2^n}$ and $k \in [0, n)$, $\alpha \overset{def}{=} 1[a^L_k + x^L_k \geq 2^k]$, $\alpha' \overset{def}{=} 1[a'^L_k + x^L_k \geq 2^k]$, $\beta \overset{def}{=} 1[a^R_k + x^R_k + \alpha \geq 2^{n-k}]$, $\beta' \overset{def}{=} 1[a'^R_k + x^R_k + \alpha \geq 2^{n-k}]$ as defined in Lemma 1, then $P(\alpha = \alpha', \beta = \beta') = \frac{4}{9}$.*

Proof. Consider α and α' first, $P(\alpha = \alpha' = 1) = P(a^L_k + x^L_k \geq 2^k, a'^L_k + x^L_k \geq 2^k)$. This is equal to $P(x^L_k \geq (2^k - min\{a^L_k, a'^L_k\}))$ what in turns can be rewritten as $P(a^L_k \geq a'^L_k)P(x^L_k \geq 2^k - a'^L_k) + P(a'^L_k > a^L_k)P(x^L_k \geq 2^k - a^L_k) = \frac{1}{2} \cdot \frac{1}{3} + \frac{1}{2} \cdot \frac{1}{3} = \frac{1}{3}$.

Similarly we can prove $P(\alpha = \alpha' = 0) = \frac{1}{3}$, so $P(\alpha = \alpha') = \frac{2}{3}$. Note the definitions of β and β' contain α and α', but $\alpha, \alpha' \in \{0, 1\}$, which is generally much smaller than 2^{n-k}, so the effect of α to β is negligible. We can roughly say $P(\beta = \beta') = \frac{2}{3}$. So $P(\alpha = \alpha', \beta = \beta') = P(\alpha = \alpha')P(\beta = \beta') = \frac{4}{9}$.

Lemma 6. *There exist an algorithm (Al1) for finding all the solutions for the equation of the form $(X \wedge C) + A = (X + B) \wedge C$. The complexity of Al1 depends only on the constant C.*

Lemma 7. *There exist an algorithm (Al2) for finding all the solutions for the equation of the form $(X \vee C) + A = (X + B) \vee C$. The complexity of Al2 depends only on the constant C.*

Proof. The proofs for the two facts are very similar with some minor changes, so we will prove only Lemma 6.

Let $X = x_{31} \ldots x_1 x_0, A = a_{31} \ldots a_1 a_0, B = b_{31} \ldots b_1 b_0, C = c_{31} \ldots c_1 c_0$. Then for each i we have:

$$(x_i \wedge c_i) \oplus a_i \oplus F_i = (x_i \oplus b_i \oplus r_i) \wedge c_i, \qquad (43)$$

where $F_i = m(x_{i-1} \wedge c_{i-1}, a_{i-1}, F_{i-1})$ is the carry at the $(i-1)$th position of $(X \wedge C + A)$, $r_i = m(x_{i-1}, b_{i-1}, r_{i-1})$ is the carry at the $(i-1)$th position of $X + B$, and $m(x, y, z) = xy \oplus xz \oplus yz$.

Equation (43), simplifies to $a_i \oplus F_i = 0$ when $c_i = 0$ and when $c_i = 1$ we get $a_i \oplus F_i = b_i \oplus r_i$.

Let us assume that we have found the values for F_i and r_i for some i. We find the smallest $j > 0$ such that $c_{i+j} = 0$. Then from the fact that $a_i \oplus F_i = 0$ and the definition of F_i we get:

$$a_{i+j} = F_{i+j} = m(x_{i+j-1}, a_{i+j-1}, F_{i+j-1}) =$$
$$= m(x_{i+j-1}, a_{i+j-1}, m(x_{i+j-2}, a_{i+j-2}, F_{i+j-2})) = \ldots$$
$$= m(x_{i+j-1}, a_{i+j-1}, m(x_{i+j-2}, a_{i+j-2}, m(\ldots, m(x_i, a_i, F_i)) \ldots))$$

In the above equation, only $x_i, x_{i+1}, \ldots x_{i+j-1}$ are unknown. So we can try all the possibilities, which are 2^j, and find all the solutions. Let us denote by $\tilde{X}$ the set of all solutions.

Now, let us find the smallest $l > 0$ such that $c_{i+j+l} = 1$. Notice that we can easily find F_{i+j+1} if considering $c_{i+j+F_0} = 0$ for $F_0 \in (0, l)$ and using $a_i \oplus F_i = 0$:

$$F_{i+j+1} = m(0, a_{i+j}, F_{i+j}) = m(0, a_{i+j}, a_{i+j}) = a_{i+j}$$
$$F_{i+j+2} = m(0, a_{i+j+1}, F_{i+j+1}) = m(0, F_{i+j+1}, F_{i+j+1}) = m(0, a_{i+j}, a_{i+j}) = a_{i+j}$$

$$\cdots$$

$$F_{i+j+l} = m(0, a_{i+j+l-1}, F_{i+j+l-1}) = a_{i+j}$$

From the relationship $a_i \oplus F_i = b_i \oplus r_i$ and definition of r_i we get:

$$a_{i+j+l} \oplus F_{i+j+l} \oplus b_{i+j+l} = r_{i+j+l} = m(x_{i+j+l-1}, b_{i+j+l-1}, r_{i+j+l-1}) =$$
$$= m(x_{i+j+l-1}, b_{i+j+l-1}, m(x_{i+j+l-2}, b_{i+j+l-2}, r_{i+j+l-2})) = \cdots$$
$$= m(x_{i+j+l-1}, b_{i+j+l-1}, m(\ldots, m(x_i, b_i, r_i) \ldots))$$

In the above equation, only $x_i, x_{i+1}, \ldots, x_{i+j+l-1}$ are unknown. So we check all the possibilities by taking $(x_i, x_{i+1}, \ldots, x_{i+j-1})$ from the set $\tilde{X}$ and the rest of the variables take all the possible values. If the equation has a solution, then this means we have fixed another F_{i+j+l}, r_{i+j+l}, and we can continue searching using the same algorithm.

The complexity of the algorithms is 2^q, where q is size of the longest consecutive sequence of ones followed by consecutive zero sequence (in the case above $q = j + l$) in the constant C. Taking into consideration the value of the constant C_1 used in the compression function of LAKE-256, we get that complexity of our algorithm for this special case is 2^8. Yet, the average complexity can be decreased additionally if first the necessary conditions are checked. For example, if we have two consecutive zeros in the constant C_1 at positions i and $i+1$ then it has to hold $a_{i+1} = a_i$. If we check for all zeros, then only with probability of 2^{-10} a constant A can pass this sieve. Therefore, the math expectancy of the complexity for a random A is less than 2^1. Note that when $\vee$ function is used instead of $\wedge$, than 0 and 1 change place. Therefore, our algorithm has a complexity of 2^6 when C_0 is used as a constant. Yet, same as for $\wedge$, early break-up strategies significantly decrease these complexities for the case when solution does not exist. Again, the average complexity is less than 2^1.   □

Lemma 8. *There exist an algorithm (Al3) for finding a solution for the following equation:* $((X + A) \oplus X) \gg 1 = (Y + B) \oplus Y$.

Proof. Instead of finding a solution w.r.t. X and Y we split the equation into a system

$$(X + A) \oplus X = -1, \quad (Y + B) \oplus Y = -1 . \tag{44}$$

We can do this because the value of -1 is invariant of any rotation. We may loose some solutions, but further we will prove that if such a solution exist then our algorithm will find it with probability 2^{-2}.

We will analyze only left equation of (44); the second one can be solved analogously. Let $X = x_{31} \ldots x_0, A = a_{31} \ldots a_0$. Then for ith bit we get: $(x_i \oplus a_i \oplus c_i) \oplus x_i = 1$, where c_i is the carry at $(i - 1)$ position of $X + A$, i.e. $c_i = m(x_{i-1}, a_{i-1}, c_{i-1})$. Obviously, this equation can be rewritten as $a_i = c_i \oplus 1$. For the $(i+1)$th bit we get $a_{i+1} = c_{i+1} \oplus 1 = m(x_i, a_i, c_i) \oplus 1 = m(x_i, a_i, a_i \oplus 1) \oplus 1 = x_i a_i \oplus x_i(a_i \oplus 1) \oplus a_i(a_i \oplus 1) \oplus 1 = x_i \oplus 1$. So, we can easily find the value of x_i for each i. When $i = 31$, x_{31} can be arbitrary. For the case when $i = 0$, considering that $c_0 = 0$, from $a_i = c_i \oplus 1$ we get $a_0 = 1$. Therefore, if $a_0 = 1$ then (44) is solvable in constant time. The solutions are $X = \overline{A \ggg 1} + i2^{32}, i = 0, 1$. Finally, for the whole system, we have that solution exist if $a_0 = b_0 = 1$, which means with probability 2^{-2}. $\qquad\square$

Lemma 9. *There exists an algorithm (Al4) for finding all the solutions for equations of the type* $(X \oplus C) + A = (X + B) \oplus K$.

Proof. We base our algorithm fully on the results of [14]. There, Paul and Preneel show, in particular, how to solve equations of the form: $(x + y) \oplus ((x \oplus \alpha) + (y \oplus \beta)) = \gamma$. Let us XOR to the both sides of the initial equation the expression $A \oplus B \oplus C$ and denote $\tilde{K} = K \oplus A \oplus B \oplus C$. Then, the equation gets the following form: $((X \oplus C) + A) \oplus A \oplus B \oplus C = (X + B) \oplus \tilde{K}$. For the $(i+1)$th bit position, we have $\tilde{k}_{i+1} = s_{i+1} \oplus F_{i+1}$, where s_i is the carry at the ith position of $(X \oplus C) + A$, and F_i is the carry at ith position of $X + B$. From the definition of s_i we get $s_{i+1} = (x_i \oplus c_i)a_i \oplus (x_i \oplus c_i)s_i \oplus a_i s_i = (x_i \oplus c_i)a_i \oplus (x_i \oplus c_i \oplus a_i)s_i = (x_i \oplus c_i)a_i \oplus (x_i \oplus c_i \oplus a_i)(\tilde{k}_i \oplus F_i)$.

From the definition of F_i we get $F_{i+1} = x_i b_i \oplus x_i F_i \oplus b_i F_i$. This means that $\tilde{k}_{i+1}$ can be computed from x_i, a_i, b_i, c_i, F_i, and $\tilde{k}_i$. Further, we apply the algorithm demonstrated in [14]. The only difference is that for each bit position we have only two unknowns x_i and F_i, whereas in [14] have three unknowns. Yet, this difference is not crucial, and the algorithm can be applied.

Our experimental results (Monte-Carlo with 2^{32} trials), show that the probability that a solution exists, when A, B, C and K are randomly chosen is around 2^{-12}. $\qquad\square$

New Cryptanalysis of Block Ciphers with Low Algebraic Degree

Bing Sun[1], Longjiang Qu[1], and Chao Li[1,2]

[1] Department of Mathematics and System Science, Science College of National
University of Defense Technology, Changsha, China, 410073
[2] State Key Laboratory of Information Security, Graduate University of Chinese
Academy of Sciences, China, 10039
happy_come@163.com, ljqu_happy@hotmail.com

Abstract. Improved interpolation attack and new integral attack are
proposed in this paper, and they can be applied to block ciphers using
round functions with low algebraic degree. In the new attacks, we can
determine not only the degree of the polynomial, but also coefficients of
some special terms. Thus instead of guessing the round keys one by one,
we can get the round keys by solving some algebraic equations over finite
field. The new methods are applied to $\mathcal{PURE}$ block cipher successfully.
The improved interpolation attacks can recover the first round key of
8-round $\mathcal{PURE}$ in less than a second; r-round $\mathcal{PURE}$ with $r \leq 21$ is
breakable with about 3^{r-2} chosen plaintexts and the time complexity is
3^{r-2} encryptions; 22-round $\mathcal{PURE}$ is breakable with both data and time
complexities being about 3×3^{20}. The new integral attacks can break
$\mathcal{PURE}$ with rounds up to 21 with 2^{32} encryptions and 22-round with
3×2^{32} encryptions. This means that $\mathcal{PURE}$ with up to 22 rounds is
breakable on a personal computer.

Keywords: block cipher, Feistel cipher, interpolation attack, integral
attack.

1 Introduction

For some ciphers, the round function can be described either by a low degree
polynomial or by a quotient of two low degree polynomials over finite field with
characteristic 2. These ciphers are breakable by using the interpolation attack,
which was first introduced by Jakobsen and Knudsen at FSE'97[2]. This attack
was generalized by K. Aoki at SAC'99[3], which is called the linear sum attack, and a method was presented that can efficiently evaluate the security of
byte-oriented ciphers against interpolation attack. In [4], the authors pointed
some mistakes in [2], and introduced a new method, root finding interpolation
attack, to efficiently find all the equivalent keys of the cipher, and this attack
can decrease the complexity of interpolation attack dramatically. To apply the
interpolation attack, a finite field should be constructed first, in [5], the effect
of the choice of the irreducible polynomial used to construct the finite field was
studied and an explicit relation between the Lagrange interpolation formula and
the Galois Field Fourier Transform was presented.

O. Dunkelman (Ed.): FSE 2009, LNCS 5665, pp. 180–192, 2009.

Interpolation attack can be applied to some ciphers which have provable securities against differential and linear cryptanalysis[15,16]. For example, in [2], a provable secure block cipher $\mathcal{PURE}$ was introduced, however, it can be broken by using interpolation attack. Later, interpolation attack was successfully applied to some simplified version of SNAKE[17,18]. However, the complexity of interpolation attack on 6-round $\mathcal{PURE}$ is 2^{36}, and it will increase when the round of the cipher becomes 7,8 and so on. In another word, it is not a real-world attack.

Integral cryptanalysis[7,8] considers the propagation of sums of (many) values. Thus it can be seen as a dual to differential cryptanalysis which considers the propagation of sums of only two values. It was first proposed in [6] but under a different name, that is square attack. A number of these ideas have been exploited, such as square attack[19,20], saturation attack[9], multiset attack[12,10], and higher order differential attack[11,13]. Integrals have a number of interesting features. They are especially well-suited to analysis of ciphers with primarily bijective components. Moreover, they exploit the simultaneous relationship between many encryptions, in contrast to differential cryptanalysis where one considers only pairs of encryptions. Consequently, integrals apply to a number of ciphers not vulnerable to differential and linear cryptanalysis. These features have made integrals an increasingly popular tool in recent cryptanalysis work.

Integral attacks are well-known to be effective against byte-oriented block ciphers. In [14], the authors outlined how to launch integral attacks against bit-based block ciphers. The new type of integral attack traces the propagation of the plaintext structure at bit-level by incorporating bit-pattern based notations. The new integral attack is applied to Noekeon, Serpent and Present reduced up to 5, 6 and 7 rounds, respectively.

In this paper, by using an algebraic method, an improved interpolation attack and a new integral attack are proposed. The complexity of interpolation attack can be decreased dramatically which leads to a real-world attack against $\mathcal{PURE}$ with up to 22 rounds. There are two improvements in this paper. The first one is an improvement of the original interpolation attack. Instead of guessing the last round key one by one, we find some algebraic equations that can efficiently find the round key. Another one is an extended integral cryptanalysis and it is somewhat like the square attack. In a square attack, value of $\sum_x f(x)$ is computed. And in our attack, value of $\sum_x x^i f(x)$ for some integer i is computed and this value can be either a constant or strongly related with only a few round-keys. Thus instead of guessing the last round key one by one, we can get the round keys by solving some algebraic equation $f_C(K) = 0$ over finite field, where C is an arbitrarily chosen constant.

The paper is organized as follows: Feistel Structure and basic attacks are presented in section 2. In section 3, we introduce the basic mathematical foundations that can efficiently improve the attacks. And the improved interpolation attack is presented in section 4. Then, in section 5, new integral cryptanalysis is presented. Results of attack against $\mathcal{PURE}$ are given in section 6. Section 7 makes the conclusion of this paper.

2   Feistel Structure and Basic Attacks

2.1   Feistel Structure

A Feistel network consists of r rounds, each of which is defined as follows. Denote by (L, R) the $2n$-bit input, set $\alpha_0 = L$ and $\beta_0 = R$, let $(\alpha_{i-1}, \beta_{i-1})$ be the input to the ith round, (α_i, β_i) and k_i be the output and the round key of the ith round, respectively. Then $(\alpha_i, \beta_i) = Round(\alpha_{i-1}, \beta_{i-1})$ is defined as:

$$\begin{cases} \alpha_i = \beta_{i-1}, \\ \beta_i = f(\beta_{i-1}, k_i) \oplus \alpha_{i-1}, \end{cases}$$

where f is the round function and in this paper, we always assume that $f(\beta_{i-1}, k_i) = f(\beta_{i-1} \oplus k_i)$. See Fig.1. After iterating $Round$ r times, the ciphertext (C_L, C_R) is defined as (β_r, α_r).

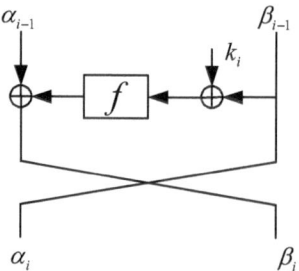

Fig. 1. Feistel Structure

2.2   Interpolation Attack on Block Ciphers

Let F be a field. Given $2t$ elements $x_1, \ldots, x_t, y_1, \ldots, y_t \in F$, where the x_is are distinct. According to Lagrange interpolation formula,

$$f(x) = \sum_{i=1}^{t} y_i \prod_{1 \le j \le t, j \ne i} \frac{x - x_j}{x_i - x_j}$$

is the only polynomial over F with degree at most $t - 1$ such that $f(x_i) = y_i$ for $i = 1, \ldots, t$.

In an interpolation attack to an r-round Feistel cipher, we construct polynomials by using pairs of plaintexts and ciphertexts. The attacker first computes the degree of the output of $(r - 1)$-th round, say N. Then he chooses $N + 2$ plaintexts P_i and encrypts them, denote by C_i the corresponding ciphertexts. By guessing the last round key k^*, the attacker partially decrypts C_i one round back and gets D_i. Now, he uses (P_i, D_i) for $1 \le i \le N + 1$ and applies the Lagrange interpolation formula to get the only polynomial $h(x)$ with degree at most N such that $h(P_i) = D_i (1 \le i \le N + 1)$. If $h(P_{N+2}) = D_{N+2}$, then put k^* as a candidate of the right key, otherwise, k^* is rejected. This process is repeated until the k^* is uniquely determined.

Assume k^* is an n-bit word, then the complexity of the interpolation attack is at least $(N + 2) \times 2^n$, since to get the ciphertexts, it needs $N + 2$ encryptions and 2^n partially decryptions for each ciphertext.

2.3 Integral Cryptanalysis

Let $(G, +)$ be a finite group and S be a subgroup of G. An integral over S is defined as the sum of all elements in S. That is,

$$\int S = \sum_{v \in S} v,$$

where the summation is defined in terms of the group operation for G.

In an integral attack, one tries to predict the values in the integral after a certain number of rounds. To be more exact, assume the input is x, and part or all of the output is $c(x)$, by computing $\sum_{x \in S} c(x)$, where S always denotes the finite field $\mathbb{F}_{2^t}$ for some integer t, one can distinguish the cipher from a random permutation. For example, in square attack, one adopts $\sum_{x \in S} c(x) = 0$ to efficiently find the round keys of a given cipher. But, if $\sum_{x \in S} c(x) = 0$, and let $h(x)$ be a nonlinear transformation, can we predict the value of $\sum_{x \in S} h(c(x))$? It seems that this is a difficult question if we cannot analyze h carefully.

Besides, most of the known integrals have the following form

$$\int (S, c) = \sum_{x \in S} c(x),$$

where x denotes the plaintext and c is the map from plaintext to ciphertext. However, in this paper, a new integral

$$\int (S, c, i) = \sum_{x \in S} x^i c(x)$$

for some integer i is proposed. This definition will facilitate our discussions in cryptanalysis.

3 Mathematical Foundation

3.1 Notations

The following notations will be used in this paper:

m : degree of the round function
r : rounds of the cipher
$2n$: size of the plaintext/ciphertext
r_0 : $\lfloor \log_m (2^n - 1) \rfloor + 1$, the largest integer $\le \log_m (2^n - 1) + 1$
$\deg(f)$: degree of a polynomial f

To simplify the discussion, let the leading coefficient of $f(x)$ be 1:

$$f(x) = x^m \oplus \sum_{i=0}^{m-1} a_i x^i \in \mathbb{F}_{2^n}[x].$$

If $m = 1$ or $m = 2$, $f(x)$ is an affine function, thus we always assume $m \geq 3$.

3.2 Algebraic Analysis of Outputs of Feistel Cipher

By interpolation, an encryption algorithm can be seen as a polynomial function with the plaintext/ciphertext as its input/output. Thus, properties of this polynomial can be studied in order to get the information of the keys. If the round function has a low algebraic degree, then, the degree and some coefficients of special terms of the polynomial function between plaintexts and ciphertexts can be computed exactly.

Proposition 1. *Let $P = (C, x)$ be the input to an r-round Feistel cipher, where $C \in \mathbb{F}_{2^n}$ is a constant, $(\alpha_t, \beta_t) = (\alpha_t(x), \beta_t(x))$ be the output of the t-th round, if $1 \leq t \leq r - 1$ and $m^{t-1} \leq 2^n - 1$, then*

$$\begin{cases} \deg \alpha_t = m^{t-1}, \\ \deg \beta_t = m^t, \end{cases}$$

where m is the degree of the round function. Furthermore, the leading coefficients of both $\alpha_t(x)$ and $\beta_t(x)$ are 1.

Proof. We can prove this proposition by induction.

If the input to the cipher is of the form $(\alpha_0, \beta_0) = (C, x)$ where C is a constant, then after the first round, $(\alpha_1, \beta_1) = (x, C \oplus f(x \oplus k_1))$. Therefore $\deg \alpha_1 = 1$, $\deg \beta_1 = \deg f = m$.

Assume $\deg \alpha_t = m^{t-1}, \deg \beta_t = m^t$, then

$$(\alpha_{t+1}, \beta_{t+1}) = (\beta_t, \alpha_t \oplus f(\beta_t \oplus k_t)),$$

thus $\deg \alpha_{t+1} = \deg \beta_t = m^t, \deg \beta_{t+1} = \deg \beta_t \times \deg f = m^{t+1}$.   □

According to Proposition 1, (α_t, β_t) can be written in the following form:

$$(\alpha_t, \beta_t) = \left(x^{m^{t-1}} \oplus g_{t-1}(x), x^{m^t} \oplus g_t(x) \right), \tag{1}$$

where $g_i(x)$ is a polynomial with degree $< m^i$.

Proposition 1 determines the degree and leading coefficients of $\alpha_t(x)$ and $\beta_t(x)$. Now let's compute the coefficient of the term x^{m^t-1} in β_t, or equivalently, the leading coefficient of $g_t(x)$. This coefficient plays a very important role in the improvement of our new attacks. By induction, the following Proposition holds:

Proposition 2. *Assume $g_t(x) = \sum_{i=0}^{m^t-1} v_i x^i \in \mathbb{F}_{2^n}[x]$ is a polynomial defined as in (1), $m \equiv 1 \mod 2$ and $t \leq r_0 - 2$, then*

$$v_{m^t-1} = k_1 \oplus a_{m-1}.$$

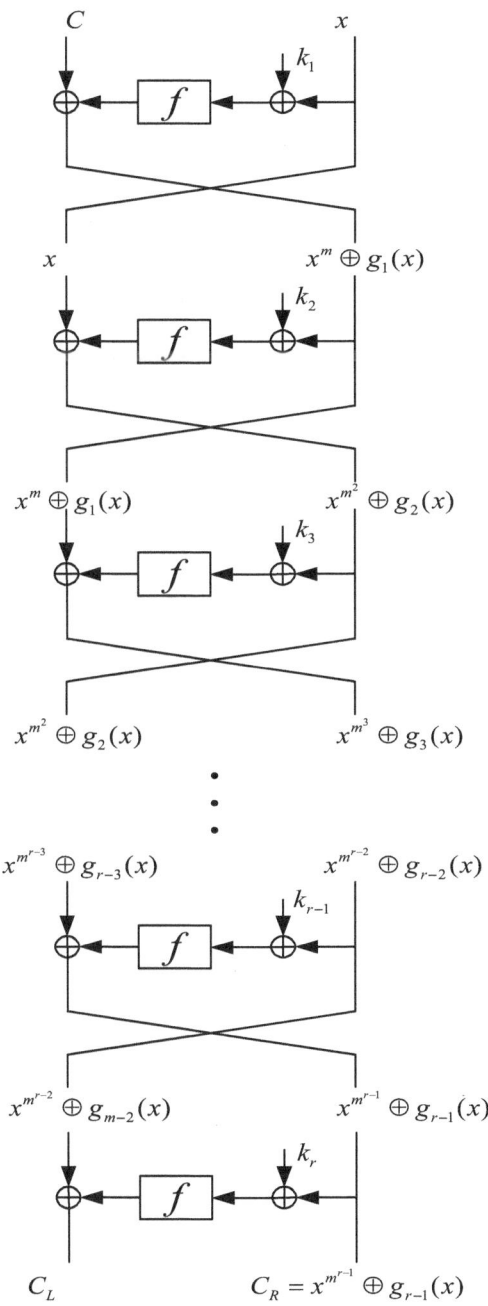

Fig. 2. Degrees of the output of each round

Proof. By computation, when $t = 1$:

$$x^m \oplus g_1(x) = C \oplus f(x \oplus k_1)$$
$$= (x \oplus k_1)^m \oplus a_{m-1}(x \oplus k_1)^{m-1} \cdots$$
$$= x^m \oplus (k_1 \oplus a_{m-1})x^{m-1} \oplus \cdots$$

Thus it is true for $t = 1$.

Assume $v_{m^t-1} = k_1 \oplus a_{m-1}$, then

$$x^{m^{t+1}} \oplus g_{t+1}(x) = \alpha_t(x) \oplus f(\beta_t \oplus k_t)$$
$$= x^{m^{t-1}} \oplus g_{t-1}(x) \oplus f(x^{m^t} \oplus g_t(x) \oplus k_t)$$
$$= (x^{m^t} \oplus g_t(x) \oplus k_t)^m \oplus a_{m-1}(x^{m^t} \oplus g_t \oplus k_t)^{m-1} \oplus \cdots$$
$$= (x^{m^t} \oplus v_{m^t-1}x^{m^t-1} \oplus \cdots)^m \oplus \cdots$$
$$= x^{m^{t+1}} \oplus (m \times v_{m^t-1})x^{m^{t+1}-1} \oplus \cdots$$

Thus $v_{m^{t+1}-1} = m \times v_{m^t-1} = k_1 \oplus a_{m-1}$, which ends our proof. $\square$

4   Improved Interpolation Attack on Feistel Ciphers

4.1   Basic Properties of the Output of A Feistel Cipher

According to Proposition 1 and 2, we can determine not only the degree of the polynomial, but also coefficients of some special terms.

Theorem 1. *For an r-round 2n-bit Feistel cipher, let the algebraic degree of the round function be an odd integer m, $r_0 = \lfloor \log_m (2^n - 1) \rfloor + 1$ and $r \leq r_0$. If the input to the cipher is of the form $P = (C, x)$ where $C \in \mathbb{F}_{2^n}$ is a constant, then the right half of the ciphertext is of the form $C_R(x) = x^{m^{r-1}} \oplus (k_1 \oplus a_{m-1})x^{m^{r-1}-1} \oplus q(x)$ where $q(x) \in \mathbb{F}_{2^n}[x]$ is a polynomial with degree $< m^{r-1} - 1$.*

Similar with Theorem 1, we can get the explicit expression of the output of an $(r_0 + 1)$-round Feistel cipher:

Theorem 2. *Let $r_0 = \lfloor \log_m (2^n - 1) \rfloor + 1$ and $r = r_0 + 1$, then for an r-round 2n-bit Feistel cipher with the algebraic degree of round function being an odd integer m, if the input to the cipher is of the form $P = (x, C)$ where $C \in \mathbb{F}_{2^n}$ is a constant, then the right half of the ciphertext is of the form $C_R(x) = x^{m^{r-2}} \oplus (f(k_1 \oplus C) \oplus k_2 \oplus a_{m-1})x^{m^{r-2}-1} \oplus p(x)$ where $p(x) \in \mathbb{F}_{2^n}[x]$ is a polynomial with degree $< m^{r-2} - 1$.*

The above two Theorems have already been used in the original interpolation attack on $\mathcal{PURE}$, however, we use them in a different manner.

To improve the interpolation attack on Feistel ciphers with low algebraic degree, we always assume that the degree of the round function is an odd integer, that is $m \equiv 1 \mod 2$.

4.2 Improved Attack

In an interpolation attack, the attacker needs to guess the last round key, thus the complexity of the attack is at least $(N+2) \times 2^n$. In our improved interpolation attack, we can compute the first round key k_1 by only using the plaintexts and corresponding ciphertexts.

Feistel cipher with r round can be broken by the following attack:

Algorithm 1: Attack on Block Ciphers with $r \leq r_0(\mathbf{I})$

Step 1: Encrypt $P = (C, x)$ for $m^{r-1} + 1$ different $x \in \mathbb{F}_{2^n}$ where $C \in \mathbb{F}_{2^n}$ is a constant. The corresponding ciphertexts are $(C_L(x), C_R(x))$;

Step 2: Compute $g(x) = ax^{m^{r-1}} \oplus sx^{m^{r-1}-1} \oplus \cdots \in \mathbb{F}_{2^n}[x]$ by interpolation such that $g(x) = C_R(x)$. According to Theorem 1, $k_1 = s \oplus a_{m-1}$.

Algorithm 1 needs $m^{r-1}+1$ encryptions, and to compute the interpolation polynomial, it needs $2 \times (m^{r-1} + 1)$ word-memories to store (P_i, C_i). It is infeasible to mount a real-world attack when m^{r-1} is too large that a computer cannot store so many plaintexts/ciphertexts.

Algorithm 2 finds the first and second round keys by solving some algebraic equations over finite field instead of guessing the keys one by one.

Algorithm 2: Attack on Block Ciphers with $r \leq r_0 + 1(\mathbf{I})$

Step 1: Encrypt $P^{(1)} = (x, C_1)$ for $m^{r-2}+1$ different $x \in \mathbb{F}_{2^n}$ where $C_1 \in \mathbb{F}_{2^n}$ is a constant. The corresponding ciphertexts are $(C_L^{(1)}(x), C_R^{(1)}(x))$;

Step 2: Compute $g(x) = ax^{m^{r-2}} \oplus s_1 x^{m^{r-2}-1} \oplus \cdots \in \mathbb{F}_{2^n}[x]$ by interpolation such that $g(x) = C_R^{(1)}(x)$; thus $s_1 = f(k_1 \oplus C_1) \oplus k_2 \oplus a_{m-1}$;

Step 3: Choose another two constants C_2 and C_3, repeat Step 1 and Step 2, then we get $s_2 = f(k_1 \oplus C_2) \oplus k_2 \oplus a_{m-1}$, $s_3 = f(k_1 \oplus C_3) \oplus k_2 \oplus a_{m-1}$;

Step 4: By finding the common roots of the following equations, we get k_1 and k_2.

$$\begin{cases} s_1 = f(k_1 \oplus C_1) \oplus k_2 \oplus a_{m-1} \\ s_2 = f(k_1 \oplus C_2) \oplus k_2 \oplus a_{m-1} \\ s_3 = f(k_1 \oplus C_3) \oplus k_2 \oplus a_{m-1} \end{cases} \qquad (2)$$

To find the solution of (2), set $h_{ij}(k_1) = f(k_1 \oplus C_i) \oplus f(k_1 \oplus C_j) \oplus s_i \oplus s_j$ for $1 \leq i < j \leq 3$. Compute $d(k_1) = \gcd(h_{12}(k_1), h_{13}(k_1), h_{23}(k_1))$, the greatest common divisor of $h_{12}(k_1), h_{13}(k_1)$ and $h_{23}(k_1)$, with great probability, $d(k_1) = k_1 \oplus K^*$ where K^* is a constant in $\mathbb{F}_{2^n}$. Thus $k_1 = K^*$, therefore, $k_2 = s_1 \oplus f(k_1 \oplus C_1) \oplus a_{m-1}$.

Comparing with the original interpolation attack, Algorithms 1 and 2 do not need to guess the key candidates. Thus the complexity of these attacks are $m^{r-1}+1$ for Algorithm 1 and $3 \times m^{r-2}+3$ for Algorithm 2, number of plaintexts to be encrypted.

5   New Integral Cryptanalysis of Block Ciphers

For 2^n pairs $(x_i, y_i) \in \mathbb{F}_{2^n}^2$ where x_is are distinct, to find the polynomial $f(x)$ of degree $\leq 2^n - 1$ such that $y_i = f(x_i)$, we can use the Lagrange interpolation formula. However, there is another way to compute $f(x)$.

Theorem 3. [1] Let $f(x) = \sum_{i=0}^{2^n-1} a_i x^i \in \mathbb{F}_{2^n}[x]$ be a polynomial with degree at most $2^n - 1$, then

$$
a_i = \begin{cases}
\sum_{x \in \mathbb{F}_{2^n}} x^{2^n-1-i} f(x) & \text{if } i \neq 0 \mod 2^n - 1, \\
f(0) & \text{if } i = 0, \\
\sum_{x \in \mathbb{F}_{2^n}} f(x) & \text{if } i = 2^n - 1.
\end{cases}
$$

If m^{r-1} or m^{r-2} is too large that the computer can not store $m^{r-1} + 1$ or $m^{r-2} + 1$ pairs of plaintext and ciphertext, we can use the following methods. The two new methods below need almost no memories to compute the round keys of a Feistel cipher. However, they need more plaintexts/ciphertexts.

Algorithm 3: Attack on Block Ciphers with $r \leq r_0$(II)

Step 1: Encrypt $P^{(1)} = (C, x)$ for all $x \in \mathbb{F}_{2^n}$ where $C \in \mathbb{F}_{2^n}$ is a constant. The corresponding ciphertexts are $(C_L(x), C_R(x))$;

Step 2: Compute $s = \sum_{x \in \mathbb{F}_{2^n}} x^{2^n-m^{r-1}} C_R(x)$, thus $k_1 = s \oplus a_{m-1}$.

Algorithm 4: Attack on Block Ciphers with $r \leq r_0 + 1$(II)

Step 1: Encrypt $P^{(1)} = (x, C_1)$ for all $x \in \mathbb{F}_{2^n}$ where $C_1 \in \mathbb{F}_{2^n}$ is a constant. The corresponding ciphertexts are $(C_L^{(1)}(x), C_R^{(1)}(x))$;

Step 2: Compute $s_1 = \sum_{x \in \mathbb{F}_{2^n}} x^{2^n-m^{r-2}} C_R^{(1)}(x)$;

Step 3: Choose another two constants $C_2, C_3 \in \mathbb{F}_{2^n}$, repeat step 1 and step 2, and compute $s_2 = \sum_{x \in \mathbb{F}_{2^n}} x^{2^n-m^{r-2}} C_R^{(2)}(x)$, $s_3 = \sum_{x \in \mathbb{F}_{2^n}} x^{2^n-m^{r-2}} C_R^{(3)}(x)$;

Step 4: Find the solution of

$$
\begin{cases}
s_1 = f(k_1 \oplus C_1) \oplus k_2 \oplus a_{m-1}, \\
s_2 = f(k_1 \oplus C_2) \oplus k_2 \oplus a_{m-1}, \\
s_3 = f(k_1 \oplus C_3) \oplus k_2 \oplus a_{m-1}.
\end{cases}
$$

Comparing Algorithms 3 and 4 with the original interpolation attack, there are some merits of the improved attacks:

(1) There is no need to store plaintexts and corresponding ciphertexts while these data should be stored in the original interpolation attack[2] as well as Algorithms 1 and 2;

(2) There is no need to guess the key candidates. Thus the complexity of these attacks are 2^n and 3×2^n respectively, number of plaintexts to be encrypted.

When applying square attack, one adopts $\sum_x y(x) = 0$. However, in the above attack, we analysis the cipher by computing $\sum_x x^i y(x)$ for some integer i. Thus square attack can be seen as a special case of the new integral attack introduced above.

NOTE 1: In Algorithm 1, for $(x_i, y_i) = (x_i, C_R(x_i))$, $1 \leq i \leq m^{r-1} + 1$, by using the Lagrange interpolation formula and computing the coefficient s of the second highest term, we get:

$$s = \sum_{1 \leq i \leq m^{r-1}+1} \frac{y_i \sum_{1 \leq j \leq m^{r-1}+1, j \neq i} x_j}{\prod_{1 \leq j \leq m^{r-1}+1, j \neq i} (x_i - x_j)}. \tag{3}$$

Instead of interpolation, k_1 can be computed by (3), and this can be seen as another extension of integrals.

6 Results of Attack on $\mathcal{PURE}$

$\mathcal{PURE}$ is a Feistel cipher with $2n = 64$ and $f(x) = x^3 \in \mathbb{F}_{2^{32}}[x]$. Though it has a provable security against differential and linear cryptanalysis, it is breakable by interpolation attack with up to 32 rounds in [2]. However, it is very difficult to mount a real-world attack by the method presented in [2].

6.1 Improved Attacks on $\mathcal{PURE}$

If $r \leq 21$, there are two cases in consideration:

1) If 3^{r-1} is too large, it is impossible to store so many data, thus by Theorem 1, the following equation holds:

$$\sum_{x \in \mathbb{F}_{2^n}} x^{2^{32}-3^{r-1}} C_R(x) = k_1. \tag{4}$$

So, k_1 can be recovered with both data and time complexities being 2^{32} respectively by using Algorithm 3. We implemented 15-round attack by using Algorithm 3, and the round key was recovered in less than 31 hours.

2) If 3^{r-1} is not too large, then the data an interpolation needs is not too large. For this case, we use Algorithm 1 by interpolation, it only needs $3^{r-1} + 1$ plaintexts to recover k_1, with some more memories to store plaintexts/ciphertexts. We implemented 10-round attack by using Algorithm 1, and the round key was recovered in less than 5 minutes.

If $r = 22$, $\mathcal{PURE}$ is breakable with 3×2^{32} encryptions by using Theorem 2, and $3 \times 3^{r-2} + 3$ by using Theorem 1:

Step 1. Encrypt $P = (P^L, P_1^R)$ where $P^L \in \mathbb{F}_{2^{32}}$ takes all values of $\mathbb{F}_{2^{32}}$ and $P_1^R \in \mathbb{F}_{2^{32}}$ is a constant;

Step 2. For the corresponding ciphertexts $C = (C_L, C_R)$, compute
$$s_1 = \sum_{P^L} (P^L)^{2^{32}-3^{r-2}} C_R;$$

Step 3. For P_2^R and P_3^R, do step 1 and step 2, then compute the corresponding s_2 and s_3.

Step 4. Solve the following equations to get k_1 and k_2:

$$\begin{cases} s_1 = (P_1^R \oplus k_1)^3 \oplus k_2 \\ s_2 = (P_2^R \oplus k_1)^3 \oplus k_2 \\ s_3 = (P_3^R \oplus k_1)^3 \oplus k_2 \end{cases} \tag{5}$$

and the solution is

$$\begin{cases} k_1 = \dfrac{s_1 \left(P_2^R \oplus P_3^R\right) \oplus s_2 \left(P_3^R \oplus P_1^R\right) \oplus s_3 \left(P_1^R \oplus P_2^R\right)}{\left(P_1^R \oplus P_2^R\right)\left(P_2^R \oplus P_3^R\right)\left(P_3^R \oplus P_1^R\right)} \oplus \left(P_1^R \oplus P_2^R \oplus P_3^R\right) \\ k_2 = s_1 \oplus \left(P_1^R \oplus k_1\right)^3 \end{cases}$$

6.2   Experimental Results

Table 1 shows the results of the attack on reduced-round $\mathcal{PURE}$, these results are computed by using the algebraic software Magma.

Table 1. Experimental Results of Attack on Reduced-round $\mathcal{PURE}$

Round	Algorithm	Data	Memory	Time	CPU
8	1	$3^7 + 1$	$3^7 + 1$	3.5 seconds	Pentium(R)4,3.06GHz
8	2	$3^6 + 1$	$3^6 + 1$	1 second	Pentium(R)4,3.06GHz
10	1	$3^8 + 1$	$3^8 + 1$	4.5 minutes	Pentium(R)4,3.06GHz
10	2	$3^9 + 1$	$3^9 + 1$	1.5 minutes	Pentium(R)4,3.06GHz
15	3	2^{32}	neglectable	31 hours	Pentium(R)4,3.06GHz
22	4	3×2^{32}	neglectable	148 hours	Pentium(R)4,3.06GHz

7   Conclusion

Both interpolation and integral attacks are improved in this paper. If the cipher can be described as a low degree polynomial, the new attacks can decrease the complexity of the original interpolation attack dramatically, which sometimes leads to a real-world attack. For example, 20-round $\mathcal{PURE}$ is not breakable on a personal computer if one uses the original method introduced in [2], while our method can do so. There are some interesting problems, for example, the square attack can be seen as a special case of this attack, since $\sum_x y$ is a special case of $\sum_x x^i y$. So can we use similar method to analyze AES? Another question is, how to extend this attack to the case of rational polynomials, that is, if the cipher can be described as $g_1(x)/g_2(x)$, how to apply this attack?

Acknowledgment

The authors wish to thank Ruilin Li, Shaojing Fu, Wentao Zhang and the anonymous reviewers for their useful comments.

The work in this paper is partially supported by the Natural Science Foundation of China (No: 60573028, 60803156), and the open research fund of State Key Laboratory of Information Security(No: 01-07).

References

1. Lidl, R., Niederreiter, H.: Finite Fields. Encyclopedia of Mathematics and Its Applications, vol. 20. Cambridge University Press, Cambridge (1997)
2. Jakobsen, T., Knudsen, L.R.: The Interpolation Attack on Block Cipher. In: Biham, E. (ed.) FSE 1997. LNCS, vol. 1267, pp. 28–40. Springer, Heidelberg (1997)
3. Aoki, K.: Efficient Evaluation of Security against Generalized Interpolation Attack. In: Heys, H.M., Adams, C.M. (eds.) SAC 1999. LNCS, vol. 1758, pp. 135–146. Springer, Heidelberg (1999)
4. Kurosawa, K., Iwata, T., Quang, V.D.: Root Finding Interpolation Attack. In: Stinson, D.R., Tavares, S. (eds.) SAC 2000. LNCS, vol. 2012, pp. 303–314. Springer, Heidelberg (2001)
5. Youssef, A.M., Gong, G.: On the Interpolation Attacks on Block Ciphers. In: Schneier, B. (ed.) FSE 2000. LNCS, vol. 1978, pp. 109–120. Springer, Heidelberg (2001)
6. Daemen, J., Knudsen, L.R., Rijmen, V.: The Block Cipher Square. In: Biham, E. (ed.) FSE 1997. LNCS, vol. 1267, pp. 149–165. Springer, Heidelberg (1997)
7. Knudsen, L.R., Wagner, D.: Integral Cryptanalysis. In: Daemen, J., Rijmen, V. (eds.) FSE 2002. LNCS, vol. 2365, pp. 112–127. Springer, Heidelberg (2002)
8. Hu, Y., Zhang, Y., Xiao, G.: Integral Cryptanalysis of SAFER+. Electronics Letters 35(17), 1458–1459 (1999)
9. Lucks, S.: The Saturation Attack — A Bait for Twofish. In: Matsui, M. (ed.) FSE 2001. LNCS, vol. 2355, pp. 1–15. Springer, Heidelberg (2002)
10. Nakahara Jr., J., Freitas, D., Phan, R.: New Multiset Attacks on Rijndael with Large Blocks. In: Dawson, E., Vaudenay, S. (eds.) Mycrypt 2005. LNCS, vol. 3715, pp. 277–295. Springer, Heidelberg (2005)
11. Knudsen, L.R.: Truncated and Higher Order Differentials. In: Preneel, B. (ed.) FSE 1995. LNCS, vol. 1008, pp. 196–211. Springer, Heidelberg (1995)
12. Biryukov, A., Shamir, A.: Structural Cryptanalysis of SASAS. In: Pfitzmann, B. (ed.) EUROCRYPT 2001. LNCS, vol. 2045, pp. 394–405. Springer, Heidelberg (2001)
13. Lai, X.: Higher Order Derivations and Differential Cryptanalysis. Communications and Cryptography: Two Sides of One Tapestry, pp. 227–233. Kluwer Academic Publishers, Dordrecht (1994)
14. Z'aba, M.R., Raddum, H., Henricksen, M., Dawson, E.: Bit-Pattern Based Integral Attack. In: Nyberg, K. (ed.) FSE 2008. LNCS, vol. 5086, pp. 363–381. Springer, Heidelberg (2008)
15. Biham, E., Shamir, A.: Differential Cryptanalysis of the Data Encryption Standard. Springer, Heidelberg (1993)
16. Matsui, M.: Linear Cryptanalysis Method for DES Cipher. In: Helleseth, T. (ed.) EUROCRYPT 1993. LNCS, vol. 765, pp. 386–397. Springer, Heidelberg (1993)
17. Lee, C., Cha, Y.: The Block Cipher: SNAKE with Provable Resistance against DC and LC Attacks. In: Proceedings of 1997 Korea-Japan Joint Workshop on Information Security and Cryptology (JW–ISC 1997), pp. 3–17 (1997)

18. Morial, S., Shimoyama, T., Kaneko, T.: Interpolation Attacks of the Block Cipher: SNAKE. In: Knudsen, L.R. (ed.) FSE 1999. LNCS, vol. 1636, pp. 275–289. Springer, Heidelberg (1999)
19. Daemen, J., Rijmen, V.: The Design of Rijndael: AES — The Advanced Encryption Standard (Information Security and Cryptography). Springer, Heidelberg (2002)
20. Ferguson, N., Kelsey, J., Lucks, S., Schneier, B., Stay, M., Wagner, D., Whiting, D.: Improved Cryptanalysis of Rijndael. In: Schneier, B. (ed.) FSE 2000. LNCS, vol. 1978, pp. 213–230. Springer, Heidelberg (2001)

Algebraic Techniques in Differential Cryptanalysis

Martin Albrecht* and Carlos Cid

Information Security Group,
Royal Holloway, University of London
Egham, Surrey TW20 0EX, United Kingdom
{M.R.Albrecht,carlos.cid}@rhul.ac.uk

Abstract. In this paper we propose a new cryptanalytic method against block ciphers, which combines both algebraic and statistical techniques. More specifically, we show how to use algebraic relations arising from differential characteristics to speed up and improve key-recovery differential attacks against block ciphers. To illustrate the new technique, we apply algebraic techniques to mount differential attacks against round reduced variants of PRESENT-128.

1 Introduction

The two most established cryptanalytic methods against block ciphers are linear cryptanalysis [22] and differential cryptanalysis [3]. These attacks are statistical in nature, in which the attacker attempts to construct probabilistic patterns through as many rounds of the cipher as possible, in order to distinguish the cipher from a random permutation, and ultimately recover the key. Due to their very nature, these attacks require a very large number of plaintext–ciphertext pairs, ensuring that (usually) they rapidly become impractical. In fact, most modern ciphers have been designed with these attacks in mind, and therefore do not generally have their security affected by them.

A new development in block cipher cryptanalysis are the so-called algebraic attacks [14,23,9]. In contrast to linear and differential cryptanalysis, algebraic attacks attempt to exploit the algebraic structure of the cipher. In its most common form, the attacker expresses the encryption transformation as a large set of multivariate polynomial equations, and subsequently attempts to solve the system to recover information about the encryption key.

The proposal of algebraic attacks against block ciphers has been the source of much speculation; while a well-established technique against some stream ciphers constructions [13], the viability of algebraic attacks against block ciphers remains subject to debate. On one hand these attack techniques promise to allow the cryptanalyst to recover secret key bits given only one or very few plaintext–ciphertext pairs. On the other hand, the runtime of algebraic attacks against

---

* This author was supported by the Royal Holloway Valerie Myerscough Scholarship.

O. Dunkelman (Ed.): FSE 2009, LNCS 5665, pp. 193–208, 2009.

block ciphers is not well understood, and it is so far not clear whether algebraic attacks can break any proposed block cipher faster than other techniques.

A promising approach however is to combine both statistical and algebraic techniques in block cipher cryptanalysis. In fact, many proposed algebraic approaches already involve statistical components. For instance, the equation systems usually considered for the AES [9,23], use the *inversion equation* $xy = 1$ for the S-Box. While this equation only holds with probability $p = 255/256$, it may well offer some advantages when compared with the correct equation $x^{254} = y$ representing the S-Box (which due to its very high degree, is usually considered impractical). Further recent examples include key bit guesses [11], the use of SAT-solvers [1] and the Raddum-Semaev algorithm [24] for solving polynomial equations. In this paper we propose a new attack technique that combines results from algebraic and differential cryptanalysis.

The paper is structured as follows. First, we briefly describe differential and algebraic cryptanalysis and give the basic idea of the attack in Section 2. We then describe the block cipher PRESENT in Section 3 and existing attacks against a reduced round version of PRESENT (Section 3.1). In Section 4 we describe the application of our new attack technique against reduced round versions of PRESENT. We present a brief discussion of the attack and possible extensions in Section 5.

2   Overview of the New Attack Technique

Since our approach combines differential and algebraic cryptanalysis, we briefly describe both techniques below.

2.1   Differential Cryptanalysis

Differential cryptanalysis was formally introduced by Eli Biham and Adi Shamir at Crypto'90 [4], and has since been successfully used to attack a wide range of block ciphers. In its basic form, the attack can be used to distinguish a n-bit block cipher from a random permutation. By considering the distribution of output *differences* for the non-linear components of the cipher (e.g. the S-Box), the attacker may be able to construct *differential characteristics* $P' \oplus P'' = \Delta P \rightarrow \Delta C_N = C'_N \oplus C''_N$ for a number of rounds N that are valid with probability p. If $p \gg 2^{-n}$, then by querying the cipher with a large number of plaintext pairs with prescribed difference ΔP, the attacker may be able to distinguish the cipher by counting the number of pairs with the output difference predicted by the characteristic. A pair for which the characteristic holds is called a *right pair*.

By modifying the attack, one can use it to recover key information. Instead of characteristics for the full N-round cipher, the attacker considers characteristics valid for r rounds only ($r = N - R$, with $R > 0$). If such characteristics exist with non-negligible probability the attacker can guess some key bits of the last rounds, partially decrypt the known ciphertexts, and verify if the result matches the one predicted by the characteristic. Candidate (last round) keys are counted, and as random noise is expected for wrong key guesses, eventually a peak may be observed in the candidate key counters, pointing to the correct round key[1].

Note that due to its statistical nature, differential cryptanalysis requires a very large number of plaintext–ciphertext pairs (for instance, approximately 2^{47} chosen plaintext pairs are required to break DES [5]). Many extensions and variants of differential cryptanalysis exist, such as the Boomerang attack [26] and truncated and higher-order differentials [21]. The technique is however very well understood, and most modern ciphers are designed to resist to differential cryptanalysis. This is often achieved by carefully selecting the cipher's non-linear operations and diffusion layer to make sure that if such differential characteristics exist, then $r \ll N$ which ensures that backward key guessing is impractical. The AES is a prime example of this approach [15].

2.2   Algebraic Cryptanalysis

Algebraic cryptanalysis against block ciphers is an attack technique that has recently received much attention, particularly after it was proposed in [14] against the AES and Serpent block ciphers. In its basic form, the attacker attempts to express the cipher as a set of low degree (often quadratic) equations, and then solve the resulting system. As these systems are usually very sparse, overdefined, and structured, it is conjectured that they may be solved much faster than generic non-linear equation systems. Several algorithms have been used and/or proposed to solve these systems including the Buchberger algorithm, XL and variants [12,29,14] , the F_4 and F_5 algorithm [17,18], and the Raddum-Semaev algorithm [24]. Another approach is to convert these equations to Boolean expressions in Conjunctive Normal Form (CNF) and use off-the-shelf SAT-solvers [2]. However, these methods have had so far limited success in targeting modern block ciphers, and no public modern block cipher, with practical relevance, has been successfully attacked using algebraic cryptanalysis faster than with other techniques.

2.3   Algebraic Techniques in Differential Cryptanalysis

A first idea in extending algebraic cryptanalysis is to use more plaintext–ciphertext pairs to construct the equation system. Given two equation systems F' and F'' for two plaintext–ciphertext pairs (P', C') and (P'', C'') under the same encryption key K, we can combine these equation systems to form a system $F = F' \cup F''$. Note that while F' and F'' share the key and key schedule variables, they do not share most of the state variables. Thus the cryptanalyst gathers almost twice as many equations, involving however many new variables. Experimental evidence indicates that this technique may often help in solving a system of equations at least up to a certain number of rounds [19]. The second step is to consider probabilistic relations that may arise from differential cryptanalysis, giving rise to what we call *Attack-A*.

---

[1] In some variants, as described in [5], no candidate key counters are required; see Section 5 for a brief discussion of this attack.

Attack-A. For the sake of simplicity, we assume the cipher is an Substitution-Permutation-Network (SP-network), which iterates layers of non-linear transformations (e.g. S-Box operations) and affine transformations. Now consider a differential characteristic $\Delta = (\delta_0, \delta_1, \ldots, \delta_r)$ for a number of rounds, where $\delta_{i-1} \to \delta_i$ is a one-round difference arising from round i and valid with probability p_i. If we assume statistical independence of one-round differences, the characteristic Δ is valid with probability $p = \prod p_i$. Each one-round difference gives rise to equations relating the input and output pairs for active S-Boxes. Let $X'_{i,j}$ and $X''_{i,j}$ denote the j-th bit of the input to the S-Box layer in round i for the systems F' and F'', respectively. Similarly, let $Y'_{i,j}$ and $Y''_{i,j}$ denote the corresponding output bits. Then we have that the expressions

$$X'_{i,j} + X''_{i,j} = \Delta X_{i,j} \to \Delta Y_{i,j} = Y'_{i,j} + Y''_{i,j},$$

where $\Delta X_{i,j}, \Delta Y_{i,j}$ are known values predicted by the characteristic, are valid with some non-negligible probability q for bits of active S-Boxes. Similarly, for non-active S-Boxes (that are not involved in the characteristic Δ and therefore have input/output difference zero), we have the relations

$$X'_{i,j} + X''_{i,j} = 0 = Y'_{i,j} + Y''_{i,j}$$

also valid with a non-negligible probability.

If we consider the equation system $F = F' \cup F''$, we can combine F and all such linear relations arising from the characteristic Δ. This gives rise to an equation system $\overline{F}$ which holds with probability p. If we attempt to solve such a system for approximately $1/p$ pairs of plaintext–ciphertext, we expect at least one non-empty solution, which should yield the encryption key. For a full algebraic key recover we expect the system $\overline{F}$ to be easier to solve than the original system F' (or F''), because many linear constrains were added without adding any new variables. However, we do not know *a priori* how difficult it will be to solve the system approximately $1/p$ times. This system $\overline{F}$ may be used however to recover some key information, leading to an attack we call *Attack-B*.

Attack-B. Now, assume that we have an SP-network, a differential characteristic $\Delta = (\delta_0, \delta_1, \ldots, \delta_r)$ valid for r rounds with probability p, and (P', P'') a right pair for Δ (so that $\delta_0 = P' \oplus P''$ and δ_r holds for the output of round r). For simplicity, let us assume that only one S-Box is active in round 1, with input $X'_{1,j}$ and $X''_{1,j}$ (restricted to this S-Box) for the plaintext P' and P'' respectively, and that there is a key addition immediately before the S-Box operation, that is

$$S(P'_j \oplus K_{0,j}) = S(X'_{1,j}) = Y'_{1,j} \text{ and } S(P''_j \oplus K_{0,j}) = S(X''_{1,j}) = Y''_{1,j}.$$

The S-Box operation S can be described by a (vectorial) Boolean function, expressing each bit of the output $Y'_{1,j}$ as a polynomial function (over $\mathbb{F}_2$) on the input bits of $X'_{1,j}$ and $K_{0,j}$. If (P', P'') is a right pair, then the polynomial equations arising from the relation $\Delta Y_{1,j} = Y'_{1,j} \oplus Y''_{1,j} = S(P'_j \oplus K_{0,j}) \oplus S(P''_j \oplus K_{0,j})$ give us a very simple equation system to solve, with only the key variables

$K_{0,j}$ as unknowns (and which do not vanish identically because we are considering nonzero differences, cf. Section 5). Consequently, if we had an effective distinguisher to determine whether $\Delta Y_{1,j}$ holds, we could learn some bits of information about the round keys involved in the first round active S-Boxes.

Experimentally, we found that, for some ciphers and up to a number of rounds, *Attack-A* can be used as such a distinguisher. More specifically, we noticed that finding a contradiction (i.e. the Gröbner basis equal to $\{1\}$) was much faster than computing the full solution of the system if the system was consistent (that is, when we have a right pair). Thus, rather than fully solving the systems to eventually recover the secret key as suggested in *Attack-A*, the *Attack-B* proceeds by measuring the time t it maximally takes to find that the system is inconsistent[2], and assume we have a right pair with good probability if this time t elapsed without a contradiction. More specifically, we expect $\Delta Y_{1,j}$ to hold with good probability. One needs to be able to experimentally estimate the time t, but for some ciphers this appears to be an efficient form of attack.

An alternative form of *Attack-B* is to recover key bits from the last round. Assume that the time t passed for a pair (P', P''), i.e. that we probably found a right pair. Now, if we guess and fix some subkey bits in the last rounds, we can check whether the time t still passes without a contradiction. If this happens, we assume that we guessed correctly. However, for this approach to work we need to guess enough subkey bits to detect a contradiction quickly. An obvious choice is to guess all subkey bits involved in the last round, which effectively removes one round from the system.

Attack-C. Experimental evidence with PRESENT (cf. Section 4) indicates that *Attack-B* in fact only relies on the differential $\delta_0 \to \delta_r$ rather than the characteristic Δ when finding contradictions in the systems. The runtimes for finding contradictions for $N = 17$ and differential characteristic of length $r = 14$ did not differ significantly from the runtimes for the same task with $N = 4$ and $r = 1$ (cf. Appendix C). This indicates that the computational difficulty is mostly determined by the difference $R = N - r$, the number of "free" rounds. We thus define a new attack (*Attack-C*) where we remove the equations for rounds $\leq r$.

This significantly reduces the number of equations and variables. After these equations are removed we are left with R rounds for each plaintext–ciphertext pair to consider; these are related by the output difference predicted by the differential. As a result, the algebraic computation is essentially equivalent to solving a related cipher of $2R - 1$ rounds (from C' to C'' via the predicted difference δ_r) using an algebraic meet-in-the-middle attack [9]. This "cipher" has a symmetric key schedule and only $2R - 1$ rounds rather than $2R$ since the S-Box applications after the difference δ_r are directly connected and lack a key addition and diffusion layer application between them. Thus we can consider these two S-Box applications as one S-Box application of S-Boxes S_i defined by the known difference δ_r: $S_i(x_{i,\ldots,i+s}) = S(S^{-1}(x_{i,\ldots,i+s}) + \delta_{r,(i,\ldots,i+s)})$ for $i \in \{0, s, \ldots, n\}$ and s the size of the S-Box.

---

[2] Other features of the calculation — like the size of the intermediate matrices created by F_4 — may also be used instead of the time t.

Again, we attempt to solve the system and wait for a fixed time t to find a contradiction in the system. If no contradiction is found, we assume that the differential $\delta_0 \rightarrow \delta_r$ holds with good probability. Note that we cannot be certain about the output difference of the first round active S-Boxes. However, the attack can be adapted such that we can still recover key bits, for instance by considering multiple suggested right pairs. A second option is to attempt to solve the resulting smaller system, to recover the encryption key. Alternatively, we can execute the guess-and-verify step described above.

To study the viability of these attacks, we describe experiments with reduced-round versions of the block cipher PRESENT.

3   The Block Cipher PRESENT

PRESENT [6] was proposed by Bogdanov et al. at CHES 2007 as an ultra-lightweight block cipher, enabling a very compact implementation in hardware, and therefore particularly suitable for RFIDs and similar devices. There are two variants of PRESENT: one with 80-bit keys and one with a 128-bit keys, denoted as PRESENT-80 and PRESENT-128 respectively. In our experiments, we consider reduced round variants of both ciphers denoted as PRESENT-K_s-N, where $K_s \in \{80, 128\}$ represents the key size in bits and $1 \leq N \leq 31$ represents the number of rounds.

PRESENT is an SP-network with a blocksize of 64 bits and both versions have 31 rounds. Each round of the cipher has three layers of operations: keyAddLayer, sBoxLayer and pLayer. The operation keyAddLayer is a simple subkey addition to the current state, while the sBoxLayer operation consists of 16 parallel applications of a 4-bit S-Box. The operation pLayer is a permutation of wires.

In both versions, these three operations are repeated $N = 31$ times. On the final round, an extra subkey addition is performed. The subkeys are derived from the user-provided key in the key schedule, which by design is also quite simple and efficient involving a cyclic right shift, one ore two 4-bit S-Box applications (depending on the key size) and the addition of a round constant. We note that the difference between the 80-bit and 128-bit variants is only the key schedule. In particular, both variants have the same number of rounds (i.e. $N = 31$). The cipher designers explicitly describe in [6] the threat model considered when designing the cipher, and acknowledge that the security margin may be somewhat tight. Although they do not recommend immediate deployment of the cipher (especially the 128-bit version), they strongly encourage the analysis of both versions.

3.1   Differential Cryptanalysis of 16 Rounds of PRESENT

In the original proposal [6], the designers of PRESENT show that both linear and differential cryptanalysis are infeasible against the cipher. In [27,28] M. Wang provides 24 explicit differential characteristics for 14 rounds. These hold with probability 2^{-62} and are within the theoretical bounds provided by the PRESENT designers. Wang's attack is reported to require 2^{64} memory accesses to cryptanalyse 16 rounds of PRESENT-80. We use his characteristics (see Appendix B for an example of one of these characteristics) to mount our attack. Furthermore, we also make use of the filter function presented in [27], which we briefly describe below.

Consider for example the differential characteristic provided in Appendix B. It ends with the difference $\delta = 1001 = 9$ as input for the two active S-Boxes of round 15. According to the difference distribution table of the PRESENT S-Box, the possible output differences are 2, 4, 6, 8, C and E. This means that the least significant bit is always zero and the weight of the output difference (with the two active S-Box) is at most 6. It then follows from pLayer that at most six S-Boxes are active in round 16. Thus we can discard any pair for which the outputs of round 16 have non-zero difference in the positions arising from the output of S-Boxes other than the active ones. There are ten inactive 4-bit S-Boxes, and we expect a pair to pass this test with probability 2^{-40}.

Furthermore, it also follows from pLayer that the active S-Boxes in round 16 (which are at most six, as described above) will have input difference 1 and thus all possible output differences are 3, 7, 9, D (and 0, in case the S-Box is inactive). Thus we can discard any pair not satisfying these output differences for these S-Boxes. We expect a pair to pass this test with probability $\frac{16}{5}^{-6} = 2^{-10.07}$. Overall we expect pairs to path both tests with probability $2^{-50.07}$. We expect to be able to construct a similar filter function for all the 24 differential characteristics presented in [28].

4 Experimental Results

To mount the attacks, we generate systems of equations $\overline{F}$ as in Section 2 for pairs of encryptions with prescribed difference as described in Section 3.1, by adding linear equations for the differentials predicted by the 14-round characteristic given in the Appendix. For PRESENT this is equivalent to adding 128 linear equations per round of the form $\Delta X_{i,j} = X'_{i,j} + X''_{i,j}$ and $\Delta Y_{i,j} = Y'_{i,j} + Y''_{i,j}$ where $\Delta X_{i,j}$ and $\Delta Y_{i,j}$ are the values predicted by the characteristic (these are zero for non-active S-Boxes).

To perform the algebraic part of the attack, we use either Gröbner basis algorithms or a SAT-solver: the SINGULAR 3-0-4-4 [20] routine groebner with the monomial odering *degrevlex*, the POLYBORI 0.5rc6 [8] routine groebner_basis with the option faugere=True and the monomial ordering dp_asc, or MiniSat 2.0 beta [16]. We note the maximal time t these routines take to detect a contradiction in our experiments for a given differential length of r, and assume we have a pair satisfying the characteristic (or differential, in *Attack-C*) with good probability if this time t elapsed without a contradiction. We note that this assumption might be too optimistic in some cases. While the attack seems to perform well enough for a small number of rounds we cannot be certain that the lack of a contradiction after the time t indeed indicates a right pair. However, with t large enough (and enough computational resources) we are guaranteed to always identify the right pair.

We performed experiments for *Attack-B* and *Attack-C*. Runtimes for *Attack-B* and *Attack-C* are given in Appendix C and D respectively. We note that Attack-C requires about 1GB of RAM to be carried out. The times were obtained on a 1.8Ghz Opteron with 64GB RAM. The attack was implemented in the mathematics software Sage [25].

If a characteristic Δ is valid with probability p, then after approximately $1/p$ attempts we expect to find a right pair and can thus set up our smaller systems for each first round active S-Box. These equations are given in Appendix A. After substitution of $P_i', P_i'', \Delta Y_i$ and elimination of the variables X_i', X_i'' in the system in Appendix A, we get an equation system with four equations in the four key variables. If we compute the reduced Gröbner basis for this system we recover two relations of the form $K_i + K_j(+1) = 0$ for two key bits K_i, K_j per S-Box, i.e. we recover 2 bits of information per first round active S-Box[3].

In order to study the behaviour of the attack, we ran simulations with small numbers of rounds to verify that the attack indeed behaves as expected. For instance, when using a 3R *Attack-C* against PRESENT-80-6 and PRESENT-80-7 we found right pairs with the expected number of trials. However, we saw false positives, i.e. the attack suggested wrong information. Yet, a majority vote on a small number of runs (e.g., 3) always recovered the correct information. We are of course aware that it is in general difficult to reason from small scale examples to bigger instances.

4.1   PRESENT-80-16

To compare with the results of [27], we can apply *Attack-C* against reduced round versions of PRESENT-80. Using this approach and under the assumption above we expect to learn 4 bits of information about the key for PRESENT-80-16 in about $2^{62-50.07} \cdot 6$ seconds to perform the consistency checks using about 2^{62} chosen plaintext–ciphertext pairs, where 6 seconds represents the highest runtime to find a contradiction we have encountered in our experiments when using POLYBORI. Even if there are instances that take longer to check, we assume that this is a safe margin because the majority should be shorter runtimes. This time gives a complexity of about 2^{62} ciphertext difference checks and about $2^{11.93} \cdot 6 \cdot 1.8 \cdot 10^9 \approx 2^{46}$ CPU cycles to find a right pair on the given 1.8 Ghz Opteron CPU. We assume that a single encryption costs at least two CPU cycles per round – one for the S-Box lookup and one for the key addition – such that a brute force search would require approximately $16 \cdot 2 \cdot 2^{80} = 2^{85}$ CPU cycles and two plaintext–ciphertext pairs due to the small blocksize.

In [28], 24 different 14-round differentials were presented, involving the 0th, 1st, 2nd, 12th, 13th and 14th S-Boxes in the first round, each having either 7 or 15 as plaintext difference restricted to one active S-Box. From these we expect to recover 18 bits of key information by repeating the attack for those S-Box configurations. We cannot recover 24 bits because we learn some redundant information. However, we can use this redundancy to verify the information

---

[3] This is as expected, since the probability of the differential used in the first round S-Box is 2^{-2}; see Lemma 1.

recovered so far. We can then guess the remaining $80 - 18 = 62$ bits, and the complete attack has a complexity of about $6 \cdot 2^{62}$ filter function applications, about $6 \cdot 2^{46}$ CPU cycles for the consistency checks and 2^{62} PRESENT applications to guess the remaining key bits[4]. (Alternatively, we may add the 18 learned linear key bit equations to any equation system for the related cipher and attempt to solve this system.) The attack in [27] on the other hand requires 2^{64} memory accesses. While this is a different metric — memory access — from the one we have to use in this case — CPU cycles — we can see that our approach has roughly the same time complexity, since the 2^{62} filter function applications cost at least 2^{62} memory accesses. However, our attack seems to have a slightly better data complexity because overall six right pairs are sufficient. When applying the attack against PRESENT-128-16, we obtain a similar complexity. We note however that for PRESENT-K_s-16, we can also make use of backward key guessing to recover more key bits. Because we assume to have distinguished a right pair already we expect the signal to noise ratio to be quite high and thus expect relatively few wrong suggestions for candidate keys.

4.2 PRESENT-128-17

Note that we cannot use the filter function for 17 rounds, thus the attack against PRESENT-80-17 gives worse performance when compared to exhaustive key search. However, it may still be applied against PRESENT-128-17. Indeed, we expect to learn 4 bits of information for PRESENT-128-17 in about $2^{62} \cdot 18$ seconds using about 2^{62} chosen plaintext–ciphertext pairs. This time is equivalent to about $2^{62} \cdot 18 \cdot 1.8 \cdot 10^9 \approx 2^{97}$ CPU cycles. If this approach is repeated 6 times for the different active S-Boxes in the PRESENT differentials, we expect to learn 18 bits of information about the key. We can then guess the remaining $128 - 18 = 110$ bits and thus have a complexity in the order of 2^{110} for the attack.

A better strategy is as follows. We identify one right pair using $2^{62} \cdot 18 \cdot 1.8 \cdot 10^9 \approx 2^{97}$ CPU cycles. Then, we guess 64 subkey bits of the last round and fix the appropriate variables in the equation system for the consistency check. Finally, we attempt to solve this system again, which is equivalent to the algebraic part of the $2R$ attack. We repeat this guess-and-verify step until the right configuration is found, i.e. the system is not inconsistent. This strategy has a complexity of 2^{97} CPU cycles for identifying the right pair and $2^{64} \cdot 6 \cdot 1.8 \cdot 10^9 \approx 2^{98}$ CPU cycles to recover 64 subkey bits. Finally, we can either guess the remaining bits or repeat the guess-and-verify step for 1R to recover another 64 subkey bits.

4.3 PRESENT-128-18

We can also attack PRESENT-128-18 using *Attack-C* as follows. First note that the limiting factor for the attack on PRESENT-128-18 is that we run out of

---

[4] Note that the attack can be improved by managing the plaintext–ciphertext pairs more intelligently and by using the fact that we can abort a PRESENT trial encryption if it does not match the known differential.

plaintext–ciphertext pairs due to the small blocksize. On the other hand, we have not yet reached the time complexity of 2^{128} for 128-bit keysizes. One way to make use of this fact is to again consider the input difference for round 15 and iterate over all possible output differences. As discussed in Section 3.1, we have six possible output differences and two active S-Boxes in round 15, which result in 36 possible output differences in total. We expect to learn 4 bits of information about the key for PRESENT-128-18 in about $36 \cdot 2^{62} \cdot 18$ seconds using about 2^{62} chosen plaintext–ciphertext pairs. This time is equivalent to about $36 \cdot 2^{62} \cdot 18 \cdot 1.8 \cdot 10^9 \approx 2^{102}$ CPU cycles. Again, we can iterate this process six times to learn 18 bits of information about the key and guess the remaining information with a complexity of approximately 2^{110} PRESENT applications.

However, this strategy might lead to false positives for each guessed output difference. To address this we need to run the brute-force attack for the remaining 110 bits for each possible candidate. Thus the overall complexity of the attack is in the order of $36 \cdot 2^{110}$ PRESENT applications. The final brute-force run will require for 2-3 plaintext-ciphertext pairs due to the large key size compared to the blocksize. This hardly affects the time complexity since only candidates passing the first plaintext-ciphertext pair need to be tested against a second and potentially third pair and these candidates are few compared to 2^{110}.

The best approach appears to be the guess-and-verify step from the $3R$ attack, which results in an overall complexity of about $36 \cdot 1.8 \cdot 10^9 (2^{62} \cdot 18 + 2^{64} \cdot 6) \approx 2^{103}$ CPU cycles.

Note that we were unable to reliably detect contradictions directly if $R = N - r \geq 4$ within 24 hours (compared to 18 seconds for $R = 3$).

4.4   PRESENT-128-19

Similarly, we can use the filter function to mount an attack against PRESENT-128-19 by iterating our attack $2^{64-50.07} = 2^{13.93}$ times (instead of 36) for all possible output differences of round 16. The overall complexity of this attack is about $2^{13.97} \cdot 1.8 \cdot 10^9 \cdot (18 \cdot 2^{62} + 6 \cdot 2^{64}) \approx 2^{113}$ CPU cycles.

5   Discussion of the Attack

While the attack has many similarities with conventional differential cryptanalysis, such as the requirement of a high probability differential Δ valid for r rounds and the use of filter functions to reduce the workload, there are however some noteworthy differences. First, *Attack-C* requires fewer plaintext–ciphertext pairs for a given differential characteristic to learn information about the key than conventional differential cryptanalysis, because the attacker does not need to wait for a peak in the partial key counter. Instead one right pair is sufficient. Second, one flavour of the attack recovers more key bits if many S-Boxes are active in the first round. This follows from its reliance on those S-Boxes to recover key information. Also note that while a high probability differential characteristic is required, the attack recovers more bits per S-Box if the differences for the active S-Box in the first round are of low probability. This is a consequence of the simple Lemma below:

Lemma 1. *Given a differential Δ with a first round active S-Box with a difference that is true with probability 2^{-b}, then* Attack-B *and* Attack-C *can recover b bits of information about the key from this S-Box.*

Finally, key-recovery differential cryptanalysis is usually considered infeasible if the differential Δ is valid for r rounds, and r is much less than the full number of rounds N, since backward key guessing for $N-r$ rounds may become impractical. In that case the *Attack-C* proposed here could *possibly* still allow the successful cryptanalysis of the cipher. However, this depends on the algebraic structure of the cipher, as it may be the case that the time required for the consistency check is such that the overall complexity remains below the one required for exhaustive key search.

We note that *Attack-C* shares many properties with the differential cryptanalysis of the full 16-round DES [5]. Both attacks are capable of detecting a right pair without maintaining a candidate key counter array. Also, both attacks use active S-Boxes of the outer rounds to recover bits of information about the key once such a right pair is found. In fact, one could argue that *Attack-C* is a generalised algebraic representation of the technique presented in [5]. From this technique *Attack-C* inherits some interesting properties: first, the attack can be carried out fully in parallel because no data structures such as a candidate key array need to be shared between the nodes. Also, we allow the encryption keys to change during the data collection phase because exactly one right pair is sufficient to learn some key information. However, if we try to learn further key bits by repeating the attack with other characteristics we require the encryption key not to change. We note however that while the attack in [5] seems to be very specific to the target cipher DES, *Attack-C* can in principle be applied to any block cipher. Another way of looking at *Attack-C* is to realise that it is in fact is a quite expensive but thorough filter function: we invest more work in the management of the outer rounds using algebraic techniques.

In the particular case of PRESENT-80-N, our attack seems to offer only marginal advantage when compared with the differential attack presented in [27]: it should require slightly less data to distinguish a right pair and similar overall complexity. On the other hand, for PRESENT-128-N this attack seems to perform better than the one in [27]. As in this case the limiting factor is the data and not the time complexity of the attack, i.e. we run out of plaintext–ciphertext pairs before running out of computation time, the attack has more flexibility.

The use of Gröbner bases techniques to find contradictions in propositional systems is a well known idea [10]. In the context of cryptanalysis, it is also a natural idea to try to detect contradictions to attack a cipher. However, in probabilistic approaches used in algebraic attacks, usually key bits are guessed. This is an intuitive idea because polynomial systems tend to be easier to solve the more overdefined they are and because the whole system essentially depends on the key. Thus guessing key bits is a natural choice. However this simplification seems to bring few benefits to the attacker, and more sophisticated probabilistic approaches seem so far to have been ignored. The method proposed in this paper can thus highlight the advantages of combining conventional (statistical) cryptanalysis and algebraic cryptanalysis. By considering differential cryptanalysis we showed how to construct an equation system for a structurally weaker and shorter related "cipher" which can then be studied independently. To attack this "cipher" algebraic attacks seem to be the natural choice since very few "plaintext–ciphertext" pairs are available but the "cipher" has few rounds (i.e. $2R - 1$). However, other techniques might also be considered.

Future research might also investigate the use of other well established (statistical) cryptanalysis techniques in combination with algebraic cryptanalysis such as linear cryptanalysis (defining a version of *Attack-A* in this case is straightforward), higher order and truncated differentials, the Boomerang attack or impossible differentials.

We note that this attack may also offer a high degree of flexibility for improvements. For example, the development of more efficient algorithms for solving systems of equations (or good algebraic representation of ciphers that may result in more efficient solving) would obviously improve the attacks proposed. For instance, by switching from SINGULAR to POLYBORI for *Attack-B*, we were able to make the consistency check up to 60 times faster[5]. As an illustration of the forementioned flexibility, if for instance an attacker could make use of an optimised method to find contradictions in $t \ll 2^{128-62} = 2^{66}$ CPU cycles for PRESENT-128-20, this would allow the successful cryptanalysis of a version of PRESENT with 6 more rounds than the best known differential, which is considered "a situation without precedent" by the cipher designers [6]. This task is equivalent to mount a meet-in-the-middle attack against an 11 round PRESENT-like cipher with a symmetric key schedule. Unfortunately with the available computer resources, we are not able to verify whether this is currently feasible.

Finally, as our results depend on experimental data and the set of data we evaluated is rather small due to the time consuming nature of our experiments, we make our claims verifiable by providing the source code of the attack online `http://bitbucket.org/malb/algebraic_attacks/src/tip/present.py`.

6   Conclusion

We propose a new cryptanalytic technique combining differential cryptanalysis and algebraic techniques. We show that in some circumstances this technique can be effectively used to attack block ciphers, and in general may offer some advantages when compared to differential cryptanalysis. As an illustration, we applied it against reduced versions of PRESENT-80 and PRESENT-128. While this paper has no implications for the security of either PRESENT-80 or PRESENT-128, it was shown that the proposed techniques can improve upon existing differential cryptanalytic methods using the same difference characteristics. Also, we pointed out promising research directions for the field of algebraic attacks.

---

[5] We did not see any further speed improvement by using e.g. MAGMA 2.14 [7].

Acknowledgements

The work described in this paper has been supported in part by the European Commission through the IST Programme under contract ICT-2007-216646 ECRYPT II. We would like to thank William Stein for allowing the use of his computers[6]. We also would like to thank Sean Murphy, Matt Robshaw, Ludovic Perret, Jean-Charles Faugère and anonymous referees for helpful comments.

References

1. Bard, G.V.: Algorithms for Solving Linear and Polynomial Systems of Equations over Finite Fields with Applications to Cryptanalysis. PhD thesis, University of Maryland (2007)
2. Bard, G.V., Courtois, N.T., Jefferson, C.: Efficient Methods for Conversion and Solution of Sparse Systems of Low-Degree Multivariate Polynomials over GF(2) via SAT-Solvers. IACR ePrint Archive, Report 2007/024 (2007),
   http://eprint.iacr.org/2007/024
3. Biham, E., Shamir, A.: Differential Cryptanalysis of DES-like Cryptosystems. Journal of Cryptology 4(1), 3–72 (1991)
4. Biham, E., Shamir, A.: Differential Cryptanalysis of DES-like Cryptosystems. In: Menezes, A., Vanstone, S.A. (eds.) CRYPTO 1990. LNCS, vol. 537, pp. 2–21. Springer, Heidelberg (1991)
5. Biham, E., Shamir, A.: Differential Cryptanalysis of the Full 16-round DES. In: Brickell, E.F. (ed.) CRYPTO 1992. LNCS, vol. 740, pp. 487–496. Springer, Heidelberg (1993)
6. Bogdanov, A., Knudsen, L.R., Leander, G., Paar, C., Poschmann, A., Robshaw, M., Seurin, Y., Vikkelsoe, C.: PRESENT: An ultra-lightweight block cipher. In: Paillier, P., Verbauwhede, I. (eds.) CHES 2007. LNCS, vol. 4727, pp. 450–466. Springer, Heidelberg (2007),
   http://www.crypto.rub.de/imperia/md/content/texte/publications/conferences/present_ches2007.pdf
7. Bosma, W., Cannon, J., Playoust, C.: The MAGMA Algebra System I: The User Language. Journal of Symbolic Computation 24, 235–265 (1997)
8. Brickenstein, M., Dreyer, A.: PolyBoRi: A framework for Gröbner basis computations with Boolean polynomials. In: Electronic Proceedings of MEGA 2007 (2007),
   http://www.ricam.oeaw.ac.at/mega2007/electronic/26.pdf
9. Cid, C., Murphy, S., Robshaw, M.: Algebraic Aspects of the Advanced Encryption Standard. Springer, Heidelberg (2006)
10. Clegg, M., Edmonds, J., Impagliazzo, R.: Using the Groebner basis algorithm to find proofs of unsatisfiability. In: Proceedings of the 28th ACM Symposium on Theory of Computing, pp. 174–183 (1996),
    http://www.cse.yorku.ca/~jeff/research/proof_systems/grobner.ps
11. Courtois, N.T., Bard, G.V.: Algebraic Cryptanalysis of the Data Encryption Standard. In: Galbraith, S.D. (ed.) Cryptography and Coding 2007. LNCS, vol. 4887, pp. 152–169. Springer, Heidelberg (2007); IACR ePrint Archive, Report 2006/402,
    http://eprint.iacr.org/2006/402

---

[6] Purchased under National Science Foundation Grant No. 0555776 and National Science Foundation Grant No. DMS-0821725.

12. Courtois, N.T., Klimov, A., Patarin, J., Shamir, A.: Efficient Algorithms for Solving Overdefined Systems of Multivariate Polynomial Equations. In: Preneel, B. (ed.) EUROCRYPT 2000. LNCS, vol. 1807, pp. 392–407. Springer, Heidelberg (2000)

13. Courtois, N.T., Meier, W.: Algebraic attacks on stream ciphers with linear feedback. In: Biham, E. (ed.) EUROCRYPT 2003. LNCS, vol. 2656, pp. 345–359. Springer, Heidelberg (2003)

14. Courtois, N.T., Pieprzyk, J.: Cryptanalysis of Block Ciphers with Overdefined Systems of Equations. IACR ePrint Archive, Report 2002/044 (2002), http://eprint.iacr.org/2002/044

15. Daemen, J., Rijmen, V.: The design of Rijndael: AES - the Advanced Encryption Standard. Springer, Heidelberg (2002)

16. Een, N., Sörensson, N.: An extensible SAT-solver. In: Giunchiglia, E., Tacchella, A. (eds.) SAT 2003. LNCS, vol. 2919, pp. 502–518. Springer, Heidelberg (2003), http://www.cs.chalmers.se/Cs/Research/FormalMethods/MiniSat/

17. Faugère, J.-C.: A New Efficient algorithm for Computing Gröbner Basis, F4 (1999), http://modular.ucsd.edu/129-05/refs/faugere_f4.pdf

18. Faugère, J.-C.: A New Efficient Algorithm for Computing Gröbner Bases without Reduction to Zero (F5). In: Proceedings of ISSAC, pp. 75–83. ACM Press, New York (2002)

19. Faugère, J.-C.: Gröbner bases: Applications in Cryptology. FSE 2007 – Invited Talk (2007), http://fse2007.uni.lu/v-misc.html

20. Greuel, G.-M., Pfister, G., Schönemann, H.: Singular 3.0. A Computer Algebra System for Polynomial Computations, Centre for Computer Algebra, University of Kaiserslautern (2005), http://www.singular.uni-kl.de

21. Knudsen, L.R.: Truncated and higher order differentials. In: Preneel, B. (ed.) FSE 1995. LNCS, vol. 1008, pp. 196–211. Springer, Heidelberg (1995)

22. Matsui, M.: Linear Cryptanalysis Method for DES Cipher. In: Helleseth, T. (ed.) EUROCRYPT 1993. LNCS, vol. 765, pp. 386–397. Springer, Heidelberg (1993), http://homes.esat.kuleuven.be/~abiryuko/Cryptan/matsui_des.PDF

23. Murphy, S., Robshaw, M.: Essential Algebraic Structure Within the AES. In: Yung, M. (ed.) CRYPTO 2002. LNCS, vol. 2442, pp. 1–16. Springer, Heidelberg (2002), http://www.isg.rhul.ac.uk/~mrobshaw/rijndael/aes-crypto.pdf

24. Raddum, H., Semaev, I.: New technique for solving sparse equation systems. IACR ePrint Archive, Report 2006/475 (2006), http://eprint.iacr.org/2006/475

25. The SAGE Group. SAGE Mathematics Software (Version 3.3) (2008), http://www.sagemath.org

26. Wagner, D.: The boomerang attack. In: Knudsen, L.R. (ed.) FSE 1999. LNCS, vol. 1636, pp. 156–170. Springer, Heidelberg (1999), http://www.cs.berkeley.edu/~daw/papers/boomerang-fse99.ps

27. Wang, M.: Differential Cryptanalysis of reduced-round PRESENT. In: Vaudenay, S. (ed.) AFRICACRYPT 2008. LNCS, vol. 5023, pp. 40–49. Springer, Heidelberg (2008)

28. Wang, M.: Private communication: 24 differential characteristics for 14-round present we have found (2008)

29. Yang, B.-Y., Chen, J.-M., Courtois, N.T.: On Asymptotic Security Estimates in XL and Gröbner Bases-Related Algebraic Cryptanalysis. In: López, J., Qing, S., Okamoto, E. (eds.) ICICS 2004. LNCS, vol. 3269, pp. 401–413. Springer, Heidelberg (2004)

A   Small Key Bit Recovery System

$$X_0' = K_0 + P_0', \quad X_1' = K_1 + P_1', \quad X_2' = K_2 + P_2', \quad X_3' = K_3 + P_3',$$
$$Y_0' = X_0'X_1'X_3' + X_0'X_2'X_3' + X_0' + X_1'X_2'X_3' + X_1'X_2' + X_2' + X_3' + 1,$$
$$Y_1' = X_0'X_1'X_3' + X_0'X_2'X_3' + X_0'X_2' + X_0'X_3' + X_0' + X_1' + X_2'X_3' + 1,$$
$$Y_2' = X_0'X_1'X_3' + X_0'X_1' + X_0'X_2'X_3' + X_0'X_2' + X_0' + X_1'X_2'X_3' + X_2',$$
$$Y_3' = X_0' + X_1'X_2' + X_1' + X_3',$$
$$X_0'' = K_0 + P_0'', \quad X_1'' = K_1 + P_1'', \quad X_2'' = K_2 + P_2'', \quad X_3'' = K_3 + P_3'',$$
$$Y_0'' = X_0''X_1''X_3'' + X_0''X_2''X_3'' + X_0'' + X_1''X_2''X_3'' + X_1''X_2'' + X_2'' + X_3'' + 1,$$
$$Y_1'' = X_0''X_1''X_3'' + X_0''X_2''X_3'' + X_0''X_2'' + X_0''X_3'' + X_0'' + X_1'' + X_2''X_3'' + 1,$$
$$Y_2'' = X_0''X_1''X_3'' + X_0''X_1'' + X_0''X_2''X_3'' + X_0''X_2'' + X_0''X_1''X_2''X_3'' + X_2'',$$
$$Y_3'' = X_0'' + X_1''X_2'' + X_1'' + X_3'',$$
$$\Delta Y_0 = Y_0' + Y_0'', \quad \Delta Y_1 = Y_1' + Y_1'', \quad \Delta Y_2 = Y_2' + Y_2'', \quad \Delta Y_3 = Y_3' + Y_3'',$$

where ΔY_i are the *known* difference values predicted by the characteristic.

B   14-Round Differential Characteristic for PRESENT

Rounds		Differences	Pr	Rounds		Difference	Pr
I		$x_2 = 7, x_{14} = 7$	1				
R1	S	$x_2 = 1, x_{14} = 1$	2^{-4}	R8	S	$x_8 = 9, x_{10} = 9$	2^{-4}
R1	P	$x_0 = 4, x_3 = 4$	1	R8	P	$x_2 = 5, x_{14} = 5$	1
R2	S	$x_0 = 5, x_3 = 5$	2^{-4}	R9	S	$x_2 = 1, x_{14} = 1$	2^{-6}
R2	P	$x_0 = 9, x_8 = 9$	1	R9	P	$x_0 = 4, x_3 = 4$	1
R3	S	$x_0 = 4, x_8 = 4$	2^{-4}	R10	S	$x_0 = 5, x_3 = 5$	2^{-4}
R3	P	$x_8 = 1, x_{10} = 1$	1	R10	P	$x_0 = 9, x_8 = 9$	1
R4	S	$x_8 = 9, x_{10} = 9$	2^{-4}	R11	S	$x_0 = 4, x_8 = 4$	2^{-4}
R4	P	$x_2 = 5, x_{14} = 5$	1	R11	P	$x_8 = 1, x_{10} = 4$	1
R5	S	$x_2 = 1, x_{14} = 1$	2^{-6}	R12	S	$x_8 = 9, x_{10} = 9$	2^{-4}
R5	P	$x_0 = 4, x_3 = 4$	1	R12	P	$x_2 = 5, x_{14} = 5$	1
R6	S	$x_0 = 5, x_3 = 5$	2^{-4}	R13	S	$x_2 = 1, x_{14} = 1$	2^{-6}
R6	P	$x_0 = 9, x_8 = 9$	1	R13	P	$x_0 = 4, x_3 = 4$	1
R7	S	$x_0 = 4, x_8 = 4$	2^{-4}	R14	S	$x_0 = 5, x_3 = 5$	2^{-4}
R7	P	$x_8 = 1, x_{10} = 1$	1	R14	P	$x_0 = 9, x_8 = 9$	1

C   Times in Seconds for *Attack-B*

N	K_s	r	p	#trials	SINGULAR	#trials	POLYBORI
4	80	4	2^{-16}	20	$11.92 - 12.16$	50	$0.72 - 0.81$
4	80	3	2^{-12}	10	$106.55 - 118.15$	50	$6.18 - 7.10$
4	80	2	2^{-8}	10	$119.24 - 128.49$	50	$5.94 - 13.30$
4	80	1	2^{-4}	10	$137.84 - 144.37$	50	$11.83 - 33.47$
8	80	5	2^{-22}	0	N/A	50	$18.45 - 63.21$
10	80	8	2^{-34}	0	N/A	20	$21.73 - 38.96$
10	80	7	2^{-30}	0	N/A	10	$39.27 - 241.17$
10	80	6	2^{-26}	0	N/A	20	$56.30 - > 4$ hours
16	80	14	2^{-62}	0	N/A	20	$43.42 - 64.11$
16	128	14	2^{-62}	0	N/A	20	$45.59 - 65.03$
16	80	13	2^{-58}	0	N/A	20	$80.35 - 262.73$
16	128	13	2^{-58}	0	N/A	20	$81.06 - 320.53$
16	80	12	2^{-52}	0	N/A	5	> 4 hours
17	80	14	2^{-62}	10	$12,317.49 - 13,201.99$	20	$55.51 - 221.77$
17	128	14	2^{-62}	10	$12,031.97 - 13,631.52$	20	$94.19 - 172.46$
17	80	13	2^{-58}	0	N/A	5	> 4 hours
17	128	13	2^{-58}	0	N/A	5	> 4 hours

D   Times in Seconds for *Attack-C*

N	K_s	r	p	#trials	SINGULAR	#trials	POLYBORI	#trials	MINISAT2
4	80	4	2^{-16}	10	$0.07 - 0.09$	50	$0.05 - 0.06$	0	N/A
4	80	3	2^{-12}	10	$6.69 - 6.79$	50	$0.88 - 1.00$	50	$0.14 - 0.18$
4	80	2	2^{-8}	10	$28.68 - 29.04$	50	$2.16 - 5.07$	50	$0.32 - 0.82$
4	80	1	2^{-4}	10	$70.95 - 76.08$	50	$8.10 - 18.30$	50	$1.21 - 286.40$
16	80	14	2^{-62}	10	$123.82 - 132.47$	50	$2.38 - 5.99$	0	N/A
16	128	14	2^{-62}	0	N/A	50	$2.38 - 5.15$	0	N/A
16	80	13	2^{-58}	10	$301.70 - 319.90$	50	$8.69 - 19.36$	0	N/A
16	128	13	2^{-58}	0	N/A	50	$9.58 - 18.64$	0	N/A
16	80	12	2^{-52}	0	N/A	5	> 4 hours	0	N/A
17	80	14	2^{-62}	10	$318.53 - 341.84$	50	$9.03 - 16.93$	50	$0.70 - 58.96$
17	128	14	2^{-62}	0	N/A	50	$8.36 - 17.53$	50	$0.52 - 8.87$
17	80	13	2^{-58}	0	N/A	5	> 4 hours	5	> 4 hours

Multidimensional Extension of Matsui's Algorithm 2

Miia Hermelin[1], Joo Yeon Cho[1], and Kaisa Nyberg[1,2]

[1] Helsinki University of Technology
[2] Nokia Research Center, Finland

Abstract. Matsui's one-dimensional Alg. 2 can be used for recovering bits of the last round key of a block cipher. In this paper a truly multidimensional extension of Alg. 2 based on established statistical theory is presented. Two possible methods, an optimal method based on the log-likelihood ratio and a χ^2-based goodness-of-fit test are compared in theory and by practical experiments on reduced round Serpent. The theory of advantage by Selçuk is generalised in multiple dimensions and the advantages and data, time and memory complexities for both methods are derived.

1 Introduction

Linear cryptanalysis was introduced by Matsui in [1]. The method uses a one-dimensio- nal linear relation for recovering information about the secret key of a block cipher. Matsui presented two algorithms, Algorithm 1 (Alg. 1) and Algorithm 2 (Alg. 2). While Alg.1 extracts one bit of information about the secret key, Alg. 2 ranks several candidates for a part of the last round key of a block cipher according to a test statistic such that the right key should be ranked highest. Using the recovered last round key, it is then possible to extract one bit of information about the other round keys.

Since then researchers have been puzzled by the question how the linear cryptanalysis method could be enhanced by making use of multiple linear approximations simultaneously. In [2] Kaliski and Robshaw used several linear relations involving the same key bits in an attempt to reduce the data complexities of Matsui's algorithms. Multiple linear relations were also used by Biryukov, et al., [3] for extracting several bits of information about the key in an Alg. 1 type attack. This basic attack was also extended to an Alg. 2 type attack. However, both [2] and [3] depend on theoretical assumptions about the statistical properties of the one-dimensional linear relations that may not hold in the general case as was shown in [4].

The statistical linear distinguisher presented by Baignères, et al., in [5] does not suffer from this limitation. It has also another advantage over the previous approaches [2] and [3]: it is based on a well established statistical theory of log-likelihood ratio, LLR, see also [6]. In [7] it was further shown how to distinguish one known probability distribution from a set of other distributions.

O. Dunkelman (Ed.): FSE 2009, LNCS 5665, pp. 209–227, 2009.

The purpose of this paper is to present two new multidimensional extensions of Matsui's Alg. 2 including an effective ranking method for the key candidates based on Selçuk's concept of advantage [8]. First a straightforward solution for Alg. 2 based on goodness-of-fit test using χ^2-statistic will be presented. We will then discuss a χ^2-based version of Alg. 1 [9] and show that the method of Biryukov, et al., is related to a combination of the χ^2-based Alg. 1 and Alg. 2. We will then present a method based on LLR which actually combines Alg. 1 and Alg. 2 and outperforms the χ^2-based method in theory and practice. In the practical experiments the data, memory and time complexity for achieved advantage is determined and compared with the values given by the theoretical statistical models developed in this paper.

The structure of this paper is as follows: In Sect. 2 the basic statistical theory and notation is given. The advantage and the generalisation of Selçuk's theory is presented in Sect. 3. The multidimensional Alg. 2 is described in Sect. 4 and the different methods based on the two test statistics are described in Sect. 5 and Sect. 6. The time, memory and data complexities of both methods are examined in Sect. 7. The experimental results are given in Sect. 8. Finally, Sect. 9 draws conclusions.

2   Boolean Function and Probability Distribution

We will denote the space of n-dimensional binary vectors by V_n. A function $f : V_n \rightarrow V_1$ is called a Boolean function. A function $f : V_n \rightarrow V_m$ with $f = (f_1, \ldots, f_m)$, where f_i are Boolean functions is called a vector Boolean function of dimension m. A linear Boolean function from V_n to V_m is represented by an $m \times n$ binary matrix U. The m rows of U are denoted by $u_1, \ldots, u_m$, where each u_i is a binary vector of length n.

The correlation between a Boolean function and zero is

$$c(f) = c(f, 0) = 2^{-n} \left(\#\{\xi \in V_n \mid f(\xi) = 0\} - \#\{\xi \in V_n \mid f(\xi) \neq 0\} \right)$$

and it is also called the correlation of f.

We say that the vector $p = (p_0, \ldots, p_M)$ is a probability distribution (p.d.) of random variable (r.v.) X and denote $X \sim p$, if $\Pr(X = \eta) = p_\eta$, for all $\eta = 0, \ldots, M$. We will denote the uniform p.d. by θ. Let $f : V_n \rightarrow V_m$ and $X \sim \theta$. We call the p.d. p of the r.v. $Y = f(X)$ the p.d. of f.

Let us study some general properties of p.d.'s. Let $p = (p_0, \ldots, p_M)$ and $q = (q_0, \ldots, q_M)$ be some p.d.'s of r.v.'s taking on values in a set with $M + 1$ elements. The Kullback-Leibler distance between p and q is defined as follows:

Definition 1. *The* relative entropy *or* Kullback-Leibler distance *between p and q is*

$$D(p \,\|\, q) = \sum_{\eta=0}^{M} p_\eta \log \frac{p_\eta}{q_\eta}, \tag{1}$$

with the conventions $0 \log 0/b = 0$, $b \neq 0$ and $b \log b/0 = \infty$.

The following property usually holds for p.d.'s related to any real ciphers, so it will be frequently used throughout this work:

Property 1. *We say that distribution p is close to q if $|p_\eta - q_\eta| \ll q_\eta$, for all $\eta = 0, 1, \ldots, M$.*

If p is close to q then we can approximate the Kullback-Leibler-distance between p and q by its Taylor series. We call the first term of the series the capacity of p and q and it is defined as follows:

Definition 2. *The capacity between two p.d.'s p and q is defined by*

$$C(p, q) = \sum_{\eta=0}^{M} \frac{(p_\eta - q_\eta)^2}{q_\eta}. \tag{2}$$

If q is the uniform distribution, then $C(p, q)$ will be denoted by $C(p)$ and called the capacity of p.

The normed normal distribution with mean 0 and variance 1 is denoted by $\mathcal{N}(0, 1)$. Its probability density function (p.d.f.) is

$$\phi(x) = \frac{1}{\sqrt{2\pi}} e^{-x^2/2} \tag{3}$$

and the cumulative distribution function (c.d.f.) is

$$\Phi(x) = \int_{-\infty}^{x} \phi(t)\, dt\,. \tag{4}$$

The normal distribution with mean μ and variance σ^2 is denoted by $\mathcal{N}(\mu, \sigma^2)$ and its p.d.f. and c.d.f. are ϕ_{μ, σ^2} and Φ_{μ, σ^2}, respectively.

The χ_M^2-distribution with M degrees of freedom has mean M and variance $2M$. The non-central $\chi_M^2(\lambda)$-distribution with M degrees of freedom has mean $\lambda + M$ and variance $2(M + 2\lambda)$. If $M > 30$, we may approximate $\chi_M^2(\lambda) \sim \mathcal{N}(\lambda + M, 2(M + 2\lambda))$ [10].

Let $X_1, \ldots, X_n$ be a sequence independent and identically distributed (i.i.d.) random variables where either $X_i \sim p$, for all $i = 1, \ldots, N$ (corresponding to null hypothesis H_0) or $X_i \sim q \neq p$, for all $i = 1, \ldots, N$ (corresponding to alternate hypothesis H_1) and let $\hat{x}_1, \ldots, \hat{x}_N$ be the empirical data. The hypothesis testing problem is then to determine whether to accept or reject H_0. The Neyman-Pearson lemma [11] states that an optimal statistic for solving this problem, or distinguishing between p and q, is the log-likelihood ratio defined by

$$\mathrm{LLR}(\hat{q}, p, q) = \sum_{\eta=0}^{M} N\hat{q}_\eta \log \frac{p_\eta}{q_\eta}, \tag{5}$$

where $\hat{q} = (\hat{q}_0, \ldots, \hat{q}_M)$ is the empirical p.d. calculated from the data $\hat{x}_1, \ldots, \hat{x}_N$ by

$$\hat{q}_\eta = \frac{1}{N} \#\{i = 1, \ldots, N \mid \hat{x}_i = \eta\}.$$

The distinguisher accepts H_0, that is, outputs p (respectively rejects H_0 or outputs q) if $\text{LLR}(\hat{q}, p, q) \geq \gamma$ $(< \gamma)$ where γ is the threshold that depends on the level and the power of the test. If the power and the level of the test are equal (as is often the case) then $\gamma = 0$.

The proof for the following result can be found in [11], see also [5].

Proposition 1. *The LLR-statistic calculated from i.i.d. empirical data $\hat{x}_i$, $i = 1, \ldots, N$ using (5) is asymptotically normal with mean and variance $N\mu_0$ and $N\sigma_0^2$ ($N\mu_1$ and $N\sigma_1^2$, resp.) if the data is drawn from p (q, resp.). The means and variances are given by*

$$\mu_0 = D(p \,\|\, q) \quad \mu_1 = -D(q \,\|\, p)$$

$$\sigma_0^2 = \sum_{\eta=0}^{M} p_\eta \log^2 \frac{p_\eta}{q_\eta} - \mu_0^2 \quad \sigma_1^2 = \sum_{\eta=0}^{M} q_\eta \log^2 \frac{p_\eta}{q_\eta} - \mu_1^2. \tag{6}$$

Moreover, if p is close to q, we have

$$\mu_0 \approx -\mu_1 \approx \frac{1}{2} C(p, q) \quad \sigma_0^2 \approx \sigma_1^2 \approx C(p, q). \tag{7}$$

3   Advantage in Key Ranking

In a key recovery attack one is given a set of key candidates, and the problem is to determine which key is the right one. Usually the keys are searched from the set V_n of all 2^n strings of n bits. The algorithm consists of four phases, the *counting phase, analysis phase, sorting phase* and *searching phase* [12]. In the counting phase one collects data from the cipher, for example, plaintext-ciphertext pairs. In the analysis phase a real-valued statistic T is used in calculating a rank (or "mark" [12]) $T(\kappa)$ for all candidates $\kappa \in V_n$.

In the sorting phase the candidates κ are sorted, i.e., ranked, according to the statistic T. Optimally, the right key, denoted by κ_0, should be at the top of the list. If this is not the case, then one must also run through a search phase, testing the keys in the list until κ_0 is found. The goal of this paper is to find a statistic $T(\kappa)$ that is easy to compute and that is also reliable and efficient in finding the right key.

The time complexity of the search phase, given amount N of data, was measured using a special purpose quantity "gain" in [3]. A similar but more generally applicable concept of "advantage" was introduced by Selçuk in [8], where it was defined as follows:

Definition 3. *We say that a key recovery attack for an n-bit key achieves an advantage of a bits over exhaustive search, if the correct key is ranked among the top $r = 2^{n-a}$ out of all 2^n key candidates.*

Statistical tests for key recovery attacks are based on the Wrong-key Hypothesis [13]. We state it as follows:

Assumption 1 (Wrong-key Hypothesis). *There are two p.d.'s q and q', $q \neq q'$ such that for the right key κ_0, the data is drawn from q and for a wrong key $\kappa \neq \kappa_0$ the data is drawn from $q' \neq q$.*

A real-valued statistic T is computed from q and q', where one of these p.d.'s may be unknown, and the purpose of a statistic T is to distinguish between q and q'. We use D_R to denote the p.d. such that $T(\kappa_0) \sim D_R$. We will assume $D_R = \mathcal{N}(\mu_R, \sigma_R^2)$, with parameters μ_R and σ_R, as this will be the case with all statistics in this paper. Then μ_R and σ_R are determined with the help of linear cryptanalysis. We denote by D_W the p.d. known based on the Wrong-key Hypothesis such that $T(\kappa) \sim D_W$ for all $\kappa \neq \kappa_0$. The p.d.f. and c.d.f. of D_W are denoted by f_W and F_W, respectively.

Ranking the keys κ according to T means rearranging the 2^n r.v.'s $T(\kappa), \kappa \in V_n$, in decreasing order of magnitude. Writing the ordered r.v.'s as $T_0 \geq T_1 \geq \cdots \geq T_M$, we call T_i the ith order statistic. Let us fix the advantage a such that the right key should be among the $r = 2^{n-a}$ highest ranking keys. Hence, the right key should be at least as high as the rth wrong key corresponding to T_r. By Theorem 1. in [8] we get that the r.v. T_r is distributed as

$$T_r \sim \mathcal{N}(\mu_a, \sigma_a^2), \text{ where}$$

$$\mu_a = F_W^{-1}(1 - 2^{-a}) \text{ and } \sigma_a \approx \frac{2^{-(n+a)/2}}{f_W(\mu_a)}. \tag{8}$$

If we now define the success probability P_S of having κ_0 among the r highest ranking keys we have

$$P_S = \Pr(T(\kappa_0) - T_r > 0) = \Phi\left(\frac{\mu_R - \mu_a}{\sqrt{\sigma_R^2 + \sigma_a^2}}\right), \tag{9}$$

since $T(\kappa_0) - T_r \sim \mathcal{N}(\mu_R - \mu_a, \sigma_R^2 + \sigma_a^2)$.

As the data complexity N depends on the parameters $\mu_R - \mu_a$ and $\sigma_R^2 + \sigma_a^2$, we can solve N from (9) as a function of a and vice versa. Hence, (9) describes the trade-off between the data complexity N and the complexity of the search phase.

In a block cipher, the unknown key is divided into a number of round keys not necessarily disjoint or independent. In [3], the keys of the last round (or first and last round) were called the outer keys and the rest of the round keys were called inner keys. The unknown key κ may consist of outer keys, the parity bits of inner keys or both. Traditionally, in Matsui's Alg. 1 key parity bit(s) of the inner keys are searched, whereas in Alg. 2. the main goal is to determine parts of the outer keys.

4   Algorithm 2

4.1   Multidimensional Linear Approximation

Let us study a block cipher with t rounds. Let $x \in V_n$ be the plaintext, $y \in V_n$ the ciphertext, $K \in V_\nu$ the fixed round key data (the inner key) used in all but the last round and $z = f_t^{-1}(y, k)$, $k \in V_l$, the input to the last round function f_t, obtained from y by decrypting with the last round key data k (outer key). Let $m \leq n$ be an integer. Using m-dimensional linear cryptanalysis one can determine an approximation p of the p.d. of the Boolean function

$$x \mapsto Ux + Wz + VK, \tag{10}$$

which defines an m-dimensional linear approximation, where U and W are $m \times n$ matrices and V is an $m \times \nu$ matrix. A way of obtaining p from the one-dimensional correlations was presented in [4]. The linear mapping V divides the inner key space to 2^m equivalence classes $g = VK \in V_m$. Let the right last round key be denoted by k_0. Denote $M = 2^m - 1$ from now on.

In the counting phase we draw N data pairs $(\hat{x}_i, \hat{y}_i), i = 1, \ldots, N$. In the analysis phase, for each last round key k, we first calculate $\hat{z}_i^k = f_t^{-1}(\hat{y}_i, k), i = 1, \ldots, N$. Then, for each k, we calculate the empirical p.d. $\hat{q}^k = (\hat{q}_0^k, \ldots, \hat{q}_M^k)$, where

$$\hat{q}_\eta^k = \frac{1}{N} \#\{i = 1, \ldots, N \,|\, U\hat{x}_i + W\hat{z}_i^k = \eta\}. \tag{11}$$

If we use the wrong key $k \neq k_0$ to decrypt $\hat{y}_i, i = 1, \ldots, N$, it means we essentially encrypt over one more round and the resulting data will be more uniformly distributed. This heuristics is behind the original Wrong-key Randomisation Hypothesis [14], which in our case means that the data $U\hat{x}_i + W\hat{z}_i^k$, $i = 1, \ldots, N$, $k \neq k_0$ is drawn i.i.d. from the uniform distribution.

When decrypting with the correct key k_0 the data $U\hat{x}_i + W\hat{z}_i^{k_0} + g$, $i = 1, \ldots, N$, where g is an unknown inner key class, is drawn i.i.d. from p. This means that the data $U\hat{x}_i + W\hat{z}_i^{k_0}$, $i = 1, \ldots, N$ is drawn i.i.d. from a fixed permutation of p denoted by p^g. These permuted p.d.'s have the property that $p_{\eta \oplus h}^g = p_\eta^{g \oplus h}$, for all $g, \eta, h \in V_m$, and consequently

$$D(p^g \,||\, \theta) = D(p \,||\, \theta) \text{ and } C(p) = C(p^g) \text{ for all } g \in V_m. \tag{12}$$

Moreover, $D(p \,||\, p^h) = D(p^g \,||\, p^{h \oplus g})$, for all $h, g \in V_m$, from which it follows that

$$\min_{g' \neq g} D(p^g \,||\, p^{g'}) = \min_{h \neq 0} D(p \,||\, p^h), \tag{13}$$

which is a constant value for all $g \in V_m$. We will denote this value by $D_{\min}(p)$ and assume in the sequel that $D_{\min}(p) \neq 0$ without restriction: We can unite the key classes for which the Kullback-Leibler distance is zero. Then we just have $m' < 2^m$ key classes whose Kullback-Leibler distance from each other is non-zero. The corresponding minimum capacity $\min_{h \neq 0} C(p, p^h)$ is denoted by $C_{\min}(p)$.

4.2   Key Ranking in One-Dimensional Alg. 2

Key ranking and advantage in the one-dimensional case, $m = 1$, of Alg. 2 was studied in [8]. We will present it here briefly for completeness. Let $c > 0$ be the correlation of (10) (the calculations are similar if $c < 0$) and let $\hat{c}^k$ be the empirical correlation calculated from the data. The statistic used in ranking the keys is then $s(k) = |\hat{c}^k|$. The r.v. $\hat{c}^{k_0}$ is binomially distributed with mean $\mu_R = c$ and variance $\sigma_R^2 = (1 - c^2)/N \approx 1/N$. The wrong key r.v.'s $\hat{c}^k$, $k \neq k_0$, are binomially distributed with mean $\mu_W = 0$ (following Assumption 1) and variance $\sigma_W^2 = \sigma_R^2$. Since N is large, we can approximate $s(k_0) \sim \mathcal{N}(\mu_R, \sigma_R^2)$ and $s(k) \sim \mathcal{FN}(\mu_W, \sigma_W^2)$, where $\mathcal{FN}$ is the folded normal distribution, see Appendix A in [8]. Now we can proceed as in [8]. We get that, with given success probability P_S and advantage a, the data complexity is

$$N = \frac{(\Phi^{-1}(P_S) + \Phi^{-1}(1 - 2^{-a-1}))^2}{c^2}. \tag{14}$$

4.3 Different Scenarios in Multiple Dimensions

When considering generalisation of Alg.2 to the case, where multiple linear approximations are used, different approaches are possible. In a previous work by Biryukov, et al., [3], a number of selected one-dimensional linear approximations with high bias are taken into account simultaneously under the assumption that they are statistically independent. As we will show later in Sect. 5.3, the statistic used in [3] is essentially a goodness-of-fit test based on least squares and searches simultaneously the key parts k_0 and g_0 which give the best fit with the theoretically estimated correlations.

The approaches taken in [5] for linear distinguishing and later in [4] for Alg. 1 do not need assumptions about independence of the linear approximations as they are based on the p.d. of the multidimensional linear approximation (10). When using the multidimensional p.d., basically two different standard statistical methods can be used:

- Goodness-of-fit (usually based on χ^2-statistic) and
- Distinguishing of an unknown p.d. from a given set of p.d.'s (usually based on LLR-statistic)

The goodness-of-fit approach is a straightforward generalisation of one-dimensional Alg. 2. It can be used in searching for $\kappa = k$. It measures whether the data is drawn from the uniform (wrong) distribution, or not, by measuring the deviation from the uniform distribution. It ranks highest the key candidate whose empirical distribution is farthest away from the uniform distribution. The statistic does not depend on the inner key class g. Information about p.d. p is required only for measuring the strength of the test. We will study this method in Sect. 5.1. After the right round key k is found, one can use the data derived in Alg. 2 in any form of Alg. 1 for finding the inner key class g. In this manner, the χ^2-approach allows separating between Alg. 1 and Alg. 2.

The LLR-method uses the information about the p.d. related to the inner key class also in Alg. 2. In this sense, it is similar to the method of [3], where the Alg. 1 and Alg. 2 were combined together for finding both the outer and inner round keys. As we noted in Sect. 2, the LLR-statistic is the optimal distinguisher between two known p.d.'s. If we knew the right inner key class g_0, we could simply use the empirical p.d.'s $\hat{q}^k$ for distinguishing p^{g_0} and the uniform distribution and then choose the k for which this distinguisher is strongest [5]. In practice, the correct inner key class g_0 is unknown when running Alg. 2 for finding the last round key.

Our approach is the following. In [7] it was described how one can use LLR to distinguish one known p.d. from a set of p.d.'s. We will use this distinguisher for distinguishing θ from the given set p^g, $g \in V_m$. In the setting of Alg. 2, we can expect that for the right k_0, it should be possible to clearly conclude that the data $(\hat{x}_i, \hat{y}_i)$, $i = 1, \ldots, N$, yields data $(\hat{x}_i, \hat{z}_i^{k_0})$, $i = 1, \ldots, N$, which follows a p.d. p^g, for some $g \in V_m$, rather than the uniform distribution. On the other hand, for the wrong $k \neq k_0$, the data follows the uniform distribution, rather than any p^g, $g \in V_m$.

To distinguish k_0 from the wrong key candidates we determine, for each round key candidate k, the inner key class g, for which the LLR-statistic is the largest with the given data. The right key k_0 is expected to have g_0 such that the LLR-statistic with this pair (k_0, g_0) is larger than for any other pair $(k, g) \neq (k_0, g_0)$. In this manner, we also recover g_0 in addition to k_0. The LLR-method is studied in Sect. 6.

5   The χ^2-Method

This method separates the Alg. 1. and Alg. 2 such that the latter does not need any information of p. Both methods are interpreted as goodness-of-fit problems, for which the natural choice of ranking statistic is χ^2. We will show how to find the last round key k with Alg. 2 first.

5.1   Algorithm 2 with χ^2

Given empirical p.d. $\hat{q}^k$, we can calculate the χ^2-statistic from the data as

$$S(k) = 2^m N \sum_{\eta=0}^{M} (\hat{q}_\eta^k - 2^{-m})^2, \tag{15}$$

where $M = 2^m - 1$ is the number of degrees of freedom. The statistic can be interpreted as the l_2-distance between the empirical p.d. and the uniform distribution. By Assumption 1, the right round key should produce data that is farthest away from the uniform distribution and we will choose the round key k for which the statistic (15) is largest. Obviously, if $m = 1$, we get the statistic $(\hat{c}^k)^2$.

According to [15] the r.v. $S(k_0)$ is distributed approximately as

$$S(k_0) \sim \chi_M^2(NC(p^{g_0})) = \chi_M^2(NC(p)), \tag{16}$$

because of the symmetry property (12). Hence, we may approximate the distribution by a normal distribution with $\mu_R = M + NC(p)$ and $\sigma_R^2 = 2(M + 2NC(p))$.

The parameters do not depend on g_0 or k_0. For the wrong keys $k \neq k_0$, we obtain by [15] that

$$S(k) \sim \chi_M^2(0) = \chi_M^2, \tag{17}$$

so that $\mu_W = M$ and $\sigma_W^2 = 2M$. The mean and variance in (8) are $\mu_a = \sigma_W b + M = \sqrt{2M}b + M$ and $\sigma_a^2 = 2^{-(l+a)/2}\sigma_W^2/\phi(b) \ll \sigma_0^2$. Now we can solve N from (9) and get that the data complexity is proportional to

$$N_{\chi^2} = \frac{\beta(M, b, P_S)}{C(p)}, \; b = \Phi^{-1}(1 - 2^{-a}), \tag{18}$$

where $\beta(M, b, P_S)$ is a parameter that depends on M, b and P_S. Assuming large b, that is, large advantage a and large P_S, we can approximate β by

$$\beta = 2\sqrt{M}b + 4\Phi^{-2}(2P_S - 1). \tag{19}$$

Note that the normal approximation of the wrong-key distribution is valid only when $m > 5$, that is, when the approximation of χ^2-distribution by a normal distribution is valid. It is not possible to perform the theoretical calculations for small m as the χ^2-distribution does not have a simple asymptotic form in that case and we cannot determine f_W and F_W in (8). Since our χ^2-statistic reduces to the square of $s(k)$ that was used by Selçuk, the theoretical distributions differ from our calculations and we get a slightly different formula for the advantage. Despite this difference, the methods are equivalent for $m = 1$.

Keeping the capacity constant, it seems that the data complexity increases exponentially as $2^{m/2}$ as the dimension m of the linear approximation increases and is sufficiently large. Hence, in order to strengthen the attack, the capacity should increase faster than $2^{m/2}$ when the m is increased. This is a very strong condition and it suggests that in applications, only approximations with small m should be used with χ^2-attack. The experimental results for different m presented in Sect. 8 as well as the theoretical curves depicted in Fig. 5(a) suggest that increasing m in the χ^2-method does not necessarily mean improved performance for Alg. 2.

Since $2^{-a} = \Phi(-b) \approx 1/\sqrt{2\pi}e^{-b^2/2}$, we can solve a from (18) as a function of N and we have proved the following theorem that can be used in describing the relationship between the data complexity and the search phase:

Theorem 1. *Suppose the cipher satisfies Assumption 1 where $q' = \theta$ and the p.d.'s p^g, $g \in V_m$ and θ are close to each other. Then the advantage of the χ^2-method using statistic (15) is given by*

$$a_{\chi^2} = \frac{(NC(p) - 4\varphi)^2}{4M}, \; \varphi = \Phi^{-2}(2P_S - 1), \; M = 2^m - 1, \tag{20}$$

where $P_S (> 0.5)$ is the probability of success, N is the amount of data used in the attack and $C(p)$ and $m (\geq 5)$ are the capacity and the dimension of the linear approximation (10), respectively.

While (20) and (18) depend on the theoretical distribution p, the actual χ^2-statistic (15) is independent of p. Hence, we do not need to know p accurately to realise the attack, we only need to find an approximation (10) that deviates as much as possible from the uniform distribution. On the other hand, if we use time and effort for computing an approximation of the theoretical p.d. and if we may assume that the approximation is accurate, we would also like to exploit this knowledge for finding the right inner key class with Alg. 1. As noted in [9], there are several ways to realising a multidimensional Alg. 1. Next we discuss Alg. 1 as a χ^2-based goodness-of-fit problem.

5.2   Algorithm 1 with χ^2

Suppose that we have obtained an empirical distribution $\hat{q}$ of data that can be used for determining the inner key class g_0 using Alg. 1. For example, we have successfully run Alg. 2 and found the correct last round key k_0 and set $\hat{q} = \hat{q}^{k_0}$.

One approach is to consider Alg. 1 as a goodness-of-fit problem, where one determines, for each g, whether the empirical p.d. $\hat{q}$ follows p^g or not. The χ^2-based ranking statistic is then

$$S_{\text{Alg1}}(g) = N \sum_{\eta=0}^{M} \frac{(\hat{q}_\eta^{k_0} - p_\eta^g)^2}{p_\eta^g}, \tag{21}$$

which should be small for g_0 and large for the wrong inner key classes $g \neq g_0$. In [9] it is shown that the data complexity of finding g_0 with given success probability P_S is

$$N_{\text{Alg } 1, \chi^2} = \frac{4m - 4\gamma_S + 2\sqrt{2M(m - \gamma_S)}}{C_{\min}(p)}, \tag{22}$$

where $\gamma_S = \ln(\sqrt{2\pi} \ln P_S^{-1})$.

5.3   Combined Method and Discussion

The sums of squares of correlations used in [3] are closely related to the sums of squares (15) and (21). Indeed, we could define a combined χ^2-statistic B by considering the sum of the statistics from (15) and (21) and setting

$$B(k, g) = \sum_{k' \neq k} S(k) + S_{\text{Alg } 1}(k, g), \tag{23}$$

where $S_{\text{Alg } 1}(k, g)$ is the statistic (21) calculated from the empirical p.d. $\hat{q}^k$, $k \in V_l$. If we approximate the denominators in (21) by 2^{-m} and scaling by $2^m N$ we obtain from $B(k, g)$ the statistic

$$B'(k, g) = \sum_{k' \neq k} ||\hat{q}^{k'} - \theta||_2^2 + ||\hat{q}^k - p^g||_2^2. \tag{24}$$

This statistic is closely related to the one used in [3].

$$\sum_{k' \neq k} ||\hat{c}^{k'}||_2^2 + ||\hat{c}^k - c^g||_2^2. \tag{25}$$

Indeed, if in (25) all correlation vectors $\hat{c}^k$ and c^g contain correlations from all linear approximations then (25) becomes the same as $2^m B'(k, g)$ as can be seen using Parseval's theorem. Initially, in the theoretical derivation of (25) only linearly and statistically independent approximations were included in the correlation vectors. However, in Sect. 3.4 of [3] it was proposed to take into account all linear approximations with strong correlations when forming the statistic (25) in practice. In practical experiments by Collard, et al. [16] this heuristic enhancement was demonstrated to improve the results. In this paper, we have shown how to remove the assumption about independence of the linear approximations and that all linear approximations that have sufficient contribution to the capacity (cf. discussion in Sect. 5.1) can and should be included.

Other possibilities for combining Alg. 1 and Alg. 2 based on χ^2 or its variants are also possible, with different weights on the terms of the sum in (24), for instance. However, the mathematically more straightforward way is to use the pure χ^2-method defined by (15) and (21), as its statistical behaviour is well-known. An even more efficient method can be developed based on LLR as will be shown next.

6   The LLR-Method

This method is also based on the same heuristic as the Wrong-key Hypothesis: For $k \neq k_0$, the distribution of the data should look uniform and for k_0 it should look like p^{g_0}, for some g_0. Hence, for each k, the problem is to distinguish the uniform distribution from the discrete and known set p^g, $g \in V_m$. Let us use the notation $L(k, g) = \mathrm{LLR}(\hat{q}^k, p^g, \theta)$. We propose to use the following ranking statistic

$$L(k) = \max_{g \in V_m} L(k, g). \tag{26}$$

Now k_0 should be the key for which this maximum over g's is the largest and ideally, the maximum should be achieved when $g = g_0$. While the symmetry property (12) allows one to develop statistical theory without knowing g_0, in practice one must search through V_l for k_0 and V_m for g_0 even if we are only interested in determining k_0.

We assume that the p.d.'s p^g and θ are all close to each other. Using Theorem 1 and property (12) we can state Assumption 1 as follows: For the right pair k_0 and g_0

$$L(k_0, g_0) \sim \mathcal{N}(N\mu_R, N\sigma_R^2), \text{ where } \mu_R = \frac{1}{2}C(p) \text{ and } \sigma_R^2 = C(p), \tag{27}$$

and for $k \neq k_0$ and any $g \in V_m$

$$L(k, g) \sim \mathcal{N}(N\mu_W, N\sigma_W^2), \text{ where } \mu_W = -\frac{1}{2}C(p) \text{ and } \sigma_W^2 = C(p). \tag{28}$$

Hence, μ_R, σ_R^2, μ_W and σ_W^2 do not depend on $g \in V_m$. For fixed $k \neq k_0$, the r.v.'s $L(k, g)$ for $k \neq k_0$ are identically normally distributed with mean μ_W and variance σ_W^2. We will assume that they are statistically independent to simplify calculations. In particular, the assumption about statistical independence of $L(k, g)$ for different g does not mean that the linear approximations should be statistically independent. The statistic itself does not depend on this assumption[1]. Moreover, the theoretical results obtained this way are a little more pessimistic that those obtained by empirical tests, as shown in Sect. 8. Hence, these calculations give a theoretical model that can be used in describing how the method behaves especially compared to other methods. Assuming that for each $k \neq k_0$, the r.v.'s $L(k, g)$'s are independent, we obtain that the c.d.f. of their maximum is given by [17]

$$F_W(x) = \Phi_{N\mu_W, N\sigma_W^2}(x)^{M+1} \tag{29}$$

and p.d.f. is

$$f_w(x) = (M+1)\Phi_{N\mu_W, N\sigma_W^2}(x)^M \phi_{\mu_W, \sigma_W^2}(x). \tag{30}$$

Let us fix the advantage a such that $r = 2^{l-a}$. The mean μ_a of the rth wrong key statistic L_r can now be calculated from (8) to be

$$\mu_a = N\mu_W + \sqrt{N}\sigma_W b = -1/2NC(p) + \sqrt{NC(p)}b,$$
$$b = \Phi^{-1}(\sqrt[M+1]{1 - 2^{-a}}), \tag{31}$$

and the variance is

$$\sigma_a^2 = \frac{2^{-l-a}\sigma_W^2}{(M+1)^2(1 - 2^{-a})^{2(1-1/(M+1))}\phi^2(b)} \ll \sigma_0^2. \tag{32}$$

Let

$$P_1 = \Pr(L(k_0, g_0) > \max_{g \neq g_0} L(k_0, g)) \tag{33}$$

be the the probability that given k_0, we choose g_0, i.e., the probability of success of Alg. 1. Let

$$P_2 = \Pr(L(k_0) > L_r) \tag{34}$$

be the probability that we rank k_0, paired with *any* $g \in V_m$, among the r highest ranking keys. Finally, let

$$P_{12} = \Pr(L(k_0) > L_r \mid L(k_0, g_0) > \max_{g \neq g_0} L(k_0, g)) \tag{35}$$

be the probability that we rank k_0 among the r highest ranking keys provided that we pair g_0 with k_0. Then

$$P_2 = P_{12}P_1 + \Pr(L(k_0) > L_r \mid L(k_0) = \mathrm{LLR}(k_0, p^g, \theta), g \neq g_0)(1 - P_1)$$
$$\geq P_{12}P_1. \tag{36}$$

---

[1] See for example [17] for calculating the c.d.f. of the maximum of dependent and identically distributed r.v.'s, when $M \geq 100$. The theoretical predictions calculated that way are slightly more pessimistic than the ones obtained in Theorem 2.

If we pair k_0 with $g \neq g_0$ then $L(k_0) \geq L(k_0, g_0)$ for a fixed empirical p.d. $\hat{q}^{k_0}$, so that k_0 gets ranked *higher* than by using the correct g_0. Hence, assuming that k_0 gets paired with g_0 only decreases P_2 so the corresponding estimate of the data complexity gets larger. Let N_1, N_2 and N_{12} be the data complexities needed to achieve success probabilities P_1, P_2 and P_{12}, respectively.

We can calculate P_{12} using (27), (28) and (9) to obtain

$$P_{12} = \Phi(\frac{\mu_R - \mu_W - \sigma_w b}{\sigma_R}) = \Phi(\sqrt{N_{12}C(p)} - b), \, b = \Phi^{-1}(\sqrt[M+1]{1 - 2^{-a}}). \quad (37)$$

Hence, the data complexity is proportional to

$$N_{12} = \left(\Phi^{-1}(P_{12}) + b\right)^2 / C(p), \quad (38)$$

which can be used in approximating an upper bound for N_2. We can approximate $\Phi(b) = \sqrt[M+1]{1 - 2^{-a}} \approx 1 - 2^{-m-a}$ such that $a \approx b^2/2 - m$ and we can solve the advantage a as a function of $N_{12} \approx N_2$ from (38). We get the following theorem:

Theorem 2. *Suppose the cipher satisfies Assumption 1 where $q' = \theta$ and the p.d.'s p^g, $g \in V_m$ and θ are close to each other. Then the advantage of the LLR-method for finding the last round key k_0 is given by*

$$a_{\mathrm{LLR}} = (\sqrt{NC(p)} - \Phi^{-1}(P_{12}))^2 / 2 - m \approx NC(p) - m. \quad (39)$$

Here N is the amount of data used in the attack, $P_{12} (> 0.5)$ is the probability of success and $C(p)$ and m are the capacity and the dimensions of the linear approximation (10), respectively.

Theorem 2 now gives the trade-off between the search phase and the data complexity of the algorithm. With fixed N and capacity $C(p)$, the advantage decreases linearly with m whereas in (20) the logarithm of advantage decreases linearly with m. For fixed m and p, the advantage of the LLR-method seems to be larger than the advantage of the χ^2-method. The experimental comparison of the methods is presented Sect. 8

In [4] it is shown that the data complexity of Alg. 1 for finding the right inner key class g_0 is proportional to

$$N_1 = \frac{16m \ln 2 - 16P_1'}{C(p)}, \quad (40)$$

where $P_1' = \ln(\sqrt{2\pi} \ln P_1^{-1})$. If we want to be certain that we have paired the right inner key class g_0 with k_0, the data complexity is given by

$$N_{\mathrm{LLR}} = \max(N_1, N_2) \propto \frac{m}{C(p)}. \quad (41)$$

The data complexity N_1 is an overestimate for the actual data complexity of Alg. 1 [9] so in practice, N_2 dominates.

7 Algorithms and Complexities

For comparing the two methods, LLR and χ^2, we are interested in the complexities of the first two phases of the Alg. 2 since the sorting and searching phase do not depend on the chosen statistic. The counting phase is done on-line and all the other phases can be done off-line. However, we have not followed this division [12] in our implementation, as we do part of the analysis phase on-line. We will divide the algorithm in two phases as follows: In the *on-line phase*, depicted in Fig. 1, we calculate the empirical p.d.'s for the round key candidates. The marks $S(k)$ for the χ^2-method and $L(k)$ for the LLR-method are then assigned to the keys in the *off-line phase*. The off-line phases for χ^2-method and LLR-method are depicted in Fig. 2 and Fig. 4, respectively. After the keys k are each given the mark, they can be ranked according to the mark. If we wish to recover g_0 with χ^2-method, we also need to store, in addition to the marks, the empirical p.d.'s q^k. Given q^{k_0}, one can use the multidimensional Alg. 1 described in Fig. 3 for finding g_0 off-line. The version of Alg. 1 is based on LLR. Obviously, one could use some other method, e.g. use the χ^2-based ranking statistic (21), which gives similar results in practice even if the LLR-based method is more powerful in theory [9].

---

initialise $2^l \times 2^m$ counters $F(k,\eta)$, $k = 0, \ldots, 2^l - 1$, $\eta = 0, \ldots, M$;
for $i = 1, \ldots, N$ **do**
    for *candidates* $k = 0, \ldots, 2^l - 1$ **do**
        decrypt the ciphertext partially: $\hat{z}_i^k = f^{-1}(\hat{y}_i, k)$;
        for $j = 1, \ldots, m$ **do**
            calculate bit $\eta_j = u_j \cdot \hat{x}_i \oplus w_j \cdot \hat{z}_i^k$;
        end
        increment counter $F(k,\eta) = \#\{i \mid U\hat{x}_i + W\hat{z}_i^k = \eta\}$, where η is the
        vector $(\eta_1, \ldots, \eta_m)$ interpreted as an integer;
    end
end

Fig. 1. On-line phase of Matsui's Alg. 2 in multiple dimensions

---

Input: table $F(k,\eta), k = 0, \ldots, 2^l - 1, \eta = 0, \ldots, M$;
for $k = 0, \ldots, 2^l - 1$ **do**
    compute $S(k) = \sum_{\eta=0}^{M}(F(k,\eta)/N - 2^{-m})^2$;
    if *wish to recover* g_0 **then**
        store $(S(k), F(k,0), \ldots, F(k,M))$;
    else
        store $S(k)$;
    end
end

Fig. 2. Off-line phase of Alg. 2 using χ^2-method

Table 1. Data, time and memory complexities of the χ^2- and LLR-method

	On-line			Off-line		
	χ^2 for k_0	χ^2 for k_0, g_0	LLR	χ^2 for k_0	χ^2 for k_0, g_0	LLR
Data	N_{χ^2}	N_{χ^2}	N_{LLR}	$-$	$-$	$-$
Time	$N_{\chi^2} 2^l m$	$N_{\chi^2} 2^l m$	$N_{LLR} 2^l m$	2^{l+m}	2^{l+m}	2^{l+m}
Memory	2^{l+m}	2^{l+m}	2^{l+m}	2^l	$2^m \max(2^l, 2^m)$	$2^m \max(2^l, 2^m)$

Input: counter values $F(k_0, 0), \ldots, F(k_0, M)$;
compute the theoretical distribution of m-dimensional approximations for each
value of 2^m inner key classes and store them in a $2^m \times 2^m$ table
$P(g, \eta), g = 0, \ldots, M, \eta = 0, \ldots, M$;
for *inner key classes* $g = 0, \ldots, M$ **do**
    calculate $G(g) = \sum_{\eta=0}^{M} F(k_0, \eta) \log P(g, \eta)$;
end
Output: g_0 such that $\max_{g \in V_m} G(g) = G(g_0)$

Fig. 3. Matsui's Alg. 1 in multiple dimensions (using LLR)

Input: table $F(k, \eta), k = 0, \ldots, 2^l - 1, \eta = 0, \ldots, M$;
compute the theoretical distribution of m-dimensional approximations for each
value of 2^m inner key classes and store them in a $2^m \times 2^m$ table
$P(g, \eta), g = 0, \ldots, M, \eta = 0, \ldots, M$;
for $k = 0, \ldots, 2^l - 1$ **do**
    for $g = 0, \ldots, M$ **do**
        $L(k, g) = \text{LLR}(\hat{q}^k, p^g, \theta)$, where $\hat{q}_\eta^k = F(k, \eta)/N$;
    end
    store $L(k) = \max_{g \in V_m} L(k, g)$;
end

Fig. 4. Off-line phase of Alg. 2 using LLR-method

The data, time and memory complexities for on-line and off-line phase for both
methods are shown in Table 1. Given success probability P_S and advantage a,
the data complexity N_{χ^2} is given by (18). If we want to recover g_0 also, then
theoretically, data complexity N_1 given by (40) is needed to successfully run
Alg. 1 given in Fig. 3. As noted in [9], the theoretical value N_1 is an overestimate
and the total data complexity in practice is probably dominated by the data
complexity N_{χ^2} of ranking k_0 high enough. Nevertheless, the data complexity of
the LLR-method is smaller than the χ^2-method.

Otherwise, the complexities for the LLR-method are mostly the same as for
χ^2-method provided that m is not much larger than l which is usually the case.
Thus, we recommend using the LLR-method rather than χ^2-method unless there
is great uncertainty about the validity of the approximative p.d p of the linear
relation (10).

In some situations it may also be advantageous to combine the different methods. For example, one may want to first find, say, r best round keys by χ^2, such that the data complexity N_{χ^2} is given by (18), where the advantage is $a = l - r$. Then one can proceed by applying the LLR-method to the remaining r keys, thus reducing the size of the round key space to be less than 2^l. Other similar variants are possible. Their usefulness depends on the cipher that is being studied.

8   Experiments

The purpose of the experiments was to test the accuracy of the derived statistical models and to demonstrate the better performance of the LLR-based method in practice. Similarly as in previous experiment on multiple linear cryptanalysis, see [16] and [3], the Serpent block cipher was used as a test-bed. The structure of Serpent is described, for example, in [18]. We have searched for a 12-bit part of the fifth round key based on m linear approximations with different m. Each experiment was performed for 16 different keys.

We calculated the capacities for the approximation (10) over 4-round Serpent for different m. Practical experiments were used in confirming that $C_{\min}(p) \approx C(p)$ and especially $C_{\min}(p) \neq 0$. We also saw that $|p_\eta^g - p_\eta^{g'}| < \frac{1}{150}p_\eta^g$, for all g, g' and $\eta \in V_m$. Hence, p^g's can be considered to be close to each other and θ.

The theoretical advantage of the χ^2-method predicted in (20) has been plotted as a function of data complexity in Fig. 5(a). The figure shows that increasing m larger than 4, the attack is weakened. This suggests using $m = 4$ base approximations in the χ^2-attack. Since we should have m at least 5 for the normal approximation of χ_M^2 to hold, the theoretical calculations do not necessarily hold for small m. However, the experiments, presented in Fig. 5(b), seem to confirm the theory for $m = 1$ and $m = 4$, too. The most efficient attack is obtained by using $m = 4$ equations. Increasing m (and hence, the time and memory complexities of the attack, see Table 1) actually weakens the attack. The optimal choice of m depends on the cipher. However, the theoretical calculations suggest that using $m \geq 5$ is usually not advantageous.

The reason is the χ^2-squared statistic itself: it only measures if the data follows a certain distribution, the uniform distribution in this case. The more approximations we use, the larger the distributions become and the more uncertainty we have about the "fitting" of the data. Small errors in experiments generate large errors in χ^2 as the fluctuations from the relative frequency 2^{-m} become more significant.

The theoretical advantage of the LLR-method (39) is plotted against the data complexity in Fig. 5(c) for different m. The empirical advantages for several different m are shown in Fig. 5(d). Unlike for χ^2 we see that the method can be strengthened by increasing m, until the increase in the capacity $C(p)$ becomes negligible compared to increase in m. For 4-round Serpent, this happens when $m \approx 12$.

Experimental results presented in Figures 5(d) and 5(b) confirm the theoretical prediction that the LLR-method is more powerful than the χ^2-method. Also

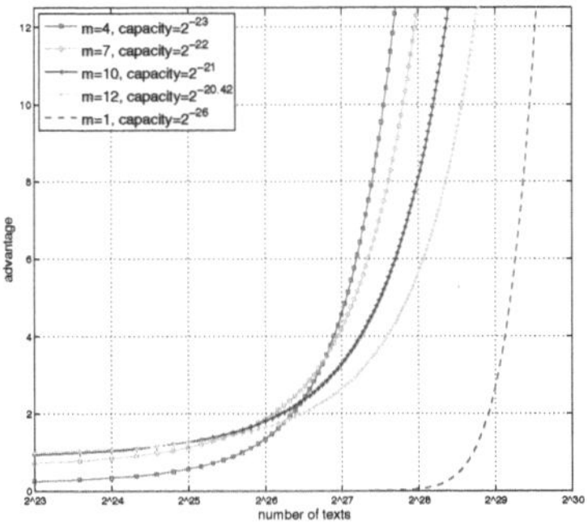

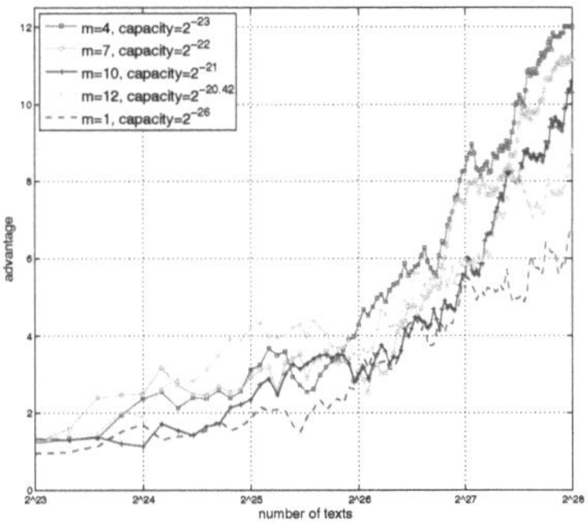

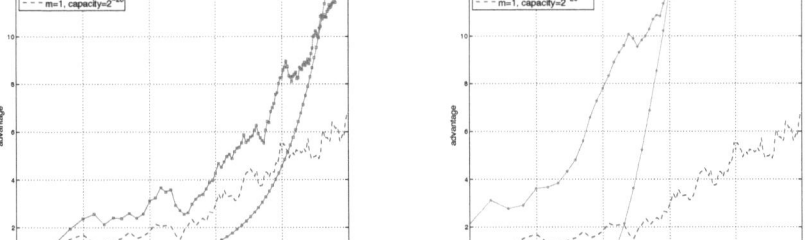

(a) Theoretical advantage for χ^2-method   (b) Empirical advantage for χ^2-method

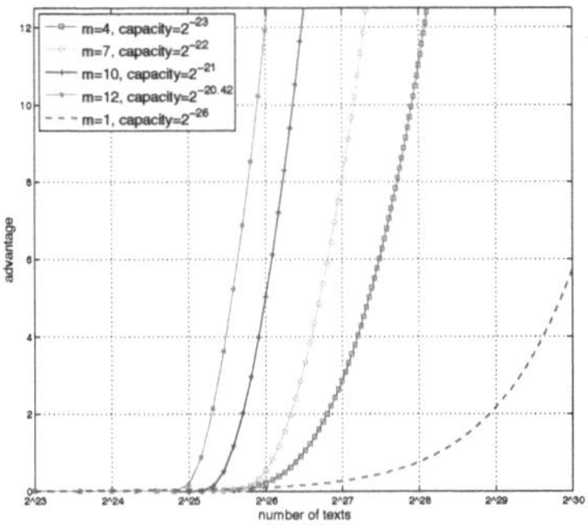

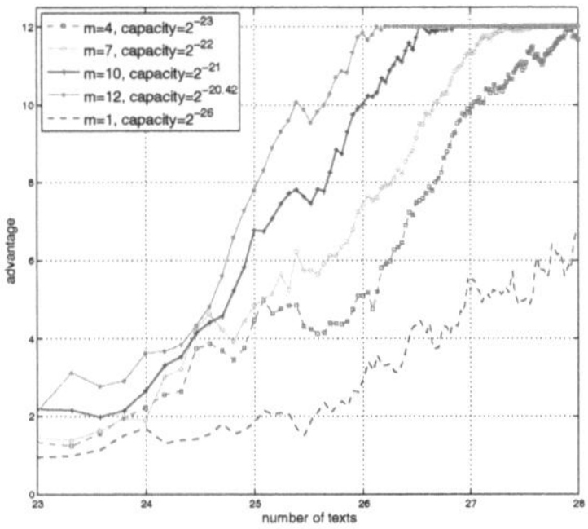

(c) Theoretical advantage for LLR-method (d) Empirical advantage for LLR-method

(e) Empirical and theoretical advantage for (f) Empirical and theoretical advantage for
χ^2 for $m = 1$ and $m = 4$ LLR for $m = 1$ and $m = 12$

Fig. 5. Theoretical and empirical advantages for χ^2- and LLR-method for different m and $P_S = P_{12} = 0.95$

the theoretical and empirical curves seem to agree nicely. For example, the full advantage of 12 bits with $m = 7$ achieved at $\log N = 27.5$ for LLR whereas χ^2-method needs about $\log N = 28$. Moreover, the LLR can be strengthened by increasing m. For $m = 12$, the empirical logarithmic data complexity is about 26.5.

9 Conclusions

There are several approaches of realising Matsui's Alg. 2 using multiple linear approximations. In this paper, methods based on two standard statistics, LLR and χ^2, were studied. Selçuk's theory of advantage describing the trade-off between data complexity and search phase was extended to multiple dimensions. The advantages of the two methods in key ranking were then determined. A description of the multidimensional Alg. 2 for both methods was given so that their performance measured in time, memory and data could be compared.

The χ^2-statistic, based on the classic goodness-of-fit test, was observed to perform poorly for large dimensions m of linear approximation, whereas the LLR-statistic, an optimal statistic for testing two known hypotheses, was shown to improve with the dimension m of the linear approximation much further. In particular, the advantage of using multiple linear approximations instead of just one is significant and of real practical importance if LLR-statistic is used in Alg. 2. In general, it was shown that the LLR-method is usually more advantageous compared to the χ^2-method. As long as there is no significant error, stemming from the linear hull-effect, for example, in determining the approximate p.d. of the multidimensional linear approximation, we recommend to use the LLR-method proposed in this paper rather than the χ^2-method.

Acknowledgements

We would like to thank Christophe de Cannière for insightful discussions and the anonymous referees for comments that helped us to improve the presentation of this paper.

References

1. Matsui, M.: Linear Cryptanalysis Method for DES Cipher. In: Helleseth, T. (ed.) EUROCRYPT 1993. LNCS, vol. 765, pp. 386–397. Springer, Heidelberg (1994)
2. Kaliski Jr., B.S., Robshaw, M.J.B.: Linear Cryptanalysis Using Multiple Approximations. In: Desmedt, Y.G. (ed.) CRYPTO 1994. LNCS, vol. 839, pp. 26–39. Springer, Heidelberg (1994)
3. Biryukov, A., Cannière, C.D., Quisquater, M.: On Multiple Linear Approximations. In: Franklin, M. (ed.) CRYPTO 2004. LNCS, vol. 3152, pp. 1–22. Springer, Heidelberg (2004)
4. Hermelin, M., Nyberg, K., Cho, J.Y.: Multidimensional Linear Cryptanalysis of Reduced Round Serpent. In: Mu, Y., Susilo, W., Seberry, J. (eds.) ACISP 2008. LNCS, vol. 5107, pp. 203–215. Springer, Heidelberg (2008)

5. Baignères, T., Junod, P., Vaudenay, S.: How Far Can We Go Beyond Linear Crypt-analysis? In: Lee, P.J. (ed.) ASIACRYPT 2004. LNCS, vol. 3329, pp. 432–450. Springer, Heidelberg (2004)

6. Junod, P.: On the optimality of linear, differential and sequential distinghers. In: Biham, E. (ed.) EUROCRYPT 2003. LNCS, vol. 2656, pp. 17–32. Springer, Heidelberg (2003)

7. Baignères, T., Vaudenay, S.: The Complexity of Distinguishing Distributions (Invited Talk). In: Safavi-Naini, R. (ed.) ICITS 2008. LNCS, vol. 5155, pp. 210–222. Springer, Heidelberg (2008)

8. Selçuk, A.A.: On probability of success in linear and differential cryptanalysis. Journal of Cryptology 21(1), 131–147 (2008)

9. Hermelin, M., Cho, J.Y., Nyberg, K.: Statistical Tests for Key Recovery Using Multidimensional Extension of Matsui's Algorithm 1. In: EUROCRYPT 2009 - poster session (2009)

10. Cramér, H.: Mathematical Methods of Statistics, 7th edn. Princeton Mathematical Series. Princeton University Press, Princeton (1957)

11. Cover, T.M., Thomas, J.A.: Elements of Information Theory, 2nd edn. Wiley Series in Telecommunications and Signal Processing. Wiley-Interscience, Hoboken (2006)

12. Vaudenay, S.: An experiment on DES statistical cryptanalysis. In: CCS 1996: Proceedings of the 3rd ACM conference on Computer and communications security, pp. 139–147. ACM, New York (1996)

13. Harpes, C., Kramer, G.G., Massey, J.L.: A Generalization of Linear Cryptanalysis and the Applicability of Matsui's Piling-Up Lemma. In: Guillou, L.C., Quisquater, J.-J. (eds.) EUROCRYPT 1995. LNCS, vol. 921, pp. 24–38. Springer, Heidelberg (1995)

14. Junod, P., Vaudenay, S.: Optimal Key Ranking Procedures in a Statistical Cryptanalysis. In: Johansson, T. (ed.) FSE 2003. LNCS, vol. 2887, pp. 235–246. Springer, Heidelberg (2003)

15. Drost, F., Kallenberg, W., Moore, D.S., Oosterhoff, J.: Power Approximations to Multinomial Tests of Fit. Journal of the American Statistican Association 84(405), 130–141 (1989)

16. Collard, B., Standaert, F.X., Quisquater, J.J.: Experiments on the Multiple Linear Cryptanalysis of Reduced Round Serpent. In: Nyberg, K. (ed.) FSE 2008. LNCS, vol. 5086, pp. 382–397. Springer, Heidelberg (2008)

17. David, H.A.: Order Statistics, 1st edn. A Wiley Publication in Applied Statistics. John Wiley & Sons, Inc., Chichester (1970)

18. Biham, E., Dunkelman, O., Keller, N.: Linear Cryptanalysis of Reduced Round Serpent. In: Matsui, M. (ed.) FSE 2001. LNCS, vol. 2355, pp. 219–238. Springer, Heidelberg (2001)

Meet-in-the-Middle Attacks on SHA-3 Candidates

Dmitry Khovratovich, Ivica Nikolić, and Ralf-Philipp Weinmann

University of Luxembourg
{dmitry.khovratovich,ivica.nikolic,ralf-philipp.weinmann}@uni.lu

Abstract. We present preimage attacks on the SHA-3 candidates Boole, EnRUPT, Edon-R, and Sarmal, which are found to be vulnerable against a meet-in-the-middle attack. The idea is to invert (or partially invert) the compression function and to exploit its non-randomness. To launch an attack on a large internal state we manipulate the message blocks to be injected in order to fix some part of the internal state and to reduce the complexity of the attack. To lower the memory complexity of the attack we use the memoryless meet-in-the-middle approach proposed by Morita-Ohta-Miyaguchi.

1 Introduction

Recent attacks on widely used hash functions standards [2,16] drew much attention to the hash function design not only from cryptographers, but also from the institutions responsible for the standardization. After several workshops and discussions had been held, NIST started the so-called SHA-3 competition [7], which called for new designs by the end of October 2008.

Since most attacks on hash functions have been differential-based collision attacks, the majority of the designs we investigated so far claimed to be resistant to differential cryptanalysis while to the resistance against other attacks were given less attention. The subject of this paper is meet-in-the-middle attacks and their application to preimage search.

A meet-in-the-middle attack on a cryptographic primitive is applicable if the execution can be expressed as a sequence of transformations all of which have at least one input that is independent of the other transformations. Providing the invertibility of the last transformation, the full execution can be divided into independent parts, which are connected using the birthday paradox.

One of the first such attack was the attack on Double-DES [3]. Double-DES, being composed of two consecutive iterations of single DES with different keys, was found to be vulnerable to the following meet-in-the-middle attack: given a pair (*plaintext, ciphertext*) one can find a Double-DES key (a pair of single DES keys), which is valid for this pair, with complexity of about 2^{32} encryptions. A full attack on Double-DES, which gives the real key, is based on this approach as well and it is faster than the brute-force.

Meet-in-the-middle attacks on hash functions based on the Merkle-Damgård construction are hard to apply since the compression function is usually assumed

O. Dunkelman (Ed.): FSE 2009, LNCS 5665, pp. 228–245, 2009.

to be non-invertible. The alternative sponge construction [1] allows invertible transformations, but requires the internal state to be large so that the meet-in-the-middle approach can not be applied.

Surprisingly, several SHA-3 proposals are vulnerable to this type of attack. In this paper we describe meet-in-the-middle based preimage attacks on Boole, EnRUPT, Edon-R, and Sarmal. Two ideas are common for all the attacks. First, all the functions have invertible (or partially invertible) transformations, which allows us to execute the meet-in-the-middle. Secondly, we reduce the intermediate state space exploiting the non-random behavior of the round transformations.

This paper is composed as follows. First, we describe the meet-in-the-middle preimage attack in general and remind how it can be maintained with little memory. Then we show how preimages for Boole, EnRUPT, Edon-R, and Sarmal can be found. We also discuss possible computation-memory tradeoffs.

2   Meet-in-the-Middle Attacks on Hash Functions

Hash functions with invertible compression functions become susceptible to preimage attacks if the size of the internal state is too small. Preimages can be obtained by performing a meet-in-the-middle attack on the compression function. In this section we will describe this generic scenario in more details.

Let $F : D \rightarrow D$ and $G : D \rightarrow D$ be two random permutations and $H = G \circ F$ the composition of these permutations. In our setting, the function H is the hash function, F is defined as the compression function with a fixed IV and G is the inverse of the compression function for a fixed target value. Furthermore, we define auxilliary functions $\pi_{1,2} : D \times D \rightarrow D$ that map tuples to their first, respectively second component.

Assume we want to perform a meet-in-the-middle attack on h. The standard technique is to compute two sets

$$S_1 = \{(F(x), x) : x \in_R D\} \quad \text{and} \quad S_2 = \{(G^{-1}(y), y) : y \in_R D\}$$

such that $|S_1| \cdot |S_2| = |D|$. Either sorting these two sets in their first component or computing them in such a way that they are already ordered in this component allows us to easily find colliding values

$$\pi_1\left((F(x), x)\right) = \pi_1\left((G^{-1}(y), y)\right)$$

by comparing the elements of the two sets in linear time. Each collision gives us a pair (x, y) such that $H(x) = y$. How to balance the size of the sets S_1 and S_2 depends on the relative cost of the function G^{-1} compared to an evaluation of the function F. It may for instance be that G is easily invertible, meaning an evaluation of G^{-1} costs about the same number of operations as an evaluation of the function F. In this case we choose the sets S_1 and S_2 to be of equal size $\left\lceil \sqrt{|D|} \right\rceil$. However, if the evaluation of G^{-1} is k times more expensive than the evaluation of F, we should choose the set $|S_1|$ to be of size $\sqrt{k \cdot |D|}$ and S_2 of size $\sqrt{k^{-1} \cdot |D|}$ to obtain a minimum number of overall operations. The memory complexity of this naive approach is non-neglible however: We need to store a total of $2 \cdot \left(\sqrt{|D|}(\sqrt{k} + \sqrt{k^{-1}}) \right)$ elements of the domain D to carry it out. Storing both sets is not really necessary: Only the smaller should be stored, the values of the larger can be computed on the fly and compared against the elements of the smaller set.

In some cases the memory requirement can be completely eliminated by a technique based on Floyd cycle finding first described in an article by Morita, Ohta and Miyaguchi [6]. Although several works on hash functions refer to memoryless variants of meet-in-the-middle attacks [10,5], all of them cite either one or both papers by Quisquater and Delescaille on collision search for DES [12,11]. These two papers however do not directly deal with meet-in-the-middle attacks, but describe the technique of using distinguished points for collision search. Oorschot and Wiener describe the same technique for memoryless meet-in-the-middle later in [14].

2.1   Eliminating the Memory Requirement

Assume we are given another function $r : D \to \{0, 1\}$ which maps elements of the domain D to a single bit in a random fashion. Using this *switching function* we can define a step function s that evaluates x either to $F(x)$ or to $G(x)$, depending on the value of x:

$$ s : D \to D, \quad x \mapsto \begin{cases} F(x) & \text{if } r(x) = 0 \\ G(x) & \text{if } r(x) = 1 \end{cases} $$

This function s can then be used in a Floyd cycle finding algorithm: We start from a random value $x \in D$ and use just two elements $a = s(x)$ and $b = s^2(x)$. In each step we then update a by applying s to it and b by applying s^2 to it. Upon finding a cycle, we must check whether we really have found a pair $F(x) = G^{-1}(y)$ or whether we have found a cycle in F or in G. If the output of r is equidistributed, for each cycle we find $\Pr(F(x) = G^{-1}(y)) = 0.5$. In case of encountering a cycle in F or G we restart the algorithm with another random element $x \in D$.

Significant problems can arise if the output of r is not equidistributed, for instance if G is very costly to compute relative to F and we want to simulate the case of $|S_1| = k \cdot |S_2|$ with k large.

For the hash functions that we attack we define two functions F and G that are used in the memoryless approach. The F function is used for the forward direction and the G function is used for the backward one. The switching function r is defined as the parity of x.

2.2   Reduced State Principle

The meet-in-the-middle (MITM) attack needs a collision in the intermediate state. However, the state may be so large that a straightforward application of the MITM approach would require more than 2^n computations for a n-bit hash digest. Thus the generic principle we use further is to generate intermediate states only from a smaller subspace (where some bits are fixed to zero) thus reducing the birthday dimension and the complexity of the attack.

The generic framework is defined as follows. A hash function with an n-bit digest has an internal state of size k bits. We manage to get intermediate states with t bits fixed to 0. Then to get a MITM connection we need to get two states that collide in $(k-t)$ bits so that the *birthday space* D has size 2^{k-t}. This implies that we must get two sets S_1 and S_2 such that $|S_1| \cdot |S_2| = 2^{k-t}$. The exact ratio between S_1 and S_2 is defined by the complexity of inverting the compression (round) function.

For the memoryless version of the MITM attack, we need to tweak the attack slightly such that we can define the functions F and G. Each of the functions is a composition of two functions, first projecting the birthday space into the state space, the second mapping the state space into the birthday space again (fixing some bits to zero). In other words, let $F = f \circ \mu$ and $G = g \circ \nu$. When memoryless meet-in-the-middle is possible in our attacks we will define these functions accordingly.

3   Boole

Boole is a family of hash functions [13] based on a stream design. Internally, Boole has a large state $\sigma_t = (R_t[0], R_t[1], \ldots, R_t[15])$ of 16 words plus 3 additional word accumulators denoted by l_t, x_t, and r_t (t is the time). The words are 64 bits each. Hashing a message in Boole is done in three phases: 1)Input phase, where the whole message is processed word by word, and for each input word the state and the accumulators are updated, 2)mixing phase, where only the state is updated depending on the values of the accumulators, 3)output phase, where the output is produced.

The *update of the state*, referred to as a *cycle*, is defined as:

$$R_{t+1}[i] \leftarrow R_t[i+1], \text{for } i = 1 \ldots 14$$
$$R_{t+1}[15] \leftarrow f_1(R_t[12] \oplus R_t[13]) \oplus (R_t[0] \lll 1)$$
$$R_{t+1}[0] \leftarrow R_{t+1}[0] \oplus f_2(R_{t+1}[2] \oplus R_{t+1}[15]),$$

where f_1 and f_2 are some non-linear functions, intended to simulate random functions.

Let w_t be a message word. The *update of the accumulators* is defined as:

$$temp \leftarrow f_1(l_t) \oplus w_t$$
$$l_{t+1} \leftarrow temp \lll 1$$
$$x_{t+1} \leftarrow x_t \oplus w_t$$
$$r_{t+1} \leftarrow (r_t \oplus temp) \ggg 1$$

The whole message is absorbed in the input phase. Sequentially, for each message word w_t the following is done:

1. *update the accumulators*
2. $R_t[3] \leftarrow R_t[3] \oplus l_{t+1}$
3. $R_t[13] \leftarrow R_t[13] \oplus r_{t+1}$
4. *update the state (cycle)*

The mixing phase is invertible and its description is irrelevant in our attack.

Each iteration of the output phase produces one output word. One iteration is defined as:

1. *cycle*
2. Output the word $v = R[0] \oplus R[8] \oplus R[12]$

For example, the output for Boole-256 is produced in 8 iterations.

Let us present two observations about the invertibility of the update functions of the state and the accumulators.

Observation 1. The state update (*cycle*) is an invertible function. If a new state σ_{t+1} is given, then the state σ_t that produced σ_{t+1} in a single *cycle* can be found from the following equations:

$$R_t[0] = (R_{t+1}[15] \oplus f_1(R_{t+1}[11] \oplus R_{t+1}[12])) \ggg 1$$
$$R_t[1] = R_{t+1}[0] \oplus f_2(R_{t+1}[2] \oplus R_{t+1}[15])$$
$$R_t[i] = R_{t+1}[i-1], i = 2, \ldots 15$$

Observation 2. The update of the accumulators can be inverted with probability $1 - 1/e$. If the values of the new accumulators $l_{t+1}, x_{t+1}, r_{t+1}$ and the input message word w_t are fixed, the values of the previous accumulators l_t, x_t, r_t are determined as:

$$l_t = f_1^{-1}((l_{t+1} \ggg 1) \oplus w_t)$$
$$x_t = x_{t+1} \oplus w_t$$
$$r_t = r_{t+1} \lll 1 \oplus f_1(l_t \oplus w_t)$$

Moreover, if the values of l_t, l_{t+1} (or r_t, r_{t+1}) are fixed, the value of the message word w_t can be found uniquelly:

$$w_t = (l_t + 1 \ggg 1) \oplus f_1(l_t)$$
$$(\, w_t = (r_{t+1} \lll 1) \oplus r_t \oplus f_1(l_t))$$

In order to invert the function f_1 we will use a look-up table $(x, f_1(x))$ with all 2^{64} values for x, sorted by the second entry. Then, a inversion of $f_1(x)$ is equivalent to a look-up in this table.

3.1 Preimage Attack on Boole-384 and Boole-512

The intermediate state of Boole has 16 state words and 3 accumulators, hence 19 words in total. Further, we will show how to fix the values of the state words $R[3], \ldots, R[12]$ (10 words in total) to zero in forward and backward directions. This will mean that $k = 19 \cdot 64 = 1216$ and $t = 10 \cdot 64 = 640$, and the birthday space D has only 9 words (576 bits). We will also define $f(x)$ and $g(x)$ for the memoryless MITM attack.

Defining μ - fixing $R[3], R[4], \ldots, R[12]$ forwards. From the description of the input phase it follows that:

$$R_{10}[3] = R_9[4] = \ldots = R_1[12] = R_0[13] \oplus r_1$$

Note that the value of r_1 can be controlled with w_0 (Observation 2). Hence, if we take $r_1 = R_0[13]$, we will get $R_{10}[3] = 0$. Similarly, for $R_{10}[4]$ we have:

$$R_{10}[4] = R_9[5] = \ldots = R_2[12] = R_1[13] \oplus r_2$$

We can change the value of r_2 with w_1 such that $R_1[13] \oplus r_2 = 0$ holds. Then $R_{10}[4] = 0$. The same technique can be applied for fixing the values of $R_{10}[5], \ldots, R_{10}[12]$.

Note that we can not fix the values of more than these 10 words. When we control the value of r_t with the input word w_{t-1}, it means that we also change the value of l_t (which is added to $R_t[3]$). Since we can not control the value of both accumulators with a single message word, and both of them are xor-ed into the registers $R[3]$ and $R[13]$, it means that we can not control the values of more than 10 words.

Defining $f(y)$ for the memoryless MITM attack. The birthday space D has 9 words. Let $y = y_1 \| y_2 \| \ldots \| y_9$, then $f(y)$ can be defined as compression of the input words y_i with $1 \leq i \leq 9$ in the first 9 *cycles*. Thus when fixing $R[3], \ldots, R[12]$ in forward direction, we first compress y, and then we start with our technique for fixing these words to zero (function μ).

Defining ν - fixing $R[3], R[4], \ldots, R[12]$ backwards. Our backwards strategy is the following: first we invert the output and the mixing phase and obtain one valid intermediate state. Then, by changing the input words, we fix $R[3], R[4], \ldots, R[12]$.

First, let us deal with the inversion of the output phase. In each *cycle* of this phase one output word is produced. Hence, the digest is produced in 8 cycles[1]. The output word v_t is defined as $v_t = R_t[0] \oplus R_t[8] \oplus R_t[12]$. Let $H^* = (h_0, \ldots, h_7)$ be the target hash value. We have to construct a state $\sigma_t = (R_t[0], \ldots, R_t[15])$ such that $h_0 = v_t, h_1 = v_{t+1}, \ldots, h_7 = v_{t+7}$. First, we put any values in $R_t[0], R_t[9], R_t[10], \ldots, R_t[15]$. The rest of the words are undefined. Then, we find $R_t[8]$ from the equation $R_t[8] = R_t[0] \oplus R_t[12] \oplus h_0$. Obviously we get that $v_t = h_0$. After the *cycle* update we obtain a new state σ_{t+1}. Then, we determine the value of $R_t[1]$ from the equation $R_t[1] = R_{t+1}[0] = R_{t+1}[8] \oplus R_{t+1}[12] \oplus h_1$, and therefore $h_1 = v_{t+1}$. The values for $R_t[2], \ldots, R_t[7]$ are determined similarly. This way we can define the rest of the words in the state σ_t, which in the 7 sequential *cycle* updates produces the target hash value.

Let us fix the accumulators to any values. Then, inverting the mixing phase is trivial because the length of the preimage, as shown further, is known and the values of the accumulators are also known.

---

[1] In Boole-384, the output is produced in 6 cycles.

Now that we have inverted the output and mixing phase, we have the freedom of choosing the input message words. The technique for fixing is rather similar to the one used for fixing this set in forward direction. But in the backward direction, we control the values of the l_t accumulators (rather then the values of r_t as in the forward direction) with the input words w_t (Observation 2). From the description of the input phase we get:

$$R_{10}[12] = R_{11}[11] = \ldots = R_{18}[4] = R_{19}[3] \oplus l_{20}$$

Therefore if we take $l_{20} = R_{19}[3]$ we will get $R_{10}[12] = 0$. Similarly, for $R_{10}[11]$ we have:

$$R_{10}[11] = R_{11}[10] = \ldots = R_{17}[4] = R_{18}[3] \oplus l_{19}$$

If we take $l_{19} = R_{18}[3]$ we obtain $R_{10}[11] = 0$. The same technique can be used to fix the variables $R_{10}[10], \ldots, R_{10}[3]$.

One may argue that for controlling the values of the l_t registers when going backwards we have to pay an additional cost because f_1 is not always invertible. But we have to keep in mind that there are values for which f_1 has many inversions. Hence, if we start with a set of N different values, we can expect to find N different inversions for these values and thus we do not have to repeat the inversion.

Defining $g(y)$ for the memoryless MITM attack. The function $g(y)$, where $y = y_1 \| y_2 \| \ldots \| y_9$, is defined as 9 consecutive backward rounds of the input phase with inputs y_i. The starting state of these 9 rounds is the state obtained after the inversion of the output and mixing phases (as described above). Note that after the application of the function $g(y)$ a new state is obtained. Then, to this state, we apply our technique for fixing $R[3], \ldots, R[12]$ in 10 backwards rounds (function ν).

3.2   Complexity of the Attack

The preimage that we obtained has a length of at least $9 + 9 + 10 + 10 = 38$ words. The memoryless MITM attack requires about $2^{\frac{9 \cdot 64}{2}} = 2^{288}$ computations[2] and 2^{64} memory (for inverting f_1).

4   Edon-R

The hash family Edon-R [4] uses the well known Merkle-Damgård design principle. The intermediate hash value is rather large, two times the digest length[3]. For an n-bit digest the chaining value H_i of Edon-R is composed of two block of n bits each, i.e. $H_i = (H_i^1, H_i^2)$. The message input M_i for the compression

---

[2] One computation is equivalent to one round of the input phase or one round of the mixing phase.

[3] Edon-224 and Edon-384 have 512 and 1024 bits chaining values, respectively.

function is also composed of two blocks, i.e. $M_i = (M_i^1, M_i^2)$. Let Edon be the compression function. Then the new chaining value is produced as follows:

$$H_{i+1} = (H_{i+1}^1, H_{i+1}^2) = \text{Edon}(M_i^1, M_i^2, H_i^1, H_i^2)$$

The hash value of a message is the value of second block of the last chaining value.

Internally, the state of Edon-R has two n-bit blocks, A and B. The compression function of Edon-R consists of eight updates, each being an application of the quasigroup operation $Q(x, y)^4$, to one of these blocks. With A_i and B_i we will denote the values of these blocks after the i-th update in the compression function (please refer to Fig. 1). Hence, each input pair (H_i, M_i) generates internal state blocks $(A_1, B_1), (A_2, B_2), \dots, (A_8, B_8)$. The new chaining value (the output of the compression function) H_{i+1} is the value of the blocks (A_8, B_8).

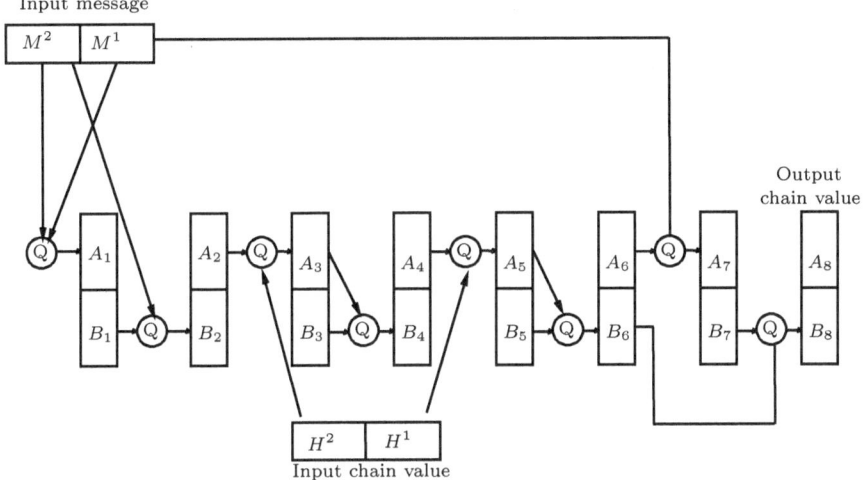

Fig. 1. Outline of the Edon-R compression function

Let us present a simple observation that is used in the attack.

Observation. The quasigroup operation $Q(x, y)$ of Edon-R is easily invertible, i.e. if A and C (B and C) are fixed then one can easily find B (C) such that $Q(A, B) = C$.

4.1 Preimage Attack on Edon-R-n

The internal state of EDON-R-n (the chaining value $H = H_1 \| H_2$) has $2n$ bits. We will show how to fix $H_1 = 0$. Then the preimage attack can be mounted using the MITM approach (Section 2), where $k = 2n$ and $t = n$. The backward step is time-consuming so we will use the memory MITM attack.

---

[4] The exact definition of the quasigroup operation can be found in [4].

Fixing H_1 in forward direction. We need only one message block to get the desired $H_1^{new} = 0$. Both initial value blocks are fixed as well. We claim that for each M_1 we can find M_2 such that this message input and the initial value blocks will produced a zero value in H_1^{new}.

Indeed, let M_1 be set to some random value. Then we obtain the value of A_6 since $A_7 = H_1^{new} = 0$ and the function Q is invertible. We consecutively obtain the values of A_5, A_4, A_3, A_2, and A_1 (keep in mind that the initial chaining value is fixed). Given A_1 and M_1, we derive M_2 by inverting the first application of Q. Finally we obtain all B's and thus a pair $(H_1^{new} = 0, H_2^{new})$.

Fixing H_1 in backward direction. We need only one step (one message block) to get a pair of the form $(0, H_2)$ from a given hash value $H = H_2^{new}$.

First, we set M_1 to some predefined value m. Then we assign to A_8 some random value and consecutively obtain the values of the following internal variables (in this order): A_7, B_7, B_6, A_6 (using M_1), $A_5, B_5, B_4, A_4, A_3, B_3$. We repeat this step 2^k times for different values of A_8 and store 2^k different pairs (A_3, B_3).

Now we set M_2 to some random value[5] and obtain the values of A_1, A_2, and B_2 using the value of M_1. If we repeat this step 2^{n-k+s} times then we will find 2^s different values of B_2 that coincide with some values of B_3 from the stored set. For each of these values we define H_2 such that $Q(A_2, H_2) = A_3$. The complexity requirements for this part are: 2^{n-k+s} computations[6], where $s - k < 65$, and $2^s + 2^k$ memory.

These 2^s pairs can be obtained using the memoryless MITM as well, where the MITM space is the value of B_2. Because of the message padding we should take any $n - 65$ bits of B_2 so that the input and the output of the MITM function F and G would have the same size. The $(n - 65)$-bit input to the function F is padded with the message padding, and the input to the function G is padded, for example, with zeros. Then, if a $(n - 65)$-bit collision between F and G is obtained, the probability that they coincide in the rest of the 65 bits is 2^{-65}. Hence, for constructing 2^s pseudo preimages with the memoryless MITM, one needs $2^s \cdot 2^{\frac{n-65}{2}+65} = 2^{\frac{n}{2}+s+32.5}$.

4.2 Complexity of the Attack

Starting from the initial value, we generate 2^{n-s} different chaining values with $H_1 = 0$. Note that we do not store these values. Then, with high probability, we can expect that one of these values will be in the set of the 2^s pseudo preimages generated in the backward direction. Under the condition $s - k < 65$ the total complexity of the attack when memory is used in the backward step is $2^{n-s} + 2^{n-k+s}$ computations and $2^s + 2^k$ memory. If only negligible memory in the backward step is used the computational complexity is $2^{n-s} + 2^{\frac{n}{2}+s+32.5}$ at the same time needing 2^s memory.

---

[5] The value is not truly random: 65 bits of the last message block are reserved for padding.

[6] Here and below, one computation is not more than one compression function call.

5   EnRUPT

The family of hash functions ENRUPT [9] is a member of a set of cryptographic primitives first presented at SASC 2008 [8].

The pseudocode of 512-bit version of ENRUPT, called ïRRUPT-512, is presented below. For details we refer the interested reader to [9].

Algorithm 1. ïRRUPT-512

Require: $p_0, \ldots, p_n$ { message blocks}
$(d_0, d_1, r) \leftarrow (0, 0, 0)$
$(x_0, \ldots, x_{47}) \leftarrow (0, \ldots, 0)$
for $i = 0$ to n **do**
    $(d_0, d_1, r, (x_0, \ldots, x_{47})) \leftarrow$ ïr8$(p_i, d_0, d_1, r, (x_0, \ldots, x_{47}))$ {squeezing}
end for
$(d_0, d_1, r, (x_0, \ldots, x_{47})) \leftarrow$ ïr8$(512, d_0, d_1, r, (x_0, \ldots, x_{47}))$
for $i = 0$ to 199 **do**
    $(d_0, d_1, r, (x_0, \ldots, x_{47})) \leftarrow$ ïr8$(0, d_0, d_1, r, (x_0, \ldots, x_{47}))$ {blank rounds}
end for
for $i = 0$ to 7 **do**
    $(d_0, d_1, r, (x_0, \ldots, x_{47})) \leftarrow$ ïr8$(0, d_0, d_1, r, (x_0, \ldots, x_{47}))$
    $z_i \leftarrow d_1$ {output}
end for
return $(z_0, \ldots, z_7)$

Algorithm 2. ïr8

Require: $p, d_0, d_1, r, (x_0, \ldots, x_{47})$
for $k = 0$ to 7 **do**
    $t \leftarrow (9 \cdot ((2 \cdot x_{(r \oplus 1) \bmod 48} \oplus x_{(r+4 \bmod 48)} \oplus d_{r\&1} \oplus r) \ggg 16)$
    $x_{(r+2) \bmod 48} \leftarrow x_{(r+2) \bmod 48} \oplus t$
    $d_{r\&1} \leftarrow d_{r\&1} \oplus t \oplus x_{r \bmod 48}$
    $r \leftarrow r + 1$
    $d_1 \leftarrow d_1 \oplus p$
end for
return $(d_0, d_1, r, x_0, \ldots, (x_0, \ldots, x_{47}))$

In the pseudo-code all indices are taken modulo 16, all multiplications are performed modulo 2^{64}, $\ggg$ stands for cyclic rotation to the right, and $\oplus$ denotes XOR. Now let us define and explain some points that are further used in the attack on ïRRUPT-512.

Equation invertibility. The accumulators d_i are updated by a non-invertible function, which can be expressed as $x \oplus g(x \oplus y)$ (see pseudo-code). Given the output of the function and the value of x a solution does not always exist. However, if we assume that the output and y are independent then the probability that the function can be inverted can be estimated by $1 - 1/e$. We did statistical tests that support this estimate.

Furthermore, while there is no solution for some input there are two (or more) solutions for other inputs (one solution on average). Thus when we perform backtracking we actually do not lose in quantity of solutions.

Look-up tables. We use look-up tables in order to find a solution for the equations arising from the round functions. All the tables used below refer to functions that have space of arguments smaller than the complexity of the attack, e.g., when we try to solve an equation $f(x \oplus C) = x$ (where C is one of 2^{64} possible constants) we use 2^{64} precomputed tables that contain values of $f(x \oplus C) \oplus x$ for all C and x.

Solving a system of equations is more complicated. Below we solve systems of form

$$\begin{cases} x = f(x, y, z, C_1); \\ y = g(x, y, z, C_2); \\ z = h(x, y, z, C_3), \end{cases}$$

where C_i are constants. We precompute for all possible x, y, z, C_i (2^{384} tuples) the sums $x \oplus f(x, y, z, C_1)$, $y \oplus g(x, y, z, C_2)$, and $z \oplus h(x, y, z, C_3)$ and then sort them so that it is easy to find a solution (or many) given C_i.

We also estimate that the time needed to find a solution is given by the complexity of the binary search which is negligible compared to the table size.

Inverting the updates in ïr8(p_i). The compression function of ïRRUPT-512 consists of the update of the state words $x_0, x_1, \ldots, x_{15}$, and the update of the accumulators d_0 and d_1. Inverting the update of the state words $x_0, x_1, \ldots, x_{15}$ is trivial:

$$x_{r+2}^{\text{old}} = x_{r+2}^{\text{new}} \oplus f.$$

The accumulator d_0 (similar formula holds for d_1) is updated by the following scheme:

$$d_0^{\text{new}} = f(x_{r \oplus 1}, x_{r+4}, d_0^{\text{old}}, r) \oplus d_0^{\text{old}} \oplus x_r$$

Instead of solving this equation for d_0^{old}, we simply use a table look-up (see above). Since the arguments of f are xored, we solve an equation of form $f(x \oplus C_1) \oplus x = C_2$. We spend $(2^{64})^2 = 2^{128}$ memory and effort to build this table for all x and C_1.

5.1   Preimage Attack on ïrRUPT-512

The preimage attack is mounted using the MITM approach (Section 2). The internal state of ïRRUPT-512 has 18 words, hence 1152 bits. We will show how to fix x_3 and x_{11} in forward and backward directions. Also, since EnRUPT does not have a message schedule and just adds the message block, we can reduce the birthday space D for an additional one word. Hence, the parameters for MITM are $k = 1152$ and $t = 192$. Getting states in both directions in not time-consuming. Therefore we will define the functions $f(x)$ and $g(x)$ and launch a memoryless MITM attack.

Defining μ - fixing x_3 and x_{11} in forward direction. We will fix the values of these two words in two consecutive application of the compression function. We will fix the value of x_3 to zero by changing the previous input message word p_0. In the following compression function iteration this value is not changed. In this iteration, we fix the value of x_{11} by setting the value of p_1.

From the definition of x_3 (notice that x_3 is updated second in the iteration but does not depend on x_2 and d_0, which has been updated before) we have:

$$x_3^{\text{new}} = 9[(2x_0 \oplus x_7 \oplus d_1) \ggg 16] \oplus x_3^{\text{old}}$$

We want to fix the value of x_3 to zero. Hence we require:

$$0 = 9[(2x_0 \oplus x_7 \oplus d_1) \ggg 16] \oplus x_3^{\text{old}}$$

In this equation the value of d_1 can be chosen freely. Simply, in the previous iteration of the compression function, the message word p, which is added to d_1 ($d_1^{\text{new}} = d_1^{\text{old}} \oplus p$) can be changed without affecting the values of the state words and d_0.

Therefore, by using a predefined table for this equation, we can find the necessary value of d_1 so that the equation holds. To build this table we spend $(2^{64})^4 = 2^{256}$ memory and computations. Notice that after the value of x_3 is fixed then, in iteration that follows, this value is not changed. In this iteration, we fix the value of x_{11} using exactly the same method. Hence, in two sequential rounds, we can fix the value of exactly two state words: x_3 and x_{11}.

Defining $f(y)$ for the memoryless MITM attack. The birthday space D has 15 words. We denote $y = y_1 || y_2 || \ldots || y_{15}$. Then $f(y)$ can be defined as compression of the input words $y_i, i = 1, \ldots, 15$ in the first 15 applications of the compression function. Thus when fixing x_3 and x_{11} in forward direction, we first compress y, and then we start with our technique for fixing these two words to zero.

Defining ν - fixing x_3 and x_{11} in backward direction. When going backwards we have to take into account two things: 1)the output hash value is produced in 8 iterations, and 2)the input message words in the last 17 iterations are fixed. Let us first address 1). When the hash value is given (as in a preimage attack), it is still hard to reconstruct the whole state of ïRRUPT-512. This is made more difficult by outputting only a small chunk of the state (the value of d_1) in each of the 8 final iterations (and not at once). So, not only we have to guess the value of the rest of the state, but we have to guess it so that in the following iterations the required values of d_1 will be output. Yet, this is possible to overcome.

Let the hash value be $H = (d_1^t, d_1^{t+1}, \ldots, d_1^{t+7})$. Consider a state where $d_1 = d_1^t$ and all of the other words of the state are left undefined. Then, we take 2^{448} different values for the rest of the state and iterate forward for 7 rounds, while producing an output word at each round. With overwhelming probability, one of these outputs will coincide with $(d_1^{t+1}, \ldots, d_1^{t+7})$. After we find the state that produces the required output, we go backwards through the blank iterations and the message length iteration. In total there are 17 iterations which is 136 rounds. The accumulators are updated non-bijectively. Therefore one may argue that the cost of inverting the accumulators through these rounds should be $(1 - 1/e)^{136}$. Yet, if in some cases solution for the accumulator doesn't exist in other cases there is more then one solution. Hence, if we start with two internal states, we can pass these iterations with a cost of two times hashing in forward direction.

Now after we have passed the output, blank rounds and message length iterations, and obtained one state, we can fix x_3 and x_{11} in two backward applications of the compression function. The following lemma holds:

Lemma 1. *Given a state $S = (x_0^{new}, \ldots, x_{15}^{new}, d_0^{new}, d_1^{new})$ one can build a state $S' = (x_0, \ldots, x_{15}, d_0, d_1)$ and a message p such that $x_3 = 0$ and $ir8(S', p_i) = S$.*

The proof is given in Appendix. The same proposition can be applied to x_{11}. Since the compression function in one application changes either x_3 or x_{11}, then in two consecutive backward applications of the compression function we can fix the values of these two words.

Defining $g(y)$ for the memoryless MITM attack. The function $g(y)$, where $y = y_1 || y_2 || \ldots || y_{15}$, is defined as 15 consecutive backward rounds of the input phase with inputs y_i. The starting state of these 9 rounds is the state obtained after the inversion of the output, blank rounds and message length iterations (as described above).

5.2   Complexity of the Attack

We spend at most 2^{384} computations to build the pre-computation tables so it is not a bottleneck. To compose a valid state after the blank rounds that gives the desired hash we need about 2^{448} trials. We also pass the blank rounds for free since the absence of solutions for some states is compensated by many of them for other ones. Thus the most computations-consuming part is the memoryless MITM attack. It requires $2^{\frac{960}{2}} = 2^{480}$ computations. The memory requirement is determined by the precomputed tables, hence it is 2^{384}.

6   Sarmal

Sarmal-n [15] is a hash family based on the HAIFA design. After the standard padding procedure, the padded message is divided into blocks of 1024 bits each, i.e. $M = M_1 || M_2 || \ldots || M_k, |M_i| = 1024, i = 1, \ldots, k$. Each block is processed by the compression functions. HAIFA design implies that the compression function f has four input arguments: the previous chain value h_{i-1}, the message block M_i, the salt s, and the block index t_i. Hence, h_i is defined as $h_i = f(h_{i-1}, M_i, s, t_i)$. The final chaining value h_k is the hash value of the whole message M. For Sarmal-n the chaining value h_i has 512 bits. Let us denote the left and the right half of

h_i as L_i and R_i respectively, i.e. $h_i = L_i||R_i$. The salt s has 256 bits (similarly let $s = s_1||s_2$), and the block index t_i has 64 bits. Then, the compression function of Sarmal-n can be defined as:

$$f(h_{i-1}, M_i, s, t_i) = \mu(L_{i-1}||s_l||c_1||t_i, M_i) \oplus \nu(R_{i-1}||s_r||c_2||t_i, M_i) \oplus h_{i-1}, \quad (1)$$

where μ and ν are functions that output 512 bit values, and c_1, c_2 are some constants. The exact definition of these functions is irrelevant for our attack.

6.1 Preimage Attack on Sarmal-512

We will show how to invert the compression function of Sarmal-512. Note that the intermediate chaining value of Sarmal has 512 bits. Then the preimage attack can be launched using the MITM approach (Section 2), where $k = 512$ and $t = 0$. The inversion of the compression function is time-consuming so we will use the memory MITM attack.

Going forward from the IV. Since we do not fix anything ($t = 0$), going forward from the IV is trivial. We simply generate a number of intermediate chaining values, by taking different random messages as an input for the first compression function.

Going backward from the target hash value. Let us explain how the compression function can be inverted.

From (1) we get:

$$\begin{aligned}
f \ (\ h_{i-1}, M_i, s, t_i) = \\
= \mu \ (L_{i-1}||s_l||c_1||t_i, M_i) \oplus \nu(R_{i-1}||s_r||c_2||t_i, M_i) \oplus h_{i-1} = \\
= \mu \ (L_{i-1}||s_l||c_1||t_i, M_i) \oplus \nu(R_{i-1}||s_r||c_2||t_i, M_i) \oplus L_{i-1}||R_{i-1} = \\
= \mu \ (L_{i-1}||s_l||c_1||t_i, M_i) \oplus \nu(R_{i-1}||s_r||c_2||t_i, M_i) \oplus L_{i-1}||0 \oplus 0||R_{i-1} = \\
= (\mu \ (L_{i-1}||s_l||c_1||t_i, M_i) \oplus L_{i-1}||0) \oplus (\nu(R_{i-1}||s_r||c_2||t_i, M_i) \oplus 0||R_{i-1})
\end{aligned}$$

Let us fix the values of $M_i, s,$ and t_i. Then, we can introduce the functions $F(L_{i-1}) = \mu(L_{i-1}||s_l||c_1||t_i) \oplus L_{i-1}||0$, and $G(R_{i-1}) = \nu(R_{i-1}||s_r||c_2||t_i) \oplus 0||R_{i-1}$. Let H^* be the target hash value. Then we get the equation:

$$F(L) \oplus G(R) = H^*$$

If we generate 2^{256} different values for $F(L)$ and the same amount for $G(R)$, then, by the birthday paradox, with high probability we can expect to get at least one pair $(F(L_l), G(R_m))$ that will satisfy the above equation and therefore obtain that $h = L_l||R_m$ is a preimage of H^*.

A memoryless version of this pseudo-preimage attack can be obtained by introducing the function $\tilde{F}(L) = F(L) \oplus H^*$, and launching the memoryless MITM attack on $\tilde{F}$ and G. This would require 2^{256} computations and negligible memory.

Table 1. Complexity of the preimage attacks described in this paper

	Computations	Memory
Boole-384/512	2^{288}	2^{64}
Edon-R-n	$2^{n-s} + 2^{n-k+s}$	$2^s + 2^k$
	$2^{n-s} + 2^{\frac{n}{2}+s+32.5}$	2^s
EnRUPT-512	2^{480}	2^{384}
Sarmal-512	$2^{512-s} + 2^{256+s}$	2^s

6.2   Complexity of the Attack

Since the backward direction, i.e. inverting the compression function, is time consuming we will use the memory version of MITM attack. Going backwards from the target hash value we create a set S_2 of 2^s different chaining values. To create this set we need $2^{256} \cdot 2^s = 2^{256+s}$ computations. Then, starting from the initial value, we generate 2^{512-s} different chaining values. Note, we do not store these values, we store only the smaller set S_2. Then, with a high probability, we can expect that these two sets coincide. The total complexity of the attack is $2^{512-s} + 2^{256+s}$ computations and 2^s memory.

7   Conclusions

We have presented meet-in-the-middle attacks on four SHA-3 candidates. These attacks became possible because we managed to invert (or partially invert) the compression functions and to reduce the birthday space so that collisions in this space can be found faster than 2^n and give a preimage.

We have also applied, when it was possible, the memoryless version of the MITM attack and thus significantly reduced the memory requirements for the attacks. For these cases we provided estimates on the computation-memory trade-offs.

The complexity of our attacks on the hash functions are summarized in the following table.

Acknowledgments

The authors would like to thank Greg Rose and the anonymous reviewers of FSE'09 for helpful comments. Dmitry Khovratovich is supported by PRP "Security & Trust" grant of the University of Luxembourg. Ivica Nikolić is supported by the Fonds National de la Recherche Luxembourg grant TR-PHD-BFR07-031. Ralf-Philipp Weinmann is supported by the ESS project of the University of Luxembourg.

References

1. Bertoni, G., Daemen, J., Peeters, M., Van Assche, G.: Sponge functions (2007), http://sponge.noekeon.org/
2. De Cannière, C., Rechberger, C.: Finding SHA-1 characteristics: General results and applications. In: Lai, X., Chen, K. (eds.) ASIACRYPT 2006. LNCS, vol. 4284, pp. 1–20. Springer, Heidelberg (2006)
3. Diffie, W., Hellman, M.E.: Exhaustive cryptanalysis of the NBS data encryption standard. Computer 10, 74–84 (1977)
4. Gligoroski, D., Ødegård, R.S., Mihova, M., Knapskog, S.J., Kocarev, L., Drápal, A.: Cryptographic hash function Edon-R. Submission to NIST (2008), http://people.item.ntnu.no/danilog/Hash/Edon-R/ Supporting_Documentation/EdonRDocumentation.pdf
5. Mendel, F., Pramstaller, N., Rechberger, C., Kontak, M., Szmidt, J.: Cryptanalysis of the GOST hash function. In: Wagner, D. (ed.) CRYPTO 2008. LNCS, vol. 5157, pp. 162–178. Springer, Heidelberg (2008)
6. Morita, H., Ohta, K., Miyaguchi, S.: A switching closure test to analyze cryptosystems. In: Feigenbaum, J. (ed.) CRYPTO 1991. LNCS, vol. 576, pp. 183–193. Springer, Heidelberg (1992)
7. National Institute of Standards and Technology. Announcing Request for Candidate Algorithm Nominations for a New Cryptographic Hash Algorithm (SHA-3) Family 72(212) of Federal Register (November 2007)
8. O'Neil, S.: EnRUPT: First all-in-one symmetric cryptographic primitive. In: SASC 2008 (2008), http://www.ecrypt.eu.org/stvl/sasc2008/
9. O'Neil, S., Nohl, K., Henzen, L.: EnRUPT hash function specification (2008), http://enrupt.com/SHA3/
10. Preneel, B.: Analysis and Design of Cryptographic Hash Functions. PhD thesis, Katholieke Universiteit Leuven, Leuven, Belgium (January 1993)
11. Quisquater, J.-J., Delescaille, J.-P.: How easy is collision search? Application to DES (extended summary). In: Quisquater, J.-J., Vandewalle, J. (eds.) EURO-CRYPT 1989. LNCS, vol. 434, pp. 429–434. Springer, Heidelberg (1989)
12. Quisquater, J.-J., Delescaille, J.-P.: How easy is collision search. new results and applications to DES. In: Brassard, G. (ed.) CRYPTO 1989. LNCS, vol. 435, pp. 408–413. Springer, Heidelberg (1990)
13. Rose, G.G.: Design and primitive specification for Boole, http://seer-grog.net/BoolePaper.pdf
14. van Oorschot, P.C., Wiener, M.J.: Parallel collision search with application to hash functions and discrete logarithms. In: ACM Conference on Computer and Communications Security, pp. 210–218 (1994)
15. Varıcı, K., Özen, O., Kocair, Ç.: Sarmal: SHA-3 proposal. Submission to NIST (2008)
16. Wang, X., Yu, H.: How to break MD5 and other hash functions. In: Cramer, R. (ed.) EUROCRYPT 2005. LNCS, vol. 3494, pp. 19–35. Springer, Heidelberg (2005)

A   Proof of the EnRUPT Lemma 1

Let $(x_0^{\mathrm{new}}, x_1^{\mathrm{new}}, x_2^{\mathrm{new}}, x_3^{\mathrm{new}}, \ldots, x_{15}^{\mathrm{new}}, d_0^{\mathrm{new}}, d_1^{\mathrm{new}})$ be our starting state. We want to invert backwards one iteration of the compression function. Hence, we want to obtain the previous state $(x_0^{new}, x_1^{new}, x_2^{old}, x_3^{old}, \ldots, x_{15}^{new}, d_0^{old}, d_1^{old})$ where $x_3^{old} = 0$. From the description of ïRRUPT-512 we get:

$$x_2^{new} = f(x_1^{new}, x_6^{old}, d_0^0, r) \oplus x_2^{old} \tag{2}$$

$$x_3^{new} = \underbrace{f(x_0^{new}, x_7^{old}, d_1^1, r+1)}_{f_3} \oplus x_3^{old}, \quad d_1^3 = f_3 \oplus d_1^1 \oplus x_1^{new} \tag{3}$$

$$x_4^{new} = f(x_3^{new}, x_8^{old}, d_0^2, r+2) \oplus x_4^{old} \tag{4}$$

$$x_5^{new} = \underbrace{f(x_2^{new}, x_9^{old}, d_1^3, r+3)}_{f_5} \oplus x_5^{old}, \quad d_1^5 = f_5 \oplus d_1^3 \oplus x_3^{new} \tag{5}$$

$$x_6^{new} = f(x_5^{new}, x_{10}^{new}, d_0^4, r+4) \oplus x_6^{old} \tag{6}$$

$$x_7^{new} = \underbrace{f(x_4^{new}, x_{11}^{new}, d_1^5, r+5)}_{f_7} \oplus x_7^{old}, \quad d_1^7 = f_7 \oplus d_1^5 \oplus x_5^{new} \tag{7}$$

$$x_8^{new} = f(x_7^{new}, x_{12}^{new}, d_0^6, r+6) \oplus x_8^{old} \tag{8}$$

$$x_9^{new} = \underbrace{f(x_6^{new}, x_{13}^{new}, d_1^7, r+7)}_{f_9} \oplus x_9^{old}, \quad d_1^{new} = f_9 \oplus d_1^7 \oplus x_7^{new} \oplus p \tag{9}$$

With d_1^i we denote the value of the accumulator d_1 used in the update of the state word x_i. We need to fix x_3^{old} to zero. Hence, from (3), we get the equation:

$$x_3^{new} = f_3 = f(x_0^{new}, x_7^{old}, d_1^1, r+1) =$$
$$= 9 \cdot ((2x_0^{new} \oplus r \oplus (x_7^{old} \oplus d_1^1)) \ggg 16).$$

In the upper equation we can denote by $X = x_7^{old} \oplus d_1^1$. Since, all the other variables are already known, a table can be built for this equation, and solution for X can be found. Let $C_1 = X = x_7^{old} \oplus d_1^1$. If we express the value of x_7^{old} from (7) then we get the following equation:

$$x_7^{new} \oplus f_7 \oplus d_1^1 = C_1. \tag{10}$$

Further, from (3), (5), (7), and (9), this equation can be rewritten as:

$$x_7^{new} \oplus f_7 \oplus f_3 \oplus f_5 \oplus f_7 \oplus f_9 \oplus x_1^{new} \oplus x_3^{new} \oplus x_5^{new} \oplus x_7^{new} \oplus p = C_1.$$

Since, $x_1^{new}, x_3^{new}, x_5^{new}, x_7^{new}$, and f_3 are all constant (the value of f_3 is equal to x_3^{new}), the upper equation can be rewritten as:

$$f_5 + f_9 + p = K, \tag{11}$$

where $K = x_3^{new} \oplus x_5^{new} \oplus f_3 \oplus C_1$. So given the values of f_5 and f_9 from (11) we can easily find the value for the message word p such that $x_3^{old} = 0$ holds. Let us try to find the values of f_5 and f_9.

The value of f_5 (from (5)) depends, in particular, on x_9^{old} and d_1^3. From (9) we get that $x_9^{old} = f_9 \oplus x_9^{new}$. From (3) and (10) we get:

$$d_1^3 = f_3 \oplus x_1^{new} \oplus d_1^1 = f_3 \oplus x_1^{new} \oplus x_7^{new} \oplus f_7 \oplus C_1. \tag{12}$$

Therefore, for the value of f_5 we get:

$$f_5 = 9 \cdot ((2x_2^{\text{new}} \oplus (r+3) \oplus x_9^{\text{old}} \oplus d_1^3) \ggg 16) =$$
$$= 9 \cdot ((K_1 \oplus f_7 \oplus f_9) \ggg 16), \quad (13)$$

where $K_1 = 2x_2^{\text{new}} \oplus (r+3) \oplus x_9^{\text{new}} \oplus f_3 \oplus x_1^{\text{new}} \oplus x_7^{\text{new}} \oplus C_1$.

Similarly, for f_7 from (7), we can see that depends on d_1^5. For this variable, from (12) and (5), we get:

$$d_1^5 = f_5 \oplus x_3^{\text{new}} \oplus d_1^3 = f_5 \oplus x_3^{\text{new}} \oplus f_3 \oplus x_1^{\text{new}} \oplus x_7^{\text{new}} \oplus f_7 \oplus C_1. \quad (14)$$

Hence, for f_7 we get:

$$f_7 = 9 \cdot ((2x_4^{\text{new}} \oplus (r+5) \oplus x_{11}^{\text{new}} \oplus d_1^5) \ggg 16) =$$
$$= 9 \cdot ((K_2 \oplus f_5 \oplus f_7) \ggg 16), \quad (15)$$

where $K_2 = 2x_4^{\text{new}} \oplus (r+5) \oplus x_{11}^{\text{new}} \oplus x_3^{\text{new}} \oplus f_3 \oplus x_1^{\text{new}} \oplus x_7^{\text{new}}$.

Finally, for f_9 from (7)), we get that it depends on d_1^7. From (14) and (7), for the value of d_1^7 we get the following:

$$d_1^7 = f_7 \oplus x_5^{\text{new}} \oplus d_1^5 =$$
$$= f_7 \oplus x_5^{\text{new}} \oplus f_5 \oplus x_3^{\text{new}} \oplus f_3 \oplus x_1^{\text{new}} \oplus x_7^{\text{new}} \oplus f_7 \oplus C_1 =$$
$$= x_5^{\text{new}} \oplus f_5 \oplus x_3^{\text{new}} \oplus f_3 \oplus x_1^{\text{new}} \oplus x_7^{\text{new}} \oplus C_1.$$

For the value of f_9 we get:

$$f_9 = 9 \cdot ((2x_6^{\text{new}} \oplus (r+7) \oplus x_{13}^{\text{new}} \oplus d_1^7) \ggg 16) = 9 \cdot ((K_3 \oplus f_5) \ggg 16), \quad (16)$$

where $K_3 = 2x_6^{\text{new}} \oplus (r+7) \oplus x_{13}^{\text{new}} \oplus x_5^{\text{new}} \oplus x_3^{\text{new}} \oplus f_3 \oplus x_1^{\text{new}} \oplus x_7^{\text{new}} \oplus C_1$.

As a result, we get a system of three equations ((13),(15), and (16)) with three unknowns f_5, f_7, and f_9:

$$\begin{cases} f_5 = 9 \cdot ((K_1 \oplus f_7 \oplus f_9) \ggg 16); \\ f_7 = 9 \cdot ((K_2 \oplus f_5 \oplus f_7) \ggg 16); \\ f_9 = 9 \cdot ((K_3 \oplus f_5) \ggg 16). \end{cases}$$

We can build a table that solves this system. There are six columns in the table: three unknowns and three constants: K_1, K_2, and K_3.

After we find the exact values of f_5 and f_9 we can easily compute the value of p from (11).

Practical Collisions for EnRUPT

Sebastiaan Indesteege[1,2,*] and Bart Preneel[1,2]

[1] Department of Electrical Engineering ESAT/COSIC, Katholieke Universiteit
Leuven. Kasteelpark Arenberg 10, B-3001 Heverlee, Belgium
`sebastiaan.indesteege@esat.kuleuven.be`
[2] Interdisciplinary Institute for BroadBand Technology (IBBT), Belgium

Abstract. The EnRUPT hash functions were proposed by O'Neil, Nohl
and Henzen [5] as candidates for the SHA-3 competition, organised by
NIST [4]. The proposal contains seven concrete hash functions, each
having a different digest length.

We present a practical collision attack on each of these seven EnRUPT
variants. The time complexity of our attack varies from 2^{36} to 2^{40} round
computations, depending on the EnRUPT variant, and the memory requirements are negligible. We demonstrate that our attack is practical
by giving an actual collision example for EnRUPT-256.

Keywords: EnRUPT, SHA-3 candidate, hash function, collision attack.

1   Introduction

Cryptographic hash functions are important cryptographic primitives that are
employed in a vast number of applications, such as digital signatures and commitment schemes. They are expected to possess several security properties, one of
which is *collision resistance*. Informally, collision resistance means that it should
be hard to find two distinct messages $m \neq m'$ that hash to the same value, i.e.,
$h(m) = h(m')$.

Many popular hash functions, such as MD5, SHA-1 and SHA-2 share a common design principle. The recent advances in the cryptanalysis of these hash
functions have raised serious concerns regarding their long-term security. This
motivates the design of new hash functions, based on different design strategies.
The National Institute of Standards and Technology (NIST) has decided to hold
a public competition, the SHA-3 competition, to develop a new cryptographic
hash function standard [4].

The EnRUPT hash functions were proposed by O'Neil, Nohl and Henzen [5]
as candidates in this SHA-3 competition. The proposal contains seven concrete
EnRUPT variants, each having a different digest length.

In this paper, we analyse EnRUPT and show that none of the proposed
EnRUPT variants is collision resistant. We present a practical collision attack

---

* F.W.O. Research Assistant, Fund for Scientific Research — Flanders (Belgium).

O. Dunkelman (Ed.): FSE 2009, LNCS 5665, pp. 246–259, 2009.

requiring only 2^{36} to 2^{40} EnRUPT round computations, depending on the En-RUPT variant. This is significantly less than the approximately $2^{n/2}$ hash computations required for a generic collision attack on an n-bit hash function based on the birthday paradox.

The structure of this paper is as follows. A short description of EnRUPT is given in Sect. 2. Section 3 introduces the basic strategy we use to find collisions for EnRUPT, which is based on the work on SHA by Chabaud and Joux [2] and Rijmen and Oswald [9]. Sections 4, 5 and 6 apply this basic attack strategy to EnRUPT, step by step. Our results, including an example collision for EnRUPT-256, are presented in Sect. 7. Finally, Sect. 8 concludes.

2   Description of EnRUPT

In this section, we give a short description of the seven EnRUPT variants that were proposed as SHA-3 candidates [5]. All share the same structure and use the same round function. The only differences lie in the parameters used. Table 1 gives the values of these parameters for each EnRUPT variant.

2.1   The EnRUPT Hash Functions

The structure shared by all EnRUPT hash functions can be split into four phases: preprocessing, message processing, finalisation and output. Figure 1 contains a description of the EnRUPT hash functions in pseudocode.

In the preprocessing phase (lines 2–4) the input message is padded to be a multiple of w bits, where w is the word size. Depending on the EnRUPT variant, the word size w is 32 or 64 bits, see Table 1. The padded message is then split into an integer number of w-bit words m_i.

The internal state of EnRUPT consists of several w-bit words: H state words x_i, P 'delta accumulators' d_i, and a round counter r. All of these are initialised to zero. The parameter P is equal to 2 for all seven EnRUPT variants. The value of H depends on the digest length, as indicated in Table 1.

Table 1. EnRUPT Parameters

EnRUPT variant	digest length	word size	parallelisation level	security parameter	number of state words
	h	w	P	s	H
EnRUPT-128	128 bits	32 bits	2	4	8
EnRUPT-160	160 bits	32 bits	2	4	10
EnRUPT-192	192 bits	32 bits	2	4	12
EnRUPT-224	224 bits	64 bits	2	4	8
EnRUPT-256	256 bits	64 bits	2	4	8
EnRUPT-384	384 bits	64 bits	2	4	12
EnRUPT-512	512 bits	64 bits	2	4	16

```
1: function EnRUPT (M)
2:        /* Preprocessing */
3:        m_0, ⋯ , m_t ← M || 1 || 0^(w−(|M|+1 mod w))   s.t.   ∀i, 0 ≤ i ≤ t : |m_i| = w
4:        d_0, ⋯ , d_{P−1}, x_0, ⋯ , x_{H−1}, r ← 0, ⋯ , 0
5:        /* Message processing */
6:        for i = 0 to n do
7:              ⟨d, x, r⟩ ← round(⟨d, x, r⟩ , m_i)
8:        end for
9:        /* Finalisation */
10:       ⟨d, x, r⟩ ← round(⟨d, x, r⟩ , uint_w(|M|))
11:       for i = 1 to H do
12:             ⟨d, x, r⟩ ← round(⟨d, x, r⟩ , 0)
13:       end for
14:       /* Output */
15:       for i = 0 to h/w − 1 do
16:             ⟨d, x, r⟩ ← round(⟨d, x, r⟩ , 0)
17:             o_i ← d_0
18:       end for
19:       return o_0 || ⋯ || o_{h/w−1}
20: end function
```

Fig. 1. The EnRUPT Hash Function

Then, in the message processing phase (lines 5–8), the round function is called once for each w-bit padded message word m_i. Each call to the round function updates the internal state $\langle d, x, r \rangle$. A detailed description of the EnRUPT round function is given in the next section, Sect. 2.2.

After all message words have been processed, a finalisation is performed (lines 9–13). The EnRUPT round function is called once with the length of the (unpadded) message, represented as a w-bit unsigned integer. Then, H blank rounds, i.e., calls to the round function with a zero message word input, are performed.

Finally, in the output phase (lines 14–18), the message digest is generated one w-bit word at a time. The EnRUPT round function is called h/w times and, after each call, the content of the 'delta accumulator' d_0 is output.

2.2   The EnRUPT Round Function

The EnRUPT round function is based entirely on a number of simple operations on words of w bits, such as bit shifts, bit rotations, exclusive OR and addition modulo 2^w. Figure 2 gives a description of the EnRUPT round function in pseudocode. The round function consists of $s \cdot P$ identical steps, where s and P are parameters of the hash function. As indicated in Table 1, $s = 4$ and $P = 2$ for all seven proposed EnRUPT variants. Thus, the EnRUPT round function consists of eight steps.

In each step, several words of the state are selected (lines 4–7) and combined into an intermediate value f (lines 9–10). Note that line 10 could equally be

```
1: function round (⟨d, x, r⟩ , m)
2:     for i = 0 to s · P − 1 do          /* An iteration of this loop is a "step" */
3:         /* Compute indices */
4:         α ← r + (i + 1 mod P) mod H
5:         β ← r + i + 2P mod H
6:         γ ← r + i + P mod H
7:         ξ ← r + i mod H
8:         /* Compute intermediate f */
9:         e ← ((x_α ≪ 1) ⊕ x_β ⊕ d_{i mod P} ⊕ uint_w(r + i)) ⋙ w/4
10:        f ← (e ≪ 3) ⊞ e                 /* Multiplication with 9 modulo 2^w */
11:        /* Update state */
12:        x_γ ← x_γ ⊕ f
13:        d_{i mod P} ← d_{i mod P} ⊕ x_ξ ⊕ f
14:    end for
15:    r ← r + s · P
16:    d_{P−1} ← d_{P−1} ⊕ m              /* Message word injection */
17:    return ⟨d, x, r⟩
18: end function
```

Fig. 2. The EnRUPT Round Function

described as a multiplication with 9 modulo 2^w. The intermediate value f is then used to update one state word, x_γ, and one 'delta accumulator', $d_{i \bmod P}$ (lines 12–13).

After all steps have been performed, the round counter is incremented by the number of steps that were carried out, i.e., $s · P$ (line 15). Finally, the input message word m is injected into one word of the internal state, the 'delta accumulator' $d_{P−1}$ (line 16).

3   Basic Attack Strategy

This section gives an overview of the linearisation method for finding collision differential characteristics for a hash function, which we use to attack EnRUPT in this work. This method was introduced by Chabaud and Joux [2], who applied it to SHA-0 and simplified variants thereof. Later, it was extended further and applied to SHA-1 by Rijmen and Oswald [9].

A Linear Hash Function. Consider a hypothetical hash function that consists only of linear operations over GF(2). When the input messages are restricted to a certain length, each output bit can be written as an affine function of the input bits. The *difference* in each output bit is given by a linear function of the differences in the input bits, as the constants (if any) cancel. A message difference that leads to a collision can be found by equating the output differences to zero, and solving the resulting system of linear equations over GF(2), for instance using Gauss elimination. Any pair of messages with this difference will result in a collision.

Linearising a Nonlinear Hash Function. Actual cryptographic hash functions contain (also) nonlinear components, so this method no longer applies. However, we may still be able to *approximate* the nonlinear components by linear ones and construct a linear approximation of the entire hash function. For our purpose, a good linear approximation $\lambda(x)$ of a nonlinear function $\gamma(x)$ is such that its differential behaviour is close to that of $\gamma(x)$. More formally, the equation

$$\gamma(x \oplus \Delta) \oplus \gamma(x) = \lambda(x \oplus \Delta) \oplus \lambda(x) = \lambda(\Delta) \qquad (1)$$

should hold for a relatively large fraction of values x. For instance, an addition modulo 2^w could be approximated by a simple XOR operation, i.e., ignoring the carries.

Finding Collisions. A *differential characteristic* consists of a message difference and a list of the differences in all (relevant) intermediate values. For the linear approximation, it is easy to find a differential characteristic that leads to a collision with probability one. But for the actual hash function, this probability will be (much) lower.

If the differential behaviour of all the nonlinear components corresponds to that of the linear approximations they were replaced with, i.e., if (1) holds simultaneously for each nonlinear component, we say that the differential characteristic is followed. In this case, the message pair under consideration will not only collide for the linearised hash function, but also for the original, nonlinear hash function. Such a message pair is called a *conforming* message pair.

Hence, a procedure for finding a collision for the nonlinear hash function could be to find a differential characteristic leading to collisions for a linearised variant of the hash function. Then, a message pair conforming to the differential characteristic is searched. In order to lower the complexity of the attack, it is important to maximise the probability that the differential characteristic is followed, i.e., we need to find a good differential characteristic.

4   Linearising EnRUPT

We now apply this general strategy to EnRUPT. Recall the description of the EnRUPT round function in Fig. 2. Note that only the modular addition in line 10 is not linear over GF(2). Indeed, the computation of the indices in lines 4–7 and the update of the round counter in line 15 do not depend on the message being hashed and can thus be precomputed. The same holds for the inclusion of the round counter in line 9, i.e., this can be seen as an XOR with a constant. The other operations are all linear over GF(2).

Replacing the modular addition in line 10 with an XOR operation yields a linearised round function, which we refer to as the EnRUPT-$\mathcal{L}$ round function. The EnRUPT-$\mathcal{L}$ hash function, i.e., the hash function built on this linearised round function, also consists solely of GF(2)-linear components.

5   The Collision Search

During the collision search phase, many collisions for EnRUPT-$\mathcal{L}$ are constructed, and a collision for EnRUPT is searched among them. Since only the modular additions (line 10 of Fig. 2) were approximated by XOR, these are the only places where the propagation of differences could differ between EnRUPT-$\mathcal{L}$ and En-RUPT. Instead of checking for a collision at the output, we can immediately check if the difference at the output of each modular addition, i.e., the difference Δf in the intermediate value f, still matches the differential characteristic.

5.1   An Observation on EnRUPT

We now make an important observation on the structure of the EnRUPT hash function. It is possible to find a conforming message pair for a given differential characteristic one round at a time.

Consider the message word m_i, which is injected into the 'delta accumulator' d_{P-1} at the end of round i. In the first $(P-1)$ steps of the next round, d_{P-1} is not used, so m_i can not influence the behaviour of the modular additions in these steps. Starting from the P-th step of round $(i+1)$, however, m_i does have an influence.

We can search for a value for m_i such that the differential characteristic is followed up to and including the first $(P-1)$ steps of round $(i+2)$. Starting with the P-th step of round $(i+2)$, the next message word, m_{i+1} also influences the modular additions. Thus, we can keep m_i fixed, and use the new freedom available in m_{i+1} to ensure the differential characteristic is also followed for the next $s \cdot P$ steps.

This drastically reduces the expected number of trials required to find a collision. Let p_i denote the probability that the differential characteristic is followed in a block of $s \cdot P$ consecutive steps, starting at the P-th step of a round. Because we can construct a conforming message pair one word at a time, the expected number of trials is $\sum_i 1/p_i$ rather than $\prod_i 1/p_i$. In other words, the complexities associated with each block of $s \cdot P$ steps should be added together, rather than multiplied. This possibility was ignored in the security analysis of EnRUPT [5], leading to the wrong conclusion that attacks based on linearisation do not apply.

5.2   Accelerating the Collision Search

An simple optimisation can be made to the collision search, which will allow us to ignore the probability associated with one step in each round. This optimisation is analogous to Wang's 'single message modification', which was first introduced in the context of MD5 [11].

Consider the P-th step of a round. In this step, the 'delta accumulator' d_{P-1}, to which a new message word m was XORed at the end of the previous round, is used for the first time. More precisely, it is used in line 9 of Fig. 2 to compute the intermediate value e. Note however that these computations can be inverted. We can choose the value of e, and compute backwards to find what the message word m should be to arrive at this value of e.

The values of e which ensure that the difference propagation of the modular addition in line 10 of Fig. 2 corresponds to that of its linear approximation can be efficiently enumerated. Thus, rather than randomly picking values for m, we can efficiently sample *good* values for e in this step, and compute backwards to find the corresponding m. This ensures that the first modular addition affected by a message word m will always exhibit the desired propagation of differences. Thus, the P-th step of every round can be ignored in the estimation of the complexity of the attack.

6   Finding Good Differential Characteristics

The key to lowering the attack complexity is to find a good differential characteristic, i.e., a characteristic which is likely to be followed for the nonlinear hash function. A general approach to this problem, based on finding low weight codewords in a linear code, was proposed by Rijmen and Oswald [9] and extended by Pramstaller et al. in [8]. In this section, we show how to apply this approach to EnRUPT.

6.1   Coding Theory

As observed by Rijmen and Oswald [9], all of the differential characteristics leading to a collision for the linearised hash function can be seen as the codewords of a linear code.

Consider the EnRUPT-$\mathcal{L}$ hash function with a h-bit output length, and the message input restricted to messages of t message words. Since it is linear over $GF(2)$, it is possible to express the difference in the output as a linear function of the difference in the input message m:

$$[\Delta o]_{1 \times h} = [\Delta m]_{1 \times tw} \cdot [\mathbf{O}]_{tw \times h} \ . \tag{2}$$

As the modular additions, or rather the multiplications with 9, in the EnRUPT round function are approximated, we are also interested in the differences that enter each of these operations. For EnRUPT restricted to t message blocks, there are $t \cdot s \cdot P$ such operations in total. Hence, we can combine the input differences to these operations in a $1 \times tsPw$ bit vector Δe. Again, for the linear approximation, Δe is simply a linear function of the message difference Δm:

$$[\Delta e]_{1 \times tsPw} = [\Delta m]_{1 \times tw} \cdot [\mathbf{E}]_{tw \times tsPw} \ . \tag{3}$$

Putting this together results in a linear code described by the following generator matrix

$$\mathbf{G} = \left[\mathbf{I}_{tw \times tw} \middle| \mathbf{E}_{tw \times tsPw} \middle| \mathbf{O}_{tw \times h} \right] \ . \tag{4}$$

Each codeword contains a message difference, the input differences to all approximated modular additions, and finally the output difference.

Thus, each codeword is in fact a differential characteristic for EnRUPT-$\mathcal{L}$, and all differential characteristics for EnRUPT-$\mathcal{L}$ are codewords of this code. To restrict ourselves to collision differentials, i.e., differential characteristicss ending in a zero output difference, we can use Gauss elimination to force the h rightmost columns of the generator matrix G to zero.

It is well known that the differential behaviour of modular addition can be approximated by that of XOR when the Hamming weight of the input difference, ignoring the most significant bit, is small [2,3,8,9]. As the input differences to the modular additions are part of the codewords, we will attempt to find a codeword with a low Hamming weight in this part of the codeword.

6.2   Low Weight Codewords

To find low weight codewords, we used a simple and straightforward algorithm that is based on the assumption that a codeword of very low weight exists in the code. For our purposes, this is a reasonable assumption, as only a very low weight codeword will lead to an attack faster than a generic attack. The algorithm is related to the algorithm of Canteaut and Chabaud [1] and the algorithm used to find low weight codewords for linearised SHA-1 by Pramstaller et al. [8].

Let G be the generator matrix of the linear code as in (4). We randomly select a set I of (appropriate) columns of the generator matrix G and force them to zero using Gauss elimination, until only d rows remain, where d is a parameter of the algorithm. Then, the remaining space of 2^d codewords is searched exhaustively. This procedure is repeated until a codeword of sufficiently low weight is encountered. By replacing only the 'oldest' column(s) in I, instead of restarting from the beginning every time, the algorithm can be implemented efficiently in practice.

If a codeword of very low weight exists in the code, it is likely that all of the columns in the randomly constructed set I will coincide with zeroes in the codeword, which implies that the codeword will be found in the exhaustive search phase. In the case of the codes originating from the seven linearised EnRUPT variants we consider, this algorithm finds a codeword of very low weight in a matter of minutes on a PC. Repeated runs of the algorithm always find the same codewords, so it is reasonable to assume that these are indeed the best codewords we can find.

6.3   Estimating the Attack Complexity

Actually, the weight of a codeword is only a heuristic for the attack complexity resulting from the corresponding differential. Codewords with a lower weight are expected to result in a lower attack complexity, but we can easily enhance our algorithm to optimise the actual attack complexity, rather than just a crude heuristic.

The Differential Probability. The probability that a differential characteristic is followed, is determined by the differences that are input to each of the multiplications with 9 (line 10 in Fig. 2) that were approximated using XOR operations.

Denote by $\mathrm{DP}^{\times 9}(\Delta)$ the probability that the propagation of differences through this nonlinear operation coincides with that of its linear approximation:

$$\mathrm{DP}^{\times 9}(\Delta) = \Pr_x \left[(x \times 9) \oplus ((x \oplus \Delta) \times 9) = \Delta \oplus (\Delta \ll 3) \right] . \tag{5}$$

The differential probability of modular addition was studied by Lipmaa and Moriai [3]. Applying their results to this situation, and taking into account that the three least significant bits of $(x \ll 3)$ are always zero, we find the following estimate for $\mathrm{DP}^{\times 9}(\Delta)$:

$$\mathrm{DP}^{\times 9}(\Delta) \approx 2^{-\mathrm{wt}\left(\left(\Delta \vee (\Delta \ll 3) \right) \wedge 0111 \cdots 111000_b \right)} . \tag{6}$$

Even though this estimate ignores the dependency between x and $(x \ll 3)$, this confirms the intuition that a difference Δ with a low Hamming weight (ignoring the most significant bit and the three least significant bits) results in a large probability $\mathrm{DP}^{\times 9}(\Delta)$. We used this as a heuristic to find a good differential characteristic: we want to minimise the Hamming weight of the relevant parts of the differences that are input to the modular additions. In other words, we want to find a low weight codeword of the aforementioned linear code, where only the bits that impact $\mathrm{DP}^{\times 9}(\Delta)$ are counted.

Exact Computation of the Differential Probability. Computing the exact value of $\mathrm{DP}^{\times 9}(\Delta)$ for any given difference Δ can be done by counting all the values x for which the differences propagation is as predicted by the linear approximation. This can be done efficiently as the modular addition can be represented compactly as a trellis, where each path through the trellis corresponds to a 'good' value of x. Using a slight variant of the Viterbi algorithm [10], the number of paths in the trellis can be counted efficiently. While this is very useful for evaluating the attack complexity, it lacks the clear intuition we can gather from (6).

Computing the Attack Complexity. Let $p_{r,i}$ be the differential probability associated with the modular addition in step i of round r of the differential characteristic. Recall the observation made in Sect. 5.1, i.e., finding a conforming message pair can be done one round at a time, or rather one message word at a time, as this does not coincide precisely with the round boundaries. Taking this into account, the complexity of finding the j-th word of a conforming message pair can thus be computed as

$$C_j = \left(\prod_{i=P-1}^{sP-1} \frac{1}{p_{j+1,i}} \right) \left(\prod_{i=0}^{P-2} \frac{1}{p_{j+2,i}} \right) . \tag{7}$$

Due to the acceleration technique presented in Sect. 5.2, we are guaranteed that the differential behaviour of the modular addition in step $P-1$ of each round will be as desired. Thus, we can set $p_{P-1} = 1$. With the default EnRUPT parameters ($P = 2$ and $s = 4$, see Table 1), this then becomes

$$C_j = \frac{1}{p_{j+1,2}} \cdot \frac{1}{p_{j+1,3}} \cdot \frac{1}{p_{j+1,4}} \cdot \frac{1}{p_{j+1,5}} \cdot \frac{1}{p_{j+1,6}} \cdot \frac{1}{p_{j+1,7}} \cdot \frac{1}{p_{j+2,0}} . \tag{8}$$

Finally, as was explained in Sect. 5.1, note that each message word can be found independent of the previous ones, due to the newly available degrees of freedom in each message word. Hence, the overall attack complexity can simply be computed as the sum of these round complexities:

$$C_{\text{tot}} = \sum_{j=0}^{t} C_j \; . \tag{9}$$

Note that, given a differential characteristic, it is easy to compute the associated attack complexity. Hence, when searching for a good differential characteristic using the algorithm described in Sect. 6.2, we can use the actual attack complexity instead of the weight of the codeword. The algorithm still implicitly uses the weight of a codeword as a heuristic, but now attempts to optimise the actual attack complexity directly.

7   Results and Discussion

We constructed differential characteristics for each of the seven EnRUPT variants in the EnRUPT SHA-3 proposal [5]. Table 2 lists the attack complexity and the length of the best characteristic we found for each variant. Recall that we fixed the length of the characteristic a priori. Note however that nothing prevents our search algorithm from proposing a shorter characteristic, padded with rounds without any difference, which we also observed in practice. We experimented with (much) longer maximum characteristic lengths, but found no better long characteristics.

The time complexities vary from 2^{36} to 2^{40} round computations, depending on the EnRUPT variant, which is remarkable. It means that the collision resistance in absolute terms of each of these EnRUPT variants is more or less the same, regardless of the digest length. Relative to the expected collision resistance of approximately $2^{n/2}$ for an n-bit hash function, however, the (relative) collision

Table 2. Summary of our attacks. Only the best attack is listed for each EnRUPT variant

EnRUPT variant	estimated time complexity [EnRUPT rounds]	length of collision differential [message words]
EnRUPT-128	$2^{36.04}$	6
EnRUPT-160	$2^{37.78}$	7
EnRUPT-192	$2^{38.33}$	8
EnRUPT-224	$2^{37.02}$	6
EnRUPT-256	$2^{37.02}$	6
EnRUPT-384	$2^{39.63}$	8
EnRUPT-512	$2^{38.46}$	10

Table 3. Our Differential Characteristic for EnRUPT-256

Round	Step	Δe	$\rightarrow$	Δf	$DP^{\times 9}$	totals
inject message word difference $\Delta m_{-1} = 0000000008000000_x$						
0	0	0000000000000000_x	$\rightarrow$	0000000000000000_x	$2^{-0.00}$	$\mathbf{2^{-0.00}}$
	1	0000000000000800_x	$\rightarrow$	0000000000004800_x	$\star$	
	2	9000000000000000_x	$\rightarrow$	1000000000000000_x	$2^{-0.85}$	
	3	4800000000000800_x	$\rightarrow$	0800000000004800_x	$2^{-3.70}$	
	4	9000000000000000_x	$\rightarrow$	1000000000000000_x	$2^{-0.85}$	
	5	4800280000000800_x	$\rightarrow$	0801680000004800_x	$2^{-7.28}$	
	6	$90000002d0000000_x$	$\rightarrow$	1000001450000000_x	$2^{-6.43}$	
	7	0000280168000800_x	$\rightarrow$	$0001680a28004800_x$	$2^{-11.02}$	
inject message word difference $\Delta m_0 = 0000000228000000_x$						
1	0	$90000002d0000000_x$	$\rightarrow$	1000001450000000_x	$2^{-6.43}$	$\mathbf{2^{-36.56}}$
	1	0000280168000000_x	$\rightarrow$	$0001680a28000000_x$	$\star$	
	2	$90000002d0000000_x$	$\rightarrow$	1000001450000000_x	$2^{-6.43}$	
	3	4800280000000000_x	$\rightarrow$	0801680000000000_x	$2^{-5.43}$	
	4	$90000002d0000000_x$	$\rightarrow$	1000001450000000_x	$2^{-6.43}$	
	5	0000080000000000_x	$\rightarrow$	0000480000000000_x	$2^{-1.85}$	
	6	9000000240000000_x	$\rightarrow$	1000001040000000_x	$2^{-3.70}$	
	7	4800080120000000_x	$\rightarrow$	0800480820000000_x	$2^{-6.54}$	
inject message word difference $\Delta m_1 = 0000002288000000_x$						
2	0	9000000240000000_x	$\rightarrow$	1000001040000000_x	$2^{-3.70}$	$\mathbf{2^{-34.08}}$
	1	0000080048000000_x	$\rightarrow$	0000480208000000_x	$\star$	
	2	9000000240000000_x	$\rightarrow$	1000001040000000_x	$2^{-3.70}$	
	3	4800080168000000_x	$\rightarrow$	$0800480a28000000_x$	$2^{-9.28}$	
	4	9000000240000000_x	$\rightarrow$	1000001040000000_x	$2^{-3.70}$	
	5	0000200000000000_x	$\rightarrow$	0001200000000000_x	$2^{-1.85}$	
	6	9000000000000000_x	$\rightarrow$	1000000000000000_x	$2^{-0.85}$	
	7	4800200000000000_x	$\rightarrow$	0801200000000000_x	$2^{-3.70}$	
inject message word difference $\Delta m_2 = 0000000208000000_x$						
3	0	9000000000000000_x	$\rightarrow$	1000000000000000_x	$2^{-0.85}$	$\mathbf{2^{-23.91}}$
	1	0000280120000000_x	$\rightarrow$	0001680820000000_x	$\star$	
	2	9000000090000000_x	$\rightarrow$	1000000410000000_x	$2^{-3.70}$	
	3	4800280168000000_x	$\rightarrow$	$0801680a28000000_x$	$2^{-11.02}$	
	4	9000000090000000_x	$\rightarrow$	1000000410000000_x	$2^{-3.70}$	
	5	0000080048000000_x	$\rightarrow$	0000480208000000_x	$2^{-4.70}$	
	6	9000000090000000_x	$\rightarrow$	1000000410000000_x	$2^{-3.70}$	
	7	4800080000000000_x	$\rightarrow$	0800480000000000_x	$2^{-3.70}$	
inject message word difference $\Delta m_3 = 0000000200000000_x$						
4	0	9000000090000000_x	$\rightarrow$	1000000410000000_x	$2^{-3.70}$	$\mathbf{2^{-34.19}}$
	1	0000080000000800_x	$\rightarrow$	0000480000004800_x	$\star$	
	2	0000000000000000_x	$\rightarrow$	0000000000000000_x	$2^{-0.00}$	
	3	0000080000000800_x	$\rightarrow$	0000480000004800_x	$2^{-3.70}$	
	4	0000000000000000_x	$\rightarrow$	0000000000000000_x	$2^{-0.00}$	
	5	4800080048000800_x	$\rightarrow$	0800480208004800_x	$2^{-8.39}$	
	6	0000000000000000_x	$\rightarrow$	0000000000000000_x	$2^{-0.00}$	
	7	4800080048000800_x	$\rightarrow$	0800480208004800_x	$2^{-8.39}$	
inject message word difference $\Delta m_4 = 0000000200000000_x$						
5	0	0000000000000000_x	$\rightarrow$	0000000000000000_x	$2^{-0.00}$	$\mathbf{2^{-20.49}}$
	1	0000000000000000_x	$\rightarrow$	0000000000000000_x	$\star$	
	$\vdots$	$\vdots$	$\rightarrow$	$\vdots$	$\vdots$	
	7	0000000000000000_x	$\rightarrow$	0000000000000000_x	$2^{-0.00}$	$\mathbf{2^{-0.00}}$

resistance of EnRUPT is much lower for the variants with a longer digest length than for those with a shorter digest length.

As an example, Table 3 lists our differential characteristic for EnRUPT-256 with an associated attack complexity of 2^{37} EnRUPT round computations. Each line in the table corresponds to one step of the EnRUPT round function. The difference in the input (Δe) and the output (Δf) of the modular addition in that step is indicated. Also, the message word differences are shown at the end of each round. The table also includes the differential probabilities of each step, which were used to compute the attack complexity. A star ('$\star$') indicates that the differential probability can be ignored in that step because of the technique presented in Sect. 5.2. The product of the step probabilities is given for eight consecutive steps. Note that these do not coincide with the rounds, as was discussed in Sect. 6.3. A collision example for EnRUPT-256, obtained using this characteristic, is given in Table 4.

Table 4. A Collision Example for EnRUPT-256

M	13_x $c8_x$ $4b_x$ 45_x 62_x 70_x 17_x $6e_x$
	04_x $f9_x$ 31_x $7e_x$ $c3_x$ $6c_x$ $e7_x$ $d3_x$
	$e1_x$ 21_x 78_x $6a_x$ 34_x 74_x 11_x 19_x
	$7f_x$ 64_x $a3_x$ $c9_x$ 40_x 07_x 75_x 76_x
	$a1_x$ $4f_x$ 90_x 86_x fd_x $c7_x$ 33_x $4a_x$
	41_x $3a_x$ 76_x 91_x 96_x 06_x $2c_x$ $a1_x$.
M'	13_x $c8_x$ $4b_x$ 45_x $6a_x$ 70_x 17_x $6e_x$
	04_x $f9_x$ 31_x $5c_x$ 43_x $6c_x$ $e7_x$ $d3_x$
	$e1_x$ 21_x 78_x 48_x bc_x 74_x 11_x 19_x
	$7f_x$ 64_x $a3_x$ cb_x 48_x 07_x 75_x 76_x
	$a1_x$ $4f_x$ 90_x 84_x fd_x $c7_x$ 33_x $4a_x$
	41_x $3a_x$ 76_x 93_x 96_x 06_x $2c_x$ $a1_x$.
EnRUPT-256$(M) =$	bd_x 67_x 51_x $7c_x$ $a6_x$ $c0_x$ 41_x 20_x
EnRUPT-256$(M') =$	82_x $e0_x$ $3b_x$ 74_x $5f_x$ fc_x $4a_x$ 64_x
	$e9_x$ $f0_x$ 92_x $c2_x$ 58_x $c3_x$ 98_x $b8_x$
	44_x $9a_x$ fe_x cb_x $7f_x$ $c8_x$ $6f_x$ 72_x.

Discussion. In response to these collision attacks, the designers of EnRUPT proposed to double the s parameter to 8, or to increase it even further to be equal to the H-parameter, see Table 1 [6,7]. As a consequence of this, the number of steps between two message word injections is at least doubled. Experiments with these EnRUPT variants indicate that this tweak seems to be effective at stopping the attacks described in this paper. For EnRUPT-256 with $s = 6$, we were still able to find a differential with an associated attack complexity of about 2^{110} EnRUPT rounds, which is still below the birthday bound. For higher values of the s parameter, all the differential characteristics we could find would result in attack complexities that are far beyond than the birthday bound, and thus should not be considered to be real attacks.

Note that the failure of this heuristic attack method for $s = 8$ or $s = H$ does not preclude the possibility of attacks based on linearisation. Our experiments only show that it is unlikely that the particular attack method used in this work can be applied directly to EnRUPT with $s \geq 8$.

8   Conclusion

We presented collision attacks on all seven variants of the EnRUPT hash function [5] that were proposed as candidates to the NIST SHA-3 competition [4]. The attacks require negligible memory and have time complexities ranging from 2^{36} to 2^{40} EnRUPT round computations, depending on the EnRUPT variant. The practicality of the attacks has been demonstrated with an example collision for EnRUPT-256.

Acknowledgements

This work was supported in part by the IAP Programme P6/26 BCRYPT of the Belgian State (Belgian Science Policy), and in part by the European Commission through the ICT programme under contract ICT-2007-216676 ECRYPT II. The collision example for EnRUPT-256 was obtained utilizing high performance computational resources provided by the University of Leuven, http://ludit.kuleuven.be/hpc.

References

1. Canteaut, A., Chabaud, F.: A New Algorithm for Finding Minimum-Weight Words in a Linear Code: Application to McEliece's Cryptosystem and to Narrow-Sense BCH Codes of Length 511. IEEE Transactions on Information Theory 44(1), 367–378 (1998)
2. Chabaud, F., Joux, A.: Differential Collisions in SHA-0. In: Krawczyk, H. (ed.) CRYPTO 1998. LNCS, vol. 1462, pp. 56–71. Springer, Heidelberg (1998)
3. Lipmaa, H., Moriai, S.: Efficient Algorithms for Computing Differential Properties of Addition. In: Matsui, M. (ed.) FSE 2001. LNCS, vol. 2355, pp. 336–350. Springer, Heidelberg (2002)
4. National Institute of Standards and Technology, Announcing Request for Candidate Algorithm Nominations for a New Cryptographic Hash Algorithm (SHA-3) Family, Federal Register 72(212), pp. 62212–62220 (2007)
5. O'Neil, S., Nohl, K., Henzen, L.: EnRUPT Hash Function Specification. Submission to the NIST SHA-3 competition (2008), http://www.enrupt.com/SHA3/
6. O'Neil, S.: Personal communication (January 20, 2009)
7. O'Neil, S.: EnRUPT. In: The First SHA-3 Candidate Conference, Leuven, Belgium, February 25–29 (2009), http://csrc.nist.gov/groups/ST/hash/sha-3/Round1/Feb2009/documents/EnRUPT_2009.pdf
8. Pramstaller, N., Rechberger, C., Rijmen, V.: Exploiting Coding Theory for Collision Attacks on SHA-1. In: Smart, N.P. (ed.) Cryptography and Coding 2005. LNCS, vol. 3796, pp. 78–95. Springer, Heidelberg (2005)

9. Rijmen, V., Oswald, E.: Update on SHA-1. In: Menezes, A. (ed.) CT-RSA 2005. LNCS, vol. 3376, pp. 58–71. Springer, Heidelberg (2005)
10. Viterbi, A.J.: Error bounds for convolutional codes and an asymptotically optimum decoding algorithm. IEEE Transactions on Information Theory 13(3), 260–269 (1967)
11. Wang, X., Yu, H.: How to Break MD5 and Other Hash Functions. In: Cramer, R. (ed.) EUROCRYPT 2005. LNCS, vol. 3494, pp. 19–35. Springer, Heidelberg (2005)

The Rebound Attack: Cryptanalysis of Reduced Whirlpool and Grøstl

Florian Mendel[1], Christian Rechberger[1], Martin Schläffer[1],
and Søren S. Thomsen[2]

[1] Institute for Applied Information Processing and Communications (IAIK)
Graz University of Technology, Inffeldgasse 16a, A-8010 Graz, Austria
[2] Department of Mathematics, Technical University of Denmark
Matematiktorvet 303S, DK-2800 Kgs. Lyngby, Denmark
martin.schlaeffer@iaik.tugraz.at

Abstract. In this work, we propose the rebound attack, a new tool for the cryptanalysis of hash functions. The idea of the rebound attack is to use the available degrees of freedom in a collision attack to efficiently bypass the low probability parts of a differential trail. The rebound attack consists of an inbound phase with a match-in-the-middle part to exploit the available degrees of freedom, and a subsequent probabilistic outbound phase. Especially on AES based hash functions, the rebound attack leads to new attacks for a surprisingly high number of rounds.

We use the rebound attack to construct collisions for 4.5 rounds of the 512-bit hash function Whirlpool with a complexity of 2^{120} compression function evaluations and negligible memory requirements. The attack can be extended to a near-collision on 7.5 rounds of the compression function of Whirlpool and 8.5 rounds of the similar hash function Maelstrom. Additionally, we apply the rebound attack to the SHA-3 submission Grøstl, which leads to an attack on 6 rounds of the Grøstl-256 compression function with a complexity of 2^{120} and memory requirements of about 2^{64}.

Keywords: Whirlpool, Grøstl, Maelstrom, hash function, collision attack, near-collision.

1 Introduction

In the last few years the cryptanalysis of hash functions has become an important topic within the cryptographic community. Especially the attacks and tools for the MD4 family of hash functions (e.g. MD5, SHA-1) have reduced the security provided by these commonly used hash functions [2, 3, 4, 24, 26, 27]. Most of the existing cryptanalytic work has been published for this particular line of hash function design. In the NIST SHA-3 competition [19], whose aim is to find an alternative hash function to SHA-2, many new hash function designs have been proposed. This is the most recent and most prominent case showing that it is very important to have tools available to analyze other design variants as well. Our work contributes to this toolbox.

O. Dunkelman (Ed.): FSE 2009, LNCS 5665, pp. 260–276, 2009.

1.1   Preview of Results

Our main result is the introduction of a technique for hash function cryptanalysis, which we call the *rebound attack*. We apply it to both block cipher based and permutation based constructions. In the rebound attack, we consider the internal cipher of a hash or compression function as three sub-ciphers. Let E be a block cipher, then $E = E_{fw} \circ E_{in} \circ E_{bw}$. Alternatively, for a permutation based construction, we decompose a permutation P into three sub-permutations $P = P_{fw} \circ P_{in} \circ P_{bw}$.

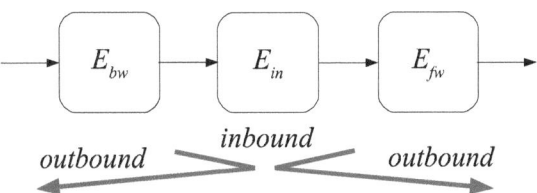

Fig. 1. A schematic view of the rebound attack. The attack consists of an inbound and two outbound phases.

The rebound attack can be described by two phases (see Fig. 1):

- **Inbound phase:** Is a meet-in-the-middle phase in E_{in} (or P_{in}), which is aided by the degrees of freedom that are available to a hash function cryptanalyst. We term the combination of meet-in-the-middle technique and exploitation of degrees of freedom leading to very efficient matches **match-in-the-middle approach.**
- **Outbound phase:** In this second phase, we use truncated differentials in both forward- and backward direction through E_{fw} and E_{bw} (or P_{fw} and P_{bw}) to obtain desired collisions or near-collisions. If the truncated differentials have a low probability in E_{fw} and E_{bw}, we can repeat the inbound phase to obtained more starting points for the outbound phase.

We apply the rebound attack on several concrete hash functions where the application on Whirlpool is probably the most relevant. Whirlpool is the only hash function standardized by ISO/IEC 10118-3:2003 (since 2000) that does not follow the MD4 design strategy. Furthermore, Whirlpool has been evaluated and approved by NESSIE [20]. Whirlpool is commonly considered to be a conservative block-cipher based design with an extremely conservative key schedule. The employed wide-trail design strategy [5] makes the application of differential and linear attacks seemingly impossible. No cryptanalytic results on the hash function Whirlpool have been published since its proposal 8 years ago.

Offsprings of Whirlpool are Maelstrom and to some extent several SHA-3 candidates, including `Grøstl`. The results of the attack on these hash functions are summarized in Table 1. For the types of attacks, we adopt the notation of [15].

Table 1. Summary of results of the attacks on reduced hash functions Whirlpool, Grøstl-256 and Maelstrom. The full versions have 10 rounds each. All attacks, except the attacks on Grøstl-256, have negligible memory requirements.

hash function	rounds	computational complexity	memory requirements	type	section
Whirlpool	4.5	2^{120}	2^{16}	collision	3
	5.5	2^{120}	2^{16}	semi-free-start collision	3
	7.5	2^{128}	2^{16}	semi-free-start near-collision	3
Grøstl-256	6	2^{120}	2^{64}	semi-free-start collision	4
Maelstrom	6.5	2^{120}	2^{16}	free-start collision	A
	8.5	2^{128}	2^{16}	free-start near-collision	A

1.2 Related Work

The rebound attack can be seen to have ancestors from various lines of research, often related to block ciphers:

- Firstly, differential cryptanalysis of block cipher based hash functions. Rijmen and Preneel [23] describe collision attacks on 15 out of 16 rounds on hash functions using DES. For the case of Whirlpool, there is an observation on the internal block cipher W by Knudsen [13]. Khovratovich *et al.* [11] studied collision search for AES-based hash functions.
- Secondly, inside-out techniques. As an application of second order differential attacks, inside-out techniques in block-cipher cryptanalysis were pioneered by Wagner in the Boomerang attack [25].
- Thirdly, truncated differentials. In the applications of the rebound technique, we used truncated differentials in the outbound parts. Knudsen [12] proposed truncated differentials as a tool in block cipher cryptanalysis, which recently have been applied to the hash function proposal Grindahl [14] by Peyrin [21].

1.3 Outline of the Paper

In the following section, we start with a description of the attacked hash functions. For the sake of presentation and concreteness, we immediately apply the rebound attack to the hash function Whirlpool in Sect. 3. In Sect. 4, we apply the rebound attack on Grøstl. The application of the attack to Maelstrom is postponed to App. A. We conclude in Sect. 5.

2   Description of the Hash Functions

In this section we give a short description of the hash functions to be analyzed in the remainder of this paper. We describe the hash function Whirlpool first, and continue with the description of the hash function Grøstl.

2.1   The Whirlpool Hash Function

Whirlpool is a cryptographic hash function designed by Barreto and Rijmen in 2000 [1]. It is an iterative hash function that processes 512-bit input message blocks with compression functions and produces a 512-bit hash value. The Whirlpool compression function basically consists of two parts: the key schedule and the state update transformation. The underlying block cipher W operates in the Miyaguchi-Preneel mode [17] as shown in Fig. 2. A detailed description of the hash function is given in [1].

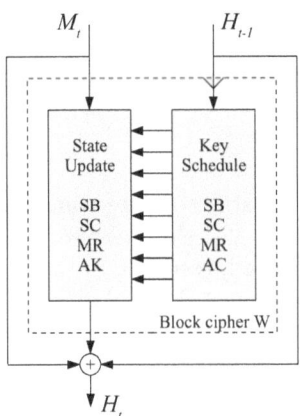

Fig. 2. A schematic view of the Whirlpool compression function. The block cipher W is used in Miyaguchi-Preneel mode.

The 512-bit block cipher W uses a 512-bit key and is similar to the Advanced Encryption Standard (AES) [18]. Both the state update transformation and the key schedule of W update an 8×8 state S of 64 bytes in 10 rounds each. The round transformations are very similar to the AES round transformations and are briefly described here:

- the non-linear layer SubBytes (SB) applies an S-Box to each byte of the state independently
- the cyclical permutation ShiftColumns (SC) rotates the bytes of column j downwards by j positions
- the linear diffusion layer MixRows (MR) multiplies the state by a constant matrix
- the key addition AddRoundKey (AK) adds the round key and/or the round constants c^r (AC) of the key schedule

In each round, the state is updated by round transformation r_i as follows:

$$r_i \equiv AK \circ MR \circ SC \circ SB.$$

In the remainder of this paper, we will use the outline of Fig. 3 for one round. We denote the resulting state of round transformation r_i by S_i and the intermediate states after SubBytes by S_i', after ShiftColums by S_i'' and after MixRows by S_i'''. The initial state prior to the first round is denoted by S_0.

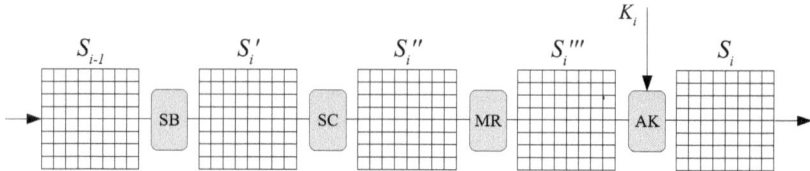

Fig. 3. One round r_i of the Whirlpool compression function with 8×8 states S_{i-1}, S_i', S_i'', S_i''', S_i and round key input K_i

After the last round of the state update transformation, the initial value or previous chaining value $H_{t-1} = S_0$, the message block M_t, and the output value of the last round S_{10} are XORed, resulting in the final output of the Whirlpool compression function, $H_t = H_{t-1} \oplus M_t \oplus S_{10}$.

2.2 The Grøstl Hash Function

Grøstl was proposed by Gauravaram et al. as a candidate for the SHA-3 competition [9], initiated by the National Institute of Standards and Technology (NIST). Grøstl is an iterated hash function with a compression function built from two distinct permutations (see Fig. 4). Grøstl is a wide-pipe design with proofs for the collision and preimage resistance of the compression function [8].

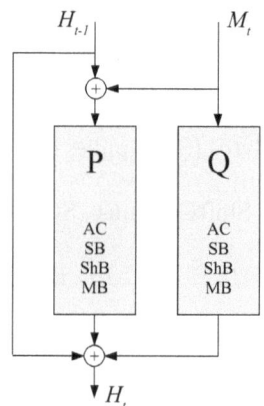

Fig. 4. The compression function of Grøstl. P and Q are $2n$-bit permutations for an n-bit hash value.

The two permutations P and Q are constructed using the wide trail design strategy and borrow components from the AES. The design of the two permutations is very similar to the block cipher W used in Whirlpool instantiated with a fixed key input. Both permutations update an 8×8 state of 64 bytes in 10 rounds each. The round transformations are very similar to the AES round transformations and are briefly described here:

- AddRoundConstant (AC) adds different one-byte round constants to the 8×8 states of P and Q
- the non-linear layer SubBytes (SB) applies the AES S-Box to each byte of the state independently
- the cyclical permutation ShiftBytes (ShB) rotates the bytes of row j left by j positions
- the linear diffusion layer MixBytes (MB) multiplies the state by a constant matrix

In each round, the state is updated by round transformation r_i as follows:

$$r_i \equiv MB \circ ShB \circ SB \circ AC$$

3   Rebound Attack on Whirlpool

In this section, we present details of the rebound attacks applied to the hash function Whirlpool. First, we will give an overview of the attack strategy which is the basis for the attacks on 4.5, 5.5 and 7.5 rounds. The main idea of the attacks is to use a 4-round differential trail [6], which has the following sequence of active S-boxes: $1 \rightarrow 8 \rightarrow 64 \rightarrow 8 \rightarrow 1$. Note that the differential probability in each round is proportional to the number of active S-boxes. Using the Rebound Attack we can cover the most expensive middle part using an efficient match-in-the-middle approach (inbound phase). In the outbound phase, the trail is extended and the two ends of the trail are linked using the feed-forward of the hash function.

3.1   Attack Overview

The core of the attack is a 4 round trail of the form $1 \rightarrow 8 \rightarrow 64 \rightarrow 8 \rightarrow 1$. This trail has the minimum number of active S-boxes and has the best differential probability according to the wide trail design strategy. In the rebound attack, we first split the block cipher W into three sub-ciphers $W = E_{fw} \circ E_{in} \circ E_{bw}$, such that most expensive part of the differential trail is covered by the efficient inbound phase E_{in}. Then, the outbound phase (E_{fw}, E_{bw}) has a relatively low probability and can be fulfilled in a probabilistic way:

$$E_{bw} = SC \circ SB \circ AK \circ MR \circ SC \circ SB$$
$$E_{in} = MR \circ SC \circ SB \circ AK \circ MR$$
$$E_{fw} = AK \circ MR \circ SC \circ SB \circ AK$$

The two phases of the rebound attack consists of basically four steps:

- **Inbound phase**
  Step 1: start with 8-byte truncated differences at the MixRows layer of round r_2 and r_3, and propagate forward and backward to the S-box layer of round r_3.
  Step 2: connect the input and output of the S-boxes of round r_3 to form the three middle states $8 \rightarrow 64 \rightarrow 8$ of the trail.
- **Outbound phase**
  Step 3: extend the trail both forward and backward to give the trail $1 \rightarrow 8 \rightarrow 64 \rightarrow 8 \rightarrow 1$ through MixRows in a probabilistic way.
  Step 4: link the beginning and the end of the trail using the feed-forward of the hash function.

If the differences in the first and last step are identical, they cancel each other through the feed-forward. The result is a collision of the round-reduced compression function of Whirlpool. See Fig. 5 for an overview of the attack.

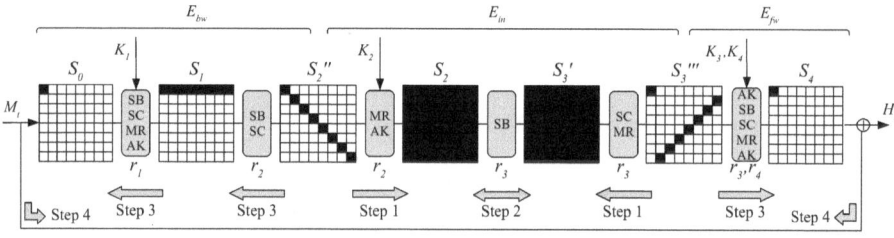

Fig. 5. A schematic view of the attack on 4 rounds of Whirlpool with round key inputs and feed-forward. Black state bytes are active.

3.2 Collision Attack for 4.5 Rounds

The collision attack on 4.5 rounds of Whirlpool is the starting point for all subsequent attacks. If the differences in the message words are the same as in the output of the state update transformation, the differences cancel each other through the feed-forward. In other words, we will construct a fixed-point (in terms of differences) for the block cipher in the state update. The outline of the attack is shown in Fig. 5 and the sequence of truncated differences has the form:

$$1 \xrightarrow{r_1} 8 \xrightarrow{r_2} 64 \xrightarrow{r_3} 8 \xrightarrow{r_4} 1 \xrightarrow{r_{4.5}} 1$$

In the following, we analyze the 4 steps of the attack in detail.

Precomputation. In the match-in-the-middle part (Step 2) we need to find a differential match for the SubBytes layer. In a precomputation step, we compute a 256×256 lookup table for each S-box differential (input/output XOR difference table). Note that only about $1/2$ of all S-box differentials exist. For each possible S-box differential, there are at least two (different) values such that the differential holds. A detailed description of the distribution of S-box differentials is given in App. B.

Step 1. We start the attack by choosing a random difference with 8 active bytes of state S_2'' prior to the MixRows layer of round r_2. Note that all active bytes have to be in the diagonal of state S_2'' (see Fig. 5). Then, the differences propagate forward to a full active state at the input of the next SubBytes layer (state S_2) with a probability of 1. Next, we start with another difference and 8 active bytes in state S_3''' after the MixRows transformation of round r_3 and propagate backwards. Again, the diagonal shape ensures that we get a full active state at the output of SubBytes of round r_3.

Step 2. In Step 2, the match-in-the-middle step, we look for a matching input/output difference of the SubBytes layer of round r_3 using the precomputed S-box differential table. Since we can find a match with a probability of $1/2$ for each byte, we can find a differential for the whole active SubBytes layer with a probability of about 2^{-64}. Hence, after repeating Step 1 of the attack about 2^{64} times, we expect to find a SubBytes differential for the whole state. Since we get at least two state values for each S-box match, we get about 2^{64} starting points for the outbound phase. Note that these 2^{64} starting points can be constructed with a total complexity of about 2^{64}. In other words, the average computational cost of each match-in-the-middle step is essentially the respective computation of the round transformations.

Step 3. In the outbound phase, we further extend the differential path backward and forward. By propagating the matching differences and state values through the next SubBytes layer, we get a truncated differential in 8 active bytes for each direction. Next, the truncated differentials need to follow a specific active byte pattern. In the case of the 4 round Whirlpool attack, the truncated differentials need to propagate from 8 to one active byte through the MixRows transformation, both in the backward and forward direction.

The propagation of truncated differentials through the MixRows transformation is modelled in a probabilistic way. The transition from 8 active bytes to one active byte through the MixRows transformation has a probability of about 2^{-56} (see App. C). Note that we require a specific position of the single active byte to find a match in the feed-forward (Step 4). Since we need to fulfill one $8 \rightarrow 1$ transitions in the backward and forward direction, the probability of the outbound phase is $2^{-2\cdot56} = 2^{-112}$. In other words, we have to repeat the inbound phase about 2^{112} times to generate 2^{112} starting points for the outbound phase of the attack.

Step 4. To construct a collision at the output of this 4 round compression function, the exact value of the input and output difference has to match. Since only one byte is active, this can be fulfilled with a probability of 2^{-8}. Hence, the complexity to find a collision for 4 rounds of Whirlpool is $2^{112+8} = 2^{120}$. Note that we can add half of a round (SB,SC) at the end for free, since we are only interested in the number of active bytes. Remember that we can construct up to 2^{128} starting points in the inbound phase of the attack, hence we have enough degrees of freedom for the attack. Note that the values of the key schedule are not influenced. Hence, the attack works with the standard IV and we can construct collisions for 4.5 rounds of the hash function of Whirlpool.

3.3 Semi-Free-Start Collision Attack for 5.5 Rounds

We can extend the collision attack on 4.5 rounds to a semi-free-start collision attack on 5.5 rounds of Whirlpool. The idea is to add another full active state in the middle of the trail. We use the additional degrees of freedom of the key schedule to fulfill the difference propagation through *two* full active S-box transformations. Note that the outbound part of the attack stays the same and the new sequence of active S-boxes is:

$$1 \xrightarrow{r_1} 8 \xrightarrow{r_2} 64 \xrightarrow{r_3} 64 \xrightarrow{r_4} 8 \xrightarrow{r_5} 1 \xrightarrow{r_{5.5}} 1$$

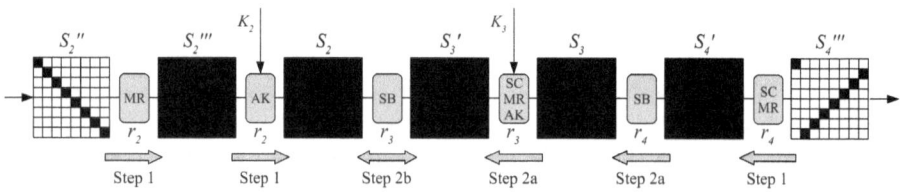

Fig. 6. In the attack on 5.5 rounds we first choose random values of the state S_4' to propagate backwards (Step 2a) and then, use the degrees of freedom from the key schedule to solve the difference propagation of the S-box in round r_3 (Step 2b).

Step 1. Figure 6 shows the inbound part of the attack in detail. Again, we can choose from up to 2^{64} initial differences with 8 active bytes at state S_2'' and S_4''' each, and linearly propagate forward to S_2 and backward to S_4' until we hit the first S-box layer. Then, we need to find a matching SubBytes differential of two consecutive S-box layers in the match-in-the-middle phase.

Step 2. To pass the S-box of round r_4 in the backward direction, we choose one of 2^{512} possible values for state S_4'. This also determines the input values and differences of the SubBytes layer (state S_3). Then, we propagate the difference further back to state S_3', which is the output of the S-box in round r_3. The 512 degrees of freedom of the key schedule input K_3 between the two S-boxes allow us to still assign arbitrary values to the state S_3'. Hence, the correct difference propagation of the S-box in round r_3 can be fulfilled by using these additional degrees of freedom to choose the state S_3' as well. The complexity of the attack does not change and is determined by the 2^{120} trials of the outbound phase.

The outbound phase (Step 3 and Step 4) of the 5.5 round attack is equivalent to the 4.5 round case. However, we cannot choose the round keys, and hence the chaining values, anymore since they are determined by the difference propagation of the S-box of round r_3. Therefore, this 5.5 round attack is only a semi-free-start collision attack on the hash function of Whirlpool.

3.4   Semi-Free-Start Near-Collision Attack for 7.5 Rounds

The collision attack on 5.5 rounds can be further extended by adding one round at the beginning and one round and at the end of the trail (see Fig. 7). The result is a semi-free-start near-collision attack on 7.5 rounds of the hash function Whirlpool with the following number of active S-boxes:

$$8 \xrightarrow{r_1} 1 \xrightarrow{r_2} 8 \xrightarrow{r_3} 64 \xrightarrow{r_4} 64 \xrightarrow{r_5} 8 \xrightarrow{r_6} 1 \xrightarrow{r_7} 8 \xrightarrow{r_{7.5}} 8$$

Since the inbound phase (Step 1 and Step 2) is identical to the attack on 5.5 rounds, we only discuss the outbound phase (Step 3 and Step 4) here.

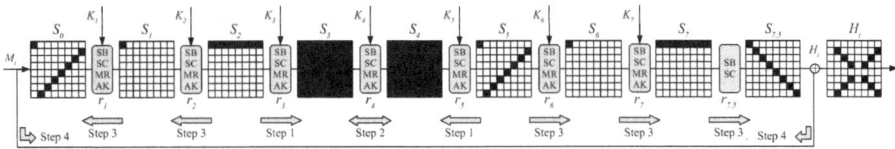

Fig. 7. In the attack on 7.5 rounds we extend the trail by one more round at the beginning and 1.5 rounds at the end to get a semi-free-start near-collision of Whirlpool

Step 3. The 1-byte difference at the beginning and end of the 4 round trail will always result in 8 active bytes after one MixRows transformation. Hence, we can go backward one round and forward 1.5 rounds with no additional costs. We add a half round at the end to get a similar pattern of 8 active S-boxes due to the ShiftColumns transformation. Note that we cannot get an exact match of active S-boxes and get therefore only a semi-free-start near-collision.

Step 4. Using the feed-forward, the position of two active S-boxes match and cancel each other with a probability of 2^{-16}. Hence, the total complexity of our semi-free-start near-collision is about $2^{112+16} = 2^{128}$. Note that the generic (birthday) complexity of a near-collision on 52 bytes is $2^{\frac{52 \cdot 8}{2}} = 2^{208}$.

4   Rebound Attack on Grøstl

In this section, we extend the attack on Whirlpool to the SHA-3 proposal Grøstl. Although the hash function is built from similar components as Whirlpool, the attack does not apply equally well. The available degrees of freedom of the second permutation cannot be used in the attack on the first permutation as in Whirlpool. Note that we can still apply the attack on 4.5 rounds of Whirlpool to the compression function of Grøstl-256 and get the same complexity of about 2^{120}.

4.1   Semi-Free-Start Collision for 5 Rounds

We can improve the Rebound Attack on Grøstl-256 by using differences in the
second permutation as well. In the attack on 5 rounds, we use the following
differential trail for both permutations:

$$8 \xrightarrow{r_1} 8 \xrightarrow{r_2} 64 \xrightarrow{r_3} 8 \xrightarrow{r_4} 8 \xrightarrow{r_5} 64$$

By using an equivalent differential trail in the second permutation one can find
a collision for the compression function of Grøstl-256 reduced to 5 rounds with
a complexity of 2^{64}, see Fig. 8.

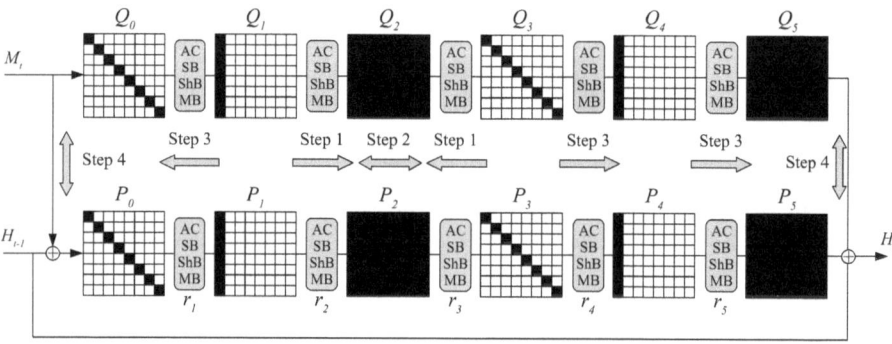

Fig. 8. Attack on Grøstl-256 reduced to 5 rounds using two equivalent trails in both
permutations P and Q

For each permutation, we can find 2^{64} inputs following this differential with a
complexity of about 2^{64} and negligible memory requirements, see Sect. 3. Hence,
the differential trail holds with probability 1 on average in both P and Q. In order
to get a semi-free-start collision of Grøstl-256 reduced to 5 rounds, we require
that the differences at the output of round 5 are equal. Since the MixBytes
transformation is linear it is sufficient that the differences before MixBytes in
round 5 are equal. Furthermore, to prevent that the feed-forward destroys the
collision again, we do not allow any differences in H. Hence, all differences are
due to differences in the message M and we require these differences at the input
of round 1 to be equal as well.

For the attack to work, differences in 16 bytes need to be equal. A straightforward implementing of the attack would result in a complexity of about 2^{128}.
However, the complexity can be significantly reduced by applying a meet-in-
the-middle attack. In detail, by generating 2^{64} differential trails for P and 2^{64}
differential trails for Q we expect to find a matching input and output. This
results in a semi-free-start collision for Grøstl-256 reduced to 5 rounds. The
attack has a total complexity of about 2^{64} evaluations of P and Q and memory
requirement of 2^{64}. Note that the memory requirements of the attack can be significantly reduced by memory less variants of the meet-in-the-middle attack [22].

4.2   Semi-Free-Start Collision for 6 Rounds

The attack can be extended to 6 rounds using an extended differential trail for
P and Q, see Fig. 9. For this attack, we use a trail with the following sequence
of active bytes:

$$8 \xrightarrow{r_1} 1 \xrightarrow{r_2} 8 \xrightarrow{r_3} 64 \xrightarrow{r_4} 8 \xrightarrow{r_5} 8 \xrightarrow{r_6} 64$$

Note that this trail holds with a probability of 2^{-56} on average. Hence, we can
find a collision for the compression function of Grøstl-256 reduced to 6 rounds
with a complexity of about $2^{56+64} = 2^{120}$, and memory requirements of 2^{64} to
match the beginning and end of each trail. In contrast to the attack on 5 rounds,
we do not see how the connection of the two permutations can be implemented
in a memory-less way.

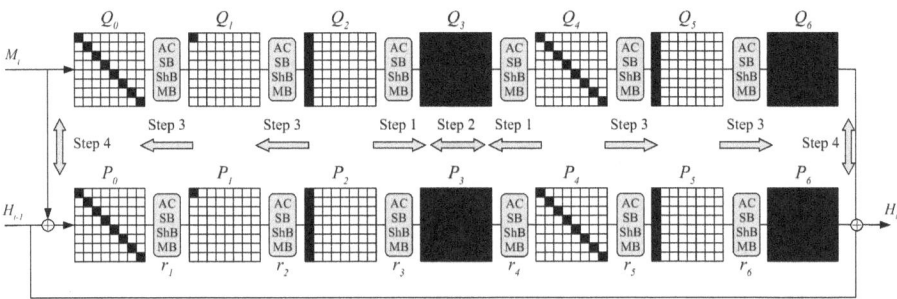

Fig. 9. Attack on Grøstl-256 reduced to 6 rounds using two equivalent trails in both
permutations P and Q

Note that we could add a half round (ShB,MB) in the beginning of Grøstl-
256, similar to the end of the trail. However, we only consider variants by reducing rounds at the end of the compression function. Trying to attack more
rounds of the Grøstl-256 compression function quickly does not leave enough
degrees of freedom to succeed, or results in a computational complexity above
2^{128}, which is above the security claims of the designers.

5   Conclusion and Open Problems

In this paper, we propose a new tool for the toolbox of hash function cryptanalysts: The rebound attack. We have successfully attacked 7.5 rounds of the
Whirlpool compression function, 6 rounds of the Grøstl-256 compression
function, and 8.5 rounds of the Maelstrom compression function (App. A).

The idea in these attacks is to use the available degrees of freedom in a collision
attack to efficiently bypass the devastating effects of the wide-trail design strategy on differential-style attacks for a feasible number of rounds. More degrees

of freedom (like the increased key-size in the Maelstrom block cipher) makes equation solving (and hence the match-in-the-middle step) easier and allows to cover even more rounds.

Most AES-based SHA-3 candidates are natural candidates for applications of the rebound attack. To this end, we can refer to preliminary results which break Twister-512 [16][1].

The idea seems applicable to a wider range of hash function constructions. For the outbound part of the rebound attack we used truncated differentials in all our examples. However, the rebound technique does not constrain the property used in the outbound part. It would be interesting to see if other non-random properties (e.g., correlations or algebraic relations) could also be used with the rebound attack.

Acknowledgments. We would like to thank Henri Gilbert, Mario Lamberger, Tomislav Nad, Vincent Rijmen and the anonymous referees for useful comments and discussions. The work in this paper has been supported in part by the Secure Information Technology Center-Austria (A-SIT), by the European Commission under contract ICT-2007-216646 (ECRYPT II) and by the IAP Programme P6/26 BCRYPT of the Belgian State (Belgian Science Policy).

References

1. Barreto, P.S.L.M., Rijmen, V.: The WHIRLPOOL Hashing Function. Submitted to NESSIE (September 2000) (Revised May 2003), http://www.larc.usp.br/~pbarreto/WhirlpoolPage.html (2008/12/11)
2. De Cannière, C., Mendel, F., Rechberger, C.: Collisions for 70-Step SHA-1: On the Full Cost of Collision Search. In: Adams, C.M., Miri, A., Wiener, M.J. (eds.) SAC 2007. LNCS, vol. 4876, pp. 56–73. Springer, Heidelberg (2007)
3. De Cannière, C., Rechberger, C.: Finding SHA-1 Characteristics: General Results and Applications. In: Lai, X., Chen, K. (eds.) ASIACRYPT 2006. LNCS, vol. 4284, pp. 1–20. Springer, Heidelberg (2006)
4. De Cannière, C., Rechberger, C.: Preimages for Reduced SHA-0 and SHA-1. In: Wagner, D. (ed.) CRYPTO 2008. LNCS, vol. 5157, pp. 179–202. Springer, Heidelberg (2008)
5. Daemen, J., Rijmen, V.: The Wide Trail Design Strategy. In: Honary, B. (ed.) Cryptography and Coding 2001. LNCS, vol. 2260, pp. 222–238. Springer, Heidelberg (2001)
6. Daemen, J., Rijmen, V.: The Design of Rijndael. In: Information Security and Cryptography. Springer, Heidelberg (2002)
7. Filho, D.G., Barreto, P.S.L.M., Rijmen, V.: The Maelstrom-0 hash function. In: SBSeg 2006 (2006)
8. Fouque, P.-A., Stern, J., Zimmer, S.: Cryptanalysis of Tweaked Versions of SMASH and Reparation. In: SAC 2008. LNCS, vol. 5381. Springer, Heidelberg (2008)

---

[1] A variant of the match-in-the-middle part of the rebound attack led to practical attacks on the compression function, which in turn led to theoretical attacks on the Twister hash function.

9. Gauravaram, P., Knudsen, L.R., Matusiewicz, K., Mendel, F., Rechberger, C., Schläffer, M., Thomsen, S.S.: Grøstl – a SHA-3 candidate (2008), http://www.groestl.info
10. Keliher, L., Sui, J.: Exact Maximum Expected Differential and Linear Probability for Two-Round Advanced Encryption Standard. Information Security, IET 1(2), 53–57 (2007)
11. Khovratovich, D., Biryukov, A., Nikolic, I.: Speeding up collision search for byte-oriented hash functions. In: Fischlin, M. (ed.) CT-RSA 2009. LNCS, vol. 5473, pp. 164–181. Springer, Heidelberg (2009)
12. Knudsen, L.R.: Truncated and Higher Order Differentials. In: Preneel, B. (ed.) FSE 1994. LNCS, vol. 1008, pp. 196–211. Springer, Heidelberg (1995)
13. Knudsen, L.R.: Non-random properties of reduced-round Whirlpool. NESSIE public report, NES/DOC/UIB/WP5/017/1 (2002)
14. Knudsen, L.R., Rechberger, C., Thomsen, S.S.: The Grindahl Hash Functions. In: Biryukov, A. (ed.) FSE 2007. LNCS, vol. 4593, pp. 39–57. Springer, Heidelberg (2007)
15. Lai, X., Massey, J.L.: Hash Functions Based on Block Ciphers. In: Rueppel, R.A. (ed.) EUROCRYPT 1992. LNCS, vol. 658, pp. 55–70. Springer, Heidelberg (1993)
16. Mendel, F., Rechberger, C., Schläffer, M.: Cryptanalysis of Twister. In: Abdalla, M., Pointcheval, D. (eds.) ACNS 2009. LNCS, vol. 5536. Springer, Heidelberg (to appear, 2009)
17. Menezes, A.J., van Oorschot, P.C., Vanstone, S.A.: Handbook of Applied Cryptography. CRC Press, Boca Raton (1997), http://www.cacr.math.uwaterloo.ca/hac/
18. National Institute of Standards and Technology. FIPS PUB 197, Advanced Encryption Standard (AES). Federal Information Processing Standards Publication 197, U.S. Department of Commerce (November 2001)
19. National Institute of Standards and Technology. Announcing Request for Candidate Algorithm Nominations for a New Cryptographic Hash Algorithm (SHA-3) Family. Federal Register 27(212), 62212–62220 (November 2007), http://csrc.nist.gov/groups/ST/hash/documents/FR_Notice_Nov07.pdf (2008/10/17)
20. NESSIE. New European Schemes for Signatures, Integrity, and Encryption. IST-1999-12324, http://cryptonessie.org/
21. Peyrin, T.: Cryptanalysis of Grindahl. In: Kurosawa, K. (ed.) ASIACRYPT 2007. LNCS, vol. 4833, pp. 551–567. Springer, Heidelberg (2007)
22. Quisquater, J.-J., Delescaille, J.-P.: How easy is collision search? Application to DES (extended summary). In: Quisquater, J.-J., Vandewalle, J. (eds.) EUROCRYPT 1989. LNCS, vol. 434, pp. 429–434. Springer, Heidelberg (1990)
23. Rijmen, V., Preneel, B.: Improved Characteristics for Differential Cryptanalysis of Hash Functions Based on Block Ciphers. In: Preneel, B. (ed.) FSE 1994. LNCS, vol. 1008, pp. 242–248. Springer, Heidelberg (1995)
24. Stevens, M., Lenstra, A.K., de Weger, B.: Chosen-Prefix Collisions for MD5 and Colliding X.509 Certificates for Different Identities. In: Naor, M. (ed.) EUROCRYPT 2007. LNCS, vol. 4515, pp. 1–22. Springer, Heidelberg (2007)
25. Wagner, D.: The Boomerang Attack. In: Knudsen, L.R. (ed.) FSE 1999. LNCS, vol. 1636, pp. 156–170. Springer, Heidelberg (1999)
26. Wang, X., Yin, Y.L., Yu, H.: Finding Collisions in the Full SHA-1. In: Shoup, V. (ed.) CRYPTO 2005. LNCS, vol. 3621, pp. 17–36. Springer, Heidelberg (2005)
27. Wang, X., Yu, H.: How to Break MD5 and Other Hash Functions. In: Cramer, R. (ed.) EUROCRYPT 2005. LNCS, vol. 3494, pp. 19–35. Springer, Heidelberg (2005)

A   Rebound Attack on Maelstrom

In this section, we apply the attack of Whirlpool to Maelstrom.

A.1   Description of the Hash Function

Maelstrom [7] is a hash function very similar to Whirlpool. It has a simpler key schedule, works on 1024-bit message blocks and uses the Davies-Meyer mode instead of Miyaguchi-Preneel. The internal block cipher of Maelstrom works on 512-bit blocks with a 1024-bit key schedule. The additional 512 degrees of freedom in the key schedule can be used to attack one more round (up to 8.5 rounds) of the compression function of Maelstrom.

A.2   Attack on 8.5 Rounds

Since Maelstrom uses the Davies-Meyer mode, we can only get a free-start collision for the hash function. However, the additional degrees of freedom of the key schedule allow us to add another round in the inbound part. The sequence of active S-boxes for the 8.5 round attack on Maelstrom is then:

$$8 \xrightarrow{r_1} 1 \xrightarrow{r_2} 8 \xrightarrow{r_3} 64 \xrightarrow{r_4} 64 \xrightarrow{r_5} 64 \xrightarrow{r_6} 8 \xrightarrow{r_7} 1 \xrightarrow{r_8} 8 \xrightarrow{r_{8.5}} 8 \qquad (1)$$

The extension is essentially the same as for the 7.5 round attack on Whirlpool. We add another state with 64 active bytes in the middle of the trail. This means, that we now have to fulfill the difference propagation of three S-box layers with 64 active bytes each. Same as in Sect. 3.3, we can fulfill one S-box propagation using the 512 degrees of freedom of the state itself. Since the second S-box difference propagation uses only 512 degrees of freedom from the key schedule, there are another 512 degrees of freedom left to fulfill the difference propagation of the third S-box. The complexity of the attack does not change and is 2^{128} for the 512-bit hash function Maelstrom. Furthermore, the semi-free-start collision attack on 5.5 rounds of Whirlpool can be extended to a 6.5 rounds free-start collision attack of Maelstrom with the same complexity of 2^{120}.

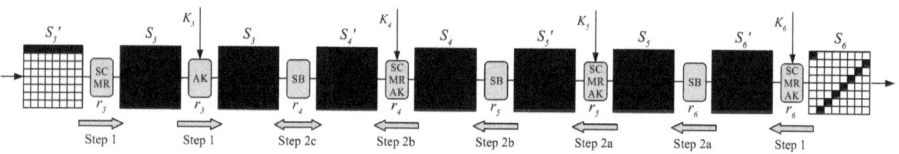

Fig. 10. In the attack on Maelstrom we use the additional degrees of freedom of the key schedule to pass three S-box layers (Step 2a,2b,2c)

B   Probability and Conditions of S-Box Differentials

In this section we analyze differentials of the Whirlpool and AES S-boxes in detail. For a fixed differential $(\Delta a, \Delta b)$ with $\Delta a = x \oplus y$ and $\Delta b = S(x) \oplus S(y)$, we get the probability $P(\Delta b = S(\Delta a)) \sim 1/2$. This can be verified by enumerating through all 256×256 input/output pairs (x, y) and $(S(x), S(y))$. Table 2 gives a distribution of possible S-box differentials for the Whirlpool and AES S-boxes [10]. Note that for each *possible* S-box differential, we get at least the two symmetric values (x, y) and (y, x). In the case of Whirlpool, we get for a small fraction of differentials even 8 possible pairs. This corresponds to the maximum probability distribution of the Whirlpool S-box, which is $8 \cdot 2^{-8} = 2^{-5}$.

Table 2. The number of differentials and possible pairs (x, y) for the Whirlpool and AES S-boxes. The first row shows the number of impossible differentials and the last row corresponds to the zero differential.

# (x, y)	Whirlpool	AES
0	39655	33150
2	20018	32130
4	5043	255
6	740	-
8	79	-
256	1	1

C   Propagation of Truncated Differentials in MixRows and MixBytes

Since the MixRows operation is a linear transformation, standard differences propagate through MixRows in a deterministic way. The propagation only depends on the values of the differences and is independent of the actual value of the state. In case of truncated differences only the position, but not the value of the difference is determined. Therefore, the propagation of truncated differences through MixRows can only be modelled in a probabilistic way. Note that the MixBytes operation of Grøstl has the same properties as MixRows.

The MDS property of the MixRows transformation ensures that the sum of the number of active input and output bytes is at least 9. Hence, a non-zero truncated difference with one active byte will propagate to a truncated difference with 8 active bytes with a probability of 1. On the other hand, a truncated difference with 8 active bytes can result in a truncated difference with one to 8 active bytes after MixRows. However, the probability of a $8 \to 1$ transition with predefined positions is only $2^{-7 \cdot 8} = 2^{-56}$ since we require 7 out of 8 truncated differences to be zero. Table 3 is similar to the table of [21] and shows the probabilities for all 81 cases with a fixed position of truncated differences. Note that the probability of any $x \to 8$ transition $(1 - \sum_{i=1}^{7} P(x \to i) \sim 2^{-0.0017})$ is approximated by 1 in this paper. Note that the probability only depends on the direction of the *propagation* of truncated differences.

Table 3. Approximate probabilities for the propagation of truncated differences through MixRows with predefined positions. D_i denotes the number of active bytes at the input and D_o the number of active bytes at the output of MixRows. Probabilities are base 2 logarithms.

$D_o \setminus D_i$	0	1	2	3	4	5	6	7	8
0	0	×	×	×	×	×	×	×	×
1	×	×	×	×	×	×	×	×	−56
2	×	×	×	×	×	×	×	−48	−48
3	×	×	×	×	×	×	−40	−40	−40
4	×	×	×	×	×	−32	−32	−32	−32
5	×	×	×	×	−24	−24	−24	−24	−24
6	×	×	×	−16	−16	−16	−16	−16	−16
7	×	×	−8	−8	−8	−8	−8	−8	−8
8	×	0	-0.0017	-0.0017	-0.0017	-0.0017	-0.0017	-0.0017	-0.0017

Revisiting the IDEA Philosophy

Pascal Junod[1,2] and Marco Macchetti[1]

[1] Nagracard SA
CH-1033 Cheseaux-sur-Lausanne, Switzerland
[2] University of Applied Sciences Western Switzerland (HES-SO/HEIG-VD)
CH-1401 Yverdon-les-Bains, Switzerland

Abstract. Since almost two decades, the block cipher IDEA has resisted an exceptional number of cryptanalysis attempts. At the time of writing, the best published attack works against 6 out of the 8.5 rounds (in the non-related-key attacks model), employs almost the whole codebook, and improves the complexity of an exhaustive key search by a factor of only two. In a parallel way, Lipmaa demonstrated that IDEA can benefit from SIMD (Single Instruction, Multiple Data) instructions on high-end CPUs, resulting in very fast implementations. The aim of this paper is two-fold: first, we describe a parallel, time-constant implementation of eight instances of IDEA able to encrypt in counter mode at a speed of 5.42 cycles/byte on an Intel Core2 processor. This is comparable to the fastest stream ciphers and notably faster than the best known implementations of most block ciphers on the same processor. Second, we propose the design of a new block cipher, named WIDEA, leveraging on IDEA's outstanding security-performance ratio. We furthermore propose a new key-schedule algorithm in replacement of completely linear IDEA's one, and we show that it is possible to build a compression function able to process data at a speed of 5.98 cycles/byte. A significant property of WIDEA is that it closely follows the security rationales defined by Lai and Massey in 1990, hence inheriting all the cryptanalysis done the past 15 years in a very natural way.

Keywords: IDEA block cipher, WIDEA compression function, Intel Core2 CPU, wordslice implementation.

1   Introduction

Finding the proper balance between security and speed has always been a non-trivial challenge for designers of block ciphers and hash functions. One possibility consists in using low-footprint arithmetical operations, like simple Boolean operators (usually AND, OR or XOR), table lookups or modular additions, to build a rather fast round function. Since the strength of such round function is usually low in cryptographic terms, one is forced to iterate it a sufficient number of times to get a proper security level. Another approach consists in using more complicated arithmetical operations, like multiplications, for instance. The inherent larger cryptographic strength of such operations naturally comes with a slower speed.

O. Dunkelman (Ed.): FSE 2009, LNCS 5665, pp. 277–295, 2009.

AES [15, 37] is maybe one of the most elegant balance between efficiency and security for a 128-bit block size. By using quite strong diffusion and confusion elements in a design having a high internal parallelism, Daemen and Rijmen have obtained a very fast cipher while keeping a reasonable security margin. However, as a matter of fact, it is interesting to note that several designs of hash functions submitted to the NIST SHA3 competition (e.g., Skein [19] and MD6 [41]) have deliberately chosen to use much simpler operations to build a "light" round and to iterate this round function a large number of times. This approach is preferred by some designers because it allows low-footprint hardware implementations as well as an easier cryptanalysis.

The IDEA block cipher [27, 26] was designed in the beginning of the 90's with the following philosophy in mind: mix three different and algebraically incompatible operations. As a result, a rather strong round function is iterated no more than 8 times to build a cipher with an outstanding security record: almost 20 years after its design, no faster attack than an exhaustive key search is known against its full flavor, despite a very intense cryptanalysis activity resulting in more than a dozen of academic papers discussing its security. However, as of today, IDEA has more been known for its security than for its speed, even if some fast implementations have been proposed by Lipmaa [29], exploiting the Intel MMX instruction set. Furthermore, it is well-known that the implementation of the so-called IDEA multiplication is rather delicate and, if not done properly, is prone to timing attacks [23].

Related Work. How to increase the block size of IDEA has been studied by Nakahara and co-authors: they proposed the MESH ciphers [36], which are ciphers relying on the same operations than in IDEA, but having block sizes of up to 128 bits as well as a stronger key-schedule algorithm. Other variants of MESH ciphers, targeting 8-bit microcontrollers, were described in [32], always exploiting the same philosophy but this times with operations working on 8-bit variables.

Our Contributions. Attacking the common belief that IDEA is rather a slow cipher, we show that, by properly exploiting modern instruction sets, it is one of the fastest block ciphers available on the market on the x86 and x86-64 architectures, beating AES by a large margin, and resulting in a formidable security-speed ratio supported by almost 20 years of unsuccessful cryptanalysis. In this paper, we revisit the IDEA philosophy (iterating only a modest number of times a relatively strong round function) at the light of the latest multimedia instruction sets SSE, SSE2 and SSE3 which are available today in virtually every PC. Our contributions are double fold: first, we exhibit a so-called *wordslice* implementation of eight parallel instances of IDEA able to encrypt in counter mode at a speed of 5.42 clock cycles per byte, according to the eSTREAM benchmarking framework, on an Intel Core2[1] CPU. Our implementation is notably more than 30% faster than the fastest known implementation of AES [22], measured

---

[1] Precisely, the CPU belongs to family 6, model 23, stepping 6.

at 7.81 cycles/byte on the same CPU, while it is able to handle as few as 64 bytes of data, compared to the 128 bytes of the implementation of Käsper and Schwabe. For the sake of completeness, we note that the fastest standard (i.e., non-bitslice) implementation of AES has been recently reported to encrypt at a rate of 10.57 cycles/byte on an Intel Core2 CPU [2]. Additionally, our implementation does not suffer from cache attacks [1, 39], is completely branch-free, and is therefore time-constant.

Our second contribution is the design of a new block cipher family named WIDEA. It relies on n parallel instances of IDEA mixed using a high-quality diffusion element. We discuss the rationales behind our design, its security and concretely specify the WIDEA-8 instance that operates on 512-bit data blocks. By applying the Davies-Meyer mode of operation, we turn WIDEA-8 into a compression function capable of processing data at 5.98 cycles/byte on the same Intel Core2 CPU.

2   The IDEA Block Cipher

In this section, we first recall the specifications of IDEA and we discuss the available literature dedicated to its security.

2.1   Overview of the Cipher

The IDEA block cipher handles 64-bit blocks of data under a 128-bit key. It consists of 8 identical rounds (Fig. 1 illustrates one round), each parametered by a 96-bit subkey, followed by a final key-whitening layer, often named *half round*. The r-th round transforms a 64-bit data input interpreted as a vector of four 16-bit words $(X_0^{(r)}, X_1^{(r)}, X_2^{(r)}, X_3^{(r)})$ to an output vector $(Y_0^{(r)}, Y_1^{(r)}, Y_2^{(r)}, Y_3^{(r)})$ having a similar shape. This process is keyed by six 16-bit subkeys denoted $Z_j^{(r)}$ with $0 \leq j \leq 5$ derived from the 128-bit master key according to a rather simple, bit-selecting (and therefore completely linear) key-schedule (see Fig. 2). The strength of IDEA is certainly due to an elegant design approach which consists in mixing three algebraically incompatible group operations: the addition over $\mathrm{GF}(2^{16})$, denoted $\oplus$, the addition over $\mathbb{Z}_{2^{16}}$, denoted $\boxplus$, and the multiplication over $\mathbb{Z}_{2^{16}+1}^*$, denoted $\odot$, where 0 represents the value 2^{16}. Round r is defined by the following operations: one first computes two intermediate values

$$\alpha^{(r)} = \left(X_0^{(r)} \odot Z_0^{(r)} \right) \oplus \left(X_2^{(r)} \boxplus Z_2^{(r)} \right) \text{ and } \beta^{(r)} = \left(X_1^{(r)} \boxplus Z_1^{(r)} \right) \oplus \left(X_3^{(r)} \odot Z_3^{(r)} \right).$$

These two values form the input of the *multiplication-addition box* (MA-box) which results in

$$\delta^{(r)} = \left(\left(\alpha^{(r)} \odot Z_4^{(r)} \right) \boxplus \beta^{(r)} \right) \odot Z_5^{(r)} \text{ and } \gamma^{(r)} = \left(\alpha^{(r)} \odot Z_4^{(r)} \right) \boxplus \delta^{(r)}.$$

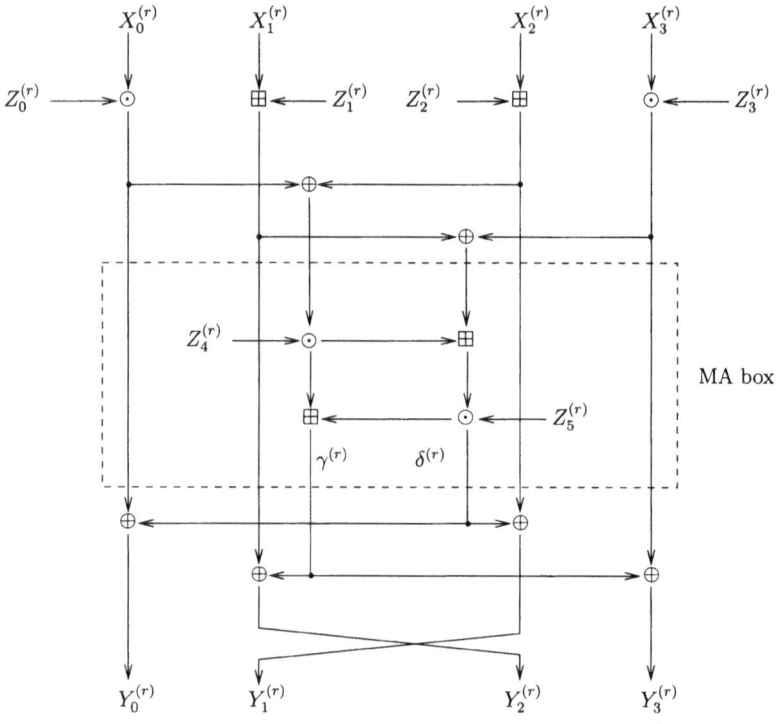

Fig. 1. Round r of IDEA

The output of i-th round is then obtained through

$$Y_0^{(r)} = \left(X_0^{(r)} \odot Z_0^{(r)}\right) \oplus \delta^{(r)}, \quad Y_1^{(r)} = \left(X_2^{(r)} \boxplus Z_2^{(r)}\right) \oplus \delta^{(r)},$$

$$Y_2^{(r)} = \left(X_1^{(r)} \boxplus Z_1^{(r)}\right) \oplus \gamma^{(r)} \text{ and } Y_3^{(r)} = \left(X_3^{(r)} \odot Z_3^{(r)}\right) \oplus \gamma^{(r)}$$

After the 8-th round, a final key-whitening layer is applied:

$$Y_0^{(9)} = X_0^{(9)} \odot Z_0^{(9)}, Y_1^{(9)} = X_2^{(9)} \boxplus Z_1^{(9)}, Y_2^{(9)} = X_1^{(9)} \boxplus Z_2^{(9)} \text{ and } Y_3^{(9)} = X_3^{(9)} \odot Z_3^{(9)}.$$

2.2 Cryptanalysis of IDEA

Designed to offer resistance to differential cryptanalysis [10], the IDEA block cipher has been subject to a very intense scrutiny by the cryptologic community since its publication in 1990. This is probably due to the fact that, at the time of writing, the best attack ever designed against IDEA in a classical scenario is able to break only 6 out of the 8.5 rounds at a $2^{126.8}$ computational cost: breaking the full version of IDEA might be considered by certain cryptanalysts as a kind of "Holy Grail". We now make a review of the available literature dedicated to the cryptanalysis of IDEA.

Round i	$Z_1^{(i)}$	$Z_2^{(i)}$	$Z_3^{(i)}$	$Z_4^{(i)}$	$Z_5^{(i)}$	$Z_6^{(i)}$
1	$Z_{[0\ldots15]}$	$Z_{[16\ldots31]}$	$Z_{[32\ldots47]}$	$Z_{[48\ldots63]}$	$Z_{[64\ldots79]}$	$Z_{[80\ldots95]}$
2	$Z_{[96\ldots111]}$	$Z_{[112\ldots127]}$	$Z_{[25\ldots40]}$	$Z_{[41\ldots56]}$	$Z_{[57\ldots72]}$	$Z_{[73\ldots88]}$
3	$Z_{[89\ldots104]}$	$Z_{[105\ldots120]}$	$Z_{[121\ldots8]}$	$Z_{[9\ldots24]}$	$Z_{[50\ldots65]}$	$Z_{[66\ldots81]}$
4	$Z_{[82\ldots97]}$	$Z_{[98\ldots113]}$	$Z_{[114\ldots1]}$	$Z_{[2\ldots17]}$	$Z_{[18\ldots33]}$	$Z_{[34\ldots49]}$
5	$Z_{[75\ldots90]}$	$Z_{[91\ldots106]}$	$Z_{[107\ldots122]}$	$Z_{[123\ldots10]}$	$Z_{[11\ldots26]}$	$Z_{[27\ldots42]}$
6	$Z_{[43\ldots58]}$	$Z_{[59\ldots74]}$	$Z_{[100\ldots115]}$	$Z_{[116\ldots3]}$	$Z_{[4\ldots19]}$	$Z_{[20\ldots35]}$
7	$Z_{[36\ldots51]}$	$Z_{[52\ldots67]}$	$Z_{[68\ldots83]}$	$Z_{[84\ldots99]}$	$Z_{[125\ldots12]}$	$Z_{[13\ldots28]}$
8	$Z_{[29\ldots44]}$	$Z_{[45\ldots60]}$	$Z_{[61\ldots76]}$	$Z_{[77\ldots92]}$	$Z_{[93\ldots108]}$	$Z_{[109\ldots124]}$
9	$Z_{[22\ldots37]}$	$Z_{[38\ldots53]}$	$Z_{[54\ldots69]}$	$Z_{[70\ldots85]}$		

Fig. 2. Complete key-schedule of IDEA. $Z_{[0..15]}$ denotes the bits 0 to 15 (inclusive) of Z, $Z_{[117..4]}$ means the bits 117-127 and 0-4 of Z, and the leftmost bit of Z has the index 0

Classical Attacks. Differential cryptanalysis [10] has been one of the first technique to be tried against IDEA by Meier [31] to break up to 2.5 rounds faster than an exhaustive search. Borst et al. [12] were able to break using a differential-linear attack and 3.5 rounds using truncated differentials. Biham et al. [5] used impossible differentials to break 4.5 rounds. Another approach to break IDEA, based on integral attacks, has been proposed by Nakahara et al. [33] against 2.5 rounds. The approach has first been pushed to 4 rounds by Demirci [17], and then to 5 rounds by Demirci et al. [18] in combination with meet-in-the-middle techniques. Inspired by a work of Nakahara et al. [35], Junod [21] presented several efficient attacks mixing the Biryukov-Demirci relation and square attacks against up to 3.5 rounds. More recently, Biham et al. [7] described a linear attack on 5-round IDEA improving the complexity of [18]. The same authors presented the first attack against 6 rounds in [8] employing almost the whole codebook and having a computational complexity of $2^{126.8}$ operations.

In the related-key setting, an attack against 6.5 rounds was proposed by Biham et al. in [6], 7.5 rounds were reached by the same authors in [8] and recently, an attack working against $4r$-round IDEA for any r has been described in [9].

Side-Channel Attacks. A few attacks exploiting side-channel information potentially leaked by implementations of IDEA have been published so far. A practical timing attack against key-dependent implementations of the IDEA multiplication has been described by Kelsey et al. [23]. Lemke et al. [28] have discussed the application of DPA-oriented techniques to attack implementations of IDEA on an 8-bit microcontroller, while Clavier et al. [14] have considered fault attacks. Protection methods have also been studied in [38].

Simplified IDEA Variants. Some authors have also attacked simplified versions of IDEA. For instance, Borisov et al. [11] have replaced all the $\boxplus$ operations by $\oplus$ ones, except for the two in the output transformation. The authors showed that for one key out of 2^{16}, there exists a multiplicative differential characteristic over eight rounds that holds with probability 2^{-32}. Raddum [40] considered at the light of differential cryptanalysis another version, called IDEA-X/2, where only half of the ⊞'s in one round are changed to ⊕ operations, namely the ⊞'s where $Z_2^{(r)}$ and $Z_3^{(r)}$ are inserted, while the MA-structure is left unchanged.

3 A Wordslice IDEA Implementation

Given the fact that IDEA does not contain S-boxes and it uses only 16-bit arithmetical operations, it is particularly suited to be optimized on those processor architectures that include SIMD multimedia extensions; nowadays practically every PC is built around the x86-64 architecture, which supports these features via the SSEx instruction sets. Moreover, this trend is also significantly showing up in the context of embedded systems, see for instance the ARC VRaptor[TM] multicore architecture, the ARM NEON[TM] technology and the new Intel Atom[TM] and VIA Nano[TM] processors.

Previous work [29] by Lipmaa has shown that a 4.35× increase in speed is achievable on Pentium processors that support the MMX instruction set. We now push the approach further, showing that IDEA can be very conveniently implemented on all those processors that implement the SSE2 instruction set, leading to encryption speed records; we also show that the multiplication modulo $2^{16} + 1$ can easily be implemented in a time-constant way, thanks to the SSE2 packed-word comparison instructions. We think that our results wipe away two common misconceptions about the IDEA block cipher, once and for all: that it is slow and difficult to secure against timing attacks [23]. On the contrary, we show that IDEA is probably one of the fastest block cipher on current microprocessor architectures, and it is completely immune to both timing attacks and cache attacks [1,39].

The SSE2 instruction set defines 144 instructions that operate on 128-bit words; the SSE2 integer arithmetic instructions operate on *packed* data, i.e. each 128-bit operand is seen as a vector of 8, 16, 32 or 64-bit words. Since IDEA is natively operating on 16-bit variables, it is clearly possible to write SSE2 code that carries out eight IDEA encryptions in parallel; we call this implementation *wordslice*, as bitslice implementations [3,30] would similarly work at bit level on 128 IDEA encryptions in parallel. The main advantage of wordslice implementations over bitslice is that to reach significant speedups it is not necessary to operate on huge amount of data, and that the orthogonalization process is straightforward.

Since the multiplication is clearly the most complex operation in the IDEA cipher, and the most critical regarding timing analysis, it deserves special care. The piece of code of Fig. 3, written using SSE2 intrinsics, is an implementation of the wordslice IDEA multiplication; it contains 11 pseudo-instructions, requires a space of four 128-bit registers and performs eight IDEA multiplications in parallel. It leverages on the unsigned multiplication instruction _mm_mulhi_epu16 , whose functionality is not available in the MMX instruction set, and the comparison instruction _mm_cmpeq_epi16, which essentially allows it to be branch-free (and thus time-constant).

```
1   t = _mm_add_epi16   (a, b);       /* t = (a + b) & 0xFFFF;        */
2   c = _mm_mullo_epi16 (a, b);       /* c = (a * b) & 0xFFFF;        */
3   a = _mm_mulhi_epu16 (a, b);       /* a = (a * b) >> 16;           */
4   b = _mm_subs_epu16  (c, a);       /* b = (c - a);                 */
                                      /* if (b & 0x80000000) b = 0;   */
5   b = _mm_cmpeq_epi16 (b, XMM_0);   /* if (b == 0) b = 0xFFFF;      */
                                      /* else b = 0;                  */
6   b = _mm_srli_epi16  (b, 15);      /* b = b >> 15;                 */
7   c = _mm_sub_epi16   (c, a);       /* c = (c - a) & 0xFFFF;        */
8   a = _mm_cmpeq_epi16 (c, XMM_0);   /* if (c == 0) a = 0xFFFF;      */
                                      /* else a = 0;                  */
9   c = _mm_add_epi16   (c, b);       /* c = (c + b) & 0xFFFF;        */
10  t = _mm_and_si128   (t, a);       /* t = t & a;                   */
11  c = _mm_sub_epi16   (c, t);       /* c = (c - t) & 0xFFFF;        */
```

Fig. 3. Eight parallel IDEA multiplications using the SSE2 instruction set

The two operands are initially contained in the a and b variables. The t and c variables are used as temporary storage and c contains the final result (the initial values of a and b are not preserved through the computation); XMM_0 is the 128-bit zero string. The algorithm takes inspiration from known efficient implementations [29, 4], but eliminates any need of comparison or branch point. The main idea behind it is that the two values that would be returned whether $a \cdot b = 0$ or not are calculated in parallel; the final choice is determined by the value of a mask, the value of a at line 8, which is also derived from the input data. The algorithm also uses the fact that the upper and the lower 16 bits of $a \cdot b$ are equal if and only if at least one of the operands is 0; this property was never observed before and can be easily checked with an exhaustive simulation. For the sake of clarity, a line-equivalent (but obviously inefficient) C implementation that performs one IDEA multiplication using unsigned integer 32-bit variables is also given. The rest of the wordslice IDEA algorithm can be implemented with packed unsigned 16-bit additions and 128-bit XORs; this part is quite straightforward to derive and will not be shown here.

After having written a complete implementation based on the SSE2 instruction set, we have proceeded to the performance tests to assess the speed level of this code. It is surely interesting to test the encryption speed in ECB mode, with pre-calculated round keys, as this gives an indication of the raw speed that can be reached. However, by running the word-slice IDEA in counter mode one can also obtain a quite efficient stream cipher. We have thus implemented some simple routines to realize a branch-free SSE2 implementation of the counter mode of operation, and obtained a stream cipher with 48-bit IV and 128-bit key.

Both codes have been compiled with the Intel compiler; speed has been measured by executing the function a high number of times and taking the average time spent by the processor. We have also integrated our code in the eSTREAM [13] benchmarking framework and found that the figures differ by no more than 1% with regards to the ones we obtained. The result is that plain ECB encryption runs on an Intel Core2 processor at 5.55 cycles per byte, while counter mode keystream generation runs at 5.42 cycles[2] per byte.

We think that the reached level of security vs. speed trade-off justifies additional research effort; the IDEA cipher appears as an extremely good building block to realize other cryptographic primitives that may be used to implement authenticated encryption and hash functions. As a first step, we proceed in the next Section with the definition of the WIDEA block cipher family.

4   The WIDEA Block Cipher Family

In this section, we first describe the rationales behind the WIDEA cipher family design, then we make short, preliminary cryptanalysis of our proposal, followed by a discussion on implementation issues and performance results.

4.1   Design Rationale

We now show how a computational skeleton composed of n IDEA instances computed in parallel can be transformed in a natural and efficient way into a $(n \cdot 64)$-bit block cipher. Basically, we need to define a minimal modification that provides diffusion over the new block size; the term *minimal* bears several meanings here:

- The modification must require minimal computational overhead.
- The modification must alter the skeleton in the least noticeable way, in particular it must not affect the achieved degree of parallelism and it must be elegant.
- The modification must follow and enforce all original IDEA design criteria.

The first problem is to define the way in which we provide full diffusion within one round; MDS matrices over $GF(2^n)$ are regarded as an optimal and efficient way to solve the problem [42], and have been extensively used in well-known constructions [16, 15]: since the IDEA structure naturally operates on 16-bit words, we choose MDS matrices over $GF(2^{16})$ as our diffusion primitive. A second problem is to identify the location in the IDEA round function to insert the diffusion block. The MA-box is an interesting place for two main reasons: it is already used to provide diffusion in the IDEA round function and it does not contain XOR operations. There are four arcs connecting the four operations inside the IDEA MA-box, but only a diffusion block inserted into the right arc

---

[2] It is noteworthy that a code with more instructions takes less time to execute. This fact is most likely due to micro-architectural optimizations automatically performed by the compiler. In other words, putting less stress on the pipeline sometimes allows a better scheduling of the micro-operations.

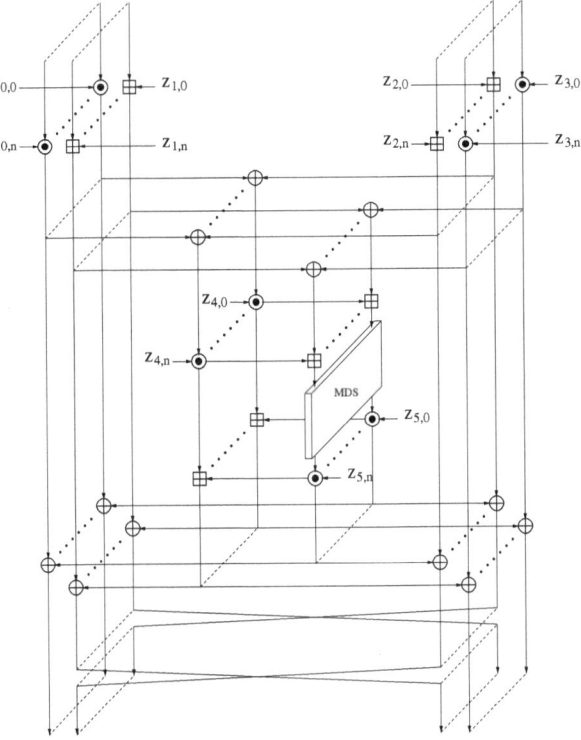

Fig. 4. The first round of WIDEA-n

would guarantee that full diffusion is still achieved. We therefore modify the n-way IDEA structure as shown in Fig. 4. The third dimension is used to represent the fact that n instances of IDEA can independently be computed and are tied together by the application of the MDS matrix.

The IDEA design criteria are effectively enforced; more specifically:

1. Full diffusion in the new block cipher is still obtained in one round, i.e. every round output bit depends on all $n \cdot 64$ input bits.
2. Every operation is still preceeded and followed by operations defined over algebraically incompatible groups.
3. No dependence on arbitrary constants is introduced, the only choices being limited to the irreducible polynomial that defines the algebraic structure of $GF(2^{16})$ and the coefficients of the $n \times n$ MDS matrix that can be chosen in order to minimize the number of operations.

We call the new core of the round function *MAD-box* (standing for Multiply-Add-Diffuse) and we refer to the global $(n \cdot 64)$-bit construction as *WIDEA* (the name providing enough hints for a "wide" block IDEA cipher); the particular members of the family obtained by fixing the value of n are identified as WIDEA-n.

Compared with AES-like constructions operating on wide blocks (such as the W block cipher instanced in the Whirlpool hash function [20]), the WIDEA structure needs only one eighth of total MDS applications, thus keeping the computational cost of diffusion quite small. Variants with $n = 4$ (256-bit) and $n = 8$ (512-bit) block sizes are easily defined for instance by taking the same MDS matrices used in the AES and in the W ciphers, but defined over $\mathrm{GF}(2^{16})$. These ciphers are very useful, as they can be used to build compression functions for 256-bit and 512-bit hash functions (this scenario also justifies the huge key sizes); as a reference, a complete specification of WIDEA-8 is given in the Appendix.

As we believe the new structure is not deviating substantially from IDEA, we think that it is not possible to exploit the bigger block size to mount attacks based on incomplete diffusion (such as integral attacks [24]) against more rounds than in IDEA, due to the fact that full diffusion is again obtained after one round. For this reason we keep the number of rounds of WIDEA at 8.5.

Instead, we focus our improvement effort on the key-schedule algorithm, which is significantly changed in order to remove the problem of weak keys and to render attacks that exploit the controllability of the key input more difficult. This is a valid scenario for related-key attacks and in case the block cipher is used to build a compression function (and a hash function). As in IDEA, we keep the key size equal to $(2n \cdot 64)$-bit (twice the block size), accepting the fact that not all key material can be used to key each round. However, to compensate for this we define a new key scheduling algorithm based on a non-linear feedback shift register, similarly to what is done in the MESH block ciphers [36]; we introduce non-linearity, diffusion and diversification in the WIDEA key scheduling, but always in a way to preserve the n-way parallelism already achieved in the cipher round function.

We denote with Z_i, $0 \leq i \leq 51$, the subkeys that are used in the 8.5 rounds of WIDEA-n; note that due to the n-way parallelism each subkey has a size of $n \cdot 16$ bits (thus each subkey Z_i can be split into the n slices $z_{i,0} \ldots z_{i,n}$). Moreover, denoting with K_i, $0 \leq i \leq 7$, the 8 words that represent the WIDEA master key, the new key scheduling algorithm is defined by the following equations:

$$Z_i = K_i \qquad\qquad\qquad\qquad\qquad\qquad\qquad 0 \leq i \leq 7$$

$$Z_i = ((((Z_{i-1} \oplus Z_{i-8}) \overset{16}{\boxplus} Z_{i-5}) \overset{16}{\lll} 5) \lll 24) \oplus C_{\frac{i}{8}-1} \quad 8 \leq i \leq 51,\ 8 \mid i$$

$$Z_i = ((((Z_{i-1} \oplus Z_{i-8}) \overset{16}{\boxplus} Z_{i-5}) \overset{16}{\lll} 5) \lll 24) \qquad\quad 8 \leq i \leq 51,\ 8 \nmid i$$

The symbols and the notation are explained more formally in the Appendix. Rotation by 5 bit positions is independently carried out on each 16-bit slice of Z_i, as suggested by the superscript; rotation by 24 bit positions (3 byte positions) is instead carried out globally on each $n \cdot 16$ bits word. The values of the constants $C_0 \div C_5$, that are injected every 8 iterations, should vary with the particular instance of the cipher.

The key scheduling is designed in a way such that it can be computed using a shift register of eight $n \cdot 16$-bit words and the same 16-bit arithmetical and logical operations used in the round function; the two rotation operations have been chosen as a practical and simple way to mix information between the n slices of the key schedule algorithm. Note that the byte level rotation is completely transparent in the context of 8-bit implementations. One could also think of fixing $n = 1$, basically obtaining IDEA with a strengthened key schedule, or $n = 2$ obtaining a cipher operating on the same block and key sizes as AES.

4.2   Preliminary Security Analysis

The fact that WIDEA is heavily based on the IDEA construction implies that all related cryptanalytic results may apply in a quite natural way, and therefore should be taken into consideration. We start this brief discussion by pointing out that considerable effort has been spent to strengthen the key-schedule part, as indicated above. Non-linearity is added by mixing XORs with integer addition, and different constants are injected every 8 iterations; thanks to this, it is very difficult to exploit repetitive patterns in the subkeys, or to find long sequences of subkeys characterized by low Hamming weight. We have verified with software simulations that thanks to the diffusion provided by the bit-level and byte-level rotations, coupled with integer additions, every bit of key material used starting from round 4 (non-linearly) depends on all the 1024 bits of the master key. For these reasons we expect that no weak keys can be found for WIDEA, and that the related-key attacks against IDEA cannot be transposed to WIDEA.

Regarding the classical attack scenarios, one may question if attacks can be based on the fact that the round function of WIDEA-n is based on n parallel instances of IDEA. Actually, the effect of the MDS diffusion matrix is to keep at 8 the number of full diffusions applied in the encryption process; to make a comparison, the AES block cipher applies a total of 5 to 7 full diffusions, depending on the key size. Recent proposals of big and efficient block ciphers, such as Threefish [19], are also quite conservatively designed to implement 7 or 8 full diffusions, depending on the digest size. Due to this property we do not expect that integral attacks, differential or linear attacks constitute a bigger threat for WIDEA than for IDEA. Obviously, independent cryptanalysis is needed to verify our claims, and we encourage further research in this direction.

4.3   WIDEA-8 Implementation Results

WIDEA-8 is certainly the most interesting member of the cipher family, since it can efficiently be computed using the XMM registers available in the x86-64 architecture[3]. The coding of WIDEA naturally takes advantage of the optimizations discussed in §3; the same code is used to perform the wordslice IDEA multiplications. Concerning the MDS matrix multiplication, we show how to perform a wordslice $GF(2^{16})$ multiplication times 2 (this is equivalent to the

---

[3] We note that in the future it may be possible to implement efficiently a WIDEA-16 instance, as Intel is planning to introduce in future micro-architectures the YMM registers, characterized by a size of 256 bits (however, only floating point instructions are planned so far for such operand size).

AES `xtime` operation [37]). We use the `_mm_cmpeq_epi16` instruction to generate a polynomial mask to be XORed to the left-shifted input basing on the value
of the MSB of each 16-bit slice of the input. The code using SSE intrinsics is
given in Fig. 5; it contains 5 pseudo instructions.

```
1    b = _mm_and_si128 (a, XMM_0x8000);
2    a = _mm_slli_epi16 (a, 1);
3    b = _mm_cmpeq_epi16 (b, XMM_0x8000);
4    b = _mm_and_si128 (b, XMM_POLY);
5    a = _mm_xor_si128 (a, b);
```

Fig. 5. The wordslice Xtime operation

The initial and final values are stored in a, while b is used as temporary
storage; `XMM_0x8000` is a vector of eight 16-bit words with only the MSB set
to 1 and `XMM_POLY` contains eight instances of the polynomial reduction mask.
This `xtime` operation is used to compute the MDS matrix multiplication; the
total number of `xtime` operations is determined by the maximum degree of the
elements in the MDS matrix (this is equal to three for WIDEA-8). We also
exploit the fact that the MDS matrix is circulant to optimize the number of
computational steps; since this technique is highly dependent on the entries of
the MDS matrix, it will not be discussed here.

Regarding the key-schedule algorithm, the operations are quite elementary
and do not deserve special mention. The Intel SSE2 instruction set can be used
to implement it easily using a bank of 9 XMM registers (8 to store the state
of the non-linear feedback shift register plus one for temporary storage); if the
SSE3 instruction set is supported by the target processor, which is the case of
all Core2 CPUs, the `_mm_shuffle_epi8` byte shuffling instruction can be used
in place of shift instructions to implement the byte-level rotation.

The WIDEA-8 cipher has been implemented by hand in Intel assembly language; this is done to exploit as much as possible the bank of XMM registers, as
their number is increased from 8 to 16 when the code is executed in 64-bit mode. In
this case it is possible to compute the key scheduling algorithm on-the-fly during
encryption. Quite amazingly, in only one point in the code the space provided by
the XMM register bank is not sufficient, and we need to save a variable in the cache
memory; thus our optimized WIDEA code is almost completely acting solely on
the processor register bank, and in a completely time-constant way.

Having on-the-fly key-scheduling is important because we want to test the
speed of WIDEA-8 in the cases when it is used to build a compression function
using the Davies-Meyer construction. From our experiments, we have determined
that such compression function is able to process data at the speed of 5.98 cycles
per byte on a Intel Core2 processor. We anticipate that such cryptographic
primitive can be used to define hash functions characterized by an outstanding
security vs. speed trade-off; we do not offer the definition of a full hash function
here, but we consider this as a very promising future work which will also benefit
from the insights about hash modes of operation obtained during the SHA-3

Hash Function	Speed (cycles per byte)
EDON-R 512	2.29
WIDEA-8	*5.98*
CubeHash8/32	6.03
Skein-512	6.10
Shabal-512	8.03
LUX	9.50
Keccak	10.00
BLAKE-64	10.00
Cheetah	13.60
Aurora	26.90
Grostl	30.45
ECHO-SP	35.70
SHAvite-3	38.20
Lesamnta	51.20
MD6	52.64
ECHO	53.50
Vortex	56.05
FUGUE	75.50

Fig. 6. Speed comparison of WIDEA used as a Davies-Meyer construction in the Merkle-Damgard mode and some SHA-3 candidates on the Intel Core2 CPU in 64-bit mode

competition. Anyway, speed comparison between some SHA3 candidates and our compression function used in a straightforward Merkle-Damgard mode is provided[4] in Table 6. The expiration of the patent protecting IDEA in a near future, and the fact that no intellectual property was applied for WIDEA, might also increase the interest in our work [5].

Acknowledgments

We would like to thank Olivier Brique, Jérôme Perrine and Corinne Le Buhan Jordan for their kind support during this work.

References

1. Bernstein, D.: Cache-timing attacks on AES (2005),
   http://cr.yp.to/papers.html
2. Bernstein, D., Schwabe, P.: New AES software speed records. In: Chowdhury, D.R., Rijmen, V., Das, A. (eds.) INDOCRYPT 2008. LNCS, vol. 5365, pp. 322–336. Springer, Heidelberg (2008), http://cr.yp.to/papers.html

---

[4] All the data are the ones provided by their respective designers in the presentation slides of the First SHA-3 Candidate Conference for the Intel Core2 x86-64 architecture.

[5] As a matter of fact, several SHA-3 submissions are re-using AES components; for instance, Vortex [25], is only about three times faster than the basic compression function we have devised here when implemented using the future Intel AES dedicated instructions.

3. Biham, E.: A fast new DES implementation in software. In: Biham, E. (ed.) FSE 1997. LNCS, vol. 1267, pp. 260–272. Springer, Heidelberg (1997)

4. Biham, E.: Optimization of IDEA. Technical report, nes/doc/tec/wp6/026/1, NESSIE Project (2002), https://www.cryptonessie.org

5. Biham, E., Biryukov, A., Shamir, A.: Miss-in-the-middle attacks on IDEA and Khufru. In: Knudsen, L.R. (ed.) FSE 1999. LNCS, vol. 1636, pp. 124–138. Springer, Heidelberg (1999)

6. Biham, E., Dunkelman, O., Keller, N.: Related-key boomerang and rectangle attacks. In: Cramer, R. (ed.) EUROCRYPT 2005. LNCS, vol. 3494, pp. 507–525. Springer, Heidelberg (2005)

7. Biham, E., Dunkelman, O., Keller, N.: New cryptanalytic results on IDEA. In: Lai, X., Chen, K. (eds.) ASIACRYPT 2006. LNCS, vol. 4284, pp. 412–427. Springer, Heidelberg (2006)

8. Biham, E., Dunkelman, O., Keller, N.: A new attack on 6-round IDEA. In: Biryukov, A. (ed.) FSE 2007. LNCS, vol. 4593, pp. 211–224. Springer, Heidelberg (2007)

9. Biham, E., Dunkelman, O., Keller, N.: A unified approach to related-key attacks. In: Nyberg, K. (ed.) FSE 2008. LNCS, vol. 5086, pp. 73–96. Springer, Heidelberg (2008)

10. Biham, E., Shamir, A.: Differential cryptanalysis of DES-like cryptosystems. Journal of Cryptology 4(1), 3–72 (1991)

11. Borisov, N., Chew, M., Johnson, R., Wagner, D.: Multiplicative differentials. In: Daemen, J., Rijmen, V. (eds.) FSE 2002. LNCS, vol. 2365, pp. 17–33. Springer, Heidelberg (2002)

12. Borst, J., Knudsen, L., Rijmen, V.: Two attacks on reduced IDEA (extended abstract). In: Fumy, W. (ed.) EUROCRYPT 1997. LNCS, vol. 1233, pp. 1–13. Springer, Heidelberg (1997)

13. De Cannière, C.: eSTREAM testing framework, http://www.ecrypt.eu.org/stream/perf/

14. Clavier, C., Gierlichs, B., Verbauwhede, I.: Fault analysis study of IDEA. In: Malkin, T.G. (ed.) CT-RSA 2008. LNCS, vol. 4964, pp. 274–287. Springer, Heidelberg (2008)

15. Daemen, J., Rijmen, V.: The Design of Rijndael. In: Information Security and Cryptography. Springer, Heidelberg (2002)

16. Damen, J., Knudsen, L., Rijmen, V.: The block cipher SQUARE. In: Biham, E. (ed.) FSE 1997. LNCS, vol. 1267, pp. 149–165. Springer, Heidelberg (1997)

17. Demirci, H.: Square-like attacks on reduced rounds of IDEA. In: Nyberg, K., Heys, H.M. (eds.) SAC 2002. LNCS, vol. 2595, pp. 147–159. Springer, Heidelberg (2003)

18. Demirci, H., Selçuk, A., Türe, E.: A new meet-in-the-middle attack on the IDEA block cipher. In: Matsui, M., Zuccherato, R.J. (eds.) SAC 2003. LNCS, vol. 3006, pp. 117–129. Springer, Heidelberg (2004)

19. Ferguson, N., Lucks, S., Schneier, B., Whiting, D., Bellare, M., Kohno, T., Callas, J., Walker, J.: The Skein hash function family – version 1.1. NIST SHA-3 Submission (2008), http://ehash.iaik.tugraz.at/wiki/The_eHash_Main_Page

20. ISO. Information technology – Security techniques – Hash-functions – Part 3: Dedicated hash-functions. ISO/IEC 10118-3:2004, International Organization for Standardization, Genve, Switzerland (2004)

21. Junod, P.: New attacks against reduced-round versions of IDEA. In: Gilbert, H., Handschuh, H. (eds.) FSE 2005. LNCS, vol. 3557, pp. 384–397. Springer, Heidelberg (2005)

22. Käsper, E., Schwabe, P.: Faster and timing-attack resistant AES-GCM. IACR ePrint Archive Report 2009/129 (2009), http://eprint.iacr.org/2009/129
23. Kelsey, J., Schneier, B., Wagner, D., Hall, C.: Side channel cryptanalysis of product ciphers. Journal of Computer Security 8(2/3) (2000)
24. Knudsen, L., Wagner, D.: Integral cryptanalysis (extended abstract). In: Daemen, J., Rijmen, V. (eds.) FSE 2002. LNCS, vol. 2365, pp. 112–127. Springer, Heidelberg (2002)
25. Kounavis, M., Gueron, S.: Vortex: A new family of one way hash functions based on Rijndael rounds and carry-less multiplication. NIST SHA-3 Submission (2008), http://ehash.iaik.tugraz.at/wiki/The_eHash_Main_Page
26. Lai, X.: On the design and security of block ciphers. ETH Series in Information Processing, vol. 1. Hartung-Gorre Verlag (1992)
27. Lai, X., Massey, J.: A proposal for a new block encryption standard. In: Damgård, I.B. (ed.) EUROCRYPT 1990. LNCS, vol. 473, pp. 389–404. Springer, Heidelberg (1991)
28. Lemke, K., Schramm, K., Paar, C.: DPA on n-bit sized Boolean and arithmetic operations and its application to IDEA, RC6, and the HMAC-construction. In: Joye, M., Quisquater, J.-J. (eds.) CHES 2004. LNCS, vol. 3156, pp. 205–219. Springer, Heidelberg (2004)
29. Lipmaa, H.: IDEA: a cipher for multimedia architectures? In: Tavares, S., Meijer, H. (eds.) SAC 1998. LNCS, vol. 1556, pp. 248–263. Springer, Heidelberg (1999)
30. Matsui, M., Nakajima, J.: On the power of bitslice implementation on Intel Core2 processor. In: Paillier, P., Verbauwhede, I. (eds.) CHES 2007. LNCS, vol. 4727, pp. 121–134. Springer, Heidelberg (2007)
31. Meier, W.: On the security of the IDEA block cipher. In: Helleseth, T. (ed.) EUROCRYPT 1993. LNCS, vol. 765, pp. 371–385. Springer, Heidelberg (1993)
32. Nakahara, J.: Faster variants of the MESH block ciphers. In: Canteaut, A., Viswanathan, K. (eds.) INDOCRYPT 2004. LNCS, vol. 3348, pp. 162–174. Springer, Heidelberg (2004)
33. Nakahara, J., Barreto, P., Preneel, B., Vandewalle, J., Kim, Y.: Square attacks on reduced-round PES and IDEA block ciphers. In: Macq, B., Quisquater, J.-J. (eds.) Proceedings of 23rd Symposium on Information Theory in the Benelux, Louvain-la-Neuve, Belgium, May 29-31, 2002, pp. 187–195 (2002)
34. Nakahara, J., Preneel, B., Vandewalle, J.: The Biryukov-Demirci attack on IDEA and MESH ciphers. Technical report, COSIC, ESAT, Katholieke Universiteit Leuven, Leuven, Belgium (2003)
35. Nakahara, J., Preneel, B., Vandewalle, J.: The Biryukov-Demirci attack on reduced-round versions of IDEA and MESH block ciphers. In: Wang, H., Pieprzyk, J., Varadharajan, V. (eds.) ACISP 2004. LNCS, vol. 3108, pp. 98–109. Springer, Heidelberg (2004)
36. Nakahara, J., Rijmen, V., Preneel, B., Vandewalle, J.: The MESH block ciphers. In: Chae, K.-J., Yung, M. (eds.) WISA 2003. LNCS, vol. 2908, pp. 458–473. Springer, Heidelberg (2004)
37. National Institute of Standards and Technology, U. S. Department of Commerce. Advanced Encryption Standard (AES), NIST FIPS PUB 197 (2001)
38. Neisse, O., Pulkus, J.: Switching blindings with a view towards IDEA. In: Joye, M., Quisquater, J.-J. (eds.) CHES 2004. LNCS, vol. 3156, pp. 230–239. Springer, Heidelberg (2004)
39. Osvik, D., Shamir, A., Tromer, E.: Cache attacks and countermeasures: the case of AES. In: Pointcheval, D. (ed.) CT-RSA 2006. LNCS, vol. 3860, pp. 1–20. Springer, Heidelberg (2006)

40. Raddum, H.: Cryptanalysis of IDEA-X/2. In: Johansson, T. (ed.) FSE 2003. LNCS, vol. 2887, pp. 1–8. Springer, Heidelberg (2003)
41. Rivest, R., Agre, B., Bailey, D., Crutchfield, C., Dodis, Y., Fleming, K., Khan, A., Krishnamurthy, J., Lin, Y., Reyzin, L., Shen, E., Sukha, J., Sutherland, D., Tromer, E., Yin, Y.: The MD6 hash function – a proposal to NIST for SHA-3. NIST SHA-3 Submission (2008), http://ehash.iaik.tugraz.at/wiki/The_eHash_Main_Page
42. Schnorr, C., Vaudenay, S.: Black box cryptanalysis of hash networks based on multipermutations. In: De Santis, A. (ed.) EUROCRYPT 1994. LNCS, vol. 950, pp. 47–57. Springer, Heidelberg (1995)

Appendix: Specification of WIDEA-8

Notation. WIDEA-8 is a block cipher having block size of 512 bits and fixed key size of 1024 bits, which is heavily based on the IDEA cipher. In the following we will indicate 128-bit words with capital letters and 16-bit words with lower-case letters. The input, state and output of the cipher can be seen as an array of four 128-bit words, where each 128-bit word can in turn be seen as an array of eight 16-bit words. Thus, each 16-bit word is indexed by two numbers: the index of the 128-bit word that contains it, followed by the index of its position in the 128-bit word (128-bit words are indexed by only one number). We adopt here a big-endian ordering, so that the index of the most significant part of a variable is equal to 0 and the index of its least significant part is the largest one. Thus, indicating with $\mathbb{X}$ the 512-bit input of the cipher, we have $\mathbb{X} = X_0 \| X_1 \| X_2 \| X_3$ and $X_0 = x_{0,0} \| x_{0,1} \| x_{0,2} \| x_{0,3} \| x_{0,4} \| x_{0,5} \| x_{0,6} \| x_{0,7}$. Different arithmetic and logic operations are used in WIDEA-8. The IDEA multiplication of two 16-bit words (multiplication over $\mathbb{Z}^*_{2^{16}+1}$ where the zero 16-bit string represents the number 2^{16}) is denoted with "$\odot$"; addition over $\mathbb{Z}_{2^{16}}$ is denoted with "$\boxplus$". Each 16-bit word can also be seen as an element of the finite field $\mathrm{GF}(2^{16})$ defined with the following irreducible polynomial $P(x) = x^{16} + x^5 + x^3 + x^2 + 1$. Addition over $\mathrm{GF}(2^{16})$, as well as bitwise logical XOR, is denoted with "$\oplus$" while multiplication over the same field is denoted with "$\cdot$"; logical left rotation of n positions is denoted with "$\lll n$". The same operators may be applied in a vectorial way over the 128-bit variables, i.e. each 16-bit slice of the operand(s) undergoes the transformations above; in this case we place the superscript "16" over the operator symbol, to distinguish the operation from the one carried out over the full 128 bits.

The Key-Schedule. The WIDEA-8 key $\mathbb{Z}$ has size equal to 1024 bits and can be seen as an array of eight 128-bit words $\mathbb{Z} = Z_0 \| Z_1 \| Z_2 \| Z_3 \| Z_4 \| Z_5 \| Z_6 \| Z_7$. This array is filled with key material starting from the most significant positions and proceeding toward the least significant ones. Once the key has been set, the subkeys can be generated. WIDEA-8, similarly to IDEA, uses a total of 52 128-bit subkeys; the first 8 are taken directly from the key, starting naturally from Z_0. The additional 44 128-bit subkeys Z_i with $8 \leq i \leq 51$ are generated using the following equations:

$$Z_i = ((((Z_{i-1} \oplus Z_{i-8}) \overset{16}{\boxplus} Z_{i-5}) \lll 5) \lll 24) \oplus C_{\frac{i}{8}-1} \qquad 8 \leq i \leq 51,\ 8 \mid i$$

$$Z_i = ((((Z_{i-1} \oplus Z_{i-8}) \overset{16}{\boxplus} Z_{i-5}) \lll 5) \lll 24) \qquad 8 \leq i \leq 51,\ 8 \nmid i$$

In practice, the same recurrence relation is used for each iteration, and a different constant is added whenever the subkey index is a multiple of 8. The six 128-bit constants are given below, in hexadecimal format:

$$C_0 = \texttt{1dea}\|\texttt{0000}\|\texttt{0000}\|\texttt{0000}\|\texttt{0000}\|\texttt{0000}\|\texttt{0000}\|\texttt{0000}$$
$$C_1 = \texttt{3825}\|\texttt{0000}\|\texttt{0000}\|\texttt{0000}\|\texttt{0000}\|\texttt{0000}\|\texttt{0000}\|\texttt{0000}$$
$$C_2 = \texttt{1dd7}\|\texttt{0000}\|\texttt{0000}\|\texttt{0000}\|\texttt{0000}\|\texttt{0000}\|\texttt{0000}\|\texttt{0000}$$
$$C_3 = \texttt{3ea4}\|\texttt{0000}\|\texttt{0000}\|\texttt{0000}\|\texttt{0000}\|\texttt{0000}\|\texttt{0000}\|\texttt{0000}$$
$$C_4 = \texttt{e57a}\|\texttt{0000}\|\texttt{0000}\|\texttt{0000}\|\texttt{0000}\|\texttt{0000}\|\texttt{0000}\|\texttt{0000}$$
$$C_5 = \texttt{f7ba}\|\texttt{0000}\|\texttt{0000}\|\texttt{0000}\|\texttt{0000}\|\texttt{0000}\|\texttt{0000}\|\texttt{0000}$$

Encryption. WIDEA-8 encryption consists in 8 full rounds followed by one half round; every round uses some subkeys calculated using the key scheduling algorithm specified above. We add an apex to each variable to indicate the round number, starting from 1; thus $\delta_{0,0}^{(i)}$ is the value of $\delta_{0,0}$ in the i-th round. The input of the i-th round is denoted as $\mathbb{X}^{(i)} = X_0^{(i)}\|X_1^{(i)}\|X_2^{(i)}\|X_3^{(i)}$ and the output $\mathbb{Y}^{(i)} = \mathbb{X}^{(i+1)}$ is calculated, for a full round, as follows. First, the inputs of the MAD-box are calculated as:

$$A^{(i)} = \left(X_0^{(i)} \overset{16}{\odot} Z_{6(i-1)}\right) \oplus \left(X_2^{(i)} \overset{16}{\boxplus} Z_{6(i-1)+2}\right)$$
$$B^{(i)} = \left(X_1^{(i)} \overset{16}{\boxplus} Z_{6(i-1)+1}\right) \oplus \left(X_3^{(i)} \overset{16}{\odot} Z_{6(i-1)+3}\right)$$

Then, the MAD-box calculation is carried out, resulting in:

$$\Delta^{(i)} = \text{MDS}\left(\left(A^{(i)} \overset{16}{\odot} Z_{6(i-1)+4}\right) \overset{16}{\boxplus} B^{(i)}\right) \overset{16}{\odot} Z_{6(i-1)+5}$$
$$\Gamma^{(i)} = \left(A^{(i)} \overset{16}{\odot} Z_{6(i-1)+4}\right) \overset{16}{\boxplus} \Delta^{(i)}$$

The MDS operation is a left-multiplication over $\text{GF}(2^{16})$ of a 128-bit string with a fixed matrix, its elements defined over $\text{GF}(2^{16})$, and is defined as follows:

$$Y = \begin{pmatrix} y_0 \\ y_1 \\ y_2 \\ y_3 \\ y_4 \\ y_5 \\ y_6 \\ y_7 \end{pmatrix} = \begin{pmatrix} 1\ 1\ 4\ 1\ 8\ 5\ 2\ 9 \\ 9\ 1\ 1\ 4\ 1\ 8\ 5\ 2 \\ 2\ 9\ 1\ 1\ 4\ 1\ 8\ 5 \\ 5\ 2\ 9\ 1\ 1\ 4\ 1\ 8 \\ 8\ 5\ 2\ 9\ 1\ 1\ 4\ 1 \\ 1\ 8\ 5\ 2\ 9\ 1\ 1\ 4 \\ 4\ 1\ 8\ 5\ 2\ 9\ 1\ 1 \\ 1\ 4\ 1\ 8\ 5\ 2\ 9\ 1 \end{pmatrix} \begin{pmatrix} x_0 \\ x_1 \\ x_2 \\ x_3 \\ x_4 \\ x_5 \\ x_6 \\ x_7 \end{pmatrix} = \text{MDS}(X)$$

Thus, we could equivalently write a set of equations looking as this:

$$y_0 = x_0 \oplus x_1 \oplus 4 \cdot x_2 \oplus x_3 \oplus 8 \cdot x_4 \oplus 5 \cdot x_5 \oplus 2 \cdot x_6 \oplus 9 \cdot x_7$$

The output of the round is finally obtained combining the outputs of the MAD-box Δ and Γ with A and B as follows:

$$Y_0^{(i)} = \left(X_0^{(i)} \overset{16}{\odot} Z_{6(i-1)} \right) \oplus \Delta^{(i)} \qquad Y_1^{(i)} = \left(X_2^{(i)} \overset{16}{\boxplus} Z_{6(i-1)+2} \right) \oplus \Delta^{(i)}$$

$$Y_2^{(i)} = \left(X_1^{(i)} \overset{16}{\boxplus} Z_{6(i-1)+1} \right) \oplus \Gamma^{(i)} \qquad Y_3^{(i)} = \left(X_3^{(i)} \overset{16}{\odot} Z_{6(i-1)+3} \right) \oplus \Gamma^{(i)}$$

On the other hand, a half round contains less operations and is defined as follows:

$$Y_0^{(i)} = X_0^{(i)} \overset{16}{\odot} Z_{6(i-1)} \qquad Y_1^{(i)} = X_2^{(i)} \overset{16}{\boxplus} Z_{6(i-1)+1}$$

$$Y_2^{(i)} = X_1^{(i)} \overset{16}{\boxplus} Z_{6(i-1)+2} \qquad Y_3^{(i)} = X_3^{(i)} \overset{16}{\odot} Z_{6(i-1)+3}$$

Decryption. WIDEA-8 decryption also consists in 8 full rounds followed by one half round; every round uses some subkeys calculated using the key scheduling algorithm specified above. The definitions of the full and half rounds for decryption is the same as that given above for encryption; the only difference is that the subkeys (previously inverted with respect to the proper law group) must be used in the inverse order. Note that the WIDEA key schedule algorithm is designed to be easily invertible, thus one may also apply on-the-fly inverse key scheduling for decryption, where the master key contains the last 8 subkeys derived for encryption.

Test vectors

PLAINTEXT

```
0000 0011 0022 0033 0044 0055 0066 0077
0088 0099 00aa 00bb 00cc 00dd 00ee 00ff
ff00 ee00 dd00 cc00 bb00 aa00 9900 8800
7700 6600 5500 4400 3300 2200 1100 0000
```

KEY

```
0000 0001 0002 0003 0004 0005 0006 0007
0008 0009 000a 000b 000c 000d 000e 000f
0000 0010 0020 0030 0040 0050 0060 0070
0080 0090 00a0 00b0 00c0 00d0 00e0 00f0
0000 0100 0200 0300 0400 0500 0600 0700
0800 0900 0a00 0b00 0c00 0d00 0e00 0f00
0000 1000 2000 3000 4000 5000 6000 7000
8000 9000 a000 b000 c000 d000 e000 f000
```

CIPHERTEXT

```
c28c 1bcf b923 65f9 d8a0 2d77 417c 3da8
f6ed 06ba 961e 3948 4162 ccaa a62a da5b
d6f2 b750 ecfb 22ce 71a3 3380 c8ef aa90
1424 67da 51fd 1d38 0978 cccc c99a 5f5a
```

Cryptanalysis of the ISDB Scrambling Algorithm (MULTI2)

Jean-Philippe Aumasson[1,*], Jorge Nakahara Jr.[2,**], and Pouyan Sepehrdad[2]

[1] FHNW, Windisch, Switzerland
jeanphilippe.aumasson@gmail.com
[2] EPFL, Lausanne, Switzerland
{jorge.nakahara,pouyan.sepehrdad}@epfl.ch

Abstract. MULTI2 is the block cipher used in the ISDB standard for scrambling digital multimedia content. MULTI2 is used in Japan to secure multimedia broadcasting, including recent applications like HDTV and mobile TV. It is the only cipher specified in the 2007 Japanese ARIB standard for conditional access systems. This paper presents a theoretical break of MULTI2 (not relevant in practice), with shortcut key recovery attacks for any number of rounds. We also describe equivalent keys and linear attacks on reduced versions with up 20 rounds (out of 32), improving on the previous 12-round attack by Matsui and Yamagishi. Practical attacks are presented on up to 16 rounds.

Keywords: ISDB, ARIB, MULTI2, block cipher, linear cryptanalysis, conditional access.

1 Introduction

MULTI2 is a block cipher developed by Hitachi in 1988 for general-purpose applications, but which has mainly been used for securing multimedia content. It was registered in ISO/IEC 9979[1] [8] in 1994, and is patented in the U.S. [13,14] and in Japan [7]. MULTI2 is the only cipher specified in the 2007 Japanese standard ARIB for conditional access systems [2]. ARIB is the basic standard of the recent ISDB (for Integrated Services Digital Broadcasting), Japan's standard for digital television and digital radio (see http://www.dibeg.org/)

Since 1995, MULTI2 is the cipher used by satellite and terrestrial broadcasters in Japan [16,18] for protecting audio and video streams, including HDTV, mobile and interactive TV. In 2006, Brazil adopted ISDB as a standard for digital-TV, and several other countries are progressively switching to ISDB (Chile, Ecuador,

---

* Supported by the Swiss National Science Foundation, project no. 113329.
** The work described in this paper has been supported in part by the European Commission through the ICT Programme under contract ICT-2007-216646 ECRYPT II. The information in this document reflects only the author's views, is provided as is and no guarantee or warranty is given that the information is fit for any particular purpose. The user thereof uses the information at its sole risk and liability.
[1] The ISO/IEC 9979, under which cryptographic algorithms were registered, was withdrawn on Feb. 2006 because of its redundancy with the ISO/IEC 18033 standard.

O. Dunkelman (Ed.): FSE 2009, LNCS 5665, pp. 296–307, 2009.
© International Association for Cryptologic Research 2009

Peru, Philippines and Venezuela). But for the moment only Japan uses the conditional access features of ISDB, thus MULTI2 is only used in Japan.

MULTI2 has a Feistel structure and encrypts 64-bit blocks using a 256-bit "system key" and a 64-bit "data key". The ISO register recommends at least 32 rounds. A previous work by Matsui and Yamagishi [11] reports attacks on a reduced version of MULTI2 with 12 rounds. Another work by Aoki and Kurokawa [1] reports an analysis of the round mappings of MULTI2, with results independently rediscovered in the present work.

Contribution. This paper presents new cryptanalytic results on MULTI2, including the description of large sets of equivalent keys, a guess-and-determine attack for any number of rounds, a linear attack on 20 rounds, and a related-key slide attack (see Table 1 for complexities). Despite no practical threat to conditional access systems, our results raise concerns on the intrinsic security of MULTI2.

Table 1. Summary of our attacks on MULTI2 (Data is given in known plaintexts)

#Rounds	Time	Data	Memory	Attack
4	$2^{16.4}$	$2^{16.4}$	—	linear distinguisher*
8	$2^{27.8}$	$2^{27.8}$	—	linear distinguisher*
12	$2^{39.2}$	$2^{39.2}$	—	linear distinguisher*
16	$2^{50.6}$	$2^{50.6}$	—	linear distinguisher*
20	$2^{93.4}$	$2^{39.2}$	$2^{39.2}$	linear key-recovery
r	$2^{185.4}$	3	2^{31}	guess-and-determine key-recovery
$r \equiv 0 \bmod 8$	$2^{128}/r$	2^{33}	2^{33}	related-key slide key-recovery

$\star$: time complexity is # of parity computations instead of # of encryptions.

2 Description of MULTI2

MULTI2 (Multi-Media Encryption Algorithm 2) is a Feistel block cipher that operates on 64-bit blocks, parametrized by a 64-bit data key and a 256-bit system key. Encryption depends only on a 256-bit key derived from the data and system keys. This encryption key is divided into eight subkeys. MULTI2 uses four key-dependent round functions π_1, π_2, π_3, and π_4, repeated in this order. The ISO register entry recommends at least 32 rounds, which is the number of rounds used in the ISDB standard. We denote MULTI2's keys as follows, parsing them into 32-bit words (see Fig. 1):

- $d = (d_1, d_2)$ is the 64-bit *data key*
- $s = (s_1, s_2, s_3, s_4, s_5, s_6, s_7, s_8)$ is the 256-bit *system key*
- $k = (k_1, k_2, k_3, k_4, k_5, k_6, k_7, k_8)$ is the 256-bit *encryption key*

MULTI2 uses no S-boxes, but only a combination of XOR ($\oplus$), modulo 2^{32} addition ($+$) and subtraction ($-$), left rotation ($\lll$) and logical OR ($\vee$). Below we denote L (resp. R) the left (resp. right) half of the encrypted data, and k_i a 32-bit encryption subkey:

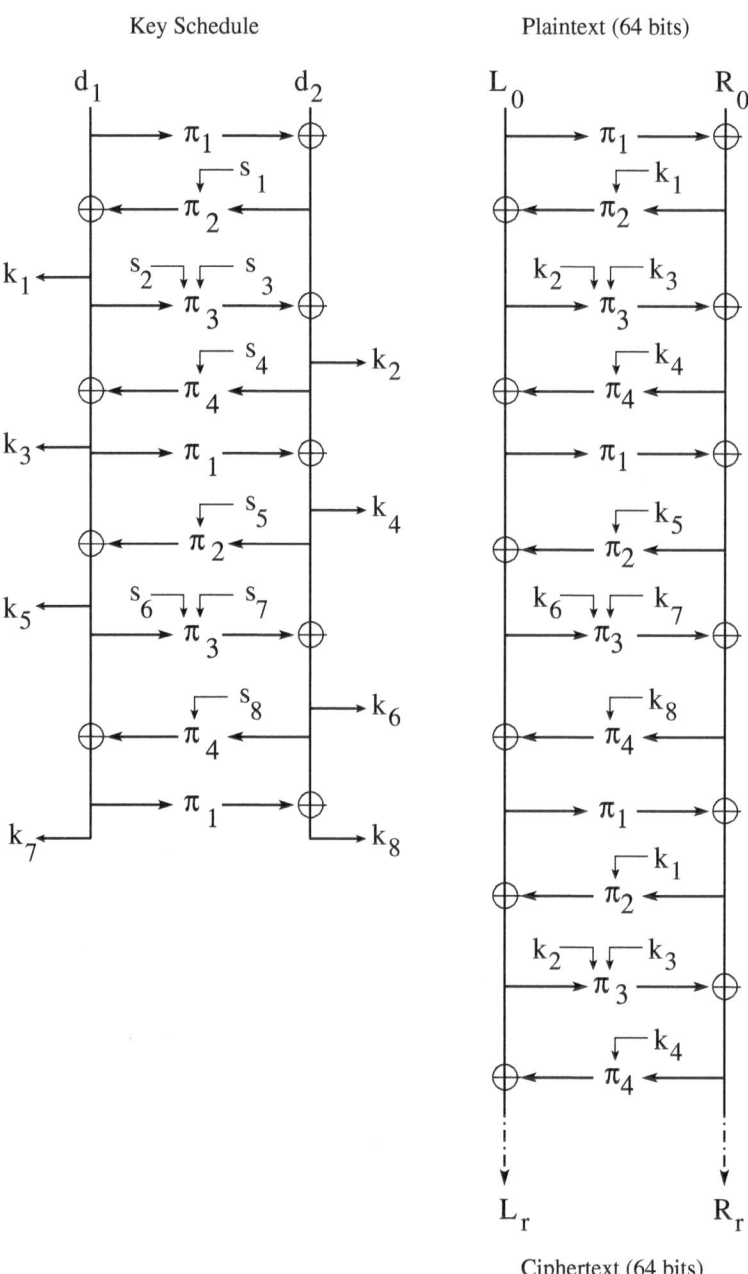

Fig. 1. Key schedule (left) and encryption (right) in MULTI2: the encryption key k is derived from the system key s and the data key d. Only k is used during the encryption.

- π_1 is the identity mapping: $\pi_1(L) = L$. It is the only surjective and key independent round transformation.
- π_2 maps 64 bits to 32 bits, and returns

$$\pi_2(R, k_i) = (x \lll 4) \oplus x \tag{1}$$

where $x = ((R + k_i) \lll 1) + R + k_i - 1$. From the definition (1) it follows that $\pi_2(R, k_i) = \pi_2(k_i, R)$, for any $k_i, R \in \{0,1\}^{32}$. Moreover, π_2 can be expressed as a function of a single value, $R + k_i$. Due to the feed forward in (1), π_2 can not be surjective. The range of π_2 contains exactly $265\,016\,655 \approx 2^{28}$ elements (only 6.2% of $\{0,1\}^{32}$, against 63% expected for a random function [12, §2.1.6]). Moreover, the set of 32 bit values output by π_2 is always the same. This follows from the observation that for fixed R if $0 \leq k_i \leq 2^{32} - 1$, then $0 \leq R + k_i \leq 2^{32} - 1$ and the same holds if k_i is fixed and $0 \leq R \leq 2^{32} - 1$.
- π_3 maps 96 bits to 32 bits, and returns

$$\pi_3(L, k_i, k_j) = (x \lll 16) \oplus (x \vee L) \tag{2}$$

where

$$x = \big(((y \lll 8) \oplus y + k_j) \lll 1\big) - \big((y \lll 8) \oplus y + k_j\big)$$

where $y = ((L + k_i) \lll 2) + L + k_i + 1$. The range of π_3 spans approximately $2^{30.8}$ values, that is, 43% of $\{0,1\}^{32}$, for a *fixed encryption key*. The fraction of the range covered by π_3 is not the same for every $k_i, k_j \in \{0,1\}^{32}$, because $\pi_3(L, k_i, k_j) \neq \pi_3(L, k_j, k_i)$.
- π_4 maps 64 bits to 32 bits, and returns

$$\pi_4(R, k_i) = ((R + k_i) \lll 2) + R + k_i + 1. \tag{3}$$

From the definition of 3, it follows that $\pi_4(R, k_i) = \pi_4(k_i, R)$ for any $k_i, R \in \{0,1\}^{32}$. The range of π_4 contains exactly $1\,717\,986\,919 \approx 2^{30.7}$ elements (i.e., 40.6% of $\{0,1\}^{32}$). The reasoning is the same as for π_2.

An additional property is the fact that these π_j functions do not commute, that is, $\pi_i \circ \pi_j \neq \pi_j \circ \pi_i$, for $i \neq j$, where $\circ$ is functional composition. Thus, the π_j mapping cannot be purposefully clustered or permuted in the cipher framework to ease cryptanalysis.

Encryption. Given subkeys $k_1, \ldots, k_8$ and a plaintext (L_0, R_0), MULTI2 computes the first eight rounds as follows (see Fig. 1):

1. $R_1 \leftarrow R_0 \oplus \pi_1(L_0)$
2. $L_1 \leftarrow L_0; L_2 \leftarrow L_1 \oplus \pi_2(R_1, k_1)$
3. $R_2 \leftarrow R_1; R_3 \leftarrow R_2 \oplus \pi_3(L_2, k_2, k_3)$
4. $L_3 \leftarrow L_2; L_4 \leftarrow L_3 \oplus \pi_4(R_3, k_4)$
5. $R_4 \leftarrow R_3; R_5 \leftarrow R_4 \oplus \pi_1(L_4)$
6. $L_5 \leftarrow L_4; L_6 \leftarrow L_5 \oplus \pi_2(R_5, k_5)$

7. $R_6 \leftarrow R_5$; $R_7 \leftarrow R_6 \oplus \pi_3(L_6, k_6, k_7)$
8. $L_7 \leftarrow L_6$; $L_8 \leftarrow L_7 \oplus \pi_4(R_7, k_8)$
9. $R_8 \leftarrow R_7$

This sequence is repeated (with suitably incremented subscripts) until the desired number of rounds r, and the ciphertext (L_r, R_r) is returned. The subkeys $k_1, \ldots, k_8$ are reused for each sequence $\pi_1, \ldots, \pi_4, \pi_1, \ldots, \pi_4$.

Key Schedule. The key schedule of MULTI2 "encrypts" a data key (d_1, d_2) (as plaintext) through nine rounds, using the system key $s_1, \ldots, s_8$. The round subkeys $k_1, \ldots, k_8$ are extracted as follows (see Fig. 1):

- $k_1 \leftarrow d_1 \oplus \pi_2(d_1 \oplus d_2, s_1)$
- $k_2 \leftarrow d_1 \oplus d_2 \oplus \pi_3(k_1, s_2, s_3)$
- $k_3 \leftarrow k_1 \oplus \pi_4(k_2, s_4)$
- $k_4 \leftarrow k_2 \oplus k_3$
- $k_5 \leftarrow k_3 \oplus \pi_2(k_4, s_5)$
- $k_6 \leftarrow k_4 \oplus \pi_3(k_5, s_6, s_7)$
- $k_7 \leftarrow k_5 \oplus \pi_4(k_6, s_8)$
- $k_8 \leftarrow k_6 \oplus k_7$

MULTI2 in ISDB. In ISDB, MULTI2 is mainly used via the B-CAS card [6] for copy control to ensure that only valid subscribers are using the service. MULTI2 encrypts transport stream packets in CBC or OFB mode. The same system key is used for all conditional-access applications, and another system key is used for other applications (DTV, satellite, etc.). The 64-bit data key is refreshed every second, sent by the broadcaster and encrypted with another block cipher. Therefore only the data key is really secret, since the system key can be obtained from the receivers. Details can be found in the ARIB B25 standard [2].

3 Equivalent Keys

The key schedule of MULTI2 maps a $(256 + 64)$-bit data-and-system key to a 256-bit encryption key (see Fig. 1). This means 64 bits of redundancy (leading to 2^{64} collisions). Further, the 256-bit encryption key $k = (k_1, \ldots, k_8)$ has entropy at most 192 bits, because the key schedule sets $k_4 = k_3 \oplus k_2$ and $k_8 = k_7 \oplus k_6$. Hence, the knowledge of two subkeys in (k_2, k_3, k_4) is sufficient to compute the third. The key schedule thus induces a loss of at least 128 bits of entropy, from the 320-bit (s, d) key. Therefore, the average size of equivalence key classes is 2^{128}.

Large sets of colliding pairs (s, d) can be found as follows: given (s, d), one just has to find s_1' such that $\pi_2(d_1 \oplus d_2, s_1') = \pi_2(d_1 \oplus d_2, s_1)$; or s_2', s_3' such that $\pi_3(k_1, s_2, s_3) = \pi_3(k_1, s_2', s_3')$; or s_4' such that $\pi_4(k_2, s_4') = \pi_4(k_2, s_4)$; or s_5' such that $\pi_2(k_4, s_5) = \pi_2(k_4, s_5')$; or s_6', s_7' such that $\pi_3(k_5, s_6, s_7) = \pi_3(k_5, s_6', s_7')$; or s_8' such that $\pi_4(k_6, s_8) = \pi_4(k_6, s_8')$. Each of these conditions are independent.

The result is a (series of) equivalent keys (s', d) that lead to the same encryption key as the pair (s, d).

However, there exist no equivalent keys with the same system key and distinct data keys. This is because the key schedule uses the data key as plaintext, hence the encryption key is trivially invertible (see Fig. 1).

Note that [8] suggests to use MULTI2 as building block for constructing hash functions. If the construction is not carefully chosen, however, equivalent keys in MULTI2 could lead to simple collisions. For example, in Davies-Meyer mode the compression function would return $E_m(h) \oplus h$, with h a chaining value and m a message block; since equivalent keys are easy to find, it is easy as well to find two (or more) distinct message block that produce the same encryption key, and thus that give multicollisions.

4   Guess-and-Determine Attack

We describe a known-plaintext attack that recovers the 256-bit encryption key in about $2^{185.4}$ r-round encryptions. The attack works for any number r of rounds, and uses only three known plaintexts/ciphertext pairs.

We recall the loss of key entropy due to redundancy in the key schedule of MULTI2 described in Sect. 3.

One recovers $k_1, \ldots, k_8$ using a guess-and-determine strategy, exploiting the non-surjectivity of the round functions π_2 and π_4 (see key schedule in Fig. 1):

1. guess k_1 and k_2 (2^{64} choices)
2. guess $\pi_4(k_2, s_4)$ ($2^{30.7}$ choices), and deduce $k_3 = k_1 \oplus \pi_4(k_2, s_4)$
3. set $k_4 = k_2 \oplus k_3$
4. guess $\pi_2(k_4, s_5)$ (2^{28} choices), and deduce $k_5 = k_3 \oplus \pi_2(k_4, s_5)$
5. guess $\pi_3(k_5, s_6, s_7)$ (2^{32} choices), and deduce $k_6 = k_4 \oplus \pi_3(k_5, s_6, s_7)$
6. guess $\pi_4(k_6, s_8)$ ($2^{30.7}$ choices), and deduce $k_7 = k_5 \oplus \pi_4(k_6, s_8)$
7. set $k_8 = k_6 \oplus k_7$

A guess of $k_1, \ldots, k_8$ is verified using three known-plaintext/ciphertext pairs (each pair gives a 64-bit condition). The total cost is about $2^{185.4}$ r-round encryptions and 2^{31} 32-bit words of memory. Note that the non-surjectivity of $\pi_3(k_5, s_6, s_7)$ cannot be exploited here, because the range depends on the system subkeys, which are unknown.

Once the encryption key k is found, one can recover all equivalent 256-bit system keys s and 64-bit data keys d as follows: starting from the end of the key schedule (Fig. 1), one iteratively searches for a valid s_8 (2^{32} π_4-computations), then a valid pair (s_6, s_7) (2^{64} π_3-computations), and so on (the complexities add up) until recovering s_4. For computing (s_2, s_3) we need the value of $d_1 \oplus d_2$. The cost for this case is $2^{32+32+32} = 2^{96}$ π_3-computations. For s_1 we need the separate values of d_1 and d_2. Since we already computed $d_1 \oplus d_2$, the cost is $2^{32+32} = 2^{64}$ π_2-computations. The final complexity is dominated by 2^{96} π_3-computations to recover all candidates pairs (s, d). The cost of computing one of the pairs (s, d) is dominated by 2^{33} π_3-computations.

5   Linear Attacks

The non-surjectivity of the round functions π_2, π_3, π_4 motivates the study of linear relations [10] for particular bitmasks. Usually, one looks for nonzero input and output bitmasks for individual cipher components, with high bias. But for MULTI2, we look for linear relations of the form $0 \xrightarrow{\pi_i} \Gamma$, $2 \leq i \leq 4$, $\Gamma \neq 0$. Because of 4-round repetition of π_i mappings in MULTI2, and to optimize the search, we looked only for *iterative* linear relations that cover all four consecutive π_i round mappings.

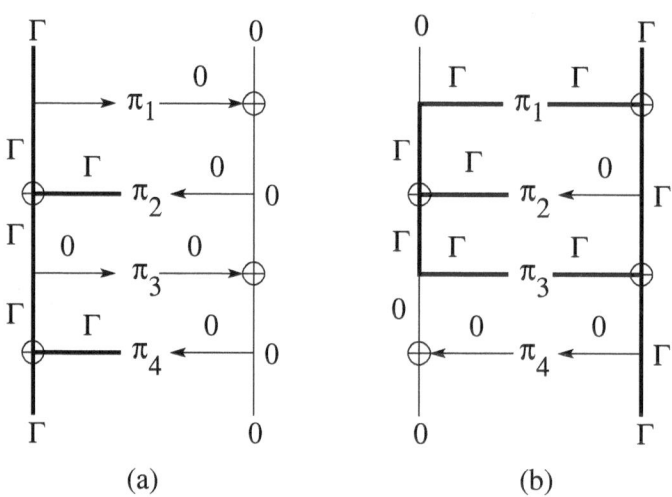

(a)   (b)

Fig. 2. Four-round iterative linear trails

In Fig. 2(a), only π_2 and π_4 are active, that is, effectively participate in the linear relation (linear trails are depicted by thick lines). In Fig. 2(b), π_2 and π_3 are active, but for π_3 the linear approximation has the form $\Gamma \xrightarrow{\pi_3} \Gamma$, where $\Gamma \neq 0$. Alternative iterative linear relations are depicted in Fig. 3. In Fig. 3(a), there is one instance of π_2 and π_4, and two instances of π_3 active along the linear relation. Fig. 3(b) is just Fig. 3(a) slided by four rounds.

We consider each round function as operating over 32-bit data (for a fixed, unknown key). Instead of deriving the bias for a given output bitmask, we have searched for promising bitmasks (with high bias), by exhaustive search over the inputs for each round function. From the linear relations listed above, the one in Fig. 2(a) is the most promising, since it involves only two active round functions for every four rounds: π_2 and π_4. This is an indication that the left half of the MULTI2 encryption/decryption framework is weaker than the right half.

Our search (over random keys) started with masks of low Hamming weight. The use of rotation suggests that masks with short repeating patterns tend to show higher biases, which our experiments confirmed: the 32-bit mask

$$\Gamma = \mathtt{AAAAAAAA}_x$$

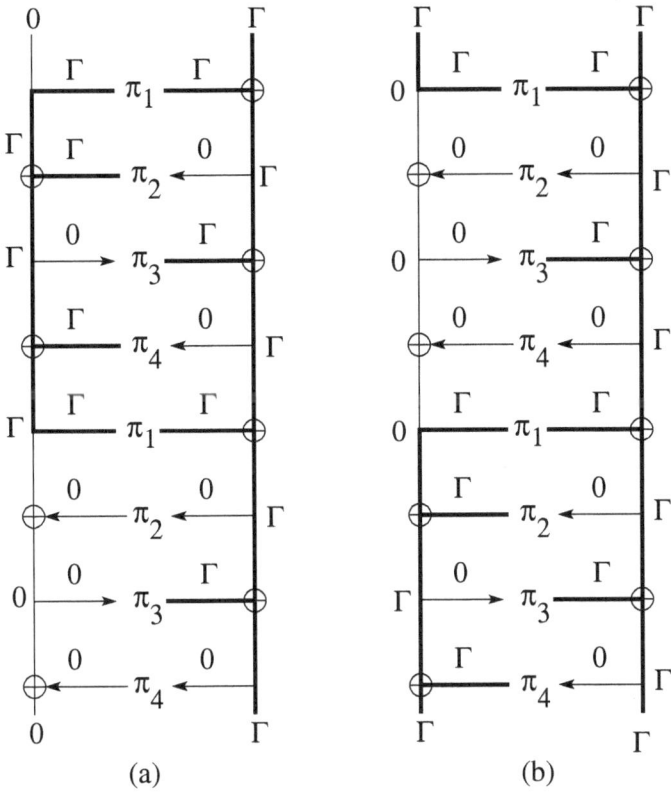

(a)   (b)

Fig. 3. Eight-round iterative linear trails

presented the highest bias: 2^{-1} for π_2, and $2^{-6.7}$ for π_4 (Fig. 2(a)). The overall bias is $p' = 2^{-6.7}$, using the piling-up lemma [10]. This bitmask was independently discovered by Aoki and Kurokawa in [1].

Comparatively, for Fig. 2(b), the best mask we have found is $\Gamma = 88888888_x$, with bias 2^{-1} for π_2, and $2^{-8.6}$ for π_3. The overall bias for four rounds is $2^{-8.6}$.

Using Fig. 2(a), one can distinguish 4-round MULTI2 from a random permutation using $8 \times (p')^{-2} = 2^{16.4}$ known plaintexts (KP), for a high success rate attack; the memory complexity is negligible and the attack effort is essentially $2^{16.4}$ parity computations. For eight rounds, the attack complexity is $8 \times (2 \times (p')^2)^{-2} = 8 \times (2^{-12.4})^{-2} = 2^{27.8}$ KP and equivalent parity computations; for twelve rounds, the data complexity becomes $8 \times (2^2 \times (p')^3)^{-2} = 8 \times (2^{-18.1})^{-2} = 2^{39.2}$ KP; for sixteen rounds, $8 \times (2^{-23.8})^{-2} = 2^{50.6}$ KP. For further rounds, more plaintexts than the codebook are required.

For key-recovery attacks on twenty rounds, we use the 12-round linear relation described in the previous paragraph. Notice that across 20-round MULTI2, the same sequence of four subkeys, $k_1, \ldots, k_4$ repeats at the first and at the last four rounds. Thus, we place the 12-round relation in the middle of 20-round MULTI2 and guess consecutively k_1, k_2 (cost 2^{32} for each of them), then k_3 (cost $2^{30.7}$), and finally k_4 (free). Time complexity is thus about $2^{94.7} + 2^{94.7} = 2^{95.7}$ 4-round decryptions, that is, $1/5 \cdot 2^{95.7} \approx 2^{93.4}$ 20-round encryptions. Storage of $2^{39.2}$ ciphertexts is necessary.

6   Related-Key Slide Attack

We present key-recovery known-plaintext related-key slide attacks [3,4,5]. These attacks exploit the partial similarity of 4-round sequences, and works for any version of MULTI2 whose number of rounds is a multiple of eight.

Let $F_{1\ldots4}$ stand for 4-round encryption involving $\pi_1, \ldots, \pi_4$ with subkeys $k_1, \ldots, k_4$. Similarly, let $F_{5\ldots8}$ stand for 4-round encryption involving $\pi_1, \ldots, \pi_4$ with subkeys $k_5, \ldots, k_8$; $F'_{1\ldots4}$ stand for $\pi_1, \ldots, \pi_4$ with subkeys $k'_1, \ldots, k'_4$, and $F'_{5\ldots8}$ stand for $\pi_1, \ldots, \pi_4$ with subkeys $k'_5, \ldots, k'_8$.

Given an unknown key pair (s, d), we consider a related-key pair (s', d') that gives k' such that

$$k'_1 = k_5 \quad k'_2 = k_6 \quad k'_3 = k_7 \quad k'_4 = k_8$$
$$k'_5 = k_1 \quad k'_6 = k_2 \quad k'_7 = k_3 \quad k'_8 = k_4 \tag{4}$$

Thus, $F'_{1\ldots4} \equiv F_{5\ldots8}$ and $F'_{5\ldots8} \equiv F_{1\ldots4}$.

For Eq. (4) to hold, it is necessary that the related key (s', d') satisfies

$$d'_1 = k_3 \quad d'_1 \oplus d'_2 = k_4$$
$$s'_1 = s_5 \quad s'_2 = s_6 \quad s'_3 = s_7 \quad s'_4 = s_8$$
$$s'_5 = s_1 \quad s'_6 = s_2 \quad s'_7 = s_3 \quad s'_8 = s_4.$$

The conditions $k'_1 = k_5$ and $k'_2 = k_6$ require

$$k_3 \oplus \pi_2(k_4, s_5) = d'_1 \oplus \pi_2(d'_1 \oplus d'_2, s'_1)$$
$$k_4 \oplus \pi_3(k_5, s_6, s_7) = d'_1 \oplus d'_2 \oplus \pi_3(k'_1, s'_2, s'_3).$$

A slid pair gives $P' = F_{1\ldots4}(P)$, which implies $C' = F'_{5\ldots8}(C) = F_{1\ldots4}(C)$, as shown below.

$$P \xrightarrow{F_{1\ldots4}} X \xrightarrow{F_{5\ldots8}} \ldots \xrightarrow{F_{5\ldots8}} C$$
$$P' \xrightarrow{F'_{1\ldots4}} \ldots \xrightarrow{F'_{1\ldots4}} Y \xrightarrow{F'_{5\ldots8}} C'$$

That is, one get two 64-bit conditions since both the plaintext and ciphertext slid pairs are keyed by the same subkeys. Thus one slid pair is sufficient to identify $k_1, \ldots, k_4$. The attack goes as follows:

1. collect 2^{32} distinct (P_i, C_i) pairs, $i = 1, \ldots, 2^{32}$ encrypted with k
2. collect 2^{32} distinct (P'_i, C'_i) pairs, $i = 1, \ldots, 2^{32}$ encrypted with k'
3. for each $(i, j) \in \{1, \ldots, 2^{32}\}^2$

4. find the value of $k_1, \ldots, k_4$ that satisfy $P'_j = F_{1 \ldots 4}(P_i)$ and $C'_j = F_{1 \ldots 4}(C_i)$
5. search exhaustively $k_5, \ldots, k_8$ (there are 2^{96} choices, exploiting the non-surjectivity of π_2 and π_4)

We cannot filter the wrong slid pairs, so we try all possible 2^{64} pairs (P_i, P_j). But each potential slid pairs provides 128-bit condition, because both the plaintext and ciphertext pairs are keyed by the same unknown subkeys. Thus, we can filter the wrong subkeys at once.

To recover $k_1, \ldots, k_4$ we use the potential slid pair (P'_j, P_i). Guess k_1 (2^{32} choices). Then, guess k_2 (2^{32} choices), and find the k_3 that yields the (known) output of π_3. Deduce k_4, as $k_2 \oplus k_3$ and finally, test whether the current choice of $k_1, \ldots, k_4$ is consistent with the second potential slid pair (C'_j, C_i).

Finding k_3 from k_2, the input of π_3, and its output one has to solve an equation of the form $(x \lll 16) \oplus (x \vee L) = b$, then an equation $(t \lll 1) - t = x$, where x and t are the unknowns. The first can be solved bit per bit, by iteratively storing the solutions for each pair (x_i, x_{i+16}). There are 16 such pairs, and for each pair there are at most two solutions. Hence in the worst case there will be 2^{16} solutions. The effort up to this point is roughly $2^{32+32} = 2^{64}$ π_2 and π_3-computations.

In total, up to this point, there are $2^{32+32+16} = 2^{80}$ possible values for (k_1, k_2, k_3, k_4). The value of $k_4 = k_2 \oplus k_3$ can be further checked using the (P_i, P'_j) pair. Let $P_i = (P_L, P_R)$ and $P'_j = (P'_L, P'_R)$. Then, $P'_L \oplus P_L \oplus \pi_2(P_R \oplus P_L, k_1) = \pi_4(P'_R, k_4)$, which is a condition on 26.7 bits, since the output of π_2 and π_4 intersect in $2^{26.7}$ distinct values. Thus, we expect only $2^{80}/2^{26.7} = 2^{53.3}$ tuples (k_1, k_2, k_3, k_4) to survive. Using (C_i, C'_j), a 64-bit condition, reduces the number of wrong key candidates to $2^{53.3}/2^{64} < 1$.

The final attack complexity is thus about $2^{64} \times 2^{64}$ 1-round computations, or $(2^{128}/r)$ r-round computations to recover $k_1, \ldots, k_4$. Further, to recover k_5, $\ldots$, k_8, we run a similar procedure, but over $r - 8$ rounds (after decrypting the top and bottom four rounds). The complexity is 2^{96} $(r - 8)$-round computations. Normalizing the complexity figures, the overall attack complexity is dominated by $(2^{128}/r)$ r-round computations. The memory complexity is 2^{33} plaintext/ciphertext pairs.

7   Conclusions

We showed that the 320-bit key of MULTI2 can be recovered in about 2^{185} trials instead of 2^{320} ideally, for any number of rounds, and using only three plaintext/ciphertext pairs. This weakness is due to the loss of entropy induced by the key schedule and the non-surjective round functions. We also described a linear (key-recovery) attack on up to 20 rounds, and a related-key slide attack in $(2^{128}/r)$ r-round computations for any number r of rounds that is a multiple of eight (thus including the recommended 32 rounds).

Although our results do not represent any practical threat when the 32-round recommendation is followed, they show that the security of MULTI2 is not as high as expected, and raise concerns on its long-term reliability. A practical break of MULTI2 would have dramatic consequences: millions of receivers would have to be replaced, a new technology and new standards would have to be designed and implemented.

Finally, note that the Common Scrambling Algorithm (CSA), used in Europe through the digital-TV standard DVB[2] also underwent some (non-practical) attacks [15,17]. For comparison, the American standard ATSC uses Triple-DES in CBC mode[3].

Acknowledgments

We wish to thank Kazumaro Aoki for communicating us a copy of his article, and also Tim Sharpe (NTT communication) and Jack Laudo for their assistance. We are also grateful to the reviewers of FSE 2009 for their valuable comments, and for pointing out reference [9].

References

1. Aoki, K., Kurokawa, K.: A study on linear cryptanalysis of Multi2 (in Japanese). In: The 1995 Symposium on Cryptography and Information Security, SCIS 1995 (1995)
2. ARIB. STD B25 v. 5.0 (2007), http://www.arib.or.jp/
3. Biham, E.: New types of cryptanalytic attacks using related keys. Journal of Cryptology 7(4), 229–246 (1994)
4. Biryukov, A., Wagner, D.: Slide attacks. In: Knudsen, L.R. (ed.) FSE 1999. LNCS, vol. 1636, pp. 245–259. Springer, Heidelberg (1999)
5. Biryukov, A., Wagner, D.: Advanced slide attacks. In: Preneel, B. (ed.) EURO-CRYPT 2000. LNCS, vol. 1807, pp. 589–606. Springer, Heidelberg (2000)
6. BS Conditional Access Systems Co., Ltd., http://www.b-cas.co.jp/
7. Hitachi: Japanese laid-open patent application no. H1-276189 (1998)
8. ISO. Algorithm registry entry 9979/0009 (1994)
9. Katagi, T., Inoue, T., Shimoyama, T., Tsujii, S.: A correlation attack on block ciphers with arithmetic operations (in Japanese). In: SCIS (2003), reference no. SCIS2003 5D-2
10. Matsui, M.: Linear cryptoanalysis method for DES cipher. In: Helleseth, T. (ed.) EUROCRYPT 1993. LNCS, vol. 765, pp. 386–397. Springer, Heidelberg (1993)
11. Matsui, M., Yamagishi, A.: On a statistical attack of secret key cryptosystems. Electronics and Communications in Japan, Part III: Fundamental Electronic Science (English translation of Denshi Tsushin Gakkai Ronbunshi) 77(9), 61–72 (1994)
12. Menezes, A.J., van Oorschot, P.C., Vanstone, S.A.: Handbook of Applied Cryptography. CRC Press, Boca Raton (1997)
13. Takaragi, K., Nakagawa, F., Sasaki, R.: U.S. patent no. 4982429 (1989)
14. Takaragi, K., Nakagawa, F., Sasaki, R.: U.S. patent no. 5103479 (1990)

---

[2] See http://www.dvb.org/

[3] See http://www.atsc.org/standards/a_70a_with_amend_1.pdf

15. Weinmann, R.-P., Wirt, K.: Analysis of the DVB common scrambling algorithm. In: 8th IFIP TC-6 TC-11 Conference on Communications and Multimedia Security (CMS). Springer, Heidelberg (2004)
16. Wikipedia. Mobaho! (accessed February 5, 2009)
17. Wirt, K.: Fault attack on the DVB common scrambling algorithm. In: Gervasi, O., Gavrilova, M.L., Kumar, V., Laganá, A., Lee, H.P., Mun, Y., Taniar, D., Tan, C.J.K. (eds.) ICCSA 2005. LNCS, vol. 3481, pp. 577–584. Springer, Heidelberg (2005)
18. Yoshimura, T.: Conditional access system for digital broadcasting in Japan. Proceedings of the IEEE 94(1), 318–322 (2006)

Beyond-Birthday-Bound Security Based on Tweakable Block Cipher

Kazuhiko Minematsu

NEC Corporation, 1753 Shimonumabe, Nakahara-Ku, Kawasaki, Japan
k-minematsu@ah.jp.nec.com

Abstract. This paper studies how to build a $2n$-bit block cipher which is hard to distinguish from a truly random permutation against attacks with $q \approx 2^{n/2}$ queries, i.e., birthday attacks. Unlike previous approaches using pseudorandom functions, we present a simple and efficient proposal using a tweakable block cipher as an internal module. Our proposal is provably secure against birthday attacks, if underlying tweakable block cipher is also secure against birthday attacks. We also study how to build such tweakable block ciphers from ordinary block ciphers, which may be of independent interest.

Keywords: Block Cipher Mode, Birthday Bound, Tweakable Block Cipher.

1 Introduction

A double-block-length cipher (DBLC), i.e. a $2n$-bit block cipher made from n-bit block components, has been one of the main research topics in the symmetric cryptography. In particular, a seminal work of Luby and Rackoff [17] proved that a 4-round Feistel permutation is computationally hard to distinguish from a truly random permutation if each round function is an n-bit pseudorandom function [11]. The proof of [17] is valid for chosen-ciphertext attacks (CCAs) using $q \ll 2^{n/2}$ queries, and is called a proof of $O(2^{n/2})$-security. As $2^{n/2}$ is related to the birthday paradox for n-bit variables, it is also called the security up to the birthday bound (for n). Then, building a DBLC having beyond-birthday-bound security, i.e., $O(2^{\omega+n/2})$-security for some $\omega > 0$, is an interesting research topic from theoretical and practical aspects. In particular, such a DBLC can improve the security of any block cipher mode that has $O(2^{n/2})$-security with an n-bit block cipher[1]. However, achieving $O(2^{\omega+n/2})$-security is generally difficult, even for a small ω. We have very few known DBLC proposals having this property. All of them were based on Feistel permutations using pseudorandom functions [22][18][20]. Although these studies indicated the great potential of Feistel permutation, we wondered if using Feistel was the only solution.

In this paper, we demonstrate how this problem can be solved using a tweakable block cipher, defined by Liskov et al.[16]. In particular, we present

---

[1] For some specific applications, such as stateful encryption and stateful authentication, block cipher modes with beyond-birthday-bound security are known [15][5].

O. Dunkelman (Ed.): FSE 2009, LNCS 5665, pp. 308–326, 2009.

how to build a DBLC based on a tweakable block cipher $\widetilde{E}$ with n-bit block and m-bit tweak for any $1 \leq m \leq n$, and prove $O(2^{(n+m)/2})$-security against CCAs. One significant fact is that it is *optimally efficient*, as it requires only two $\widehat{E}$ calls (independently of m) and some universal hash functions. Thus, assuming very fast universal hash functions (e.g., [25]), our DBLC will have almost the same throughput as that of a tweakable block cipher. This means that, the task of building a secure $2n$-bit block cipher can be efficiently reduced to that of building a secure n-bit block tweakable block cipher. We think this is an interesting application of tweakable block cipher, that has not been mentioned before. As a by-product, we provide some variants such as a pseudorandom function with $2n$-bit input and n-bit output. All variants are optimally efficient in the sense defined above.

We have to emphasize that the birthday bound here is with respect to n, and not to $n + m$. The security of our scheme is still up to the birthday bound of *input length* of the cryptographic primitive (as with Yasuda [28]). Although this makes the problem much easier in general, our result is still non-trivial and highly optimized as a solution to beyond-birthday-bound security for n.

As our DBLC requires a tweakable block cipher with beyond-birthday-bound security, we also discuss how to realize it. Specifically, we focus on constructions using n-bit block ciphers. Although known constructions [16][24] are only $O(2^{n/2})$-secure, we provide a simple solution using tweak-dependent key changes with a concrete security proof. Unfortunately, this scheme is only the first step: it can be very slow and has some severe theoretical limitations, thus is far from being perfect. Building a better scheme remains an interesting future direction of research.

2 Preliminaries

2.1 Basic Notations

A random variable will be written in capital letters and its sampled value will be written in the corresponding small letters. Let Σ^n denote $\{0,1\}^n$. The bit length of a binary sequence x is denoted by $|x|$, and $x_{[i,j]}$ denotes a subsequence of x from i-th to j-th bit, for $1 \leq i < j \leq |x|$. A uniform random function (URF) with n-bit input and ℓ-bit output, denoted by $\mathsf{R}_{n,\ell}$, is a random variable uniformly distributed over $\{f : \Sigma^n \to \Sigma^\ell\}$. Similarly, a random variable uniformly distributed over all n-bit permutations is an n-bit block uniform random permutation (URP) and is denoted by P_n. If $F_K : \mathcal{X} \to \mathcal{Y}$ is a keyed function, then F_K is a random variable (not necessarily uniformly) distributed over $\{f : \mathcal{X} \to \mathcal{Y}\}$. If F_K is a keyed permutation, F_K^{-1} will denote its inversion. We will omit K and write $F : \mathcal{X} \to \mathcal{Y}$, when K is clear from the context.

A tweakable block cipher [16] is a keyed permutation with auxiliary input called tweak. Formally, a ciphertext of a tweakable blockcipher, $\widetilde{E} : \mathcal{M} \times \mathcal{T} \to \mathcal{M}$, is $C = \widetilde{E}(M,T)$, where $M \in \mathcal{M}$ is a plaintext and $T \in \mathcal{T}$ is the tweak. The encryption, $\widetilde{E}$, must be a keyed permutation over $\mathcal{M}$ for every $T \in \mathcal{T}$, and the decryption is defined as $\widetilde{E}^{-1}(C,T) = M$ with $\widetilde{E}^{-1} : \mathcal{M} \times \mathcal{T} \to \mathcal{M}$. If $\widetilde{E}$ has n-bit block and m-bit tweak, we say it is an (n, m)-bit tweakable cipher. An (n, m)-bit tweakable URP is the set of 2^m independent URPs (i.e., an n-bit URP is used for each m-bit tweak) and is denoted by $\widetilde{\mathsf{P}}_{n,m}$. We write $\widetilde{\mathsf{P}}_n$ if m is clear from the context.

2.2   Security Notion

Consider the game in which we want to distinguish two keyed functions, G and G', using a black-box access to them. We define classes of attacks: chosen-plaintext attack (CPA), and chosen-ciphertext attack (CCA), and their tweaked versions, i.e., a tweak and a plaintext (or ciphertext) can be arbitrarily chosen. Here, (tweaked) CCA can be defined when G and G' are (tweakable) permutations. Let $\mathtt{atk} \in \{\mathtt{cpa}, \mathtt{cca}, \widetilde{\mathtt{cpa}}, \widetilde{\mathtt{cca}}\}$, where $\widetilde{\mathtt{cpa}}$ ($\widetilde{\mathtt{cca}}$) denotes tweaked CPA (CCA). The maximum advantage of adversary using $\mathtt{atk}$ in distinguishing G and G' is:

$$\mathrm{Adv}_{G,G'}^{\mathtt{atk}}(\theta) \overset{\mathrm{def}}{=} \max_{\mathcal{D}:\theta-\mathtt{atk}} \left| \Pr[\mathcal{D}^G = 1] - \Pr[\mathcal{D}^{G'} = 1] \right|, \qquad (1)$$

where $\mathcal{D}^G = 1$ denotes that $\mathcal{D}$'s guess is 1, which indicates G or G'. The parameter θ denotes the attack resource, such as the number of queries, q, and time complexity [11], τ. If θ does not contain τ, the adversary has no computational restriction. The maximum is taken for all $\mathtt{atk}$-adversaries having θ. For $G : \varSigma^n \to \varSigma^m$, we have

$$\mathrm{Adv}_G^{\mathtt{prf}}(\theta) \overset{\mathrm{def}}{=} \mathrm{Adv}_{G,\mathsf{R}_{n,m}}^{\mathtt{cpa}}(\theta), \ \ \mathrm{Adv}_G^{\mathtt{sprp}}(\theta) \overset{\mathrm{def}}{=} \mathrm{Adv}_{G,\mathsf{P}_n}^{\mathtt{cca}}(\theta), \ \ \mathrm{Adv}_G^{\mathtt{prp}}(\theta) \overset{\mathrm{def}}{=} \mathrm{Adv}_{G,\mathsf{P}_n}^{\mathtt{cpa}}(\theta),$$

where the last two equations are defined if G is an n-bit permutation, Moreover, if G is an (n, m)-bit tweakable cipher, we define

$$\mathrm{Adv}_G^{\widetilde{\mathtt{sprp}}}(\theta) \overset{\mathrm{def}}{=} \mathrm{Adv}_{G,\widetilde{\mathsf{P}}_{n,m}}^{\widetilde{\mathtt{cca}}}(\theta), \ \ \text{and} \ \ \mathrm{Adv}_G^{\widetilde{\mathtt{prp}}}(\theta) \overset{\mathrm{def}}{=} \mathrm{Adv}_{G,\widetilde{\mathsf{P}}_{n,m}}^{\widetilde{\mathtt{cpa}}}(\theta).$$

If $\mathrm{Adv}_G^{\mathtt{prf}}(\theta)$ is negligibly small for all practical θ (the definition of "practical θ" depends on users.), G is a pseudorandom function (PRF)[11]. If G is invertible, it is also called a pseudorandom permutation (PRP). In addition, if $\mathrm{Adv}_G^{\mathtt{sprp}}(\theta)$ is negligibly small, G is a strong pseudorandom permutation (SPRP). If G is a tweakable cipher, tweakable SPRP and PRP are similarly defined using $\mathrm{Adv}_G^{\widetilde{\mathtt{sprp}}}(\theta)$ and $\mathrm{Adv}_G^{\widetilde{\mathtt{prp}}}(\theta)$. Generally, we say G is secure if $\mathrm{Adv}_G^{\widetilde{\mathtt{sprp}}}(\theta)$ is negligibly small.

CCA-CPA conversion. In our security proof, it is convenient to use a conversion from a $\mathtt{cca}$-advantage into a $\mathtt{cpa}$-advantage. For this purpose, we introduce a subclass of $\mathtt{cpa}$ called $\mathtt{cpa}'$, which is as follows. First, for any keyed permutation G over $\mathcal{M}$, let $\langle G \rangle : \mathcal{M} \times \varSigma \to \mathcal{M}$ be the equivalent representation of G, where $\langle G \rangle(x, 0) = G(x)$ and $\langle G \rangle(x, 1) = G^{-1}(x)$. This expression also holds true for tweakable permutations, i.e., for any tweakable permutation $\widetilde{G}$ with message space $\mathcal{M}$ and tweak space $\mathcal{T}$, $\langle \widetilde{G} \rangle$ is an equivalent keyed function $: \mathcal{M} \times \mathcal{T} \times \varSigma \to \mathcal{M}$. The LSB of a query to $\langle G \rangle$ is called a operation indicator.

Consider $F : \mathcal{M} \times \Sigma \to \mathcal{M}$ and a cpa-adversary $\mathcal{D}$ interacting with F. Let $X_i = (M_i, W_i) \in \mathcal{M} \times \Sigma$ be the i-th query of $\mathcal{D}$ and let $Y_i \in \mathcal{M}$ be the i-th answer. For any $\mathcal{D}$, we assume $M_i \neq M_j$ always holds when $W_i = W_j$ with $i < j$. Moreover, if $Y_i \neq M_j$ holds whenever $W_i \neq W_j$ holds with $i < j$, $\mathcal{D}$ is said to follow the *invertibility condition*. A cpa-adversary following the invertibility condition is called a cpa$'$-adversary. If F corresponds to $\langle G \rangle$ for a keyed permutation G, violating the invertibility condition is clearly pointless, as outputs are predictable. Thus any cca-adversary avoiding useless queries for G can be simulated by a cpa$'$-adversary interacting with $\langle G \rangle$. In other words, for any keyed permutations, E and G, we have

$$\mathrm{Adv}_{E,G}^{\mathrm{cca}}(q, \tau) = \mathrm{Adv}_{\langle E \rangle, \langle G \rangle}^{\mathrm{cpa}}(q, \tau) = \mathrm{Adv}_{\langle E \rangle, \langle G \rangle}^{\mathrm{cpa}'}(q, \tau). \tag{2}$$

In general, cpa$'$ is weaker than cpa when at least one of two target functions is not invertible. Note that, following the invertibility condition does not exclude all collisions that can not be happened for permutations. For example, if a cpa$'$-adversary is interacting with $F = \langle G \rangle$ for some keyed permutation G, $M_i \neq Y_j$ holds true for all $i < j$ with $W_i \neq W_j$ (in addition to $Y_i \neq M_j$, which is guaranteed from the invertibility condition). However $M_i = Y_j$ can happen when (e.g.) F is a URF, as Y_j is uniform and independent of X_i for all $i < j$.

2.3 Maurer's Methodology

Our security proof will be based on a methodology developed by Maurer [19]. Here, we briefly describe it. See Maurer [19] for a more detailed description. Consider a binary random variable A_i as a (non-deterministic) function of i input/output pairs (and internal variables) of a keyed function. We denote the event $A_i = 1$ by a_i, and denote $A_i = 0$ by $\overline{a_i}$. We assume a_i is monotone; i.e., a_i never occurs if $\overline{a_{i-1}}$ occurs. For instance, a_i is monotone if it indicates that all i outputs are distinct. An infinite sequence $\mathcal{A} = a_0 a_1 \ldots$ is called a *monotone event sequence* (MES). Here, a_0 is some tautological event (i.e. $A_0 = 1$ with probability 1). Note that $\mathcal{A} \wedge \mathcal{B} = (a_0 \wedge b_0)(a_1 \wedge b_1) \ldots$ is an MES if $\mathcal{A} = a_0 a_1 \ldots$ and $\mathcal{B} = b_0 b_1 \ldots$ are both MESs. For any sequence of random variables, $X_1, X_2, \ldots$, let X^i denote $(X_1, \ldots, X_i)$. Let MESs $\mathcal{A}$ and $\mathcal{B}$ be defined for two keyed functions, $F : \mathcal{X} \to \mathcal{Y}$ and $G : \mathcal{X} \to \mathcal{Y}$, respectively. Let $X_i \in \mathcal{X}$ and $Y_i \in \mathcal{Y}$ be the i-th input and output. Let P^F be the probability space defined by F. For example, $P_{Y_i | X^i Y^{i-1}}^F(y_i, x^i, y^{i-1})$ means $\Pr[Y_i = y_i | X^i = x^i, Y^{i-1} = y^{i-1}]$ where $Y_j = F(X_j)$ for $j \geq 1$. If $P_{Y_i | X^i Y^{i-1}}^F(y_i, x^i, y^{i-1}) = P_{Y_i | X^i Y^{i-1}}^G(y_i, x^i, y^{i-1})$ for all possible (x^i, y^{i-1}), then we write $P_{Y_i | X^i Y^{i-1}}^F = P_{Y_i | X^i Y^{i-1}}^G$ and denote it by $F \equiv G$. Here, note that the definitions of $\mathcal{X}$ and $\mathcal{Y}$, and the set of possible (x^i, y^{i-1}) may depend on the target attack class. Inequalities such as $P_{Y_i | X^i Y^{i-1}}^F \leq P_{Y_i | X^i Y^{i-1}}^G$ are similarly defined.

Definition 1. *We write $F^{\mathcal{A}} \equiv G^{\mathcal{B}}$ if $P^F_{Y_i a_i | X^i Y^{i-1} a_{i-1}} = P^G_{Y_i b_i | X^i Y^{i-1} b_{i-1}}$ holds[2] for all $i \geq 1$. Moreover, we write $F|\mathcal{A} \equiv G|\mathcal{B}$ if $P^F_{Y_i | X^i Y^{i-1} a_i} = P^G_{Y_i | X^i Y^{i-1} b_i}$ holds for all $i \geq 1$.*

Definition 2. *For MES $\mathcal{A}$ defined for F, $\nu_{\mathtt{atk}}(F, \overline{a_q})$ denotes[3] the maximal probability of $\overline{a_q}$ for any $\mathtt{atk}$-adversary using q queries (and infinite computational power) that interacts with F.*

The equivalences defined by Definition 1 are crucial to information-theoretic security proofs. For example, the following theorem holds true.

Theorem 1. *(Theorem 1 (i) of [19]) If $F^{\mathcal{A}} \equiv G^{\mathcal{B}}$ or $F|\mathcal{A} \equiv G$ holds for an attack class $\mathtt{atk}$, then $\mathrm{Adv}^{\mathtt{atk}}_{F,G}(q) \leq \nu_{\mathtt{atk}}(F, \overline{a_q})$.*

We will use some of Maurer's results including Theorem 1 to make simple and intuitive proofs . For completeness, these results are cited in Appendix A.

3   Previous Constructions of DBLC

There are many $O(2^{n/2})$-secure DBLC proposals. Luby and Rackoff proved that the 4-round random Feistel cipher (denoted by ψ_4) is $O(2^{n/2})$-secure. Here, "random" means that each round function is an independent n-bit block PRFs. Later, Naor and Reingold [22] proved that the first and last round functions of ψ_4 need not necessarily be pseudorandom, but only required to be ϵ-almost XOR uniform (ϵ-AXU) for sufficiently small ϵ. Here, if H is a keyed function being ϵ-AXU, we have $\Pr[H(x) \oplus H(x') = \delta] \leq \epsilon$ for any $x \neq x$ and δ. The result of [22] stimulated many related works, e.g., [23][13][26], to name a few. Above all, what inspired us was another proposal of Naor and Reingold [22], which is so-called NR mode. Basically it is an mn-bit block cipher for arbitrarily large m, using n-bit block cipher, E. When $m = 2$, NR mode encrypts a plaintext $M \in \Sigma^{2n}$ as:

$$C = G_2^{-1} \circ \mathrm{ECB}[E] \circ G_1(M), \tag{3}$$

where $\mathrm{ECB}[E]$ is the $2n$-bit permutation from ECB mode of E. G_1 and G_2 are keyed permutations called pairwise independent permutations [22]. That is, $\Pr[G_i(x) = y, G_i(x') = y'] = 1/(2^n \cdot (2^n - 1))$ for any $x \neq x'$ and $y \neq y'$, where probability is defined by the distribution of G_i's key for $i = 1, 2$.

Compared to the vast amount of $O(2^{n/2})$-secure proposals, we have very few schemes achieving better security. A scheme of Aiello and Venkatesan [1] has some beyond-birthday-bound security but not invertible. A proposal of [22] was based on unbalanced Feistel round, where each round function has inputs longer than n-bit. The $O(2^{n/2})$-security of ψ_6 was proved by Patarin [18] and another proof of ψ_r for $r \to \infty$ was given by Maurer and Pietrzak [20], though we omit the details here.

---

[2] As a_i denotes $A_i = 1$, this equality means $P^F_{Y_i A_i | X^i Y^{i-1} A_{i-1}}(y_i, 1, x^i, y^{i-1}, 1)$ equals to $P^G_{Y_i B_i | X^i Y^{i-1} B_{i-1}}(y_i, 1, x^i, y^{i-1}, 1)$ for all (x^i, y^{i-1}) such that both $P^F_{A_{i-1} X^i Y^{i-1}}(1, x^i, y^{i-1})$ and $P^G_{B_{i-1} X^i Y^{i-1}}(1, x^i, y^{i-1})$ are positive.

[3] The original definition does not contain $\mathtt{atk}$; this is for readability.

4 Building a DBLC with Beyond-Birthday-Bound Security

4.1 Extending Naor-Reingold Approach

Our goal is to build a $O(2^{\omega+n/2})$-secure DBLC, i.e., an $2n$-bit keyed permutation which is hard to be distinguished from P_{2n} against any practical CCA using $q \ll 2^{\omega+n/2}$ queries for some $\omega > 0$ (a large ω indicates a strong security). Our initial question is if we can adopt a Mix-Encrypt-Mix structure[4] similar to Eq. (3). In the following, we provide a novel solution using tweakable block ciphers. The scheme has Mix-Encrypt-Mix structure similar to NR mode, thus we call our scheme Extended Naor-Reingold (ENR)[5]. It has a parameter $m \in \{1,\ldots,n\}$, and we will prove $O(2^{(n+m)/2})$-security.

For convenience, for any random variable X, we abbreviate $X_{[1,m]}$ to $\widehat{X}$ (i.e., $\widehat{X}$ is the first m-bit of X). If $|X| = m$, we have $\widehat{X} = X$. Let $\widetilde{E}$ be an (n,m)-bit tweakable cipher, and let $\widetilde{E}_L$ and $\widetilde{E}_R$ denote two independently-keyed instances of $\widetilde{E}$. ENR consists of $\widetilde{E}_L$, $\widetilde{E}_R$, and a $2n$-bit keyed permutation, G. For plaintext $(M_l, M_r) \in \Sigma^n \times \Sigma^n$ and ciphertext $(C_l, C_r) \in \Sigma^n \times \Sigma^n$, the encryption and decryption of ENR are defined as Fig. 1.

Algorithm 4.1: $\mathrm{ENR}[G, \widetilde{E}](M_l, M_r)$

$(S,T) \leftarrow G(M_l, M_r)$
$U \leftarrow \widetilde{E}_L(S, \widehat{T}), \quad V \leftarrow \widetilde{E}_R(T, \widehat{U})$
$(C_l, C_r) \leftarrow G_{\mathrm{rev}}^{-1}(U, V)$
return $((C_l, C_r))$

Algorithm 4.2: $\mathrm{ENR}[G, \widetilde{E}]^{-1}(C_l, C_r)$

$(U,V) \leftarrow G_{\mathrm{rev}}(C_l, C_r)$
$T \leftarrow \widetilde{E}_R^{-1}(V, \widehat{U}), \quad S \leftarrow \widetilde{E}_L^{-1}(U, \widehat{T})$
$(M_l, M_r) \leftarrow G^{-1}(S, T)$
return $((M_l, M_r))$

Fig. 1. Encryption (left) and decryption (right) procedures of ENR

Here, G_{rev} denotes the mirrored image of G, i.e., $G_{\mathrm{rev}}(x) = \mathrm{rev}(G(\mathrm{rev}(x)))$ with $\mathrm{rev}(x_1,\ldots,x_{2n}) = (x_{2n},\ldots,x_1)$. We assume G_{rev} and G use the same key. Basically, we can prove the security of ENR for a more general setting where the second mixing layer is not restricted to G_{rev}. We here focus on the use of G_{rev} because it allows us to reuse the key and implementation of G.

4.2 Security Proof of ENR

To prove the security of $\mathrm{ENR}[G, \widetilde{E}]$, we first introduce a condition for G.

---

[4] Naor and Reingold's unbalanced Feistel cipher is based on Mix-Encrypt-Mix structure and achieves $O(2^{\omega+n/2})$-security. However, as it uses PRFs with input longer than n-bit, it is not comparable to ours. Moreover, an important difference is that the number of round of their scheme is depending on the security parameter (for higher security more rounds are needed), while that of ours is constant.

[5] If our scheme is realized with non-tweakable permutation (by setting $m = 0$), it will be very close to NR mode.

Definition 3. *Let G be a $2n$-bit keyed permutation. Let $m \in \{1, \ldots, n\}$ be a parameter. If G is (ϵ, γ, ρ)-almost uniform $((\epsilon, \gamma, \rho)$-AU), we have*

$$\Pr[G(x)_{[1,n+m]} = G(x')_{[1,n+m]}] \le \epsilon, \text{ and}$$
$$\Pr[G(x)_{[n+1,2n]} = G(x')_{[n+1,2n]}] \le \gamma, \text{ and}$$
$$\Pr[G(x)_{[n+1,n+m]} = G(x')_{[n+1,n+m]}] \le \rho, \text{ for any distinct } x, x' \in \Sigma^{2n}.$$

A $2n$-bit pairwise independent permutation is $(2^{-(n+m)}, 2^{-n}, 2^{-m})$-AU. Even a more efficient construction is possible by using Feistel permutation (see Corollaries 1 and 2). The security proof of general ENR is as follows.

Theorem 2. *If G is (ϵ, γ, ρ)-AU for $m \in \{1, \ldots, n\}$ and $\widetilde{E}$ is an (n, m)-bit tweakable cipher, we have*

$$\mathrm{Adv}^{\mathrm{sprp}}_{\mathrm{ENR}[G, \widetilde{E}]}(q, \tau)$$

$$\le 2\mathrm{Adv}^{\widetilde{\mathrm{sprp}}}_{\widetilde{E}}(q, \tau + O(q)) + q^2 \left(3\epsilon + \frac{2\gamma}{2^m} + \frac{\rho}{2^n} + \max\left\{\frac{\gamma}{2^m}, \frac{\rho}{2^n}\right\} + \frac{1}{2^{n+m}} \right).$$

We also provide two instantiations of ENR with Feistel-based implementations of G.

Corollary 1. *Let $m = n$ and $\psi[H]$ be a balanced $2n$-bit (left-to-right, see the left of Fig. 2) Feistel using a round function $H : \Sigma^n \to \Sigma^n$. H is defined as $H(x) = K \cdot x$, where multiplication is defined over $\mathrm{GF}(2^n)$ and key K is uniformly random over $\mathrm{GF}(2^n)$. Then we have*

$$\mathrm{Adv}^{\mathrm{sprp}}_{\mathrm{ENR}[\psi[H], \widetilde{E}]}(q, \tau) \le 2\mathrm{Adv}^{\widetilde{\mathrm{sprp}}}_{\widetilde{E}}(q, \tau + O(q)) + \frac{5q^2}{2^{2n}}.$$

Proof. When $m = n$, every $2n$-bit keyed permutation is $(0, \gamma, \rho)$-AU for some $\gamma = \rho$. The probability of $\psi[H](x)_{[n+1,\ldots,2n]} = \psi[H](x')_{[n+1,\ldots,2n]}$ is at most γ for any $x \ne x'$, if H is γ-AXU. Here, our H is 2^{-n}-AXU, thus $\psi[H]$ is $(0, 2^{-n}, 2^{-n})$-AU. Combining this fact and Theorem 2 proves the corollary.

Corollary 2. *Let $m < n$, and K_1, K_2, and K_3 be independent and uniform over $\mathrm{GF}(2^n)$ (represented as n-bit values). We define $H_1 : \Sigma^n \to \Sigma^n$ as $H_1(x) = K_1 \cdot x$, and define $H_2 : \Sigma^{n-m} \to \Sigma^{n+m}$ as $H_2(x) = (K_2 \cdot \acute{x} \| K_3 \cdot \acute{x})_{[1,\ldots,n+m]}$, where $\acute{x} = x\|0^m$. Then,*

$$\mathrm{Adv}^{\mathrm{sprp}}_{\mathrm{ENR}[\psi[H_1, H_2], \widetilde{E}]}(q, \tau) \le 2\mathrm{Adv}^{\widetilde{\mathrm{sprp}}}_{\widetilde{E}}(q, \tau + O(q)) + q^2 \left(\frac{8}{2^{n+m}} + \frac{2}{2^{2n}} \right),$$

where $\psi[H_1, H_2]$ is a 2-round Feistel permutation with i-th round function H_i (the first round is balanced and the second is unbalanced, see the right of Fig. 2).

Proof. We show that $\psi[H_1, H_2]$ is (ϵ, γ, ρ)-AU with $\epsilon = 2^{-(n+m)}$, $\gamma = 2^{-n}$, and $\rho = 2^{-n} + 2^{-m}$. The proofs for ϵ and γ are easy, as H_2 is $2^{-(n+m)}$-AXU and H_1 is 2^{-n}-AXU. To prove ρ, let $\mathcal{E}_1$ denote the collision event on $\psi[H_1](x)_{[1,\ldots,n]}$ and let $\mathcal{E}_2$ denote the collision event on $\psi[H_1, H_2](x)_{[n+1,...,n+m]}$. Here, ρ is obtained by bounding $\Pr(\mathcal{E}_2)$ for any two distinct inputs to $\psi[H_1, H_2]$, which is as follows.

$$\Pr(\mathcal{E}_2) = \Pr(\mathcal{E}_1) \cdot \Pr(\mathcal{E}_2|\mathcal{E}_1) + \Pr(\overline{\mathcal{E}_1}) \cdot \Pr(\mathcal{E}_2|\overline{\mathcal{E}_1}) \leq \Pr(\mathcal{E}_1) + \Pr(\mathcal{E}_2|\overline{\mathcal{E}_1}) \leq 2^{-n} + 2^{-m},$$

where the last inequality follows from that H_1 is 2^{-n}-AXU and $H_2(x)_{[n+1,...,n+m]}$ is 2^{-m}-AXU. Combining this observation and Theorem 2, the proof is completed.

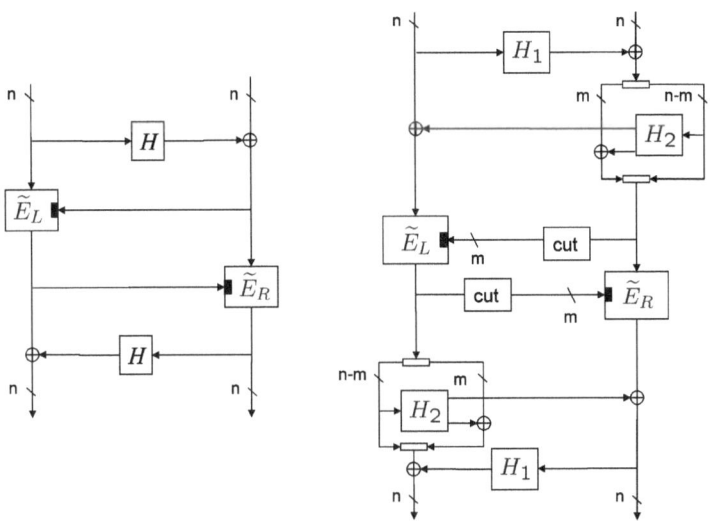

Fig. 2. Encryption of ENR. Left: the case $m = n$. Right: the case $m < n$.

4.3   Proof of Theorem 2

Setup. Let us abbreviate $\text{ENR}[G, \widetilde{\mathsf{P}}_{n,m}]$ to ENR^*, where G is (ϵ, γ, ρ)-AU. We only present the information-theoretic part, that is, the indistinguishability of ENR^* from P_{2n} against any computationally unbounded cca-adversary. The computational part is easy from the standard technique (see e.g., [2]). For convenience, we introduce some notations. For any $F : \mathcal{M} \times \Sigma \to \mathcal{M}$ with a set $\mathcal{M}$, $F[i] : \mathcal{M} \to \mathcal{M}$ is defined as $F[i](x) = F(x, i)$ for $i \in \Sigma$. For $F : \Sigma^{2n} \times \Sigma \to \Sigma^{2n}$, we define $\mathbb{G}F : \Sigma^{2n} \times \Sigma \to \Sigma^{2n}$ as

$$\mathbb{G}F(x, 0) = G_{\text{rev}}^{-1} \circ F[0] \circ G(x), \quad \text{and} \quad \mathbb{G}F(x, 1) = G^{-1} \circ F[1] \circ G_{\text{rev}}(x). \tag{4}$$

Then, $\text{DR} : \Sigma^{2n} \times \Sigma \to \Sigma^{2n}$ is defined as

$$\text{DR}((x_l, x_r), 0) = (U, \mathsf{R}_R(x_r, \widehat{U}, 0)), \quad \text{where } U = \mathsf{R}_L(x_l, \widehat{x_r}, 0),$$
$$\text{DR}((x_l, x_r), 1) = (\mathsf{R}_R(x_l, \widehat{T}, 1), T), \quad \text{where } T = \mathsf{R}_L(x_r, \widehat{x_l}, 1), \tag{5}$$

using two independent URFs, R_L, $R_R : \Sigma^n \times \Sigma^m \times \Sigma \to \Sigma^n$. Let $\widetilde{P}_L$ and $\widetilde{P}_R$ denote two independent instances of $\widetilde{P}_{n,m}$. Using them, we also define DP : $\Sigma^{2n} \times \Sigma \to \Sigma^{2n}$ in the same way as DR but R_R and R_L are substituted with $\langle \widetilde{P}_R \rangle$ and $\langle \widetilde{P}_L \rangle$, respectively. Here, note that $\mathbb{G}$DP is equivalent to $\langle \text{ENR}^* \rangle$.

The proof outline is as follows. We analyze cpa′-advantage between $\mathbb{G}$DP and $\langle P_{2n} \rangle$, which corresponds to what we want from Eq. (2). Then, using the triangle inequality, we move as $\mathbb{G}$DP $\Rightarrow$ $\mathbb{G}$DR $\Rightarrow$ $R_{2n+1,2n}$ $\Rightarrow$ $\langle P_{2n} \rangle$, that is, we evaluate the maximum cpa′-advantages for the game with $\mathbb{G}$DP and $\mathbb{G}$DR (Game 1), and the game with $\mathbb{G}$DR and $R_{2n+1,2n}$ (Game 2), and the game with $R_{2n+1,2n}$ and $\langle P_{2n} \rangle$ (Game 3). Formally, we have

$$\text{Adv}_{\text{ENR}^*}^{\text{sprp}}(q) = \text{Adv}_{\text{ENR}^*,P_{2n}}^{\text{cca}}(q) = \text{Adv}_{\langle \text{ENR}^* \rangle, \langle P_{2n} \rangle}^{\text{cpa}'} = \text{Adv}_{\mathbb{G}\text{DP}, \langle P_{2n} \rangle}^{\text{cpa}'} \tag{6}$$

$$\leq \text{Adv}_{\mathbb{G}\text{DP},\mathbb{G}\text{DR}}^{\text{cpa}'}(q) + \text{Adv}_{\mathbb{G}\text{DR},R_{2n+1,2n}}^{\text{cpa}'}(q) + \text{Adv}_{R_{2n+1,2n}, \langle P_{2n} \rangle}^{\text{cpa}'}(q). \tag{7}$$

Analysis of Game 3. By extending the well-known PRP-PRF switching lemma (e.g., Lemma 1 of [4]), we easily get

$$\text{Adv}_{R_{2n+1,2n}, \langle P_{2n} \rangle}^{\text{cpa}'}(q) \leq \binom{q}{2} \cdot \frac{1}{2^{2n}}. \tag{8}$$

Analysis of Game 2. We first observe that

$$\mathbb{G}R_{2n+1,2n} \equiv R_{2n+1,2n}, \text{ and thus } \text{Adv}_{\mathbb{G}\text{DR},R_{2n+1,2n}}^{\text{cpa}'}(q) = \text{Adv}_{\mathbb{G}\text{DR},\mathbb{G}R_{2n+1,2n}}^{\text{cpa}'}(q), \tag{9}$$

since pre- and post-processing added by $\mathbb{G}$ are permutations. We consider an adversary, $\mathcal{D}$, accessing to F which is DR or $R_{2n+1,2n}$. For each time period $i = 1, \ldots, q$, $\mathcal{D}$ can choose whether $F[0]$ or $F[1]$ is queried. This information is denoted by $W_i \in \Sigma$, and if $W_i = 0$, the input to $F[0]$ is denoted by $(SE_i, TE_i) \in \Sigma^n \times \Sigma^n$ (if $W_i = 1$, (SE_i, TE_i) is undefined), and the corresponding output is denoted by $(UE_i, VE_i) \in \Sigma^n \times \Sigma^n$. Similarly, if $W_i = 1$, the input to $F[1]$ and the output from $F[1]$ are denoted by (UD_i, VD_i) and (SD_i, TD_i), respectively (see Fig. 3). These notations will also be used for adversaries accessing to $\mathbb{G}F$. We define an MES $\mathcal{E} = e_0 e_1 \ldots$, where e_q denotes the event that

$$(SE_i, \widehat{TE}_i) \neq (SE_j, \widehat{TE}_j) \text{ and } (\widehat{UE}_i, TE_i) \neq (\widehat{UE}_j, VE_j), \text{ and} \tag{10}$$

$$(UD_i, \widehat{TD}_i) \neq (UD_j, \widehat{TD}_j) \text{ and } (\widehat{UD}_i, VD_i) \neq (\widehat{UD}_j, VD_j), \tag{11}$$

holds for all possible $i \neq j$, $i, j \in \{1, \ldots, q\}$, e.g., Eq. (10) for $i \neq j$ with $W_i = W_j = 0$. Then, we obtain the following equivalence. Its proof is in Appendix B.

$$\text{DR}^{\mathcal{E}} \equiv R_{2n+1,2n}^{\mathcal{E}}. \tag{12}$$

From Eq. (12) and Lemma 2, we have

$$\mathbb{G}\text{DR}^{\mathcal{E}} \equiv \mathbb{G}R_{2n+1,2n}^{\mathcal{E}}. \tag{13}$$

Using Eqs. (9) and (13) and Theorem 1, we obtain

$$\text{Adv}_{\mathbb{G}\text{DR},R_{2n+1,2n}}^{\text{cpa}'}(q) = \text{Adv}_{\mathbb{G}\text{DR},\mathbb{G}R_{2n+1,2n}}^{\text{cpa}'}(q) \leq \nu_{\text{cpa}'}(\mathbb{G}R_{2n+1,2n}, \overline{e_q}). \tag{14}$$

We leave the analysis of the last term of Eq. (14) for now.

Analysis of Game 1. We consider the indistinguishability between $\langle \widetilde{P}_{n,m} \rangle$ and $R_{n+m+1,n}$. We first focus on the input/output collision for the same tweak value. More precisely, let $(X_i, T_i, W_i) \in \Sigma^n \times \Sigma^m \times \Sigma$ denote the i-th input to $\langle \widetilde{P}_{n,m} \rangle$ or $R_{n+m+1,n}$, and let $Y_i \in \Sigma^n$ be the i-th output. For $b \in \Sigma$ and $t \in \Sigma^m$, let $\mathcal{X}_b^t = \{X_i : i \in \{1, \ldots, q\}, T_i = t, W_i = b\}$ and $\mathcal{Y}_b^t = \{Y_i : i \in \{1, \ldots, q\}, T_i = t, W_i = b\}$. Then, a_q denotes the event that

$$[\mathcal{X}_0^t \cap \mathcal{Y}_1^t = \emptyset] \wedge [\mathcal{X}_1^t \cap \mathcal{Y}_0^t = \emptyset] \text{ for all } t \in \Sigma^m.$$

The corresponding MES, $\mathcal{A} = a_0 a_1 \ldots$, is called the generalized collision-freeness (GCF). Then, we have

$$\langle \widetilde{P}_{n,m} \rangle^{\mathcal{A} \wedge \mathcal{C}} \equiv R_{n+m+1,n}^{\mathcal{A}}, \text{ for some MES } \mathcal{C}. \tag{15}$$

The proof of Eq.(15) is written in Appendix C. As mentioned, if we substitute R_L and R_R with $\langle \widetilde{P}_L \rangle$ and $\langle \widetilde{P}_R \rangle$, we will obtain DP. Thus, from Eq. (15), we get

$$DP^{\mathcal{AL} \wedge \mathcal{CL} \wedge \mathcal{AR} \wedge \mathcal{CR}} \equiv DR^{\mathcal{AL} \wedge \mathcal{AR}}, \text{ and } \mathbb{G}DP^{\mathcal{AL} \wedge \mathcal{CL} \wedge \mathcal{AR} \wedge \mathcal{CR}} \equiv \mathbb{G}DR^{\mathcal{AL} \wedge \mathcal{AR}}, \tag{16}$$

where $\mathcal{AL} = al_0 al_1 \ldots$ denotes the GCF for $\langle \widetilde{P}_L \rangle$ or R_L, and $\mathcal{AR} = ar_0 ar_1 \ldots$ denotes the GCF for $\langle \widetilde{P}_R \rangle$ or R_R, and $\mathcal{CL}$ and $\mathcal{CR}$ are some MESs (implied by Eq. (15)). The second equivalence follows from Lemma 2. Thus, using Theorem 1 we obtain

$$Adv_{\mathbb{G}DP,\mathbb{G}DR}^{cpa'}(q) \leq \nu_{cpa'}(\mathbb{G}DR, \overline{al_q \wedge ar_q}). \tag{17}$$

For DR and $R_{2n+1,2n}$, the occurrence of $al_q \wedge ar_q$ can be completely determined by the q inputs and outputs. From this fact and Lemma 5, we can *adjoin* $\mathcal{AL} \wedge \mathcal{AR}$ to the both sides of Eq. (12) and obtain

$$DR^{\mathcal{E} \wedge \mathcal{AL} \wedge \mathcal{AR}} \equiv R_{2n+1,2n}^{\mathcal{E} \wedge \mathcal{AL} \wedge \mathcal{AR}}. \tag{18}$$

Moreover, it is easy to see that $\mathcal{E} \wedge \mathcal{AL} \wedge \mathcal{AR} \equiv \mathcal{AL} \wedge \mathcal{AR}$ holds for DR and $R_{2n+1,2n}$. Combining this observation, Eq. (18), and Lemmas 2 and 3, we have

$$\mathbb{G}DR^{\mathcal{AL} \wedge \mathcal{AR}} \equiv \mathbb{G}R_{2n+1,2n}^{\mathcal{AL} \wedge \mathcal{AR}}, \text{ and}$$

$$Adv_{\mathbb{G}DP,\mathbb{G}DR}^{cpa'}(q) \leq \nu_{cpa'}(\mathbb{G}DR, \overline{al_q \wedge ar_q}) = \nu_{cpa'}(\mathbb{G}R_{2n+1,2n}, \overline{al_q \wedge ar_q}). \tag{19}$$

Collision Probability analysis. Combining Eqs. (7), (8), (14), and (19), and Lemma 4, we have

$$Adv_{ENR^*}^{sprp}(q) \leq \binom{q}{2} \frac{1}{2^{2n}} + \sum_{ev = \overline{e_q}, \overline{al_q}, \overline{ar_q}} \nu_{cpa'}(\mathbb{G}R_{2n+1,2n}, ev). \tag{20}$$

We need to bound $\nu_{cpa'}$ terms of Eq. (20). First, the maximum probabilities of $\overline{al_q}$ and $\overline{ar_q}$ (under $\mathbb{G}R_{2n+1,2n}$) are the same because of the symmetry from Eq. (4)

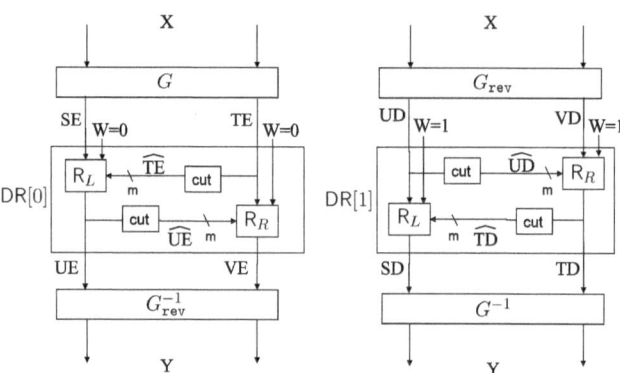

Fig. 3. $\mathbb{G}$DR function

and that G_{rev} is a mirrored image of G. Thus we only need to evaluate the maximum probabilities of $\overline{al}_q$ and $\overline{e}_q$.

As shown by Eqs. (10) and (11), $\overline{e_q}$ consists of collision events such as

$$\mathrm{type}[\mathbf{e}_1] : (SE_i, \widehat{TE}_i) = (SE_j, \widehat{TE}_j), \qquad \mathrm{type}[\mathbf{e}_2] : (\widehat{UE}_i, TE_i) = (\widehat{UE}_j, TE_j)$$
$$\mathrm{type}[\mathbf{e}_3] : (\widehat{UD}_i, VD_i) = (\widehat{UD}_j, VD_j), \quad \mathrm{type}[\mathbf{e}_4] : (UD_i, \widehat{TD}_i) = (UD_j, \widehat{TD}_j)$$

for all possible $i \neq j$, $i, j \in \{1, \ldots, q\}$. Moreover, $\overline{al}_q$ consists of collision events such as

$$\mathrm{type}[\mathbf{a}_1] : (SE_i, \widehat{TE}_i) = (SE_j, \widehat{TE}_j), \qquad \mathrm{type}[\mathbf{a}_2] : (UE_i, \widehat{TE}_i) = (UE_j, \widehat{TE}_j)$$
$$\mathrm{type}[\mathbf{a}_3] : (UD_i, \widehat{TD}_i) = (UD_j, \widehat{TD}_j), \qquad \mathrm{type}[\mathbf{a}_4] : (SD_i, \widehat{TD}_i) = (SD_j, \widehat{TD}_j)$$
$$\mathrm{type}[\mathbf{a}_5] : (SE_i, \widehat{TE}_i) = (SD_j, \widehat{TD}_j), \qquad \mathrm{type}[\mathbf{a}_6] : (UE_i, \widehat{TE}_i) = (UD_j, \widehat{TD}_j).$$

Note that $\mathrm{type}[\mathbf{a}_1]$ and $\mathrm{type}[\mathbf{a}_3]$ are the same as $\mathrm{type}[\mathbf{e}_1]$ and $\mathrm{type}[\mathbf{e}_4]$, respectively. Let $\Pr[\mathbf{x}]$ be the maximum probability of $\mathrm{type}[\mathbf{x}]$-collision for $\mathbf{x} \in \{\mathbf{e}_1, \ldots, \mathbf{e}_4, \mathbf{a}_1, \ldots, \mathbf{a}_6\}$ under $\mathbb{G}R_{2n+1,2n}$, where the maximum is taken for all q-cpa' (possibly adaptive) adversaries and for all $i, j \in \{1, \ldots, q\}$ with $i \neq j$. Using the union bound and the symmetry of $\mathbb{G}R_{2n+1,2n}[0]$ and $\mathbb{G}R_{2n+1,2n}[1]$, the R.H.S. of Eq. (20) is at most

$$\binom{q}{2} \left(\frac{1}{2^{2n}} + \sum_{i=1,\ldots4} \Pr[\mathbf{e}_i] + 2 \sum_{j=1,\ldots6} \Pr[\mathbf{a}_j] \right). \tag{21}$$

From Eq. (9), the adversary's choice must be independent of (the key of) G and G_{rev}. With this fact, each collision probability of Eq. (21) can be easily bounded for any cpa'-adversary if G is (ϵ, γ, ρ)-AU^6. The full description of our analysis

---
[6] Note that if G is (ϵ, γ, ρ)-AU, the mirrored image of G_{rev} is also (ϵ, γ, ρ)-AU.

is rather long[7], thus we here describe some typical examples. Other collision probabilities can be analyzed in a similar way. Let XE_i and YE_i be the i-th $2n$-bit input and output of $\mathbb{GR}_{2n+1,2n}$ with $W_i = 0$ (they are undefined for i satisfying $W_i = 1$). Similarly, XD_i and YD_i denote the i-th input and output with $W_i = 1$.

- type[e_1]. Here, a collision means $G(XE_i)_{[1,\dots,n+m]} = G(XE_j)_{[1,\dots,n+m]}$ for $XE_i \neq XE_j$. Moreover, XE_i and XE_j are independent of G's key. Thus we have $\Pr[e_1] \leq \epsilon$ as G is assumed to be (ϵ, γ, ρ)-AU.
- type[e_2]. Without loss of generality, we assume $i < j$. The probability of $TE_i = TE_j$ is at most γ, as G is (ϵ, γ, ρ)-AU. Since G is invertible, the inputs to $R_{2n+1,2n}[0]$ are always distinct. This implies that $\widehat{UE}_j$ is independent of previous variables (including $\widehat{UE}_i$, TE_i and TE_j) and uniform, even conditioned by the event $TE_i = TE_j$. Thus we get

$$\Pr[e_2] = \max_{i \neq j} \Pr[\widehat{UE}_i = \widehat{UE}_j | TE_i = TE_j] \cdot \Pr[TE_i = TE_j] \leq 2^{-m}\gamma. \quad (22)$$

- type[a_5]. When $i < j$, $(SD_j, \widehat{TD}_j)$ is uniform and independent of $(SE_i, \widehat{TE}_i)$, thus collision probability is exactly $2^{-(n+m)}$. When $j < i$, $XE_i \neq YD_j$ must hold as we consider cpa$'$-adversary (i.e., $XE_i = YD_j$ for $j < i$ means an intentional invertibility check). Hence

$$\Pr[(SE_i, \widehat{TE}_i) = (SD_j, \widehat{TD}_j)]$$
$$= \Pr[G(XE_i)_{[1,\dots,n+m]} = G(YD_j)_{[1,\dots,n+m]}] \leq \epsilon. \quad (23)$$

Thus we have $\Pr[a_5] \leq \max\{2^{-(n+m)}, \epsilon\}$. Here, $\epsilon \geq 2^{-(n+m)}$ as it is the collision probability over $(n + m)$ bits.

In summary, we obtain all maximum collision probabilities:

$$\Pr[e_1] \leq \epsilon, \quad \Pr[e_2] \leq 2^{-m}\gamma, \quad \Pr[e_3] \leq \epsilon, \quad \Pr[e_4] \leq 2^{-m}\gamma,$$
$$\Pr[a_1] \leq \epsilon, \quad \Pr[a_2] \leq 2^{-n}\rho, \quad \Pr[a_3] \leq 2^{-m}\gamma,$$
$$\Pr[a_4] \leq 2^{-(n+m)}, \quad \Pr[a_5] \leq \epsilon, \quad \Pr[a_6] \leq \max\{2^{-m}\gamma, 2^{-n}\rho\}. \quad (24)$$

Combining Eqs. (24) and (21) proves the theorem.

4.4   PRP and PRF Versions of ENR

Although our primary target is a DBLC secure against CCA, a slight simplification of our proposal yields a CPA-secure variant of ENR. It saves some operations from the original ENR at the cost of a weaker attack class.

Definition 4. *The simplified ENR (sENR) is defined as ENR with G_{rev} being omitted (or, substituted with the identity function).*

---

[7] This is mainly because we have to think of the cases when the adversary's choice is adaptive, even though it is independent of G.

Corollary 3. *Let G be (ϵ, γ, ρ)-AU. Then, the* cpa-*security of* $\mathrm{sENR}[G, \widetilde{E}]$ *is:*

$$\mathrm{Adv}^{\mathrm{prp}}_{\mathrm{sENR}[G,\widetilde{E}]}(q, \tau) \leq 2\mathrm{Adv}^{\widetilde{\mathrm{prp}}}_{\widetilde{E}}(q, \tau + O(q)) + q^2 \left(\epsilon + \frac{\gamma}{2^m} + \frac{\rho}{2^{n+1}} + \frac{1}{2^{n+m}}\right).$$

The proof is similar to that of Theorem 2, thus is omitted here. The reason why G_{rev} can be omitted is that, we do not have to consider some bad events (e.g., the collision of $(\widehat{UD}, VD)$) that have to be avoided by G_{rev} when decryption query is possible. Moreover, by truncating the rightmost n-bit output, we obtain a PRF : $\Sigma^{2n} \to \Sigma^n$ which is $O(2^{(n+m)/2})$-secure for any $m = 1, \ldots, n$ (the proof is trivial from Corollary 3). We emphasize that ENR, and the simplified ENR, and its truncated-output version are optimally efficient, for they need exactly c calls of $\widetilde{E}$ when the output is cn bits, for $c = 1, 2$.

5   A Simple Construction of Tweakable Block Cipher with Beyond-Birthday-Bound Security

Our proposal requires a tweakable block cipher with beyond-birthday-bound security. Then, one may naturally ask how to realize it. A straightforward approach is building from scratch, e.g., Mercy [9] and HPC [8]. Recent studies [10][21] demonstrated that adding a tweak to some internal variables of a (generalized) Feistel cipher could yield a secure tweakable block cipher. This technique, called direct tweaking, may well be applied to a concrete tweakable cipher using (e.g.) S-box and linear diffusion. Another approach, which we focus on, is building from ordinary block ciphers. There are several schemes [16][24] that turn an n-bit block cipher into an (n, n)-bit tweakable cipher. However, they only have $O(2^{n/2})$-security[8]. Building a tweakable block cipher with better security has been considered as rather difficult (Liskov et al. mentioned it as an open problem [16]).

Our solution is simple and intuitive: changing keys depending on tweaks. This idea was possibly in mind of [16]. However, to our knowledge it has not been seriously investigated[9]. Although our scheme is simple, its security proof needs some cares. Throughout this section, we occasionally write E_K instead of E, if we need to specify the key we use.

Definition 5. *For $E_K : \Sigma^n \to \Sigma^n$ with key $K \in \mathcal{K}$ and $F_{MK} : \Sigma^m \to \mathcal{K}$ with key $MK \in \mathcal{K}'$, Tweak-dependent Rekeying (TDR) is an (n, m)-bit tweakable cipher, where its encryption is $\mathrm{TDR}[E, F](x, t) = E_{F_{MK}(t)}(x)$, and decryption is $\mathrm{TDR}[E, F]^{-1}(y, t) = E^{-1}_{F_{MK}(t)}(y)$. Here, the key of $\mathrm{TDR}[E, F]$ is F's key, MK.*

---

[8] In [10], tweakable ciphers having "security against exponential attacks" are proposed. They correspond to $(2n, m)$-bit tweakable ciphers with $O(2^n)$-security, thus their security is up to the birthday bound of the block size.

[9] Liskov et al., said that a change of a tweak should be faster than a change of a key. This requirement is certainly desirable, however not mandatory one.

Theorem 3. $\mathtt{Adv}^{\widetilde{\mathtt{sprp}}}_{\mathtt{TDR}[E,F]}(q,\tau) \leq \mathtt{Adv}^{\mathtt{prf}}_{F}(\eta,\tau+O(q)) + \eta\mathtt{Adv}^{\mathtt{sprp}}_{E}(q,\tau+O(q))$, where $\eta \overset{\mathrm{def}}{=} \min\{q,2^m\}$.

Proof. Let $\mathsf{R}: \Sigma^m \to \mathcal{K}$ be the URF. We have

$$\mathtt{Adv}^{\widetilde{\mathtt{sprp}}}_{\mathtt{TDR}[E,F]}(q,\tau) \leq \mathtt{Adv}^{\widetilde{\mathtt{cca}}}_{\mathtt{TDR}[E,F],\mathtt{TDR}[E,\mathsf{R}]}(q,\tau) + \mathtt{Adv}^{\widetilde{\mathtt{sprp}}}_{\mathtt{TDR}[E,\mathsf{R}]}(q,\tau). \qquad (25)$$

The first term of R.H.S. for Eq. (25) is clearly at most $\mathtt{Adv}^{\mathtt{prf}}_{F}(\eta,\tau+O(q))$, as we can evaluate F or R on at most η points. For the second term, the adversary can produce at most η instances of E, and their keys are independent and uniform (as keys are generated from URF). For each sampled key, the adversary can query at most q times[10]. Thus, the second term is at most $\eta\mathtt{Adv}^{\mathtt{sprp}}_{E}(q,\tau+O(q))$ from the triangle inequality.

At first glance, TDR seems to provide a desirable security, since it simulates the tweakable URP in an intuitive way. However, this is not always the case. For example, when $\mathcal{K} = \Sigma^n$ and $m = n$, a simple attack using about $2^{n/2}$ queries can easily distinguish TDR from $\widetilde{\mathsf{P}}_{n,n}$: we first query a fixed plaintext with many distinct tweaks, and if a ciphertext collision is found for tweak t and t', then query a new plaintext with tweaks t and t' and see if the new ciphertexts collide again[11].

Nevertheless, this scheme can have beyond-birthday-bound security if tweak length is not longer than the half of block length:

Corollary 4. Let E_K be an n-bit block cipher with key $K \in \Sigma^n$. For $m < n/2$, let $E'': \Sigma^m \to \Sigma^n$ be defined as $E''(x) = E(x\|0^{n-m})$. Then $\mathtt{Adv}^{\widetilde{\mathtt{sprp}}}_{\mathtt{TDR}[E,E'']}(q,\tau)$ is at most $(\eta+1)\mathtt{Adv}^{\mathtt{sprp}}_{E}(q,\tau+O(q)) + \eta^2/2^{n+1}$, where $\eta = \min\{q,2^m\}$.

Here, $\mathtt{TDR}[E,E'']$ is secure if $2^{-(n-2m)}$ is sufficiently small and E is computationally secure, where "secure" means cca-advantage being much smaller than 2^{-m}. Unfortunately, Corollary 4 does not tell us how large q is admissible by itself, since the first term of the bound would not be negligible if q is large. Nonetheless, as the first term is at least $\eta\tau + O(q)/2^n \approx q/2^{n-m}$ when $q \geq 2^m$ (achieved by the exhaustive key search, see [3]), we expect that $\mathtt{TDR}[E,E'']$ is computationally secure against attacks with $q \ll 2^{n-m}$ queries, if E is sufficiently secure.

Practically, the big problem of TDR is the frequent key scheduling of E, as it may be much slower than encryption. Still, the negative impact on speed could be alleviated when on-the-fly key scheduling is possible.

Combining ENR and TDR. A combination of ENR and TDR provides a DBLC using an n-bit block cipher E. Let us consider combining the schemes

---

[10] A more elaborate analysis can significantly improve the second term of the bound. However, it requires some additional parameters to describe the adversary's strategy and thus the result would look rather complicated. We here make it simple.

[11] This does not contradict with Theorem 3: the second term of the bound is at least $\eta(\tau + O(q))/|\mathcal{K}|$, which is about $q^2/2^n$ when $|\mathcal{K}| = 2^n$, as pointed out by Bellare et al. [3].

from Corollaries 2 and 4. The resulting DBLC needs 4 calls of E and two key schedulings of E. By assuming $\mathtt{Adv}_E^{\mathtt{sprp}}(q, \tau) \approx q/2^n$, the security bound of this DBLC is about $2q/2^{n-m} + 8q^2/2^{(n+m)/2} + 2q^2/2^{2n} + 1/2^{n-2m}$. Then the choice $m \approx n/3$ achieves the security against $q \ll 2^{2n/3}$ queries for fixed n, which is the best possible for this combination. For example, if we use AES (i.e., $n = 128$) and set $m = 42$, the combined scheme's security is about 83.5-bit, assuming AES's security. Compared to the previous DBLCs having 64-bit security, the gain is not that large, but non-negligible. Of course, the security and efficiency of the resulting ENR would be greatly improved by using a better AES-based tweakable block cipher.

6   Conclusion

We described the extended Naor-Reingold (ENR), which converts an n-bit block tweakable block cipher into a $2n$-bit block cipher. ENR has the beyond-birthday-bound security (for n) if underlying tweakable block cipher does, and has almost the same throughput as that of the tweakable block cipher. Hence, we have shown that a good (i.e., fast and secure) tweakable cipher implies a good double-block-length cipher. We also described a way to convert an n-bit block cipher into tweakable one and achieves beyond-birthday-bound security based on the computational indistinguishability of the underlying block cipher. Unfortunately, this scheme has both theoretical and practical drawbacks due to its frequent rekeying. Thus, finding an efficient scheme without rekeying would be an important open problem.

Future Directions. It would be interesting to extend ENR to mn-bit block cipher for $m > 2$ and make ENR tweakable, keeping beyond-birthday-bound security for n. Both problems can be basically solved by using ENR as a module of some known block cipher modes (e.g., CMC mode [12]) as they have $O(2^n)$-security with $2n$-bit pseudorandom permutation. However, more efficient constructions may well be possible.

Acknowledgments

We would like to thank Thomas Ristenpart, Tetsu Iwata, and the anonymous referees for very helpful comments and suggestions. We also thank Debra L. Cook for suggesting references.

References

1. Aiello, W., Venkatesan, R.: Foiling Birthday Attacks in Length-Doubling Transformations - Benes: A Non-Reversible Alternative to Feistel. In: Maurer, U.M. (ed.) EUROCRYPT 1996. LNCS, vol. 1070, pp. 307–320. Springer, Heidelberg (1996)
2. Bellare, M., Desai, A., Jokipii, E., Rogaway, P.: A Concrete Security Treatment of Symmetric Encryption. In: Proceedings of the 38th Annual Symposium on Foundations of Computer Science, FOCS 1997, pp. 394–403 (1997)

3. Bellare, M., Krovetz, T., Rogaway, P.: Luby-Rackoff Backwards: Increasing Security by Making Block Ciphers Non-invertible. In: Nyberg, K. (ed.) EUROCRYPT 1998. LNCS, vol. 1403, pp. 266–280. Springer, Heidelberg (1998)
4. Bellare, M., Rogaway, P.: The Security of Triple Encryption and a Framework for Code-Based Game-Playing Proofs. In: Vaudenay, S. (ed.) EUROCRYPT 2006. LNCS, vol. 4004, pp. 409–426. Springer, Heidelberg (2006)
5. Bernstein, D.J.: Stronger Security Bounds for Wegman-Carter-Shoup Authenticators. In: Cramer, R. (ed.) EUROCRYPT 2005. LNCS, vol. 3494, pp. 164–180. Springer, Heidelberg (2005)
6. Bertoni, G., Breveglieri, L., Fragneto, P., Macchetti, M., Marchesin, S.: Efficient Software Implementation of AES on 32-Bit Platforms. In: Kaliski Jr., B.S., Koç, Ç.K., Paar, C. (eds.) CHES 2002. LNCS, vol. 2523, pp. 159–171. Springer, Heidelberg (2003)
7. Carter, L., Wegman, M.: Universal Classes of Hash Functions. Journal of Computer and System Science 18, 143–154 (1979)
8. Schroeppel, R.: Hasty Pudding Cipher (1998),
   http://www.cs.arizona.edu/rcs/hpc
9. Crowley, P.: Mercy: A Fast Large Block Cipher for Disk Sector Encryption. In: Schneier, B. (ed.) FSE 2000. LNCS, vol. 1978, pp. 49–63. Springer, Heidelberg (2000)
10. Goldenberg, D., Hohenberger, S., Liskov, M., Schwartz, E.C., Seyalioglu, H.: On Tweaking Luby-Rackoff Blockciphers. In: Kurosawa, K. (ed.) ASIACRYPT 2007. LNCS, vol. 4833, pp. 342–356. Springer, Heidelberg (2007)
11. Goldreich, O.: Modern Cryptography, Probabilistic Proofs and Pseudorandomness. Algorithms and Combinatorics, vol. 17. Springer, Heidelberg (1998)
12. Halevi, S., Rogaway, P.: A Tweakable Enciphering Mode. In: Boneh, D. (ed.) CRYPTO 2003. LNCS, vol. 2729, pp. 482–499. Springer, Heidelberg (2003)
13. Iwata, T., Yagi, T., Kurosawa, K.: On the Pseudorandomness of KASUMI Type Permutations. In: Safavi-Naini, R., Seberry, J. (eds.) ACISP 2003. LNCS, vol. 2727. Springer, Heidelberg (2003)
14. Iwata, T., Kurosawa, K.: OMAC: One-Key CBC MAC. In: Johansson, T. (ed.) FSE 2003. LNCS, vol. 2887, pp. 129–153. Springer, Heidelberg (2003)
15. Iwata, T.: New Blockcipher Modes of Operation with Beyond the Birthday Bound Security. In: Robshaw, M.J.B. (ed.) FSE 2006. LNCS, vol. 4047, pp. 310–327. Springer, Heidelberg (2006)
16. Liskov, M., Rivest, R.L., Wagner, D.: Tweakable Block Ciphers. In: Yung, M. (ed.) CRYPTO 2002. LNCS, vol. 2442, pp. 31–46. Springer, Heidelberg (2002)
17. Luby, M., Rackoff, C.: How to Construct Pseudo-random Permutations from Pseudo-random functions. SIAM J. Computing 17(2), 373–386 (1988)
18. Patarin, J.: Security of Random Feistel Schemes with 5 or More Rounds. In: Franklin, M. (ed.) CRYPTO 2004. LNCS, vol. 3152, pp. 106–122. Springer, Heidelberg (2004)
19. Maurer, U.: Indistinguishability of Random Systems. In: Knudsen, L.R. (ed.) EUROCRYPT 2002. LNCS, vol. 2332, pp. 110–132. Springer, Heidelberg (2002)
20. Maurer, U., Pietrzak, K.: The Security of Many-Round Luby-Rackoff Pseudo-Random Permutations. In: Biham, E. (ed.) EUROCRYPT 2003. LNCS, vol. 2656, pp. 544–561. Springer, Heidelberg (2003)
21. Mitsuda, A., Iwata, T.: Tweakable Pseudorandom Permutation from Generalized Feistel Structure. In: Baek, J., Bao, F., Chen, K., Lai, X. (eds.) ProvSec 2008. LNCS, vol. 5324, pp. 22–37. Springer, Heidelberg (2008)

22. Naor, M., Reingold, O.: On the Construction of Pseudorandom Permutations: Luby-Rackoff Revisited. Journal of Cryptology 12(1), 29–66 (1999)
23. Patel, S., Ramzan, Z., Sundaram, G.: Towards Making Luby-Rackoff Ciphers Optimal and Practical. In: Knudsen, L.R. (ed.) FSE 1999. LNCS, vol. 1636, pp. 171–185. Springer, Heidelberg (1999)
24. Rogaway, P.: Efficient Instantiations of Tweakable Blockciphers and Refinements to Modes OCB and PMAC. In: Lee, P.J. (ed.) ASIACRYPT 2004. LNCS, vol. 3329, pp. 16–31. Springer, Heidelberg (2004), http://www.cs.ucdavis.edu/~rogaway/papers/offsets.pdf
25. Krovetz, T.: Message Authentication on 64-Bit Architectures. In: Biham, E., Youssef, A.M. (eds.) SAC 2006. LNCS, vol. 4356, pp. 327–341. Springer, Heidelberg (2007)
26. Vaudenay, S.: On the Lai-Massey scheme. In: Lam, K.-Y., Okamoto, E., Xing, C. (eds.) ASIACRYPT 1999. LNCS, vol. 1716, pp. 8–19. Springer, Heidelberg (1999)
27. Wegman, M., Carter, L.: New Hash Functions and Their Use in Authentication and Set Equality. Journal of Computer and System Sciences 22, 265–279 (1981)
28. Yasuda, K.: A One-Pass Mode of Operation for Deterministic Message Authentication- Security beyond the Birthday Barrier. In: Nyberg, K. (ed.) FSE 2008. LNCS, vol. 5086, pp. 316–333. Springer, Heidelberg (2008)

A   Lemmas from Maurer's Methodology

We describe some lemmas developed by Maurer [19] that we used in this paper. We assume that F and G are two random functions with the same input/output size; we define MESs $\mathcal{A} = a_0 a_1 \ldots$ and $\mathcal{B} = b_0 b_1 \ldots$ for F and G. The i-th input and output are denoted by X_i and Y_i for F (or G), respectively.

Lemma 1. *(Lemma 1 (iv) of [19]) If $F|\mathcal{A} \equiv G|\mathcal{B}$ and*
$$P^F_{a_i|X^iY^{i-1}a_{i-1}} \le P^G_{b_i|X^iY^{i-1}b_{i-1}}$$ *holds for $i \ge 1$, then there exists an MES $\mathcal{C}$ defined for G such that $F^{\mathcal{A}} \equiv G^{\mathcal{B} \wedge \mathcal{C}}$.*

Lemma 2. *(Lemma 4 (ii) of [19]) Let $\mathbb{F}$ be the function of F and G (i.e., $\mathbb{F}[F]$ is a function that internally invokes F, possibly multiple times, to process its inputs). Here, $\mathbb{F}$ can be probabilistic, and if so, we assume $\mathbb{F}$ is independent of F or G. If $F^{\mathcal{A}} \equiv G^{\mathcal{B}}$ holds, $\mathbb{F}[F]^{\mathcal{A}'} \equiv \mathbb{F}[G]^{\mathcal{B}'}$ also holds, where a'_i denotes an event that $\mathcal{A}$-event is satisfied for the time period i. For example, if $\mathbb{F}[F]$ always invoke F c times for any input, then $a'_i = a_{ci}$. $\mathcal{B}'$ is defined in the same way.*

Lemma 3. *(Lemma 6 (ii) of [19]) If $F^{\mathcal{A}} \equiv G^{\mathcal{B}}$ holds for the attack class atk, then $\nu_{\mathrm{atk}}(F, \overline{a_q}) = \nu_{\mathrm{atk}}(G, \overline{b_q})$ holds.*

Lemma 4. *(Lemma 6 (iii) of [19]) $\nu_{\mathrm{atk}}(F, \overline{a_q \wedge b_q}) \le \nu_{\mathrm{atk}}(F, \overline{a_q}) + \nu_{\mathrm{atk}}(F, \overline{b_q})$.*

Lemma 5. *(An extension of Lemma 2 (ii) of [19]) If $F^{\mathcal{A}} \equiv G^{\mathcal{B}}$, then $F^{\mathcal{A} \wedge \mathcal{C}} \equiv G^{\mathcal{B} \wedge \mathcal{C}}$ holds for any MES $\mathcal{C}$ defined on the inputs and/or outputs.*

B   Proof of Equation (12)

We focus on the indistinguishability between DR[0] and $R_{2n,2n}$. Let $\text{dist}(X^i, Y^j)$ denote an event that there is no collision among $\{X_1, \ldots, X_i, Y_1, \ldots, Y_j\}$. Let i-th input (to DR[0] or $R_{2n,2n}$) be $X_i \overset{\text{def}}{=} (SE_i, TE_i) \in \Sigma^n \times \Sigma^n$, and i-th output be $Y_i \overset{\text{def}}{=} (UE_i, VE_i) \in \Sigma^n \times \Sigma^n$. We define events $il_q \overset{\text{def}}{=} \text{dist}(SE^q, \widehat{TE^q})$ and $ir_q \overset{\text{def}}{=} \text{dist}(\widehat{UE^q}, TE^q)$ and the corresponding MESs, $\mathcal{IL}$ and $\mathcal{IR}$. For DR[0], let us analyze the conditional probability of $\widehat{Y_q}$ (which equals to $\widehat{UE_q}$), given $X^q = x^q$, $Y^{q-1} = y^{q-1}$ and $il_q \wedge ir_q$. Note that the inputs to $R_L[0]$ are distinct from il_q, which means $\widehat{Y^q}$ are independent and uniform. However, if $te_i = te_j$ for some $i \neq j$, $ue_i \neq ue_j$ must hold from ir_q. Thus, $\widehat{Y_q}$ is uniform over $\widehat{\mathcal{Y}}^c = \{0,1\}^m \setminus \widehat{\mathcal{Y}}$, where $\widehat{\mathcal{Y}} \overset{\text{def}}{=} \{ue_i : te_i = te_q, i = 1, \ldots, q-1\}$. The remaining $(2n - m)$ bits of Y_q are uniform over Σ^{2n-m} from il_q and ir_q. For $R_{2n,2n}$, the set $\widehat{\mathcal{Y}}$ can be defined in the same way and Y_q (given $X^q = x^q$, $Y^{q-1} = y^{q-1}$ and $il_q \wedge ir_q$) is clearly uniformly distributed over $\widehat{\mathcal{Y}}^c \times \Sigma^{2n-m}$. Thus we have

$$P_{Y_q \mid X^q Y^{q-1} il_q ir_q}^{\text{DR}[0]} = P_{Y_q \mid X^q Y^{q-1} il_q ir_q}^{R_{2n,2n}}. \tag{26}$$

Next, we see that

$$P_{il_q ir_q \mid X^q Y^{q-1} il_{q-1} ir_{q-1}}^{\text{DR}[0]}(x^q, y^{q-1}) = \begin{cases} 0 & \text{if } il_q \text{ is contradicted by } x^q, \\ \dfrac{|\widehat{\mathcal{Y}}^c|}{2^m} & \text{otherwise.} \end{cases} \tag{27}$$

holds true, as il_q is a function of x^q and the conditional probability of ir_q given $X^q = x^q$, $Y^{q-1} = y^{q-1}$, and $il_q \wedge b_{q-1}$ is the probability of $\widehat{Y_q} \notin \widehat{\mathcal{Y}}$, where $\widehat{Y_q}$ is uniform given il_q. Therefore, the conditional probability of ir_q is $|\widehat{\mathcal{Y}}^c|/2^m$ when x^q satisfies il_q. The same analysis also holds for $R_{2n,2n}$. Thus we have

$$P_{il_q ir_q \mid X^q Y^{q-1} il_{q-1} ir_{q-1}}^{\text{DR}[0]} = P_{il_q ir_q \mid X^q Y^{q-1} il_{q-1} ir_{q-1}}^{R_{2n,2n}} \tag{28}$$

From Eqs. (26) and (28),

$$\text{DR}[0]^{\mathcal{IL} \wedge \mathcal{IR}} \equiv R_{2n,2n}^{\mathcal{IL} \wedge \mathcal{IR}}, \text{ and } \text{DR}[1]^{\mathcal{IL}' \wedge \mathcal{IR}'} \equiv R_{2n,2n}^{\mathcal{IL}' \wedge \mathcal{IR}'} \tag{29}$$

is obtained. The latter equivalence is derived by symmetry, where $\mathcal{IL}'$ and $\mathcal{IR}'$ are defined by $il_q' \overset{\text{def}}{=} \text{dist}(UD^q, \widehat{TD^q})$ and $ir_q' \overset{\text{def}}{=} \text{dist}(\widehat{UD^q}, VD^q)$ (here, i-th input is $X_i = (UD_i, VD_i)$ and output is $Y_i = (SD_i, TD_i)$). From Eq. (29) and the independence of DR[0] and DR[1], the proof is completed.

C   Proof of Equation (15)

We abbreviate $R_{n+m+1,n}$ and $\widetilde{P}_{n,m}$ to R and P, respectively. From the definition of GCF event a_q, it is easy to derive

$$\langle \widetilde{P} \rangle \mid \mathcal{A} \equiv R \mid \mathcal{A}. \tag{30}$$

Next, we have

$$P^{\langle\widetilde{\mathsf{P}}\rangle}_{a_q|X^qY^{q-1}a_{q-1}}(x^q, y^{q-1}) = 1, \text{ and } P^{\mathsf{R}}_{a_q|X^qY^{q-1}a_{q-1}}(x^q, y^{q-1}) = 1 - \frac{\theta}{2^n}, \quad (31)$$

for any x_q consistent with a_q (given x^{q-1} and y^{q-1}), since $\langle\widetilde{\mathsf{P}}\rangle$'s output always keeps GCF, while R's output is uniform and thus has a chance to violate GCF. From Eq. (31), we get $P^{\mathsf{R}}_{a_q|X^qY^{q-1}a_{q-1}} \leq P^{\langle\widetilde{\mathsf{P}}\rangle}_{a_q|X^qY^{q-1}a_{q-1}}$. The proof is completed by combining this inequality, and Eq. (30), and Lemma 1.

Enhanced Target Collision Resistant Hash Functions Revisited*

Mohammad Reza Reyhanitabar, Willy Susilo, and Yi Mu

Centre for Computer and Information Security Research,
School of Computer Science and Software Engineering
University of Wollongong, Australia
{rezar, wsusilo, ymu}@uow.edu.au

Abstract. Enhanced Target Collision Resistance (eTCR) property for a hash function was put forth by Halevi and Krawczyk in Crypto 2006, in conjunction with the randomized hashing mode that is used to realize such a hash function family. eTCR is a strengthened variant of the well-known TCR (or UOWHF) property for a hash function family (*i.e.* a dedicated-key hash function). The contributions of this paper are twofold. First, we compare the new eTCR property with the well-known collision resistance (CR) property, where both properties are considered for a dedicated-key hash function. We show there is a *separation* between the two notions, that is *in general*, eTCR property cannot be claimed to be weaker (or stronger) than CR property for any arbitrary dedicated-key hash function. Second, we consider the problem of eTCR property preserving domain extension. We study several domain extension methods for this purpose, including (Plain, Strengthened, and Prefix-free) Merkle-Damgård, Randomized Hashing (considered in dedicated-key hash setting), Shoup, Enveloped Shoup, XOR Linear Hash (XLH), and Linear Hash (LH) methods. Interestingly, we show that the only eTCR preserving method is a *nested variant* of LH which has a drawback of having high key expansion factor. Therefore, it is interesting to design a new and *efficient* eTCR preserving domain extension in the *standard model*.

Keywords: Hash Functions, CR, TCR, eTCR, Domain Extension.

1 Introduction

Cryptographic hash functions are widely used in many cryptographic schemes, most importantly as building blocks for digital signature schemes and message authentication codes (MACs). Their application in signature schemes following hash-and-sign paradigm, like DSA, requires the collision resistance (CR) property. Contini and Yin [5] showed that breaking the CR property of a hash function can also endanger security of the MAC schemes, which are based on the hash function, such as HMAC. Despite being a very essential and widely-desirable security property of a hash function, CR has been shown to be a very

---

* The full version of this paper is available from [18].

O. Dunkelman (Ed.): FSE 2009, LNCS 5665, pp. 327–344, 2009.
© International Association for Cryptologic Research 2009

strong and demanding property for hash functions from theoretical viewpoint [22, 4, 17] as well as being a practically endangered property by the recent advances in cryptanalysis of widely-used standard hash functions like MD5 and SHA-1 [25, 24]. In response to these observations in regard to the strong CR property for hash functions and its implication on the security of many applications, recently several ways out of this uneasy situation have been proposed.

The first approach is to avoid relying on the CR property in the design of new applications and instead, just base the security on other weaker than CR properties like Target Collision Resistance ("Ask less of a hash function and it is less likely to disappoint! " [4]). This is an attractive and wise methodology in the design of new applications using hash functions, but unfortunately it might be of limited use to secure an already implemented and in-use application, if the required modifications are significant and hence prohibitive (and not cost effective) in practice.

The second approach is to design new hash functions to replace current endangered hash function standards like SHA-1. For achieving this goal, NIST has started a public competition for selecting a new secure hash standard SHA-3 to replace the current SHA-1 standard [15]. It is hoped that new hash standard will be able to resist against all known cryptanalysis methods, especially powerful statistical methods like differential cryptanalysis which have been successfully used to attack MD5, SHA-1 and other hash functions [25, 24, 23].

Another methodology has also recently been considered as an intermediate step between the aforementioned two approaches in [10, 9]. This approach aims at providing a "safety net" by fixing the current complete reliance on endangered CR property without having to change the internals of an already implemented hash function like SHA-1 and instead, just by using the hash function in some black-box modes of operation. Based on this idea, Randomized Hashing mode was proposed in [10] and announced by NIST as Draft SP 800-106 [16]. In a nutshell, Randomized Hashing construction converts a keyless hash function H (e.g. SHA-1) to a dedicated-key hash function $\tilde{H}$ defined as $\tilde{H}_K(M) = H(K||(M_1 \oplus K)||\cdots||(M_L \oplus K))$, where H is an iterated Merkle-Damgård hash function based on a compression function h. ($M_1||\cdots||M_L$ is the padded message after applying strengthening padding.) Note that Randomized Hashing keys the entire iterated hash function H at once, by using it as a black-box function and just preprocessing the input message M with the key K (*i.e.* the *random salt*).

Although the main motivation for the design of a randomized hashing mode in [10] was to free reliance on collision resistance assumption on the underlying hash function (by making off-line attacks ineffective by using a random key), in parallel to this aim, a new security property was also introduced and defined for hash functions, namely *enhanced* Target Collision Resistance (eTCR) property. Having $\tilde{H}$ as the first example of a construction for eTCR hash functions in hand, we also note that an eTCR hash function is an interesting and useful new primitive. In [10], the security of the specific example function $\tilde{H}$ in eTCR sense is based on some new assumptions (called c-SPR and e-SPR) about keyless compression function h. However, this example function $\tilde{H}$, may be threatened as a result of future cryptanalysis results, but the notion of eTCR hashing will still remain useful independently from this specific function. By using an eTCR hash function family $\{H_K\}$ in a hash-and-sign digital signature scheme, one does not need to sign the key K used for the hashing. It is only required to sign $H_K(M)$ and the key K is sent in public to the verifier as part of the signed message [10]. This is an improvement compared to using a TCR (UOWHF) hash function family where one needs to sign $H_K(M)\|K$ [4].

Our Contributions

Our aim in this paper is to investigate the eTCR hashing as a new and interesting notion. Following the previous background on the CR notion, the first natural question that arises is whether eTCR is weaker than CR in general. It is known that both CR and eTCR imply TCR property (*i.e.* are stronger notions than TCR) [14, 20, 10], but the relation between CR and eTCR has not been considered yet. As our first contribution in this paper, we compare the eTCR property with the CR property, where both properties are considered formally for a dedicated-key hash function. We show that there is a separation between eTCR and CR notions, that is *in general*, eTCR property cannot be claimed to be weaker (or stronger) than CR property for any arbitary dedicated-key hash function. Although our separation result does not rule out the possibility of designing specific dedicated-key hash functions in which eTCR might be easier to achieve compared to CR, it emphasizes the point that any such a construction should explicitly show that this is indeed the case.

As our second contribution, we consider the problem of eTCR preserving domain extension. Assuming that one has been able to design a dedicated-key compression function which possesses eTCR property, the next step will be how to extend its domain to obtain a full-fledged hash function which also provably possesses eTCR property and is capable of hashing any variable length message. In the case of CR property the seminal works of Merkle [12] and Damgård [7] show that Merkle-Damgård (MD) iteration with strengthening (length indicating) padding is a CR preserving domain extender. Analysis and design of (multi-)property preserving domain extenders for hash function has been recently attracted new attention in several works considering several different security properties, such as [4, 3, 2, 1]. We investigate eight domain extension transforms for this purpose; namely Plain MD [12, 7], Strengthened MD [12, 7], Prefix-free MD [6, 11], Randomized Hashing [10] (considered in dedicated-key hash setting), Shoup [21], Enveloped Shoup [2], XOR Linear Hash (XLH) [4], and a variant of Linear Hash (LH) [4] methods. Interestingly, we show that the only eTCR preserving method among these methods is a *nested variant* of LH (defined based on a variant proposed in [4]) which has the drawback of having high key expansion factor. The overview of constructions and the properties they preserve are shown in Table 1. The symbol "✓" means that the notion is provably preserved by the construction; "×" means that it is not preserved. Underlined entries related to eTCR property are the results shown in this paper.

Table 1. Overview of constructions and the properties they preserve

Scheme	CR	TCR	eTCR
Plain MD	× [12, 7]	× [4]	×
Strengthened MD	✓ [12, 7]	× [4]	×
Prefix-free MD	× [2]	× [2]	×
Randomized Hashing	✓ [1]	× [1]	×
Shoup	✓ [21]	✓ [21]	×
Enveloped Shoup	✓ [2]	✓ [2]	×
XOR Linear Hash (XLH)	✓ [1]	✓ [4]	×
Nested Linear Hash (LH)	✓ [4]	✓ [4]	✓

2 Preliminaries

2.1 Notations

If A is a probabilistic algorithm then by $y \xleftarrow{\$} A(x_1, \cdots, x_n)$ it is meant that y is a random variable which is defined from the experiment of running A with inputs $x_1, \cdots, x_n$ and assigning the output to y. To show that an algorithm A is run without any input (*i.e.* when the input is an empty string) we use the notation $y \xleftarrow{\$} A()$. By time complexity of an algorithm we mean the running time, relative to some fixed model of computation (e.g. RAM) plus the size of the description of the algorithm using some fixed encoding method. If X is a finite set, by $x \xleftarrow{\$} X$ it is meant that x is chosen from X uniformly at random. Let $x||y$ denote the string obtained from concatenating string y to string x. Let 1^m and 0^m, respectively, denote a string of m consecutive 1 and 0 bits. For a binary string M, let $M_{1...n}$ denote the first n bits of M, $|M|$ denote its length in bits and $|M|_b \triangleq \lceil |M|/b \rceil$ denote its length in b-bit blocks. For a positive integer m, let $\langle m \rangle_b$ denotes binary representation of m by a string of length exactly b bits. If S is a finite set we denote size of S by $|S|$. The set of all binary strings of length n bits (for some positive integer n) is denoted as $\{0,1\}^n$, the set of all binary strings whose lengths are variable but upper-bounded by N is denoted by $\{0,1\}^{\leq N}$ and the set of all binary strings of arbitrary length is denoted by $\{0,1\}^*$.

2.2 Two Settings for Hash Functions

In a formal study of cryptographic hash functions and their security notions, two different but related settings can be considered. The first setting is the traditional keyless hash function setting where a hash function refers to a single function H (e.g. H=SHA-1) that maps variable length messages to fixed length output hash value. In the second setting, by a hash function it is meant a family of hash functions $H : \mathcal{K} \times \mathcal{M} \to \{0,1\}^n$, also called a dedicated-key hash function [2], which is indexed by a key space $\mathcal{K}$. A key $K \in \mathcal{K}$ acts as an index to select a specific member function from the family and often the key argument is denoted as a subscript, that is $H_K(M) = H(K, M)$, for all $M \in \mathcal{M}$. In a formal treatment of hash functions and the study of relationships between different security properties, one should clarify the target setting, namely whether keyless or dedicated-key setting is considered. This is worth emphasizing as some security properties like TCR and eTCR are inherently defined and make sense for a dedicated-key hash function [20, 10]. Regarding CR property there is a well-known foundational dilemma, namely CR can only be *formally* defined for a dedicated-key hash function, but it has also been used widely as a security *assumption* in the case of keyless hash functions like SHA-1. We will briefly review this formalization issue for CR in Subsection 2.3 and for a detailed discussion we refer to [19].

2.3 Definition of Security Notions: CR, TCR and eTCR

In this section, we recall three security notions directly relevant to our discussions in the rest of the paper; namely, CR, TCR, and eTCR, where these properties are formally defined for a dedicated-key hash function. We also recall the well-known definitional dilemma regarding CR assumption for a keyless hash function.

A dedicated-key hash function $H : \mathcal{K} \times \mathcal{M} \to \{0,1\}^n$ is called (t, ϵ)-x secure, where $x \in \{CR, TCR, eTCR\}$ if the advantage of *any* adversary, having time complexity at most t, is less than ϵ, where the advantage of an adversary A, denoted by $\mathrm{Adv}_H^x(A)$, is defined as the probability that a specific winning condition is satisfied by A upon finishing the game (experiment) defining the property x. The probability is taken over all randomness used in the defining game as well as that of the adversary itself. The advantage functions for an adversary A against the CR, TCR and eTCR properties of the hash function H are defined as follows, where in the case of TCR and eTCR, adversary is denoted by a two-stage algorithm $A = (A_1, A_2)$:

$$\mathrm{Adv}_H^{CR}(A) = \Pr\left\{ K \xleftarrow{\$} \mathcal{K}; (M, M') \xleftarrow{\$} A(K) : M \neq M' \wedge H_K(M) = H_K(M') \right\}$$

$$\mathrm{Adv}_H^{TCR}(A) = \Pr\left\{ \begin{array}{l} (M, State) \xleftarrow{\$} A_1(); \\ K \xleftarrow{\$} \mathcal{K}; \\ M' \xleftarrow{\$} A_2(K, State); \end{array} : M \neq M' \wedge H_K(M) = H_K(M') \right\}$$

$$\mathrm{Adv}_H^{eTCR}(A) = \Pr\left\{ \begin{array}{ll} (M, State) \xleftarrow{\$} A_1(); & (K, M) \neq (K', M') \\ K \xleftarrow{\$} \mathcal{K}; & \wedge \\ (K', M') \xleftarrow{\$} A_2(K, State); & H_K(M) = H_{K'}(M') \end{array} \right\}$$

CR for a Keyless Hash Function. Collision resistance as a security property cannot be formally defined for a keyless hash function $H : \mathcal{M} \to \{0,1\}^n$. *Informally*, one would say that it is *"infeasible"* to find two distinct messages M and M' such that $H(M) = H(M')$. But it is easy to see that if $|\mathcal{M}| > 2^n$ (*i.e.* if the function is compressing) then there are many colliding pairs and hence, trivially there *exists* an *efficient* program that can always output a colliding pair M and M', namely a simple one with M and M' included in its code. That is, *infeasibility* cannot be formalized by an statement like "there exists no efficient adversary with non-negligible advantage" as clearly there are many such adversaries as mentioned before. The point is that *no human being knows* such a program [19], but the latter concept cannot be formalized mathematically. Therefore, in the context of keyless hash functions, CR can only be treated as a strong *assumption* to be used in a constructive security reduction following human-ignorance framework of [19]. We will call such a CR assumption about a keyless hash function as **keyless-CR assumption** to distinguish it from formally definable CR notion for a dedicated-key hash function. We note that as a result of recent collision finding attacks, it is shown that keyless-CR assumption is completely invalid for MD5 [25] and theoretically endangered assumption for SHA-1 [24].

3   eTCR Property *vs.* CR Property

In this Section, we show that there is a separation between CR and eTCR, that is none of these two properties can be claimed to be weaker or stronger than the other *in general* in dedicated-key hash function setting. We emphasize that we consider relation between CR and eTCR as formally defined properties for a dedicated-key hash function following the methodology of [20]. The CR property considered in this section should not be mixed with the strong keyless-CR assumption for a keyless hash function.

3.1   CR $\nRightarrow$ eTCR

We want to show that the CR property does not imply the eTCR property. That is, eTCR as a security notion for a dedicated-key hash function is not weaker than the CR property. This is done by showing as a counterexample, a dedicated-key hash function which is secure in CR sense but completely insecure in eTCR sense.

Lemma 1 (CR does not imply eTCR). *Assume that there exists a dedicated-key hash function* $H : \{0,1\}^k \times \{0,1\}^m \to \{0,1\}^n$ *which is* $(t, \epsilon)-CR$. *Select (and fix) an arbitrary message* $M^* \in \{0,1\}^m$ *and an arbitrary key* $K^* \in \{0,1\}^k$ *(e.g.* $M^* = 1^m$ *and* $K^* = 1^k$*). The dedicated-key hash function* $G : \{0,1\}^k \times \{0,1\}^m \to \{0,1\}^n$ *shown in this lemma is* $(t', \epsilon')-CR$, *where* $t' = t - cT_H$ *and* $\epsilon' = \epsilon + 2^{-k}$, *but it is completely insecure in* $eTCR$ *sense.* T_H *denotes the time for one computation of* H *and* c *is a small constant.*

$$
G_K(M) = \begin{cases} M^*_{1\cdots n} & \text{if } M = M^* \bigvee K = K^* & (1) \\ H_K(M^*) \text{ if } M \neq M^* \bigwedge K \neq K^* \bigwedge H_K(M) = M^*_{1\cdots n} & (2) \\ H_K(M) \quad \text{otherwise} & (3) \end{cases}
$$

The proof is valid for any arbitrary selection of parameters $M^* \in \{0,1\}^m$ and $K^* \in \{0,1\}^k$, and hence, this construction actually shows 2^{m+k} such counterexample functions, which are CR but not eTCR.

Proof. Let's first demonstrate that G as a dedicated-key hash function is not secure in eTCR sense. This can be easily shown by the following simple adversary $A = (A_1, A_2)$ playing eTCR game against G. In the first stage of eTCR attack, A_1 outputs the target message as $M = M^*$. In the second stage of the attack, A_2, after receiving the first randomly selected key K (where $K \xleftarrow{\$} \{0,1\}^k$), outputs a different message $M' \neq M^*$ and selects the second key as $K' = K^*$. It can be seen easily that the adversary $A = (A_1, A_2)$ wins the eTCR game, as $M' \neq M^*$ implies that $(M^*, K) \neq (M', K^*)$ and by the construction of G we have $G_K(M^*) = G_{K^*}(M') = M^*_{1...n}$; that is both of the conditions for winning eTCR game are satisfied. Therefore, the hash function family G is completely insecure in eTCR sense.

To complete the proof, we need to show that the hash function family G inherits the CR property of H. This is done by reducing CR security of G to that of H. Let A be an adversary that can win CR game against G with probability ϵ' using time complexity t'. We construct an adversary B against CR property of H with success probability of at least $\epsilon = \epsilon' - 2^{-k}$ ($\approx \epsilon'$, for large k) and time $t = t' + cT_H$ as stated in the lemma. The construction of B and the complete analysis can be found in the full version of this paper in [18].   □

3.2   eTCR $\not\Rightarrow$ CR

We want to demonstrate that the eTCR property does not imply the CR property. That is, the CR property as a security notion for a dedicated-key hash function is not a weaker than the eTCR property. This is done by showing as a counterexample, a dedicated-key hash function which is secure in eTCR sense but completely insecure in CR sense.

Lemma 2 (eTCR does not imply CR). *Assume that there exists a dedicated-key hash function* $H : \{0,1\}^k \times \{0,1\}^m \to \{0,1\}^n$, *where* $m > k \geq n$, *which is* $(t, \epsilon) - eTCR$. *The dedicated-key hash function* $G : \{0,1\}^k \times \{0,1\}^m \to \{0,1\}^n$ *shown in this lemma is* $(t', \epsilon') - eTCR$, *where* $t' = t - c$, $\epsilon' = \epsilon + 2^{-k+1}$, *but it is completely insecure in CR sense. (c is a small constant.)*

$$G_K(M) = \begin{cases} H_K(0^{m-k}||K) & \text{if } M = 1^{m-k}||K \\ H_K(M) & \text{otherwise} \end{cases}$$

Proof. We firstly demonstrate that G as a dedicated-key hash function is not secure in CR sense. This can be easily shown by the following simple adversary A that plays CR game against G. On receiving the key K, the adversary A outputs two different messages as $M = 1^{m-k}||K$ and $M' = 0^{m-k}||K$ and wins the CR game as we have $G_K(1^{m-k}||K) = H_K(0^{m-k}||K) = G_K(0^{m-k}||K)$.

It remains to show that that G indeed is an eTCR secure hash function family. Let $A = (A_1, A_2)$ be an adversary which wins the eTCR game against G with probability ϵ' and using time complexity t'. We construct an adversary $B = (B_1, B_2)$ which uses A as a subroutine and wins eTCR game against H with success probability of at least $\epsilon = \epsilon' - 2^{-k+1} (\approx \epsilon'$, for large k) and having time complexity $t = t' + c$ where small constant c can be determined from the description of algorithm B. The description of the algorithm B and the complete analysis can be found in the full version of this paper in [18]. □

3.3 The Case for Randomized Hashing

Randomized Hashing method is a simple method to obtain a dedicated-key hash function $\tilde{H} : \mathcal{K} \times M \to \{0,1\}^n$ from an iterated (keyless) hash function H as $\tilde{H}(K, M) \triangleq H(K||(M_1 \oplus K)|| \cdots ||(M_L \oplus K))$, where $\mathcal{K} = \{0,1\}^b$ and H itself is constructed by iterating a keyless compression function $h : \{0,1\}^{n+b} \to \{0,1\}^n$ and using a *fixed* initial chaining value IV. The analysis in [10] reduces the security of $\tilde{H}$ in eTCR sense to some assumptions, called c-SPR and e-SPR, on the keyless compression function h which are weaker than the keyless-CR assumption on h.

Here, we are interested in a somewhat different question, namely whether (formally definable) CR for this specific design of dedicated-key hash function $\tilde{H}$ implies that it is eTCR or not. Interestingly, we can gather a strong evidence that CR for $\tilde{H}$ implies that it is also eTCR, by the following argument. First, from the construction of $\tilde{H}$ it can be seen that CR for $\tilde{H}$ implies keyless-CR for a hash function H^* which is identical to the H except that its initial chaining value is a random and *known* value $IV^* = h(IV||K)$ instead of the prefixed IV (Note that K is selected at random and is provided to the adversary at the start of CR game). This is easily proved, as any adversary that can find collisions for H^* (*i.e.* breaks it in keyless-CR sense) can be used to construct an adversary that can break $\tilde{H}$ in CR sense. Second, from recent cryptanalysis methods which use differential attacks to find collisions [25, 24], we have a strong evidence that finding collisions for H^* under known IV^* would not be harder than finding collisions for H under IV, for a practical hash function like MD5 or SHA-1. That is, we argue that if H^* is keyless-CR then H is also keyless-CR. Finally, we note that keyless-CR assumption on H in turn implies that $\tilde{H}$ is eTCR as follows. Consider a successful eTCR attack against $\tilde{H}$ where on finishing the attack we will have $(K, M) \neq (K', M')$ and $\tilde{H}(K, M) = \tilde{H}(K', M')$, where $M = M_1|| \cdots ||M_L$ and $M' = M'_1|| \cdots ||M'_L$. Referring to the construction of $\tilde{H}$ this is translated to $H(K||(M_1 \oplus K)|| \cdots ||(M_L \oplus K)) = H(K||(M'_1 \oplus K)|| \cdots ||(M'_L \oplus K))$ and from $(K, M) \neq (K', M')$ we have that $(K||(M_1 \oplus K)|| \cdots ||(M_L \oplus K)) \neq (K||(M'_1 \oplus K)|| \cdots ||(M'_L \oplus K))$. Hence, we have found a collision for H and this contradicts the assumption that H is keyless-CR. Therefore, for the case of the specific dedicated-key hash function $\tilde{H}$ obtained via Randomized Hashing mode, it can be argued that CR implies eTCR.

4   Domain Extension and eTCR Property Preservation

In this section we investigate the eTCR preserving capability of eight domain extension transforms, namely Plain MD [12, 7], Strengthened MD [12, 7], Prefix-free MD [6, 11], Randomized Hashing [10], Shoup [21], Enveloped Shoup [2], XOR Linear Hash (XLH)[4], and Linear Hash (LH) [4] methods.

Assume that we have a compression function $h : \{0,1\}^k \times \{0,1\}^{n+b} \rightarrow \{0,1\}^n$ that can only hash messages of fixed length $(n + b)$ bits. A domain extension transform can use this compression function (as a black-box) to construct a hash function $H : \mathcal{K} \times \mathcal{M} \rightarrow \{0,1\}^n$, where the message space $\mathcal{M}$ can be either $\{0,1\}^*$ or $\{0,1\}^{<2^m}$, for some positive integer m (e.g. $m = 64$). The key space $\mathcal{K}$ is determined by the construction of a domain extender. Clearly $log_2(|\mathcal{K}|) \geq k$, as H involves at least one invocation of h. The difference between $log_2(|\mathcal{K}|)$ (i.e. the key length of H) and k (i.e. the key length of h) is called the 'key expansion' of domain extension transform and is a measure of its efficiency: the less key expansion is, the more efficient the domain extension transform will be.

A domain extension transform comprises of two functions: an injective padding function Pad and an iteration function f_I. First, the padding function $Pad : \mathcal{M} \rightarrow D_I$ is applied to an input message $M \in \mathcal{M}$ to convert it to the padded message $Pad(M)$ in a domain D_I. Then, the iteration function $f_I : \mathcal{K} \times D_I \rightarrow \{0,1\}^n$ uses the compression function h as many times as required, and outputs the final hash value. The full-fledged hash function H is obtained by combining the two functions.

The padding functions used in the eight domain extension transforms that we consider in this paper are defined as follows:

- **Plain:** $pad : \{0,1\}^* \rightarrow \bigcup_{L \geq 1} \{0,1\}^{Lb}$, where $pad(M) = M \| 10^p$ and p is the minimum number of 0's required to make the length of $pad(M)$ a multiple of block length.
- **Strengthening:** $pad_s : \{0,1\}^{<2^m} \rightarrow \bigcup_{L \geq 1} \{0,1\}^{Lb}$, where $pad_s(M) = M \| 10^p \| \langle |M| \rangle_m$ and p is the minimum number of 0's required to make the length of $pad_s(M)$ a multiple of block length.
- **Prefix-free:** $padPF : \{0,1\}^* \rightarrow \bigcup_{L \geq 1} \{0,1\}^{Lb}$, where $padPF$ transforms the input message space $\{0,1\}^*$ to a prefix-free message space, $i.e. padPF(M)$ is not a prefix of $padPF(M')$ for any two *distinct* messages M and M'. An example of a Prefix-free padding function, which we consider in this paper, is as follows. Append 10^p to the message where p is the minimum number of 0's required to make the length of the resulted message a multiple of $b - 1$ bits. Parse the resulted message into blocks of $b - 1$ bits and prepend a '0' to all blocks but the final block where a '1' must be prepended.
- **Strengthened Chain Shift:** $padCS_s : \{0,1\}^{<2^m} \rightarrow \bigcup_{L \geq 1} \{0,1\}^{Lb+b-n}$, where $padCS_s(M) = M \| 10^r \| \langle |M| \rangle_m \| 0^p$, and parameters p and r are defined in two ways depending on the block length b. If $b \geq n + m$ then $p = 0$, otherwise $p = b - n$. Then r is the minimum number of 0's required to make the padded message a member of $\{0,1\}^{Lb+b-n}$, for some positive integer L.

The iteration functions for MD, Randomized Hashing, Shoup, Enveloped Shoup, XLH and LH are shown in Fig. 1.

4.1 Merkle-Damgård Does Not Preserve eTCR

MD iteration function as shown in Fig. 1 can be used together with Plain (pad), Strengthening(pad_s), or Prefix-free$(padPF)$ padding function to construct

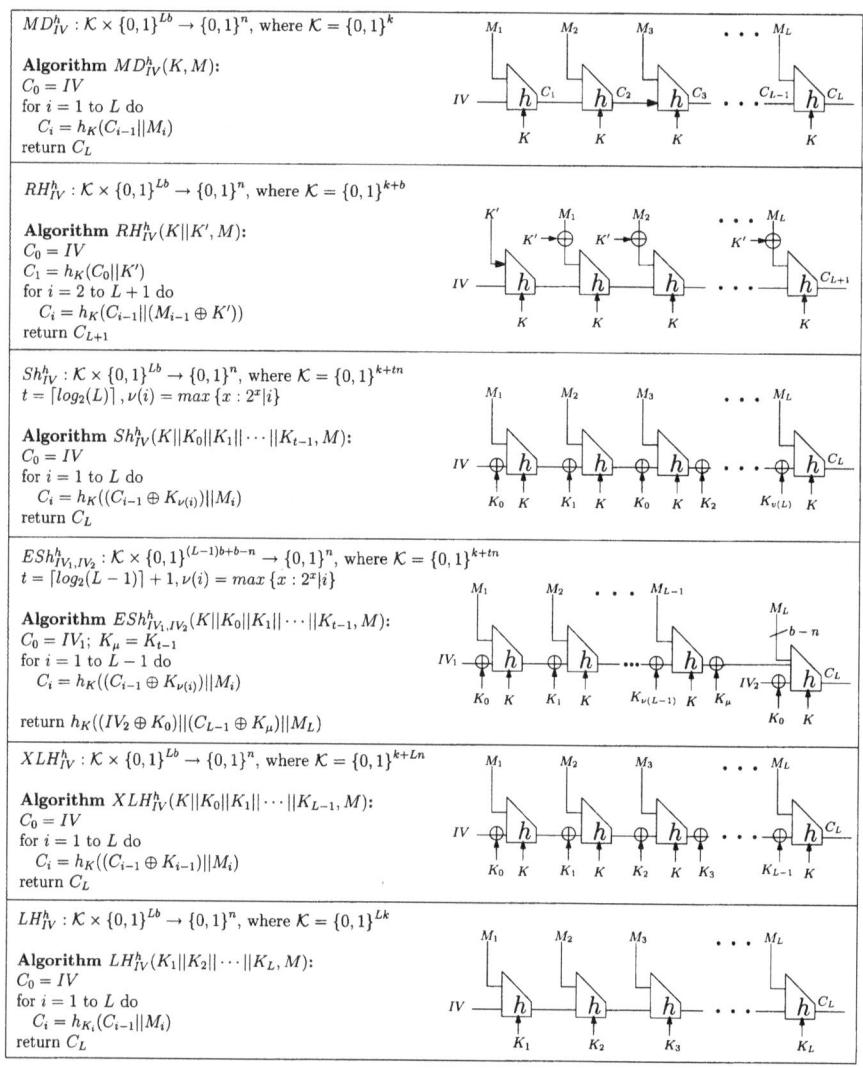

Fig. 1. Iteration functions used in domain extension transforms: Merkle-Damgård (MD), Randomized Hashing (RH), Shoup (Sh), Enveloped Shoup (ESh), XLH and LH. The iteration functions are ordered top-down based on their efficiency in terms of key expansion, MD iteration does not expand the key length of underlying compression function and is the most efficient transform and LH is the least efficient transform.

a domain extension transform, which is called Plain MD, Strengthened MD, or Prefix-free MD, respectively. In this section we show that none of these three domain extension transforms can be used as an eTCR preserving domain extender.

Theorem 1 (Negative Result). *Plain MD, Strengthened MD, and Prefix-free MD do not preserve eTCR.*

Proof. We borrow the construction of the following counterexample from [4] where it was used in the context of TCR property. Assume that there is a dedicated-key compression function $g : \{0,1\}^k \times \{0,1\}^{n+b} \to \{0,1\}^n$ with $b > k$ which is (t, ϵ)-eTCR secure. Set $b = k + b'$ where $b' > 0$ by the assumption that $b > k$. Consider the following dedicated-key compression function $h : \{0,1\}^k \times \{0,1\}^{(n+k)+b'} \to \{0,1\}^{n+k}$:

$$h(K, X\|Y\|Z) = h_K(X\|Y\|Z) = \begin{cases} g_K(X\|Y\|Z)\|K & \text{if } K \neq Y \\ 1^{n+k} & \text{if } K = Y \end{cases}$$

where $K \in \{0,1\}^k$, $X \in \{0,1\}^n$, $Y \in \{0,1\}^k$, $Z \in \{0,1\}^{b'}$ ($n + k$ is chaining variable length and b' is block length for h).

To complete the proof, we first show in Lemma 3 that h_K inherits the eTCR property from g_K. Note that this cannot be directly inferred from the proof in [4] that h_K inherits the weaker notion TCR from g_K. Then, we show a simple attack in each case to show that the hash function obtained via either of Plain, Strengthened, or Prefix-free MD transform by extending domain of h_K is completely insecure in eTCR sense.

Lemma 3. *The dedicated-key compression function h is (t', ϵ')-eTCR secure, where $\epsilon' = \epsilon + 2^{-k+1} \approx \epsilon$ and $t' = t - c$, for a small constant c.*

Proof. Let $A = (A_1, A_2)$ be an adversary which wins the eTCR game against h_K with probability ϵ' and using time complexity t'. We construct an adversary $B = (B_1, B_2)$ which uses A as a subroutine and wins eTCR game against g_K with success probability of at least $\epsilon = \epsilon' - 2^{-k+1} (\approx \epsilon'$, for large k) and spending time complexity $t = t' + c$ where small constant c can be determined from the description of algorithm B. Algorithm B is as follows:

Algorithm $B_1()$	**Algorithm $B_2(K_1, M_1, State)$**
$(M_1 = X_1\|Y_1\|Z_1, State) \overset{\$}{\leftarrow} A_1()$;	Parse M_1 as $M_1 = X_1\|Y_1\|Z_1$
return $(M_1, State)$;	if $[K_1 = Y_1 \bigvee K_1 = 1^k]$ **return** 'Fail';
	$(M_2 = X_2\|Y_2\|Z_2, K_2) \overset{\$}{\leftarrow} A_2(K_1, M_1, State)$;
	return (M_2, K_2);

At the first stage of eTCR attack, B_1 just merely runs A_1 and returns whatever it returns as the first message (*i.e.* $M_1 = X_1\|Y_1\|Z_1$) and any possible state information to be passed to the second stage algorithm. At the second stage of the attack, let **Bad** be the event that $[K_1 = Y_1 \bigvee K_1 = 1^k]$. If **Bad** happens then algorithm B_2 (and hence B) will fail in eTCR attack; otherwise (*i.e.* if $\overline{\textbf{Bad}}$ happens) we show that B will be successful in eTCR attack against g whenever A succeeds in eTCR attack against h.

Assume that the event $\overline{\textbf{Bad}}$ happens; that is, $[K_1 \neq Y_1 \bigwedge K_1 \neq 1^k]$. We claim that in this case if A succeeds then B also succeeds. Referring to the construction of (counterexample) compression function h in this lemma, it can be seen that if A succeeds, *i.e.*, whenever $(M_1, K_1) \neq (M_2, K_2) \bigwedge h_{K_1}(M_1) = h_{K_2}(M_2)$, it must be the case that $g_{K_1}(M_1)||K_1 = g_{K_2}(M_2)||K_2$ which implies that $g_{K_1}(M_1) = g_{K_2}(M_2)$ (and also $K_1 = K_2$). That is, (M_1, K_1) and (M_2, K_2) are also valid a colliding pair for the eTCR attack against g. (Remember that $M_1 = X_1||Y_1||Z_1$ and $M_2 = X_2||Y_2||Z_2$.)

Now note that $\Pr[\mathbf{Bad}] \leq \Pr[K_1 = Y_1] + \Pr[K_1 = 1^k] = 2^{-k} + 2^{-k} = 2^{-k+1}$, as K_1 is selected uniformly at random just after the message M_1 is fixed in the eTCR game. Therefore, we have $\epsilon = \Pr[B \text{ succeeds}] = \Pr[A \text{ succeeds} \wedge \overline{\mathbf{Bad}}] \geq \Pr[A \text{ succeeds}] - \Pr[\mathbf{Bad}] \geq \epsilon' - 2^{-k+1}$.

To complete the proof of Theorem 1, we need to show that MD transforms cannot preserve eTCR while extending the domain of this specific compression function h_K. For this part, the same attacks that used in [4, 2] against TCR property also work for our purpose here as clearly breaking TCR implies breaking its strengthened variant eTCR. The eTCR attacks are as follows:

The Case of Plain MD and Strengthened MD:
Let's denote the Plain MD and Strengthened MD domain extension transforms, respectively, by **pMD** and **sMD**. The following adversary $A = (A_1, A_2)$ can break the hash function obtained using either of **pMD** or **sMD** transforms, in the eTCR sense. A_1 outputs $M_1 = 0^{b'}||0^{b'}$ and A_2, on receiving the first key K, outputs a different message as $M_2 = 1^{b'}||0^{b'}$ together with the same key K as the second key. Considering that the initial value $IV = IV_1||IV_2 \in \{0,1\}^{n+k}$ is fixed before adversary starts the attack game and K is chosen at random afterward in the second stage of the game, we have $\Pr[K = IV_2] = 2^{-k}$. If $K \neq IV_2$ which is the case with probability $1 - 2^{-k}$ then adversary becomes successful as we have:

$$MD_{IV}^h(K, 0^{b'}||0^{b'}) = h_K(h_K(IV_1||IV_2||0^{b'})||0^{b'}) = h_K(g_K(IV_1||IV_2||0^{b'})||K||0^{b'})$$
$$= 1^{n+k}$$
$$MD_{IV}^h(K, 1^{b'}||0^{b'}) = h_K(h_K(IV_1||IV_2||1^{b'})||0^{b'}) = h_K(g_K(IV_1||IV_2||1^{b'})||K||0^{b'})$$
$$= 1^{n+k}$$

$$\mathbf{pMD} : \begin{cases} MD_{IV}^h(K, pad(0^{b'}||0^{b'})) = h_K(MD_{IV}^h(K, 0^{b'}||0^{b'})||10^{b'-1}) \\ \qquad = h_K(1^{n+k}||10^{b'-1}) \\ \\ MD_{IV}^h(K, pad(1^{b'}||0^{b'})) = h_K(MD_{IV}^h(K, 1^{b'}||0^{b'})||10^{b'-1}) \\ \qquad = h_K(1^{n+k}||10^{b'-1}) \end{cases}$$

$$\mathbf{sMD} : \begin{cases} MD_{IV}^h(K, pad_s(0^{b'}||0^{b'})) = h_K(MD_{IV}^h(K, 0^{b'}||0^{b'})||10^{b'-m-1}|| \langle 2b' \rangle_m) \\ \qquad = h_K(1^{n+k}||10^{b'-m-1}|| \langle 2b' \rangle_m) \\ \\ MD_{IV}^h(K, pad_s(1^{b'}||0^{b'})) = h_K(MD_{IV}^h(K, 1^{b'}||0^{b'})||10^{b'-m-1}|| \langle 2b' \rangle_m) \\ \qquad = h_K(1^{n+k}||10^{b'-m-1}|| \langle 2b' \rangle_m) \end{cases}$$

The Case of Prefix-free MD: Denote Prefix-free MD domain extension transform by **preMD**. The full-fledged hash function $H : \{0,1\}^k \times \mathcal{M} \to \{0,1\}^{n+k}$ will be defined as $H(K, M) = \mathbf{preMD}^h_{IV}(K, M) = MD^h_{IV}(K, padPF(M))$. Note that we have $\mathcal{M} = \{0,1\}^*$ due to the application of $padPF$ function. The following adversary $A = (A_1, A_2)$ which is used for TCR attack against Prefix-free MD in [2], can also break H in eTCR sense, as clearly any TCR attacker against H is an eTCR attacker as well. Here, we provide the description of the attack for eTCR, for completeness. A_1 outputs $M_1 = 0^{b'-1}||0^{b'-2}$ and A_2 on receiving the first key K outputs a different message as $M_2 = 1^{b'-1}||0^{b'-2}$ together with the same key K as the second key. Considering that the initial value $IV = IV_1||IV_2 \in \{0,1\}^{n+k}$ is fixed before the adversary starts the attack game and K is chosen at random afterward, we have $\Pr[K = IV_2] = 2^{-k}$. If $K \neq IV_2$ which is the case with probability $1 - 2^{-k}$ then the adversary becomes successful as we have:

$$
\begin{aligned}
MD^h_{IV}(K, padPF(0^{b'-1}||0^{b'-2})) &= MD^h_{IV}(K, 0^{b'}||10^{b'-2}1) \\
&= h_K(h_K(IV_1||IV_2||0^{b'})||10^{b'-2}1) \\
&= h_K(g_K(IV_1||IV_2||0^{b'})||K||10^{b'-2}1) \\
&= 1^{n+k}
\end{aligned}
$$

$$
\begin{aligned}
MD^h_{IV}(K, padPF(1^{b'-1}||0^{b'-2})) &= MD^h_{IV}(K, 01^{b'-1}||10^{b'-2}1) \\
&= h_K(h_K(IV_1||IV_2||01^{b'-1})||10^{b'-2}1) \\
&= h_K(g_K(IV_1||IV_2||01^{b'-1})||K||10^{b'-2}1) \\
&= 1^{n+k}
\end{aligned}
$$

4.2   Randomized Hashing Does Not Preserve eTCR

Our aim in this section is to show that Randomized Hashing (RH) construction, *if considered as a domain extension for a dedicated-key compression function,* does not preserve eTCR property. Note that (this dedicated-key variant of) RH method as shown in Fig. 1 expands the key length of the underlying compression function by only a constant additive factor of b bits, that is $log_2(|\mathcal{K}|) = k + b$. This characteristic, *i.e.* a small and message-length-independent key expansion could have been considered a stunning advantage from efficiency viewpoint, if RH had been able to preserve eTCR. Nevertheless, unfortunately we shall show that randomized hashing does not preserve eTCR.

Following the specification of the original scheme for Randomized Hashing in [10], we assume that the padding function is the strengthening padding pad_s. The full-fledged hash function $H : \{0,1\}^k \times \mathcal{M} \to \{0,1\}^{n+k}$ will be defined as $H(K||K', M) = RH^h_{IV}(K||K', pad_s(M))$. Note that we have $\mathcal{M} = \{0,1\}^{<2^m}$ due to the application of pad_s function.

Theorem 2 (Negative Result). *The Randomized Hashing transform does not preserve eTCR.*

Proof. We use the same counterexample as used in the proof of Theorem 1 to show that Randomized Hashing transform does not preserve eTCR property.

As we have previously shown in Lemma 3 that the constructed h_K inherits the eTCR property of g_K, it just remains to show that RH^h_{IV} cannot extend the domain of h_K while preserving its eTCR property. Consider an adversary $A = (A_1, A_2)$ that plays the eTCR game against the hash function H, obtained via Randomized Hashing, as follows. A_1 outputs a one-block long target message $M_1 = 0^{b'}$ (note that for the counterexample compression function h_K, b' is the block length and $n + k$ is the chaining variable length). A_2 on getting the first key $K||K'$ for H (in the second stage of eTCR attack), outputs the second message as $M_2 = 1^{b'}$ and puts the second key the same as the first key. As $M_2 \neq M_1$, we just need to show that these two messages collide under the same key, *i.e.* $K||K'$. Considering that the initial value $IV = IV_1||IV_2 \in \{0,1\}^{n+k}$ for RH^h_{IV} is (selected and) fixed before the adversary starts the attack game and $K||K'$ is chosen at random latter in the second stage of the game, we have $\Pr[K = IV_2] = 2^{-k}$. If $K \neq IV_2$ (which is the case with probability $1 - 2^{-k}$) then the adversary $A = (A_1, A_2)$ becomes successful as we have:

$$RH^h_{IV}(K||K', pad_s(0^{b'})) = RH^h_{IV}(K||K', 0^{b'}||10^{b'-1-m} \langle b' \rangle_m)$$
$$= h_K \Big(h_K \big(h_K(IV_1||IV_2||K')||(K' \oplus 0^{b'}) \big)||(K' \oplus 10^{b'-1-m} \langle b' \rangle_m) \Big)$$
$$= h_K \Big(h_K \big(g_K(IV_1||IV_2||K')||K||(K' \oplus 0^{b'}) \big)||(K' \oplus 10^{b'-1-m} \langle b' \rangle_m) \Big)$$
$$= h_K(1^{n+k}||(K' \oplus 10^{b'-1-m} \langle b' \rangle_m))$$

$$RH^h_{IV}(K||K', pad_s(1^{b'})) = RH^h_{IV}(K||K', 1^{b'}||10^{b'-1-m} \langle b' \rangle_m)$$
$$= h_K \Big(h_K \big(h_K(IV_1||IV_2||K')||(K' \oplus 1^{b'}) \big)||(K' \oplus 10^{b'-1-m} \langle b' \rangle_m) \Big)$$
$$= h_K \Big(h_K \big(g_K(IV_1||IV_2||K')||K||(K' \oplus 1^{b'}) \big)||(K' \oplus 10^{b'-1-m} \langle b' \rangle_m) \Big)$$
$$= h_K(1^{n+k}||(K' \oplus 10^{b'-1-m} \langle b' \rangle_m)) \qquad \square$$

4.3 Shoup, Enveloped Shoup and XLH Do Not Preserve eTCR

In previous subsections, we have shown that neither MD nor RH are eTCR preserving transforms. The next three most efficient candidates from key expansion viewpoint that we consider are Shoup (Sh), Enveloped Shoup (ESh) and XLH transforms.

Theorem 3 (Negative Results). *Sh, ESh, and XLH transforms do not preserve eTCR.*

Proof. The proof is quite simple but the results are stronger than previous counterexample based proofs, as here the negative results hold for any arbitrary compression function (irrespective of how secure the compression function h is), and not only for some specific counterexamples. That is, *these XOR masking based domain extension transforms are structurally insecure in eTCR sense.* Intuitively, the inability if these domain extenders to preserve eTCR is due to the fact that they use XOR operation to add the key to the internal state (*i.e.* chaining variable), and hence an eTCR adversary will be able to cancel internal differences by taking advantage of its ability to select the value of the second key in the second stage of eTCR attack. (We note that this is also the case for the XTH scheme of [4].)

For the formal proof, we provide the following simple eTCR attack against Shoup construction. The attacks for the cases of ESh and XLH are quite similar and can be found in the full version of this paper in [18].

Consider the hash function obtained via Shoup domain extension transform, *i.e.* pad_s padding function followed by Sh_{IV}^h iteration method. The following simple adversary $A = (A_1, A_2)$ can break it in the eTCR sense. At the first stage of the eTCR attack, A_1 outputs a two-block message $M = M_1 || M_2$ as the target message which after applying pad_s will become a three-block message $M_1 || M_2 || (10^{b-1-m} \langle 2b \rangle_m)$ to be input to the three-round Sh_{IV}^h iteration. In the second stage of eTCR game, A_2, after receiving the first key as $K || K_0 || K_1 || K_0$ from the challenger, chooses the second two-block message as $M' = M_1' || M_2$ which after padding becomes $M_1' || M_2 || (10^{b-1-m} \langle 2b \rangle_m)$. A_2 also puts the second key as $K || K_0 || K_1' || K_0$, where the value of K_1' is computed as $K_1' = K_1 \oplus h_K \big((IV \oplus K_0) || M_1 \big) \oplus h_K \big((IV \oplus K_0) || M_1' \big)$. It is easy to see (referring to Fig. 1) that this value for K' cancel the introduced difference in chaining variable which was created due to the different message blocks M_1 and M_1'. So, $(K || K_0 || K_1, M)$ and $(K || K_0 || K_1', M')$ constitute a colliding pair for H in eTCR sense. (Note that the key sequence used for iteration function Sh_{IV}^h is $K || K_0 || K_1 || K_0$ because padded message $pad_s(M)$ has an extra third block containing the length information.) $\square$

4.4   LH Transform and Its Nested Variant

Up to know we have shown that neither of MD, RH, Sh, or XLH transforms can preserve eTCR property. Henceforth, we have lost all efficient methods from key expansion viewpoint and now we have reached to the same starting-point (and the least efficient) transform for TCR preserving scenario as in [4], *i.e.* the LH method whose key expansion is linear in the message length. We now consider whether at least (but hopefully not the last) this LH transform or its variants can be used for eTCR preserving domain extension or not. Fortunately, we gather a positive answer for this. The proof for this positive result is a straightforward extension of the methodology used in [4] for the case of TCR, but with some necessary adaptations required for considering eTCR attack scenario where adversary has more power in second stage by getting to choose a different key as well as a different message. Firstly, in Theorem 4 we show that if the compression function h is eTCR secure then the hash function LH_{IV}^h will be secure against a restricted class of eTCR adversaries which only find equal-length colliding pairs. Let's denote this equal-length eTCR notion by eTCR*. Secondly, it is shown in Theorem 5 that a *nested variant* of LH can be made eTCR secure, *i.e.* against any arbitrary adversary. The proofs for these two theorems can be found in the full version of this paper in [18].

Assume that the input messages have length a multiple of block length and the maximum length in blocks is some positive integer N, i.e. $|M| \leq Nb$ where b is the length of one block in bits. This restriction of message space to a domain with messages of variable but multiple-block length can be easily removed by using any proper injective padding function like plain padding function pad. LH_{IV}^h iteration function can be used to define a hash function as $H(K_1||\cdots||K_N, M) \triangleq LH_{IV}^h(K_1||\cdots||K_m, M)$, where m is the length of M in blocks.

Theorem 4 (Positive Result: eTCR*). *Assume that the compression function $h : \{0,1\}^k \times \{0,1\}^{n+b} \to \{0,1\}^n$ is (t, ϵ)-eTCR. Then the hash function $H : \{0,1\}^{Nk} \times \{0,1\}^{\leq Nb} \to \{0,1\}^n$ obtained using LH_{IV}^h iteration of h, will be (t', ϵ')-eTCR*, where $\epsilon' = N\epsilon$, $t' = t - \Theta(N)(T_h + n + b + k)$, where T_h is the time for one computation of the compression function h.*

Theorem 5 (From eTCR* to eTCR). *Assume that $H_1 : \{0,1\}^{k_1} \times \mathcal{M} \to \{0,1\}^n$ is (t_1, ϵ_1)-eTCR* hash function and $h : \{0,1\}^{k_2} \times \{0,1\}^{n+b} \to \{0,1\}^n$ is (t_2, ϵ_2)-eTCR compression function, where $b \geq \lceil log_2(|M|) \rceil$, for any $M \in \mathcal{M}$. Then the composition function $H : \{0,1\}^{k_1+k_2} \times \mathcal{M} \to \{0,1\}^n$, defined as $H(K1||K2, M) = h(K2, H_1(K1, M)||\langle |M| \rangle_b)$, will be (t, ϵ)-eTCR; where $\epsilon = \epsilon_1 + 2\epsilon_2$, and $t = min\{t_1 - k_2, t_2 - k_1 - 2T_{H_1} - 2b\}$.*

Nested Linear Hash: Let H_1 be the equal-length eTCR hash function obtained via LH transform as stated in Theorem 4. From Theorem 5 we can obtain a variant of LH which is eTCR secure. This variant which we call it Nested LH is obtained by the composition of H_1 with an eTCR compression function h, that is, LH nested by this final application of the compression function in the way stated in Theorem 5 (*i.e.* final block is just $\langle |M| \rangle_b$). Theorem 5 and Theorem 4 show that this Nested LH will be eTCR if the compression function is eTCR. Alternatively, this Nested LH construction can be seen as obtained using a *variant* of strengthening padding followed by LH iteration on the compression function h. This variant of strengthening padding, which can be called full-final-block strengthening, acts as follows. On input a message M, append the message by 10^r to make its length a multiple of block length and then append another full block which only contains the representation of length of M in an exactly b-bit string, i.e. $\langle |M| \rangle_b$.

5   Conclusion

The introduction of the eTCR property by Halevi and Krawczyk [10] has been proven to be very useful to enrich the notions of hash functions, in particular with its application to construct the Randomized Hashing mode which has been announced by NIST as Draft SP 800-106. Nonetheless, the relationships between eTCR with the existing properties of hash functions need to be further studied. In this paper, we showed that there is a *separation* between the new eTCR property with the well-known collision resistance (CR) property, where both properties are considered for a dedicated-key hash function. Furthermore, when considering the problem of eTCR property preserving domain extension, we found that the only eTCR preserving method is a nested variant of LH which has a drawback of having high key expansion factor. Therefore, it is interesting to design a new eTCR preserving domain extension in the *standard model*, which is *efficient*. We left this as an open problem in this paper.

Acknowledgments. We would like to thank the anonymous reviewers of FSE 2009 for their insightful comments and suggestions.

References

1. Andreeva, E., Neven, G., Preneel, B., Shrimpton, T.: Seven-Property-Preserving Iterated Hashing: ROX. In: Kurosawa, K. (ed.) ASIACRYPT 2007. LNCS, vol. 4833, pp. 130–146. Springer, Heidelberg (2007)
2. Bellare, M., Ristenpart, T.: Hash Functions in the Dedicated-Key Setting: Design Choices and MPP Transforms. In: Arge, L., Cachin, C., Jurdziński, T., Tarlecki, A. (eds.) ICALP 2007. LNCS, vol. 4596, pp. 399–410. Springer, Heidelberg (2007)
3. Bellare, M., Ristenpart, T.: Multi-Property-Preserving Hash Domain Extension and the EMD Transform. In: Lai, X., Chen, K. (eds.) ASIACRYPT 2006. LNCS, vol. 4284, pp. 299–314. Springer, Heidelberg (2006)
4. Bellare, M., Rogaway, P.: Collision-Resistant Hashing: Towards Making UOWHFs Practical. In: Kaliski Jr., B.S. (ed.) CRYPTO 1997. LNCS, vol. 1294, pp. 470–484. Springer, Heidelberg (1997)
5. Contini, S., Yin, Y.L.: Forgery and Partial Key-Recovery Attacks on HMAC and NMAC using Hash Collisions. In: Lai, X., Chen, K. (eds.) ASIACRYPT 2006. LNCS, vol. 4284, pp. 37–53. Springer, Heidelberg (2006)
6. Coron, J.S., Dodis, Y., Malinaud, C., Puniya, P.: Merkle-Damgård Revisited: How to Construct a Hash Function. In: Shoup, V. (ed.) CRYPTO 2005. LNCS, vol. 3621, pp. 430–448. Springer, Heidelberg (2005)
7. Damgård, I.: A Design Principle for Hash Functions. In: Brassard, G. (ed.) CRYPTO 1989. LNCS, vol. 435, pp. 416–427. Springer, Heidelberg (1990)
8. den Boer, B., Bosselaers, A.: Collisions for the Compressin Function of MD5. In: Helleseth, T. (ed.) EUROCRYPT 1993. LNCS, vol. 765, pp. 293–304. Springer, Heidelberg (1993)
9. Dodis, Y., Puniya, P.: Getting the Best Out of Existing Hash Functions; or What if We Are Stuck with SHA? In: Bellovin, S.M., Gennaro, R., Keromytis, A.D., Yung, M. (eds.) ACNS 2008. LNCS, vol. 5037, pp. 156–173. Springer, Heidelberg (2008)
10. Halevi, S., Krawczyk, H.: Strengthening Digital Signatures Via Randomized Hashing. In: Dwork, C. (ed.) CRYPTO 2006. LNCS, vol. 4117, pp. 41–59. Springer, Heidelberg (2006)
11. Maurer, U.M., Sjödin, J.: Single-Key AIL-MACs from Any FIL-MAC. In: Caires, L., Italiano, G.F., Monteiro, L., Palamidessi, C., Yung, M. (eds.) ICALP 2005. LNCS, vol. 3580, pp. 472–484. Springer, Heidelberg (2005)
12. Merkle, R.C.: One Way Hash Functions and DES. In: Brassard, G. (ed.) CRYPTO 1989. LNCS, vol. 435, pp. 428–446. Springer, Heidelberg (1990)
13. Mironov, I.: Hash Functions: From Merkle-Damgård to Shoup. In: Pfitzmann, B. (ed.) EUROCRYPT 2001. LNCS, vol. 2045, pp. 166–181. Springer, Heidelberg (2001)

14. Naor, M., Yung, M.: Universal One-Way Hash Functions and Their Cryptographic Applications. In: STOC 1989, pp. 33–43. ACM, New York (1989)
15. National Institute of Standards and Technology. Cryptographic Hash Algorithm Competition, http://csrc.nist.gov/groups/ST/hash/sha-3/index.html
16. National Institute of Standards and Technology. Draft NIST SP 800-106: Randomized Hashing for Digital Signatures (August 2008),
    http://csrc.nist.gov/publications/PubsDrafts.html#SP-800-106
17. Preneel, B.: The State of Cryptographic Hash Functions. In: Damgård, I.B. (ed.) EEF School 1998. LNCS, vol. 1561, pp. 158–182. Springer, Heidelberg (1998)
18. Reyhanitabar, M.R., Susilo, W., Mu, Y.: Enhanced Target Collision Resistant Hash Functions Revisited. IACR ePrint Archive, Report 2009/051 (2009),
    http://eprint.iacr.org/2009/051
19. Rogaway, P.: Formalizing Human Ignorance: Collision-Resistant Hashing without the Keys. In: Nguyên, P.Q. (ed.) VIETCRYPT 2006. LNCS, vol. 4341, pp. 211–228. Springer, Heidelberg (2006)
20. Rogaway, P., Shrimpton, T.: Cryptographic Hash-Function Basics: Definitions, Implications, and Separations for Preimage Resistance, Second-Preimage Resistance, and Collision Resistance. In: Roy, B., Meier, W. (eds.) FSE 2004. LNCS, vol. 3017, pp. 371–388. Springer, Heidelberg (2004)
21. Shoup, V.: A Composition Theorem for Universal One-Way Hash Functions. In: Preneel, B. (ed.) EUROCRYPT 2000. LNCS, vol. 1807, pp. 445–452. Springer, Heidelberg (2000)
22. Simon, D.R.: Finding Collisions on a One-Way Street: Can Secure Hash Functions be Based on General Assumptions? In: Nyberg, K. (ed.) EUROCRYPT 1998. LNCS, vol. 1403, pp. 334–345. Springer, Heidelberg (1998)
23. Wang, X., Lai, X., Feng, D., Chen, H., Yu, X.: Cryptanalysis of the Hash Functions MD4 and RIPEMD. In: Cramer, R. (ed.) EUROCRYPT 2005. LNCS, vol. 3494, pp. 1–18. Springer, Heidelberg (2005)
24. Wang, X., Yin, Y.L., Yu, H.: Finding Collisions in the Full SHA-1. In: Shoup, V. (ed.) CRYPTO 2005. LNCS, vol. 3621, pp. 17–36. Springer, Heidelberg (2005)
25. Wang, X., Yu, H.: How to Break MD5 and Other Hash Functions. In: Cramer, R. (ed.) EUROCRYPT 2005. LNCS, vol. 3494, pp. 19–35. Springer, Heidelberg (2005)

MAC Reforgeability

John Black[1] and Martin Cochran[2]

[1] Dept. of Computer Science, University of Colorado, Boulder CO 80309, USA
jrblack@cs.colorado.edu
www.cs.colorado.edu/~jrblack
[2] Google, Inc. 1600 Amphitheatre Parkway, Mountain View, CA 94043, USA
martin.cochran@gmail.com

Abstract. Message Authentication Codes (MACs) are core algorithms deployed in virtually every security protocol in common usage. In these protocols, the integrity and authenticity of messages rely entirely on the security of the MAC; we examine cases in which this security is lost.

In this paper, we examine the notion of "reforgeability" for MACs, and motivate its utility in the context of {power, bandwidth, CPU}-constrained computing environments. We first give a definition for this new notion, then examine some of the most widely-used and well-known MACs under our definition in a variety of adversarial settings, finding in nearly all cases a failure to meet the new notion. We examine simple counter-measures to increase resistance to reforgeabiliy, using state and truncating the tag length, but find that both are not simultaneously applicable to modern MACs. In response, we give a tight security reduction for a new MAC, WMAC, which we argue is the "best fit" for resource-limited devices.

Keywords: Message Authentication Codes, Birthday Attacks, Provable Security.

1 Introduction

MESSAGE AUTHENTICATION CODES. Message authentication codes (MACs) are the most efficient algorithms to guarantee message authenticity and integrity in the symmetric-key setting, and as such are used in nearly all security protocols. They work like this: if Alice wishes to send a message M to Bob, she processes M with an algorithm MAC using her shared key K and possibly some state or random bits we denote with s. This produces a short string Tag and she then sends (M, s, Tag) to Bob. Bob runs a verification algorithm VF with key K on the received tuple and VF outputs either ACCEPT or REJECT. The goal is that Bob should virtually never see ACCEPT unless (M, s, Tag) was truly generated by Alice; that is, an imposter should not be able to impersonate Alice and forge valid tuples.

There are a large number of MACs in the literature. Most have a proof of security where security is expressed as a bound on the probability that an attacker

O. Dunkelman (Ed.): FSE 2009, LNCS 5665, pp. 345–362, 2009.

will succeed in producing a forgery after making q queries to an oracle that produces MAC tags on messages of his choice. The bound usually contains a term $q^2/2^t$ where q is the total number of tags generated under a given key and t is the tag length in bits. This quadratic term typically comes from the probability that two identical tags were generated by the scheme for two different messages; this event is typically called a "collision" and once it occurs the analysis of the scheme's security no longer holds. The well-known birthday phenomenon is responsible for the quadratic term: if we generate q random uniform t-bit strings independently, the expected value of q when the first collision occurs is about $\sqrt{\pi 2^{t-1}} = \Theta(2^{t/2})$.

REFORGEABILITY. The following is a natural question: if a forgery is observed or constructed by an adversary, what are the consequences? One possibility is that this forgery does not lead to any additional advantage for the adversary: a second forgery requires nearly as much effort to obtain as the first one did. We might imagine using a random function $f : \Sigma^* \to \{0,1\}^t$ as a stateless MAC. Here, knowing a forgery amounts to knowing distinct $M_1, M_2 \in \Sigma^*$ with $f(M_1) = f(M_2)$. However it is obvious this leads to no further advantage for the adversary: the value of f at points M_1 and M_2 are independent of the values of f on all remaining unqueried points.

Practical MAC schemes, however, usually do not come close to truly random functions, even when implemented as pseudorandom functions (PRFs). Instead they typically contain structure that allows the adversary to use the obtained collision to infer information about the inner state of the algorithm. This invariably leads to further forgeries with a minimum of computation.

APPLICATIONS. One might reasonably ask why we care about reforgeability. After all, aren't MACs designed so that the first forgery is extremely improbable? They are, in most cases, and for many scenarios this is the correct approach, but there are several settings where we might want to think about reforgeability nonetheless:

- In sensor nodes, where radio power is far more costly than computing power, short tag-length MACs might be employed to reduce the overhead of sending tags.
- Streaming video applications might use a low-security MAC with the idea that forging one frame would hardly be noticeable to the viewer; our concern would be that the attacker would be unable to efficiently forge arbitrarily many frames, thereby taking over the video transmission.
- VOIP is another setting where reforgeability is arguably more appropriate than current MAC security models. In this setting, a forged packet probably only corresponds to a fraction of a second of sound and is relatively harmless.

In all cases, if parameters are chosen correctly so that an attacker's best strategy is to guess tags, the overwhelming number of incorrect guesses can be used to inform users in situations where a forged packet could potentially have serious consequences.

MAC scheme	Expected queries for j forgeries	Succumbs to padding attack	Succumbs to other attack	Message freedom
CBC MAC	$C_1 + j$		$\checkmark$	$m - 2$
EMAC	$C_1 + j$	$\checkmark$	$\checkmark$	$m - 2$
XCBC	$C_1 + j$	$\checkmark$	$\checkmark$	$m - 2$
PMAC	$C_1 + j$		$\checkmark$	1
ANSI retail MAC	$C_1 + j$	$\checkmark$	$\checkmark$	$m - 2$
HMAC	$\sum_i C_i/2^i + j$	$\checkmark$		$m - 1$

Fig. 1. Summary of Results. The upper table lists each well-known MAC scheme we examined, along with its resistance to reforgeability attacks. Here n is the output length (in bits) of each scheme, and m is the length (in n-bit blocks) of the queries to the MAC oracle; the i-th collision among the tags is denoted by event C_i. For most schemes, the first forgery is made after the first collision among the tags, and each subsequent forgery requires only one further MAC query. With a general birthday attack, the first collision is expected at around $2^{n/2}$ MAC queries, although the exact number for each scheme can differ somewhat. The last column gives the number of freely-chosen message blocks in the forgery.

MAIN RESULTS. In this paper we conduct a systematic study of reforgeability, treated first in the literature by McGrew and Fluhrer [23].We first give a definition of reforgeability, both in the stateless and stateful settings. We then examine a variety of well-known MAC schemes and assess their resistance to reforgeability attacks. We find that for all stateless schemes and many stateful schemes there exists an attack that enables efficient generation of forgeries given knowledge of an existing collision in tags. In some cases this involves fairly constrained modification of just the final block of some fixed message; in other cases we obtain the MAC key and have free rein. For each stateful scheme where we could not find an attack, we then turned our attentions to another related problem: nonce misuse. That is, if nonces are reused with the same key, can we forge multiple times? The answer is an emphatic "yes." For many of these MACs only a single protocol error is required to break the security; querying to the birthday bound is unnecessary.

Figure 1 and Figure 2 give a synopsis of our findings. In most cases, our attack is based on finding collisions and this in turn leads to a substantial number of subsequent forgeries; the degree to which each scheme breaks is noted in the table. For some Wegman-Carter-Shoup (WCS) [7, 27] MACs, the attack is more severe: nonce misuse yields the universal hash family instance almost immediately.

After an earlier draft of this paper appeared on eprint, many of the attacks in Figure 1 and Figure 2 were subsequently improved in [18] by Handschuh and Preneel. In light of this, we include no attacks within this version of the paper. For attack details we refer the interested reader to the full version of this paper [8] and the other literature on this subject [10, 18, 23, 24].

These attacks were sufficient to make us wonder if there exists an efficient and practical MAC scheme resistant to reforgeability attacks. A natural first try is

UHF in FH mode	Expected queries for j forgeries	Reveals key	Queries for key recovery
hash127/Poly1305	$C_1 + \log m + j$	$\checkmark$	$C_1 + \log m$
VMAC	$C_1 + 2j$		
Square Hash	$C_1 + 2j$	$\checkmark$	mC_1
Topelitz Hash	$C_1 + 2j$		
Bucket Hash	$C_1 + 2j$		
MMH/NMH	$C_1 + 2j$		

UHF in WCS mode with nonce misuse	Expected queries for j forgeries	Repeated nonce	Reveals key	Queries for key recovery
hash127/Poly1305	$2 + \log m + j$	1	$\checkmark$	$2 + \log m$
VMAC	$C_1 + 2j$	$C_1 + j$		
Square Hash	$3m + j$	m	$\checkmark$	$3m$
Topelitz Hash	$2j + 2$	1		
Bucket Hash	$2j + 2$	1		
MMH/NMH	$2m + j$	m	$\checkmark$	$2m$

Fig. 2. Results for Carter-Wegman MACs. The top table lists 6 well-known universal hash families, each made into a MAC via the FH construction [11, 29] where the hash family is composed with a pseudorandom function to produce the MAC tag. These similarly succumb to reforgeability attacks after a collision in the output tags, with hash127/Poly1305 and Square-Hash surrendering their key in the process. The last column gives the expected number of queries for key recovery, where possible. The bottom table considers the same hash families in the Wegman-Carter-Shoup (WCS) [7, 27] paradigm (the most prominent MAC paradigm for ϵ-AU hash families), but where nonces are misused and repeated. With many families, only one repeated nonce query is enough to render the MAC totally insecure. Others reveal the key with a few more queries using the same nonce. See [18] for further attacks on these and other hash families in a similar setting.

to add state, in the form of a nonce inserted in a natural manner, to the schemes above. We show, however, that this approach can be insufficient or insecure when subtly misused. Another approach would be to use a stateless MAC such as HMAC, and truncate the output so a collision in tags does not expose some exploitable internal information. However, this is also somewhat unsatisfactory because all the fastest MACs are stateful WCS-style MACs where trucation severely reduces the security.

We therefore devised a new (stateful) scheme, WMAC, that allows nonce reuse and where for most parameter sizes guessing the tag is the best reforgeability strategy. The scheme is described fully in Section 3 but briefly it works as follows.

Let $\mathcal{H}$ be some $\epsilon-$AU hash family $\mathcal{H} = \{h : D \rightarrow \{0,1\}^l\}$, and $\mathcal{R}$ a set of functions $\mathcal{R} = \text{Rand}(l+b, n)$. Let $\rho \xleftarrow{\$} \mathcal{R}$ and $h \xleftarrow{\$} \mathcal{H}$; the shared key is (ρ, h). Let $\langle cnt \rangle_b$ denote the encoding of cnt using b bits. To MAC a message (M, cnt), the signer first ensures that $cnt < 2^b - 1$ and if so sends $(cnt, \rho(\langle cnt \rangle_b \parallel h(M)))$. To verify a received message M with tag (i, Tag), the verifier computes $\rho(\langle i \rangle_b \parallel h(M))$ and ensures it equals Tag.

WHY WMAC? There are essentially four parameters which much be balanced when choosing a suitable MAC: speed, security, tag length, and deployment feasibility. WCS MACs provide excellent performance on the first two items, but require long tags and absolutely non-repeatable nonces (which also increases the tag length), a potential deployment problem where the state might have to be consistent across several machines. Stateless MACs, whose tags may be truncated without degrading security and therefore tend to do well on the last two items, lag behind on the first two.

WMAC can be seen as a compromise between the two sets of MACs. It has the speed of the fastest WCS MACs but the tag length may be truncated appropriately and nonces may be reused. A fixed nonce may be used for all queries if desired, effectively yielding the FH [11, 29] scheme as a special case. At the other extreme end, nonces are never repeated and WMAC retains a high degree of security comparable to the WCS setting. For most real-world applications that may already have implicit nonces (via the underlying networking protocol, eg) and that could use the added security benefits from nonces but do not want to enforce nonce uniqueness, WMAC is the best solution.

As an example, consider the following concrete WMAC instantiation. Let $\epsilon \leq 2^{-82}$, $b = 8$, and our PRF will be AES truncated to 24 bits. Then after 2^{32} signing queries and 2^{24} verification queries, one forgery is expected (from guessing the output of the PRF). The hash family can be a variant of the VHASH used in VMAC-128, so that the speed of the family is comparable to VMAC-128.[1] Moreover, the total tag length, including the nonce is only 32 bits. There is no efficient MAC which, using 32 bits for both the tag and nonce, can safely MAC as many messages with so few expected forgeries. (Note that the nonce greatly helps the security in this case; without it an expected 64 forgeries would be possible.)

We stress that although WMAC offers good tradeoffs for resource-constrained environments where some forgeries may be acceptable, it is still susceptible to attacks that exploit some bad event that occurs during operation, usually related to the value of ϵ for the ϵ-almost universal hash family used. To be clear, the attacks from [18] still apply and indeed come within a constant factor of matching the bound given in our security reduction.[2]

RELATED WORK. David McGrew and Scott Fluhrer have also done some work [23] on a similar subject, produced concurrently with our work but published earlier. They examine MACs with regard to multiple forgeries, although they view the subject from a different angle. They show that for HMAC, CBC MAC, and GMAC from the Galois Counter Mode (GCM) of operation for blockciphers

---

[1] Dan Bernstein has recently proposed [5] an almost-universal hash family which should be as fast or faster than VMAC-64, but which uses a much smaller key than VMAC. Bernstein's hash would use fewer multiplications and additions than VMAC-128, although those operations are done in some field $\mathcal{F}$, not modulo 2^n.

[2] Our bound also highlights interesting behavior with a verification query-only attack when the length of the tag is much smaller than $\lg(\epsilon^{-1})$. This case is also matched by essentially the attacks from [18].

[21], reforgeability is possible. However, they examine reforgeability in terms of the number of expected forgeries (parameterized by the number of queries) for each scheme, which is dependent on the precise security bounds for the respective MACs. Although our focus is somewhat different, our work complements their paper by showing their techniques and bounds apply to all major MACs.

Handschuh and Preneel investigated attacks on ϵ-almost universal hash families used in Wegman-Carter-Shoup mode MACs, and found new classes of attacks [18]. Their attacks improve on ours in several ways, probably the most significant of which is that they do not require misuse of nonce values to work.

OUTLINE OF THE PAPER. In the next section we cover the basic notation and security models used. After that, we jump right in to the discussion of WMAC and its security reduction, our main contribution, deferring the attacks that motivated its construction to the full version [8].

2   Preliminaries

Let $\{0,1\}^n$ denote the set of all binary strings of length n. For an alphabet Σ, let Σ^* denote the set of all strings with elements from Σ. Let $\Sigma^+ = \Sigma^* - \{\epsilon\}$ where ϵ denotes the empty string. For strings s, t, let $s \| t$ denote the concatenation of s and t. For set S, let $s \xleftarrow{\$} S$ denote the act of selecting a member s of S according to a probability distribution on S. Unless noted otherwise, the distribution is uniform. For a binary string s let $|s|$ denote the length of s. For a string s where $|s|$ is a multiple of n, let $|s|_n$ denote $|s|/n$. Unless otherwise noted, given binary strings s, t such that $|s| = |t|$, let $s \oplus t$ denote the bitwise XOR of s and t. For a string M such that $|M|$ is a multiple of n, $|M|_n = m$, then we will use the notation $M = M_1 \| M_2 \| \ldots \| M_m$ such that $|M_1| = |M_2| = \ldots = |M_m|$. Let $\mathrm{Rand}(l, L) = \{f \mid f : \{0,1\}^l \rightarrow \{0,1\}^L\}$ denote the set of all functions from $\{0,1\}^l$ to $\{0,1\}^L$.

UNIVERSAL HASH FAMILIES. Universal hash families are used frequently in the cryptographic literature. We now define several notions needed later.

Definition 1. *(Carter and Wegman [11]) Fix a domain $\mathcal{D}$ and range $\mathcal{R}$. A finite multiset of hash functions $\mathcal{H} = \{h : \mathcal{D} \rightarrow \mathcal{R}\}$ is said to be* **Universal** *if for every $x, y \in \mathcal{D}$ with $x \neq y$, $Pr_{h \in \mathcal{H}}[h(x) = h(y)] = 1/|\mathcal{R}|$.*

Definition 2. *Let $\epsilon \in \mathbb{R}^+$ and fix a domain $\mathcal{D}$ and range $\mathcal{R}$. A finite multiset of hash functions $\mathcal{H} = \{h : \mathcal{D} \rightarrow \mathcal{R}\}$ is said to be ϵ-**Almost Universal** (ϵ-AU) if for every $x, y \in \mathcal{D}$ with $x \neq y$, $Pr_{h \in \mathcal{H}}[h(x) = h(y)] \leq \epsilon$.*

Definition 3. *(Krawczyk [20], Stinson [28]) Let $\epsilon \in \mathbb{R}^+$ and fix a domain $\mathcal{D}$ and range $\mathcal{R} \subseteq \{0,1\}^r$ for some $r \in \mathbb{Z}^+$. A finite multiset of hash functions $\mathcal{H} = \{h : \mathcal{D} \rightarrow \mathcal{R}\}$ is said to be ϵ-**Almost XOR Universal** (ϵ-AXU) if for every $x, y \in \mathcal{D}$ and $z \in \mathcal{R}$ with $x \neq y$, $Pr_{h \in \mathcal{H}}[h(x) \oplus h(y) = z] \leq \epsilon$.*

Throughout the paper we assume that a given value of ϵ for an ϵ-AU or ϵ-AXU family includes a parameter related to the length of the messages. If we speak of a fixed value for ϵ, then we implicitly specify an upper bound on this length.

MESSAGE AUTHENTICATION. Formally, a stateless message authentication code is a pair of algorithms, (MAC, VF), where MAC is a 'MACing' algorithm that, upon input of key $K \in \mathcal{K}$ for some key space $\mathcal{K}$, and a message $M \in \mathcal{D}$ for some domain $\mathcal{D}$, computes a τ-bit tag Tag; we denote this by Tag $=$ MAC$_K(M)$. Algorithm VF is the 'verification' algorithm such that on input $K \in \mathcal{K}$, $M \in \mathcal{D}$, and Tag $\in \{0,1\}^\tau$, outputs a bit. We interpret 1 as meaning the verifier *accepts* and 0 as meaning it *rejects*. This computation is denoted VF$_K(M, \text{Tag})$. Algorithm MAC can be probabilistic, but VF typically is not. A restriction is that if MAC$_K(M) = \text{Tag}$, then VF$_K(M, \text{Tag})$ must output 1. If MAC$_K(M) = $ MAC$_K(M')$ for some K, M, M', we say that messages M and M' *collide* under that key.

The common notion for MAC security is resistance to adaptive chosen message attack [3]. This notion states, informally, that an adversary *forges* if he can produce a new message along with a valid tag after making some number of queries to a MACing oracle. Because we are interested in *multiple* forgeries, we now extend this definition in a natural way.

Definition 4 (MAC Security—j Forgeries). *Let $\Pi = (\text{MAC}, \text{VF})$ be a message authentication code, and let A be an adversary. We consider the following experiment:*

> *Experiment* $\mathbf{Exmt}_\Pi^{juf\text{-}cma}(A, j)$
> $K \xleftarrow{\$} \mathcal{K}$
> *Run* $A^{\text{MAC}_K(\cdot), \text{VF}_K(\cdot, \cdot)}$
> *If A made j distinct verification queries (M_i, Tag_i), $1 \le i \le j$, such that*
> — VF$_K(M_i, \text{Tag}_i) = 1$ *for each i from 1 to j*
> — *A did not, prior to making verification query (M_i, Tag_i), query its* MAC$_K$ *oracle at M_i*
> *Then return 1 else return 0*

The juf-cma advantage of A in making j forgeries is defined as

$$\mathbf{Adv}_\Pi^{juf\text{-}cma}(A, j) = \Pr\left[\mathbf{Exmt}_\Pi^{juf\text{-}cma}(A, j) = 1\right].$$

For any q_s, q_v, μ_s, μ_v, Time ≥ 0 we overload the above notation and define

$$\mathbf{Adv}_\Pi^{juf\text{-}cma}(t, q_s, \mu_s, q_v, \mu_v, j) = \max_A \{\mathbf{Adv}_\Pi^{juf\text{-}cma}(A, j)\}$$

where the maximum is over all adversaries A that have time-complexity at most Time, *make at most q_s MAC-oracle queries, the sum of those lengths is at most μ_s, and make at most q_v verification queries where the sum of the lengths of these messages is at most μ_v.*

The special case where $j = 1$ corresponds to the regular definition of MAC security. If, for a given MAC, $\mathbf{Adv}_{\Pi}^{juf\text{-}cma}(t, q_s, \mu_s, q_v, \mu_v, j) \leq \epsilon$, then we say that MAC is (j, ϵ)-secure. For the case $j = 1$, the scheme is simply ϵ-secure.

It is worth noting that the adversary is allowed to adaptively query VF_K and is not penalized for queries that return 0. All that is required is for j distinct queries to VF_K return 1, subject to the restriction these queries were not previously made to the MACing oracle.

STATEFUL MACs. We will also examine stateful MACs that require an extra parameter or nonce value. Our model will let the adversary control the nonce, but limit the number of MAC queries per nonce. Setting this limit above 1 will simulate a protocol error where nonces are re-used in computing tags.

A stateful message authentication code is a pair of algorithms, (MAC, VF), where MAC is an algorithm that, upon input of key $K \in \mathcal{K}$ for some key space $\mathcal{K}$, a message $M \in \mathcal{D}$ for some domain $\mathcal{D}$, and a state value S from some prescribed set of states $\mathcal{S}$, computes a τ-bit tag Tag; we denote this by Tag $= \mathrm{MAC}_K(M, S)$. Algorithm VF is the verification algorithm such that on inputs $K \in \mathcal{K}$, $M \in \mathcal{D}$, Tag $\in \{0, 1\}^\tau$, and $S \in \mathcal{S}$, VF outputs a bit, with 1 representing accept and 0 representing reject. This computation is denoted $\mathrm{VF}_K(M, S, \mathrm{Tag})$. A restriction on VF is that if $\mathrm{MAC}_K(M, S) = \mathrm{Tag}$, then $\mathrm{VF}_K(M, S, \mathrm{Tag})$ must output 1.

As discussed later, all our attacks on stateless MACs work by examining the event of a collision in tag values, by virtue of the birthday phenomenon or otherwise. With stateful MACs an adversary may see collisions in tags, but the state mitigates, and in most cases neutralizes, any potentially damaging information leaked in such an event. With that in mind, we will consider two different security models with regard to stateful MACs. In one, we treat stateful MACs as intended: nonces are not repeated among queries, but repeated nonces may be used with verification queries. Many MACs we examine have security proofs in this model, so it is not surprising that they perform well, even with short tags. Others don't, and we provide the analysis.

We also provide analysis for a plausible and interesting protocol error: that in which nonces are reused. This can happen in several reasonable scenarios: 1) the nonce is a 16- or 32-bit variable, and overflow occurs unnoticed, and 2) the same key is used across multiple virtualized environments. This latter case may happen when MACs in differing virtualized environments are keyed with the same entropy pools, or one environment is cloned from another.

These protocol misuses are captured formally by allowing an adversary a maximum of α queries per nonce between the two oracles. For most MACs we examine, α need only be 2 for successful reforgery attacks.

Definition 5 (Stateful MAC Security—j Forgeries). *Let $\Pi = (\mathrm{MAC}, \mathrm{VF})$ be a stateful message authentication code, and let A be an adversary. We consider the following experiment:*

Experiment $\mathbf{Exmt}_{\Pi}^{jsuf\text{-}cma}(A, j, \alpha)$
$K \xleftarrow{\$} \mathcal{K}$

Run $A^{\mathrm{MAC}_K(\cdot),\mathrm{VF}_K(\cdot,\cdot)}$

If A made j distinct verification queries $(M_i, s_i, \mathrm{Tag}_i)$, $1 \leq i \leq j$, such that

— $\mathrm{VF}_K(M_i, s_i, \mathrm{Tag}_i) = 1$ for each i from 1 to j
— A did not, prior to making verification query $(M_i, s_i, \mathrm{Tag}_i)$, query its MAC oracle with (M_i, s_i)
— A did not make more than α queries to MAC_K with the same nonce.
Then return 1 else return 0

The jsuf-cma advantage of A in making j forgeries is defined as

$$\mathbf{Adv}_{\Pi}^{jsuf\text{-}cma}(A) = \Pr\left[\mathbf{Exmt}_{\Pi}^{jsuf\text{-}cma}(A, j, \alpha) = 1\right].$$

For any $q_s, q_v, \mu_s, \mu_v, \mathsf{Time}, j, \alpha \geq 0$ we let

$$\mathbf{Adv}_{\Pi}^{jsuf\text{-}cma}(t, q_s, \mu_s, q_v, \mu_v, j, \alpha) = \max_{A}\{\mathbf{Adv}_{\Pi}^{jsuf\text{-}cma}(A, j, \alpha)\}$$

where the maximum is over all adversaries A that have time-complexity at most Time, *make at most q_s MACing queries, the sum of those lengths is at most μ_s, where no more than α queries were made per nonce, and make at most q_v verification queries where the sum of the lengths of the messages involved is at most μ_v.*

If, for a given MAC, $\mathbf{Adv}_{\Pi}^{jsuf\text{-}cma}(t, q_s, \mu_s, q_v, \mu_v, j, \alpha) \leq \epsilon$, then we say that MAC is (j, ϵ)-secure. For the case $j = 1$, the scheme is simply ϵ-secure.

3   A Fast, Stateful MAC with Short Tags

For some stateful MACs we found no attack, and others are accompanied by a proof of security. Similarly, tag truncation is a simple technique which may be used to ensure that security is retained well after one starts seeing collisions in tags. Perhaps we should be satisfied and consider our search for reforgeability-resistant MACs complete. However, both of these techniques have drawbacks for the applications in mind which require very short tags. Namely, the nonce value must be transmitted with each query, and tag truncation may not be used on the fastest MACs without seriously degrading security.[3]

It is with these thoughts in mind, and with newfound knowledge of the perils associated with nonce misuse in WCS MACs, that we designed WMAC. WMAC boasts speed comparable to VMAC/Poly1305, can use much shorter tags, and is the first MAC we know of to use repeating nonces, a side effect of which is shorter tags.

---

[3] Truncating the tag of VMAC or Poly1305-AES by t bits also effectively grows ϵ for the ϵ-AU family by a multiplicative factor of 2^t. If these MACs were to be revised into FH mode, truncation would be possible, but without nonces they succumb to attacks covered in this paper, and with nonces ϵ needs to be unacceptably reduced to make room for the nonce input.

WMAC. Let $\mathcal{H} = \{h : \mathcal{D} \to \mathcal{R}\}$ be a family of ϵ-AU hash functions and let $F : \mathcal{K} \times \mathcal{T} \times \mathcal{R} \to \{0,1\}^n$ be a PRF. We define

$$\text{WMAC}[\mathcal{H}, F]^t_{h, F_K}(x) = F_K(t, h(x)),$$

where $t \in \mathcal{T}$, $h \xleftarrow{\$} \mathcal{H}$, $K \xleftarrow{\$} \mathcal{K}$, and $x \in \mathcal{D}$. Informally, once keyed with the selection of $K \in \mathcal{K}$ and AU hash instance h, WMAC accepts a message x and nonce t as inputs and returns $F_K(t, h(x))$ as the tag.

NONCES IN WMAC. WMAC's nonce use can be considered as "flexible" in the sense that the security analysis is done for different uses. To model this, we are mainly interested in an adversary of somewhat limited capability, that is, an adversary which can make at most α signing queries for each nonce $t \in \mathcal{T}$. The adversary's verification queries per nonce are not similarly bounded. We call such an adversary α-limited, and define $\mathbf{Adv}_\Pi^{jsuf\text{-}cma}(q, t, \alpha)$ be the maximum of $\mathbf{Adv}_\Pi^{jsuf\text{-}cma}(A)$ over every α-limited adversary A which makes at most $q = q_s + q_v$ oracle queries (q_s to the signing oracle and q_v to the verification oracle) and halts within time Time. We say that Π is secure as an α-limited MAC, if $\mathbf{Adv}_\Pi^{jsuf\text{-}cma}(q, t, \alpha)$ is negligibly small for any reasonably large q and Time.

As an example, the FH and FCH [11, 29] modes of operation are special cases of WMAC where α is set to q_s and 1, respectively.

Theorem 1. *For any α-limited adversary A of WMAC which makes at most $q = q_s + q_v$ queries in time Time, there exists an adversary B of F such that*

$$\mathbf{Adv}_{\text{WMAC}}^{jsuf\text{-}cma}(A) \le \mathbf{Adv}_F^{prf}(B) + \frac{\epsilon(\alpha - 1)q_s}{2} +$$

$$\frac{\epsilon}{2^{n-1}} \left(q_v^2 + q_v q_s + \max\{2^n, q^{\frac{1}{2}} 2^{\frac{n}{2}+3}\} q_v \right) + \delta(j, n, q_v).$$

and where B makes at most q queries, using time proportional to Time $+ \text{Hash}(q)$, *where* $\text{Hash}(1)$ *is the time to compute $h(M)$ for some message $M \in \mathcal{D}$ and $h \xleftarrow{\$} \mathcal{H}$. The term $\delta(j, n, q_v)$ is defined as*

$$\sum_{k=j}^{|S|} \sum_{X \in S_k} \left[\Pi_{x' \in S : x' \notin X} \left(1 - \frac{q_{v,x'}}{2^n} \right) \Pi_{x \in X} \left(\frac{q_{v,x}}{2^n} \right) \right]$$

where S is the set of distinct message-tag pairs seen in all verification queries, S_k is the set of k-tuples in S, and for an element $x \in S$, $q_{v,x}$ is the number of verification queries made for that element.

DISCUSSION OF THE BOUND AND EXPECTED NUMBER OF FORGERIES. Mc-Grew and Fluhrer discuss the expected number of forgeries for GMAC (a WCS MAC)[21], CBC MAC, and HMAC in terms of ϵ, n, and q. Our specific attacks complement their analysis by showing their methods apply to all major stateful and stateless MACs. Essentially, they show that for stateless MACs, the expected number of forgeries is $cq^3 2^{-n} + \mathcal{O}(q^4 2^{-2n})$, where n is output size of the blockcipher or hash function and c is a constant. For WCS MACs, they show the expected number of forgeries is $cq^2 \epsilon + \mathcal{O}(q^3 \epsilon^2)$.

We believe this sort of analysis should supplant the current definition of MAC security for the simple reason that it more accurately quantifies the risks for MACing q messages over the lifetime of one key and, in the case of our bound in particular, makes the bound more easily understood. Rather than giving the traditional security bound and suggesting the number of queries be "well below" a certain value ($2^{n/2}$, usually), producing a specific expected number of forgeries is much superior.

And in this spirit, we give a formula for the expected number of forgeries for WMAC, which also helps to understand the rather obtuse bound in theorem 1. For a given MAC scheme $\Pi = (\text{MAC}, \text{VF})$, let $E(\text{Forge}_\Pi, q_s, q_v)$ denote the expected number of forgeries when q_s queries are allowed to the MAC oracle and q_v queries are allowed to the VF oracle.

Following [23], we will assume WMAC uses an ideal random function as the PRF. Unless q_v is unreasonably large, the expected number of forgeries is overwhelmingly influenced by the chance that an adversary sets bad to true during one of the q_s queries to the MAC oracle. If this occurs, we give the adversary q_v forgeries. There is a small chance bad is set to true in the verification phase and to simplify the analysis we admit q_v forgeries in this case as well. Thus, we bound the expected number of forgeries as q_v times the probability that bad is set to true. Finally, we must consider the expected number of forgeries when the adversary merely guesses the correct outputs of the ideal random function, which is $q_v 2^{-n}$. Thus,

$$E(\text{Forge}_{\text{WMAC}}, q) \leq \frac{\epsilon q_v q_s (\alpha - 1)}{2} + \frac{q_v \epsilon}{2^{n-1}} \left(q_v^2 + q_v q_s + 2^{n/2+3} q_v \sqrt{q} \right) + q_v 2^{-n}.$$

It is this formula which is used to give figures in the example from section 1. Note that when $q = q_s = q_v$, letting α take on values in $\{1, q\}$ gives bounds similar to those from [23].

Proof. Without loss of generality, we may assume that A doesn't ask the same signing query twice, and that A makes all signing queries before making any verification queries.[4] Our adversary B has access to an oracle $Q(t, x)$. We construct B, which runs A as a subroutine, by directly simulating the oracles A expects. That is, in the startup phase, B randomly selects $h \xleftarrow{\$} \mathcal{H}$. It then runs A, responding to A's signing query (t, M) by querying its oracle at $(t, h(M))$ and returning the answer to A. Similarly, B responds to a verification query (t, M, Tag) by querying its oracle at $(t, h(M))$ and returning 1 if the answer is equal to Tag, 0 otherwise. After A has completed all queries, B outputs the same bit as A.

---

[4] This condition is not required by our security reduction— an adversary may make queries in any order she wishes — but for ease of notation we adopt it.

```
     ┌─────────────────────────────────────────────────────────────────────────────────┐
     │ Procedure Initialize                                                              │
  0  │ V ← ∅, h ←$ H, ρ ←$ Rand(T × R, {0,1}ⁿ),                                          │
     │ InitializeMap(Map, T × D, T × R), InitializeMap(Map_o, T × D, T × R)              │
     │ Procedure MAC(t, x)                                                                │
  1  │ v ← h(x)                                                                           │
     │                                                                                    │
  2  │ If (t, v) ∈ V then { bad ← true, │ (t, v) ←$ T × R \ V │ }                         │
  3  │ V ← V ∪ (t, v)                                                                     │
  4  │ return ρ(t, v)                                                                     │
     │ Procedure VF(t, x, Tag)                                                            │
  5  │ If Map[(t, x)] = ⊥ then {                                                          │
  6  │    v ← h(x), Map[(t, x)] ← (t, v)                                                  │
     │    ┌──────────────────────────────────────────────────────────────────────────┐   │
  7  │    │ If (t, v) ∈ V then { Map_o[(t, x)] ← (t, v), Map[(t, x)] ←$ T × R \ V,    │   │
     │    │ (t, v) ← Map[(t, x)] }                                                     │   │
  8  │    │ V ← V ∪ (t, v)                                                             │   │
     │    └──────────────────────────────────────────────────────────────────────────┘   │
     │    }                                                                               │
  9  │ If Map_o[(t, x)] ≠ ⊥ then {                                                        │
 10  │    If Tag = ρ(Map_o[(t, x)]) or Tag = ρ(Map[(t, x)]) then { bad ← true }           │
     │    }                                                                               │
 11  │ return Tag = ρ(Map[(t, x)])                                                        │
     └─────────────────────────────────────────────────────────────────────────────────┘
```

Fig. 3. Game G_0 and $\boxed{\text{Game } G_1}$

Consider the games G_0 and G_1 in figure 3, where Game G_1 includes the boxed statement. The function InitializeMap takes as arguments a map name, a domain, and a range, and initializes a map with the input name where every map lookup returns $\bot$.

Clearly, A^{G_0} corresponds to the experiment where A is given access to the signing oracle $\rho(t, h(x))$ and verification oracle $\rho(t, h(x)) = \text{Tag}$, and A^{G_1} corresponds to the experiment where the tags for A's queries (either signing or verification), are choosen as uniform random outputs. Because A doesn't ask the same signing query twice and by the way we constructed B, this is precisely the answers A will get when the signing oracle is a uniform random function and the verification oracle behaves similarly. Finally, when B's oracle is F_K, B simulates the oracle A expects exactly. Therefore,

$$\mathbf{Adv}_F^{prf}(B) = \Pr\left[1 \leftarrow A^{\text{WMAC}_{K,h}}\right] - \Pr\left[1 \leftarrow A^{G_0}\right]$$

$$= \Pr\left[1 \leftarrow A^{\text{WMAC}_{K,h}}\right] - \Pr\left[1 \leftarrow A^{G_1}\right] + \Pr\left[1 \leftarrow A^{G_1}\right] - \Pr\left[1 \leftarrow A^{G_0}\right]$$

$$\geq \Pr\left[1 \leftarrow A^{\text{WMAC}_{K,h}}\right] - \Pr\left[1 \leftarrow A^{G_1}\right] - \Pr\left[A^{G_1} \text{ sets } bad\right]$$

$$= \mathbf{Adv}_{\text{WMAC}}^{jsuf\text{-}cma}(A) - \Pr\left[1 \leftarrow A^{G_1}\right] - \Pr\left[A^{G_1} \text{ sets } bad\right],$$

since G_0 and G_1 are identical-until-bad games.

The term $\delta(j, n, q_v)$ represents the probability of A's success when presented with the oracle of game G_1. In this case, a verification query (t_i, x_i, τ_i) with a new message-nonce pair (t_i, x_i) 'succeeds' iff $\rho(t_i, h(x_i)) = \tau_i$, and this happens with probability 2^{-n}. Similarly, for ℓ verification queries made with (t_i, x_i) as the message-tag pair, the total success probability is $\ell/2^n$. By summing over all possibilities for correct and incorrect guesses, we have that

$$\sum_{k=j}^{|S|} \sum_{X \in S_k} \left[\Pi_{x' \in S : x' \notin X} \left(1 - \frac{q_{v,x'}}{2^n} \right) \Pi_{x \in X} \left(\frac{q_{v,x}}{2^n} \right) \right].$$

(A much more intuitive grasp of this term can be obtained by considering its expected value, $q_v 2^{-n}$. This can be seen by the fact that the expected number of forgeries for any one message tag pair $x \in S$ is $q_{v,x} 2^{-n}$; the value follows by linearity of expectation of independent events and the fact that $q_v = \sum_{x \in S} q_{v,x}$.)

```
  Procedure Initialize
0   V ← ∅, h ←$ H, ρ ←$ Rand(T × R, {0,1}ⁿ), InitializeMap(Map, T × D, T × R),
    InitializeMap(Map_o, T × D, T × R), InitializeMap(O, T × R, {0,1}ⁿ)
  Procedure Q(t, x)
1   v ← h(x)
2   If (t, v) ∈ V then { bad ← true, (t, v) ←$ T × R \ V }
3   V ← V ∪ (t, v), O[(t, v)] ←$ {0,1}ⁿ
4   return O[(t, v)]
  Procedure VF(t, x, Tag)
5   If Map[(t, x)] = ⊥ then {
6     v ← h(x), Map[(t, x)] ← (t, v)
7     If (t, v) ∈ V then { Map_o[(t, x)] ← (t, v), Map[(t, x)] ←$ T × R \ V, (t, v) ← Map[(t, x)] }
8     V ← V ∪ (t, v), O[(t, v)] ←$ {0,1}ⁿ
    }
9   If Map_o[(t, x)] ≠ ⊥ then {
10    If Tag = O[Map_o[(t, x)]] or Tag = O[Map[(t, x)]] then { bad ← true }
    }
11  return Tag = O[Map[(t, x)]]
```

Fig. 4. Game G_2

Now we must bound the probability that bad is set to true, but first we go through some output distribution-preserving game transitions to make the analysis easier. The difference between Game G_1 and Game G_2 is that in G_2, MAC(t, x) returns a uniform random value τ from $\{0,1\}^n$ and VF(t, x, Tag) chooses its outputs in line 8 from uniform random values from $\{0,1\}^n$. But in Game G_1, $\rho(t, v)$ is computed for all distinct (t, v) in line 4 and in line 11 $\rho(\text{Map}[(t, x)])$ is computed for all distinct values of $\text{Map}[(t, x)]$ when distinct (t, x) values are used. Therefore the two games are identical. In Game G_3, we clean things up by removing the unnecessary ρ, and removing the statement $(t, v) \xleftarrow{\$} T \times R \setminus V$. This is possible because this occurs after bad ← true.

In Game G_4, we first generate all the random answers to the queries of A, and on ith signing query, save the query and just return the ith random answer. The verification queries are handled similarly by using the saved values. We can check whether we should set bad at the finalization step, using the saved query values. Clearly, all games G_2, G_3, and G_4 preserve the probability that bad gets set. Therefore,

$$\mathbf{Adv}_{\text{WMAC}}^{jsuf\text{-}cma}(A) \leq \mathbf{Adv}_F^{prf}(B) + \Pr[A^{G_4} \text{ sets bad}] + \delta(j, n, q_v).$$

Procedure Initialize

0    $V \leftarrow \emptyset$, $O \leftarrow \emptyset$, $h \xleftarrow{\$} \mathcal{H}$, InitializeMap(Map, $\mathcal{T} \times \mathcal{D}, \mathcal{T} \times \mathcal{R}$),
     InitializeMap(Map$_o$, $\mathcal{T} \times \mathcal{D}, \mathcal{T} \times \mathcal{R}$), InitializeMap($O$, $\mathcal{T} \times \mathcal{R}, \{0,1\}^n$)

Procedure $Q(t, x)$

1    $v \leftarrow h(x)$
2   If $(t, v) \in V$ then $\{$ bad $\leftarrow$ true $\}$
3    $V \leftarrow V \cup (t, v)$, $O[(t, v)] \xleftarrow{\$} \{0, 1\}^n$
4   return $O[(t, v)]$

Procedure VF(t, x, Tag)

5   If Map$[(t, x)] = \bot$ then $\{$
6      $v \leftarrow h(x)$, Map$[(t, x)] \leftarrow (t, v)$
7   If $(t, v) \in V$ then $\{$ Map$_o[(t, x)] \leftarrow (t, v)$, Map$[(t, x)] \xleftarrow{\$} \mathcal{T} \times \mathcal{R} \setminus V$, $(t, v) \leftarrow$ Map$[(t, x)] \}$
8      $V \leftarrow V \cup (t, v)$, $O[(t, v)] \xleftarrow{\$} \{0, 1\}^n$
     $\}$
9   If Map$_o[(t, x)] \neq \bot$ then $\{$
10   If Tag $= O[\text{Map}_o[(t, x)]]$ or Tag $= O[\text{Map}[(t, x)]]$ then $\{$ bad $\leftarrow$ true $\}$
     $\}$
11   return Tag $= O[\text{Map}[(t, x)]]$

Fig. 5. Game G_3

Procedure Initialize

0    $h \xleftarrow{\$} \mathcal{H}$, $(\tau_1, \dots, \tau_{q_s + \#q_v}) \xleftarrow{\$} (\{0, 1\}^n)^{q_s + \#q_v}$, $i \leftarrow 0$,
     InitializeMap(O, $\mathcal{T} \times \mathcal{R}, \{0, 1\}^n$)

Procedure $Q(t, x)$

1    $i \leftarrow i + 1$, $t_i \leftarrow t$, $x_i \leftarrow x$, $O[(t, x)] \leftarrow \tau_i$
2   return τ_i

Procedure VF(t, x, Tag)

3   If $O[(t, x)] = \bot$ then $\{$
4      $i \leftarrow i + 1$, $t_i \leftarrow t$, $x_i \leftarrow x$, $O[(t, x)] \leftarrow \tau_i$, Tag$_i \leftarrow$ Tag
     $\}$
5   return $\tau_i = \text{Tag}_i$

Procedure Finalize

6   If $(t_i, h(x_i)) = (t_j, h(x_j))$ for some $i < j \leq q_s$, then $\{$ bad $\leftarrow$ true $\}$
7   If $(t_i, h(x_i)) = (t_j, h(x_j))$ for some $i < j$, $q_s < j$ then $\{$
8   If $O[(t_i, x_i)] = \text{Tag}_j$ or $O[(t_j, x_j)] = \text{Tag}_j$ then $\{$ bad $\leftarrow$ true $\}$
     $\}$

Fig. 6. Game G_4

We will use the fact that

$$\Pr[A^{G_4} \text{ sets } bad] \leq \Pr[A^{G_4} \text{ sets } bad \text{ in line 6}] + \Pr[A^{G_4} \text{ sets } bad \text{ in line 8}].$$

It is easy to analyze the probability $\Pr[A^{G_4} \text{ sets } bad \text{ in line 6}]$; In Game G_4, the adversary A gets no information about h at all, and the random variables t_i and x_i are independent from h. Let's enumerate all the elements of $\mathcal{T}$ as $T_1, \dots, T_{|\mathcal{T}|}$, and let $q_{s,i}$ be the number of signing queries (t, x) such that $t = T_i$. Then,

$$\Pr[A^{G_4} \text{ sets } bad \text{ in line 6}] \leq \sum_{i=1}^{|\mathcal{T}|} \epsilon \cdot \frac{q_{s,i}(q_{s,i} - 1)}{2} \leq \sum_{i=1}^{|\mathcal{T}|} \epsilon \cdot \frac{q_{s,i}(\alpha - 1)}{2}$$

$$= \frac{\epsilon(\alpha - 1)}{2} \sum_{i=1}^{|\mathcal{T}|} q_{s,i} = \frac{\epsilon(\alpha - 1)q_s}{2}.$$

We must also bound the probability $\Pr[A^{G_4}$ sets bad in line 8$]$. The adversary A still learns no information about h, but we must account for an optimal tag guessing strategy with respect to bad being set to true. We first focus on the case where A does not guess multiple tags for a message-nonce pair and then handle the general case. For each value $k \in \mathcal{T}$ let S_k be the set of indices i such that $1 \le i \le q_s$ and $t_i = k$. Similarly, let V_k be the set of indices i such that $q_s < i \le q_s + q_v$ and $t_i = k$. Let g be the number of correctly guessed tags during the verification phase. Let $X_k = \{x_i : i \in S_k \vee (i \in V_k \wedge \mathrm{Tag}_i = \tau_i)\}$ and let $X_k^\tau = \{\tau_i : x_i \in X_k\}$. (Note that $\sum_{k \in \mathcal{T}} |X_k| = q_s + g$.) For any value $\tau \in \{0,1\}^n$, let $G_k(\tau) = \{x_i : \tau_i \in X_k^\tau, \tau = \tau_i\}$. Let $C_k = \max\{|G_k(\tau)| : \tau \in X_k^\tau\}$ and $C = \max\{C_k\}$ and let E_b be the the the event that A^{G_4} sets bad in line 8. Then,

$$\Pr[E_b] \le \sum_{k \in \mathcal{T}} \sum_{i \in V_k} \left(\max_{\tau \in X_k^\tau} \Pr\left[h(x_i) = h(x) : x \in G_k(\tau)\right] \right. \tag{1}$$

$$+ \Pr\left[h(x_i) = h(x) : x \in X_k\right] \cdot \Pr[\mathrm{Tag}_i = \tau_i] \tag{2}$$

$$+ \left. \sum_{j \in V_k, j < i} \Pr[h(x_j) = h(x_i)] \cdot \Pr\left[\mathrm{Tag}_j = \tau_j \vee \mathrm{Tag}_j = \tau_i\right] \right) \tag{3}$$

$$\le \sum_{k \in \mathcal{T}} \sum_{i \in V_k} \left(\epsilon C_k + \epsilon |X_k| 2^{-n} + \sum_{j \in V_k, j < i} \epsilon 2^{-n+1} \right) \tag{4}$$

$$\le \epsilon \sum_{k \in \mathcal{T}} \sum_{j=0}^{|V_k|-1} (C_k + (\alpha + g)2^{-n} + j 2^{-n+1}) \tag{5}$$

$$\le \epsilon \sum_{k \in \mathcal{T}} \left(|V_k|(C + (\alpha + g)2^{-n}) + 2^{-n+1} \binom{|V_k|}{2} \right) \tag{6}$$

$$\le \epsilon \left(q_v(C + (\alpha + g)2^{-n}) + 2^{-n+1} \sum_{k \in \mathcal{T}} \binom{|V_k|}{2} \right) \tag{7}$$

$$\le \epsilon \left(q_v(C + (\alpha + g)2^{-n}) + 2^{-n+1} \binom{q_v}{2} \right) \tag{8}$$

On a verification query $(t_j, x_j, \mathrm{Tag}_j)$ we consider two cases where the conditional on line 7 is met: $i \le q_s$ and $q_s < i$. Also, on line 8, there are two events that may set bad to true: A's guess may be correct for the unmodified output τ_i, or it may be a correct guess for the modified output τ_j. Suppose bad is set to true on line 8, then we distinguish these four events:

- $E_{1,j} : i \le q_s$ and A's guess was correct for the unmodified output.
- $E_{2,j} : i \le q_s$ and A's guess was correct for the modified output.
- $E_{3,j} : q_s < i$ and A's guess was correct for the unmodified output.
- $E_{4,j} : q_s < i$ and A's guess was correct for the modified output.

Then $\Pr[A^{G_4}$ sets bad in line 8$]$ on the j-th query is

$$\Pr[E_{1,j} \vee E_{2,j} \vee E_{3,j} \vee E_{4,j}] \le \Pr[E_{1,j}] + \Pr[E_{2,j}] + \Pr[E_{3,j}] + \Pr[E_{4,j}].$$

Lines (1) and (2) of the set of the equations denote $\Pr[E_{1,j}]$ and $\Pr[E_{2,j}]$, respectively, and line (3) contains $\Pr[E_{3,j} \vee E_{4,j}]$. The justification for line (1) is that an adversary's best strategy when $i \leq q_s$ is to guess the most frequently occuring tag returned during the signing phase (or a tag that is known by being guessed correctly during the verification phase). In lines (2) and (3) the adversary must try to guess an independent uniform random sample from 2^n values once the conditional is met. Line (4) upper bounds the probabilities for these events to occur, lines (5-7) simplify the equation, and the last inequality is justified by the fact that the quantity is maximized by making all verification queries with the same nonce.

Finally, with a simple argument we cover the case where during the verification phase multiple tags are guessed for a particular message-nonce pair. Since A learns nothing about h during the game, A has no way of learning which of its queries caused the conditional on line 7 to be true and gains no advantage from this approach. Indeed, the optimal strategy is to only make one verification query per message-nonce pair, so that the odds of line 8 being reached are increased with each query by forcing more values to be re-mapped as in line 7 of game G_3.

The full version contains a bound for C for values of interest. In particular, $C \leq \max\{1, \frac{q_s+g}{2^n} + 15\sqrt{\frac{q_s+g}{2^{n/2}}}\}$ and the expected value of g is $q_v 2^{-n}$. Putting it together, we have

$$
\Pr[A^{G_4} \text{ sets bad}] \leq \frac{\epsilon(\alpha-1)q_s}{2} + \epsilon\left(q_v(C + (\alpha+g)2^{-n}) + 2^{-n+1}\binom{q_v}{2}\right)
$$

$$
\leq \frac{\epsilon(\alpha-1)q_s}{2} + \epsilon\left(\frac{q_v^2}{2^n} + 2q_v C\right)
$$

$$
\leq \frac{\epsilon(\alpha-1)q_s}{2} + \epsilon\left(\frac{q_v^2}{2^n} + \frac{q_v q_s}{2^{n-1}} + \frac{q_v^2}{2^{2n-1}} + \frac{15 q_v \sqrt{q}}{2^{n/2}}\right)
$$

$$
\leq \frac{\epsilon(\alpha-1)q_s}{2} + \frac{\epsilon}{2^{n-1}}\left(q_v^2 + q_v q_s + 2^{n/2+3} q_v \sqrt{q}\right).
$$

4   Conclusions

We have shown that for most MACs, forging multiple times is not much harder than forging once. We then find that two natural ways of improving resistance to reforgeability are, unfortunately, mutually exclusive when applied to common MACs. WMAC, which aims to reconcile these two methods with a modern Carter-Wegman-Shoup MAC, is introduced and the security bounds given match the best known attacks [18]. WMAC provides parameter choices that yield constructions with varying security, speed, tag length, and use of state. For this flexibility, the inputs to WMAC are longer than other Wegman-Carter style MAC constructions and therefore messages take slightly longer to process.

Acknowledgments

John Black's work was supported by NSF CAREER-0240000 and a gift from the Boettcher Foundation. Many thanks to Aaram Yum and Yongdae Kim for their thoughtful insights and suggestions on earlier draft versions.

References

1. Association, A.B.: ANSI X9.19. Financial institution retail message authentication, Washington, D. C (August 1986)
2. Bellare, M., Canetti, R., Krawczyk, H.: Keying hash functions for message authentication. In: Koblitz [19], pp. 1–15
3. Bellare, M., Kilian, J., Rogaway, P.: The security of cipher block chaining. In: Desmedt [14], pp. 341–358
4. Bernstein, D.: Floating-point arithmetic and message authentication. Draft available as, http://cr.yp.to/papers/hash127.dvi
5. Bernstein, D.: Polynomial evaluation and message authentication. Draft available as, http://cr.yp.to/papers.html#pema
6. Bernstein, D.J.: The Poly1305-AES message-authentication code. In: Gilbert and Handschuh [16], pp. 32–49
7. Bernstein, D.J.: Stronger security bounds for Wegman-Carter-Shoup authenticators. In: Cramer, R. (ed.) EUROCRYPT 2005. LNCS, vol. 3494, pp. 164–180. Springer, Heidelberg (2005)
8. Black, J., Cochran, M.: MAC reforgeability. IACR ePrint Archive, Report 2006/095 (2006), http://eprint.iacr.org/2006/095
9. Black, J., Halevi, S., Krawczyk, H., Krovetz, T., Rogaway, P.: UMAC: Fast and secure message authentication. In: Wiener [30], pp. 216–233
10. Brincat, K., Mitchell, C.J.: New CBC-MAC forgery attacks. In: Varadharajan, V., Mu, Y. (eds.) ACISP 2001. LNCS, vol. 2119, pp. 3–14. Springer, Heidelberg (2001)
11. Carter, J., Wegman, M.: Universal hash functions. Journal of Computer and System Sciences 18, 143–154 (1979)
12. Dai, W., Krovetz, T.: VHASH security. IACR ePrint Archive, Report 2007/338 (2007), http://eprint.iacr.org/2007/338
13. den Boer, B.: A simple and key-economical unconditional authentication scheme. Journal of Computer Security 2, 65–72 (1993)
14. Desmedt, Y. (ed.): CRYPTO 1994. LNCS, vol. 839. Springer, Heidelberg (1994)
15. Etzel, M., Patel, S., Ramzan, Z.: Square hash: Fast message authentication via optimized universal hash functions. In: Wiener [30], pp. 234–251
16. Gilbert, H., Handschuh, H. (eds.): FSE 2005. LNCS, vol. 3557. Springer, Heidelberg (2005)
17. Halevi, S., Krawczyk, H.: MMH: Software message authentication in the gbit/second rates. In: Biham, E. (ed.) FSE 1997. LNCS, vol. 1267, pp. 172–189. Springer, Heidelberg (1997)
18. Handschuh, H., Preneel, B.: Key-recovery attacks on universal hash function based MAC algorithms. In: Wagner, D. (ed.) CRYPTO 2008. LNCS, vol. 5157, pp. 144–161. Springer, Heidelberg (2008)
19. Koblitz, N. (ed.): CRYPTO 1996. LNCS, vol. 1109. Springer, Heidelberg (1996)
20. Krawczyk, H.: LFSR-based hashing and authentication. In: Desmedt [14], pp. 129–139

21. McGrew, D., Viega, J.: The Galois/counter mode of operation (GCM). NIST Special Publication (2005), http://cs.www.ncsl.nist.gov/groups/ST/toolkit/BCM/documents/proposedmodes/gcm/gcm-revised-spec.pdf
22. McGrew, D., Weis, B.: Requirements on fast message authentication codes. IETF Internet-Draft. Intended status: Informational (February 2008), http://www.ietf.org/internet-drafts/draft-irtf-cfrg-fast-mac-requirements-01.txt
23. McGrew, D.A., Fluhrer, S.R.: Multiple forgery attacks against message authentication codes. IACR ePrint Archive, Report 2005/161 (2005), http://eprint.iacr.org/2005/161
24. Preneel, B., van Oorschot, P.C.: MDx-MAC and building fast MACs from hash functions. In: Coppersmith, D. (ed.) CRYPTO 1995. LNCS, vol. 963, pp. 1–14. Springer, Heidelberg (1995)
25. Rogaway, P.: Bucket hashing and its application to fast message authentication. Journal of Cryptology: the journal of the International Association for Cryptologic Research 12, 2, 91–115 (1999)
26. Rogaway, P., Bellare, M., Black, J.: OCB: A block-cipher mode of operation for efficient authenticated encryption. ACM Trans. Inf. Syst. Secur. 6, 3, 365–403 (2003)
27. Shoup, V.: On fast and provably secure message authentication based on universal hashing. In: Koblitz [19], pp. 313–328
28. Stinson, D.R.: On the connections between universal hashing, combinatorial designs and error-correcting codes. Electronic Colloquium on Computational Complexity (ECCC) 2, 52 (1995)
29. Wegman, M., Carter, J.: New hash functions and their use in authentication and set equality. Journal of Computer and System Sciences 22, 265–279 (1981)
30. Wiener, M.J. (ed.): CRYPTO 1999. LNCS, vol. 1666. Springer, Heidelberg (1999)

New Distinguishing Attack on MAC Using Secret-Prefix Method*

Xiaoyun Wang[1,2], Wei Wang[2], Keting Jia[2], and Meiqin Wang[2]

[1] Center for Advanced Study, Tsinghua University, Beijing 100084, China
xiaoyunwang@mail.tsinghua.edu.cn
[2] Key Laboratory of Cryptographic Technology and Information Security,
Ministry of Education, Shandong University, Jinan 250100, China
wwang@math.sdu.edu.cn, kejia@mail.sdu.edu.cn, mqwang@sdu.edu.cn

Abstract. This paper presents a new distinguisher which can be applied to secret-prefix MACs with the message length prepended to the message before hashing. The new distinguisher makes use of a special truncated differential path with high probability to distinguish an inner near-collision in the first round. Once the inner near-collision is detected, we can recognize an instantiated MAC from a MAC with a random function. The complexity for distinguishing the MAC with 43-step reduced SHA-1 is $2^{124.5}$ queries. For the MAC with 61-step SHA-1, the complexity is $2^{154.5}$ queries. The success probability is 0.70 for both.

Keywords: MAC, secret prefix method, distinguishing attack, SHA-1.

1 Introduction

Message Authentication Code (MAC) algorithms play an important role in internet security protocols (SSL/TLS, SSH, IPsec) and the financial sector for debit and credit transaction. A MAC algorithm is a hash function with a secret key K as the secondary input, which guarantees data integrity and authenticity. The *secret prefix* method is a MAC construction which prepends a secret K to the message before the hashing operation, which is the basic design unit for HMAC/NMAC [1]. One suggestion to guarantee a secure secret prefix MAC is to prepend the message length to the message before hashing [13]. Recent work [2,3,15,16,17,19] discovered devastating collision attacks on hash functions from the MDx family. Such attacks have undermined the confidence in the most popular hash functions such as MD5 and SHA-1, and promoted the reevaluation of the actual security of the MAC algorithms based on them [4,6,8,11,12,14,18].

There are two kinds of distinguishing attacks on MACs, and they are respectively called *distinguishing-R* and *distinguishing-H* attacks [8]. Distinguishing-R attack means distinguishing a MAC from a random function, and distinguishing-H attack detects an instantiated MAC (by an underlying hash function or block

---

* Supported by the National Natural Science Foundation of China (NSFC Grant No. 60525201 and No.90604036) and 973 Project (No.2007CB807902).

O. Dunkelman (Ed.): FSE 2009, LNCS 5665, pp. 363–374, 2009.

cipher) from a MAC with a random function. Preneel and van Oorschot [10] introduced a general distinguishing-R attack on all iterated MACs using the birthday paradox, which requires about $2^{n/2}$ messages and works with a success rate 0.63, where n is the length of the hash output. Their attack can immediately be converted into a general forgery attack. For the distinguishing-H attack, its ideal complexity should be exhaustive search cost.

This paper focuses on the distinguishing-H attack that checks which cryptographic hash function is embedded in a MAC. For simplicity, we call it distinguishing attack.

Kim *et al.* [8] described two kinds of distinguishers for the HMAC structure, which are differential and rectangle distinguishers. For the differential distinguisher, it needs a collision differential path with probability higher than 2^{-n}, and the rectangle distinguisher needs a near-collision differential with probability higher than $2^{-n/2}$. For MD4, because it is easy to find a differential path with high probability [15,20], there are some successful cryptanalytic results on MACs based on MD4 [4,6,14]. For HMAC/NMAC-MD5, there is only one available differential path that is called dBB pseudo-collision path [5]. Because the dBB pseudo-collision consists of two different IVs and the same message, so all the attacks [4,6,11,14] are in the related-key setting. For HMAC/NMAC-SHA-0, there exists a partial key-recovery attack [4]. For HMAC-SHA-1, Kim *et al.* proposed a distinguishing attack on 43-step HMAC-SHA-1 with data complexity $2^{154.9}$, which was improved by Rechberger and Rijmen [12]. The improved attack detected 50-step HMAC-SHA-1 with data complexity $2^{153.5}$. The paper [12] also proposed a related-key distinguishing attack on 62-step (17-78) HMAC-SHA-1 and a full key-recovery attack on 34-step NMAC-SHA-1 in the related-key setting.

All the above attacks make use of collision or near-collision differential paths for the underlying compression function with probability higher than 2^{-n}. For MD5 and SHA-1, there are too many sufficient conditions in the collision or near-collision differential paths, which imply a complexity more than 2^n. Because most conditions focus on the first round, it is hard to analyze the MACs with MD5/SHA-1 for more steps without related keys. In this paper, we only consider the MACs with reduced SHA-1 starting from the first step.

One recent work [18] presented a new distinguishing attack on HMAC/NMAC-MD5 and MD5-MAC without related keys. Their distinguisher detects an inner near-collision in the first iteration. This motivates us to explore a similar attack on MACs based on SHA-1. For the MAC with SHA-1, the situation is more complex, because SHA-1 dose not have a differential path with high probability. If we do not consider the first round, the probability of the existing differential paths for the last three rounds is high. How to avoid the differential path in the first round, and completely explore the probability advantage in the last three rounds? We neglect the exact path in the first round, replace it with an inner near-collision, and explore the new techniques to detect the inner near-collision by a birthday attack. Our new distinguishing attack is applicable to secret-prefix MACs with the length prepended before hashing, which is denoted as LPMAC.

For LPMAC based on 43-step SHA-1, the complexity is $2^{124.5}$ queries, and for LPMAC with 61-step SHA-1, the complexity is $2^{154.5}$.

This paper is organized as follows: Section 2 gives brief descriptions of LPMAC and SHA-1. In Section 3, we present the new distinguisher which is available to LPMAC structure, and describe the details of the distinguishing attack on LPMAC with 61-step SHA-1 in Section 4. Finally, we conclude the paper in Section 5.

2   Backgrounds and Definitions

In this section, we define the notations used in this paper, and give brief descriptions of the LPMAC and SHA-1.

2.1   Notations

H	: a hash function
$\overline{H}$	: a hash function without padding and length appending
n	: the length of the hash output
b	: the length of one message block
IV	: the initial chaining value
$x\|y$	: the concatenation of the two bitstrings x and y
$x_{i,j}$	: the j-th bit of x_i, where x_i is a 32-bit word, $j = 1, \ldots, 32$, and 32 is the most significant bit
$+, -$	: addition and subtration modular 2^{32}
$\Delta^- x$	: modular difference $x - x'$, where x and x' are two 32-bit words
$\wedge, \neg, \vee, \oplus$	: bitwise AND, NOT, OR and exclusive OR
$\lll s$	: left-rotation operation by s-bit

2.2   MAC Using Secret Prefix Method

The *secret prefix* method is to append a message M to a secret key K before the hashing operation:

$$\text{Secret-Prefix-MAC}_K(M) = H(K\|M),$$

where H is an unkeyed hash function. This method was proposed in the 1980s, and suggested for MD4 independently in [7,13]. The original secret prefix MAC is insecure: given a message and its MAC, an attacker can easily append another message to the message and update the MAC accordingly, as the given MAC value can be taken as the initial chaining value for the appended message [10].

Prefixing the message length to the message before hashing is one suggestion to avoid the above attack [13]. However, our new distinguisher specifically works for this kind of MAC, which we call LPMAC. We provide that $K\|\#length\|pad$ is a full block, and this kind of MAC corresponds to a hash function $\overline{H}$ with a secret IV (K'), which is denoted as:

$$LPMAC_K(M) = \overline{H}(K\|\#length\|pad\|M) = \overline{H}_{K'}(M).$$

In the rest of this paper, LPMAC refers to a MAC with the new form $\overline{H}_{K'}(M)$.

2.3 Brief Description of SHA-1

The hash function SHA-1 was issued by NIST in 1995 as a Federal Informa-tion Processing Standard [9]. It follows the Merkle-Damgård iterative construc-tion, takes a message M with the bit-length less than 2^{64}, and produces a 160-bit digest. The compression function takes a 160-bit chaining value $h^i = (a_0, b_0, c_0, d_0, e_0)$ and a 512-bit message block M^i as inputs, and produces an-other 160-bit chaining value h^{i+1}, where h^0 is the initial value IV, and $M = M^0\| \cdots \|M^{t-1}$. By iterating all the message blocks M^i, we obtain the final 160-bit value h^t which is the hash value.

Each 512-bit block M^i is divided into sixteen 32-bit words, which is denoted as $(m_0, m_1, \ldots, m_{15})$. The message words are expanded to eighty 32-bit words $w_0, \ldots, w_{79}$:

$$w_j = \begin{cases} m_j, & \text{if } j = 0, \ldots, 15, \\ (w_{j-3} \oplus w_{j-8} \oplus w_{j-14} \oplus w_{j-16}) \lll 1, & \text{if } j = 16, \ldots, 79. \end{cases}$$

The compression function consists of 4 rounds, and each round includes 20 steps. The details for the compression function are the following:

– Input: $w_0, \ldots, w_{79}$ and $h^i = (a_0, b_0, c_0, d_0, e_0)$, where h^i is a 160-bit chaining value .
– Step update: For $j = 1, \ldots, 80$,

$$a_j = (a_{j-1} \lll 5) + f_j(b_{j-1}, c_{j-1}, d_{j-1}) + e_{j-1} + w_{j-1} + k_j,$$
$$b_j = a_{j-1}, \ c_j = b_{j-1} \lll 30, \ d_j = c_{j-1}, \ e_j = d_{j-1},$$

where the Boolean function f_j and constant k_j are described in Table 1.
– Output: $h^{i+1} = (a_0 + a_{80}, b_0 + b_{80}, c_0 + c_{80}, d_0 + d_{80}, e_0 + e_{80})$.

Table 1. Boolean Functions and Constants Involved in SHA-1

round	steps	f_j		k_j
1	1-20	IF	$: (x \wedge y) \vee (\neg x \wedge z)$	0x5a827999
2	21-40	XOR	$: x \oplus y \oplus z$	0x6ed6eba1
3	41-60	MAJ	$: (x \wedge y) \vee (x \wedge z) \vee (y \wedge z)$	0x8fabbcdc
4	61-80	XOR	$: x \oplus y \oplus z$	0xca62c1d6

3 New Distinguisher on LPMAC Structure

This section introduces a new distinguisher of the LPMAC structure. It is based on a near-collision differential path with two message blocks.

3.1 Recent Attack on HMAC/NMAC-MD5 and MD5-MAC [18]

We first recall the distinguishing attack on HMAC/NMAC-MD5 and MD5-MAC presented in [18]. This distinguisher utilizes the dBB pseudo-collision path of

MD5 [5], where the hash values collide with probability 2^{-46} when the IV difference satisfies the dBB-condition, i.e.,

$$IV \oplus IV' = (0x80000000, 0x80000000, 0x80000000, 0x80000000),$$
$$\text{MSB}(B_0) = \text{MSB}(C_0) = \text{MSB}(D_0),$$

where $(A_0, B_0, C_0, D_0) = IV$, and MSB means the most significant bit. To discard the related-key setting, and give a real distinguishing attack, the adversary firstly collects many one-block messages to guarantee enough pairs which produce a difference with the dBB-condition. Then append a fixed one-block message to the collected messages, query their MACs, and try to find out a dBB-collision. The main idea of the distinguishing attack is summarized as follows:

To maintain the appearance of a dBB-collision under the dBB-condition, i. e., the dBB pseudo-collision happens in the second iteration, the adversary selects a structure S composed of 2^{89} one-block messages P. Then a fixed 447-bit message M is appended to each $P \in S$. Query the MACs with all $P\|M$, and find all collision pairs $(P\|M, P'\|M)$ by birthday attack. A dBB -collision is detected as follows.

- If $(P\|M, P'\|M)$ collides, append another M' to (P, P'), and query two MACs for the new pair $(P\|M', P'\|M')$. If two MACs are the same, we conclude that (P, P') is an internal collision.
- If (P, P') dos not collide, append 2^{47} different M' to (P, P'), query their MACs, and search whether there is a collision. If a collision is found, $(\overline{H}(P), \overline{H}(P'))$ must satisfy the dBB-condition, and $(P\|M, P'\|M)$ is a dBB-collision.
- Otherwise, $(\overline{H}(P), \overline{H}(P'))$ are random values.

Once a dBB-collision is detected, it is concluded that the MAC is a MAC based on MD5, otherwise, based on a random function.

3.2   Description of the New Distinguisher

Our attack is motivated by the above idea, which is to distinguish an instantiated LPMAC from a random function by detecting the target inner near-collision. However, our distinguisher is a totally different one. All the previous attacks detect the collision (or near-collision) generated by a full iteration of the hash function [4,6,8,11,12,14,18], but our attack is trying to distinguish an inner near-collision occurring in the first round, and there is no published techniques to detect such a near-collision till now.

For SHA-1 reduced to 61 steps, we consider a differential path with two message blocks, and assume that $P\|M_0\|M_1$ and $P'\|M_0'\|M_1'$ are a message pair, which produces a target differential path. Here, P and P' are one-block messages, M_0 and M_0' are 448-bit (14 words) truncated messages of the second block, and M_1 and M_1' are the corresponding 64-bit messages left. Denote the output of the first iteration $\overline{H}_K(P)$ as h_P, $\overline{H}_K(P')$ as h_P', and suppose the intermediate chaining variables $(a_i, \cdots, e_i)$ of the second block as ch_i, and $(a'_i, \cdots, e'_i)$ as ch'_i. It is clear that $h_P + ch_{61}$ and $h'_P + ch'_{61}$ are hash values of the message pair.

We select a target differential path, such that the first iteration can be any differential path, and the second includes a near-collision differential path. To make the truncated differential path of the last 47 steps with high probability, we choose the same disturbance vector as [16], which produces the near-collision differential path for the second iteration (See Table 2). We divide the second differential path into two parts, the first part consists of the previous 14 steps which involves most conditions, and the second part is the last 47 steps with only 34 conditions. We neglect the special differential path in previous 14 steps, and only consider its output difference, which can be regarded as an inner near-collision. We select the specific difference $\triangle ch_{14} = ch_{14} \oplus ch'_{14}$ as

$$(0x00000000, 0x00000000, 0x80000000, 0x20000002, 0x00000040).$$

The choice of $\triangle ch_{14}$ is to cancel the message word differences $\triangle w_{14}$ and $\triangle w_{15}$. The sufficient conditions for the cancelation are as follows:

$$a_{10,9} = w_{14,7} + 1,$$
$$a_{11,4} = w_{15,2} + 1, a_{11,32} = w_{14,30},$$
$$a_{13,2} = 1, a_{13,30} = 0, a_{13,32} = 1, a_{13,4} = w_{14,2} + w_{16,2} + 1,$$
$$a_{14,4} = w_{14,2} + w_{16,2}, a_{14,32} = 0.$$

The core of our attack is to explore some mathematical properties that can be used to distinguish the inner near-collision in the 14th step. For the LPMAC, there are two obstacles to do this:

1. In the first iteration, the output difference $\varDelta^- H_P = H_P - H'_P$ is unknown, which conceals the difference of the near-collision $\varDelta^- H_P + \varDelta^- ch_{61}$. Hence, the birthday attack can not be applied directly to the second iteration like the distinguishing attacks on MACs based on MD5.
2. How to choose messages, and fulfill the birthday attack to detect the inner near-collision?

We explore the following mathematical properties of the differential path to surpass the above two obstacles:

- If the inner near-collision occurs, replace (M_1, M'_1) with another $(\overline{M}_1, \overline{M'_1})$, then $(P\|M_0\|\overline{M}_1, P'\|M'_0\|\overline{M'_1})$ follows the differential path with probability 2^{-34}.
- If two pairs $(P\|M_0\|M_1, P'\|M'_0\|M'_1)$ and $(P\|M_0\|N_1, P'\|M'_0\|N'_1)$ result in the near-collision differential path, i. e.,

$$\overline{H}_K(P\|M_0\|M_1) - \overline{H}_K(P'\|M'_0\|M'_1) = \overline{H}_K(P\|M_0\|N_1) - \overline{H}_K(P'\|M'_0\|N'_1)$$
$$= \varDelta^- H_P + \varDelta^- ch_{61} = \delta, \qquad (1)$$

then we have (See Fig. 1.)

$$\overline{H}_k(P\|M_0\|M_1) - \overline{H}_K(P\|M_0\|N_1) = \overline{H}_K(P'\|M'_0\|M'_1) - \overline{H}_K(P'\|M'_0\|N'_1) = \delta'. \qquad (2)$$

Table 2. A Differential Path for Steps 1~61 on SHA-1

step i	disturb. vector	XOR difference of the input to step i Δw_{i-1}	Δa_i	Δb_i	Δc_i	Δd_i	Δe_i	conditions
1	80000001	2, 7, 31, 32	-	-	-	-	-	-
2	2	5, 6	-	-	-	-	-	-
3	3	2, 7, 30, 31, 32	-	-	-	-	-	-
4	2	6, 7, 30	-	-	-	-	-	-
5	80000002	1, 7, 30, 31, 32	-	-	-	-	-	-
6	2	5, 7, 30	-	-	-	-	-	-
7	80000003	1, 7, 31, 32	-	-	-	-	-	-
8	0	2, 5, 6, 7, 30, 31, 32	-	-	-	-	-	-
9	80000000	1, 2, 30, 32	-	-	-	-	-	-
10	2	2, 5, 31, 32	9	-	-	-	-	-
11	80000001	1, 7, 30, 31	4, 32	9	-	-	-	-
12	0	2, 5, 6, 31, 32	2	4,32	7	-	-	-
13	0	1, 30	2	2, 30	7	-	-
14	2	2, 31, 32			32	2, 30	7	$a_{13,2} = 1, a_{13,30} = 0, a_{13,32} = 1$ $a_{11,4} = w_{15,2} + 1, a_{11,32} = w_{14,30}, a_{10,9} = w_{14,7} + 1$
15	2	2, 7, 30, 31, 32	2			32	2,30	$a_{15,2} = w_{14,2}, a_{14,32} = 1$
16	0	2, 7, 30, 31	2				32	$a_{14,4} = w_{14,2} + 16,2, a_{13,4} = w_{14,2} + w_{16,2} + 1$
17	0	2, 32			32			$a_{16,32} = 0$
18	0					32		$a_{17,32} = 1$
19	0						32	
20	0	32						
21	2	2	2					$a_{21,2} = w_{20,2}$
22	0	7		2				$a_{20,4} = a_{19,4} + w_{20,2} + w_{23,7} + 1$
23	2	2			32			$a_{23,2} = w_{23,7} + 1$
24	0	7,32		2		32		$a_{22,4} = a_{21,4} + w_{23,7} + w_{25,7}$
25	2	32	2		32		32	$a_{25,2} = w_{25,7} + 1$
26	0	7		2		32		$a_{24,4} = a_{23,4} + w_{25,7} + w_{26,1}$
27	3	1, 32	1		32		32	$a_{27,1} = w_{26,1} + 1$
28	0	6, 7		1		32		$a_{26,3} = a_{25,3} + w_{26,1} + w_{28,1} + 1$
29	0	1, 2, 32			31		32	$a_{28,31} = a_{26,1} + w_{26,1} + w_{29,31}$
30	2	2,31	2			31		$a_{30,2} = w_{29,2}, a_{29,31} = a_{28,1} + w_{26,1} + w_{30,31} + 1$
31	0	7,31, 32		2			31	$a_{29,4} = a_{28,4} + w_{29,2} + w_{31,2} + 1$
32	0	2, 31, 32			32			
33	0	32				32		
34	0	32					32	
35	2	2,32	2					$a_{35,2} = w_{34,2}$
36	0	7		2				$a_{34,4} = a_{33,4} + w_{34,2} + w_{36,2} + 1$
37	0	2			32			
38	0	32				32		
39	0	32					32	
40	0	32						
41	2	2	2					$a_{41,2} = w_{40,2}$
42	0	7		2				$a_{40,4} = a_{39,4} + 1$
43	2	2			32			$a_{43,2} = w_{41,2}, a_{42,32} = a_{40,2} + 1$
44	0	7, 32		2		32		$a_{42,4} = a_{41,4} + 1, a_{43,32} = a_{42,2} + 1$
45	0	2,32			32		32	$a_{44,32} = a_{42,2} + 1$
46	0					32		$a_{45,32} = a_{44,2} + 1$
47	0	32					32	
48	0	32						
...	0							
54	0							
55	4	3	3					$a_{55,3} = w_{54,3}$
56	0	8		3				$a_{54,5} = a_{53,5} + 1$
57	0	3			1			$a_{56,1} = a_{54,3} + 1$
58	8	1, 4	4			1		$a_{57,1} = a_{56,3} + 1, a_{58,4} = w_{57,4}$
59	4	1, 3, 9	3		4		1	$a_{57,6} = a_{56,6} + 1, a_{59,3} = w_{58,3}$
60	0	1, 4, 8			3	2		$a_{59,2} = a_{57,4} + w_{57,4} + w_{60,2} + 1,$ $a_{58,5} = a_{57,5} + w_{58,3} + w_{60,3} + 1$
61	10	2, 3, 5	5			1	2	

We utilize the equation (2) to construct a distinguisher for LPMAC with 61-step SHA-1.

- Firstly, collect enough messages $P\|M_0\|M_1$, $P\|M_0\|N_1$, $P'\|M_0'\|M_1'$, $P'\|M_0'\|N_1'$ and their MACs. Compute two structures, one structure is

$$S_1 = \{(\overline{H}_K(P\|M_0\|M_1) - \overline{H}_K(P\|M_0\|N_1)\},$$

and the other is

$$S_2 = \{\overline{H}_K(P'\|M_0'\|M_1') - \overline{H}_K(P'\|M_0'\|N_1')\}.$$

- Secondly, apply the birthday attack to the structures S_1 and S_2, and search all the collisions such that

$$\overline{H}_K(P\|M_0\|M_1) - \overline{H}_K(P\|M_0\|N_1) = \overline{H}_K(P'\|M_0'\|M_1') - \overline{H}_K(P'\|M_0'\|N_1').$$

- Finally, for each collision, detect whether the corresponding pair $(P\|M_0\|$ $M_1, P'\|M_0'\|M_1')$ satisfies the differential path in Table 2.

More details about the distinguisher are described in the following section. It is noted that, it is hard to fulfill the birthday attack directly to search the solution to equation (1), but easy to get the solution to equation (2) by birthday attack.

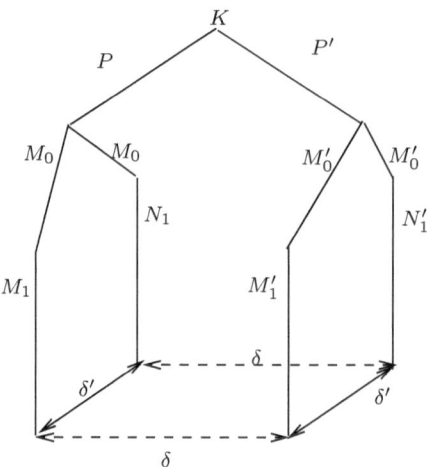

Fig. 1. The Distinguishing Attack on LPMAC

The distinguisher is applicable to LPMACs with other hash functions, such as LPMAC with reduced SHA-2.

4   New Distinguishing Attack on LPMAC Based on 61-Step SHA-1

In this section, we describe the new distinguishing attack in detail. We assume that the LPMAC algorithm is either LPMAC with 61-step SHA-1 or LPMAC with a random function.

Before we introduce the new attack, we need to make clear that how many chosen messages are needed to guarantee an inner near-collision. The total suffi-cient conditions for the near-collision is 169, where 160 conditions are from the difference $\triangle ch_{14}$, and 9 conditions (See Subsection 3.2) are deduced from the cancelation of $\triangle w_{14}$ and $\triangle w_{15}$. So, we need $2^{169/2} = 2^{84.5}$ messages to guarantee such an inner near-collision happen.

Select four messages $M_0\|M_1, M_0\|N_1, M_0'\|M_1'$ and $M_0'\|N_1'$ such that $\varDelta(M_0\|M_1)$ $= (M_0\|M_1) \oplus (M_0'\|M_1')$ and $\varDelta(M_0\|N_1) = (M_0\|N_1) \oplus (M_0'\|N_1')$ are consistent with the target message difference in Table 2, and $M_0 \| M_1$, $M_0 \| N_1$ satisfy the sufficient conditions in Table 3.

The distinguishing attack implements the following four steps:

Table 3. Conditions on Messages

$w_{14,31} = w_{14,30} + 1, w_{15,7} = w_{14,2} + 1, w_{15,30} = w_{14,30}, w_{15,31} = w_{15,30} + 1,$
$w_{21,7} = w_{20,2} + 1, w_{27,6} = w_{26,1} + 1, w_{27,7} = w_{26,1}, w_{28,2} = w_{28,1} + 1,$
$w_{30,7} = w_{29,2} + 1, w_{31,31} = w_{26,1} + 1, w_{35,7} = w_{34,2} + 1, w_{41,7} = w_{40,2} + 1,$
$w_{43,7} = w_{40,2} + 1, w_{44,2} = w_{40,2} + 1, w_{55,8} = w_{54,3} + 1, w_{56,3} = w_{54,3} + 1,$
$w_{57,1} = w_{54,3} + 1, w_{58,1} = w_{54,3} + 1, w_{58,9} = w_{57,4} + 1, w_{59,1} = w_{54,3} + 1,$
$w_{59,4} = w_{57,4} + 1, w_{59,8} = w_{58,3} + 1$

1. Randomly choose a structure S, which consists of $2^{84.5}$ different one-block messages.
2. For all $P \in S$, query the MACs with $P\|M_0\|M_1$, $P\|M_0'\|M_1'$, $P\|M_0\|N_1$ and $P\|M_0'\|N_1'$, respectively, and compute the following two structures of differences

$$S_1 = \{LPMAC(P\|M_0\|M_1) - LPMAC(P\|M_0\|N_1) \mid P \in S\},$$
$$S_2 = \{LPMAC(P\|M_0'\|M_1') - LPMAC(P\|M_0'\|N_1') \mid P \in S\}.$$

Search all the collisions between two structures by a birthday attack.
3. For each collision, compute $LPMAC(P\|M_0\|M_1) - LPMAC(P'\|M_0'\|M_1')$, and denote it as δ. Then for the message pair $(P\|M_0, P'\|M_0')$, we choose 2^{34} different message pairs $(\overline{M_1}, \overline{M_1'})$ such that $M_0\|\overline{M_1}$ satisfies the message sufficient conditions for the near-collision path in Table 3. Query the MACs for $(P\|M_0\|\overline{M_1}, P'\|M_0'\|\overline{M_1'})$. Check whether the difference $LPMAC(P\|M_0 \|\overline{M_1}) - LPMAC(P'\|M_0'\|\overline{M_1'})$ is equivalent to δ.

   - If a pair $(P\|M_0 \|\overline{M_1}, P'\|M_0'\|\overline{M_1'})$ that matches the difference δ is searched, we conclude that the LPMAC is based on 61-step SHA-1, and stop the algorithm.
   - Else, go to step 4.

4. Repeat steps 1-3. If the number of structures exceeds 2^{68}, we conclude that the LPMAC is constructed from a random function.

Complexity
Summing up the above attack, we choose $4 \cdot 2^{68} \cdot (2^{84.5} + 2^{34}) \approx 2^{154.5}$ messages in total. Since we can use the birthday attack to search collisions in step 2, a table with size of $2^{84.5}$ needs to be built. We need about $2^{68} \cdot 2^{84.5} = 2^{152.5}$ table lookups and $2^{154.5}$ queries.

Success Rate
From the above process, the success rate of our attack can be divided into two parts :

– If the LPMAC is constructed from 61-step SHA-1, once a second collision in step 3 is detected, the attack succeeds. The probability is computed as follows:

- There are 169 conditions to guarantee the inner near-collision in the step 14, and 34 conditions to follow the differential path in the last steps 15-61. According to the birthday paradox and Taylor series expansion, for $2^{68} \cdot 2^{169} = 2^{237}$ pairs among the structures S_1 and S_2, there exists an inner near-collision with probability

$$1 - (1 - \frac{1}{2^{237}})^{2^{237}} \approx 1 - e^{-1} \approx 0.63.$$

- If the first collision is captured in step 2, the second collision in step 3 is searched with probability

$$1 - (1 - \frac{1}{2^{34}})^{2^{34}} \approx 1 - e^{-1} \approx 0.63.$$

Hence, when the LPMAC is based on 61-step SHA-1, the distinguishing attack successes with probability

$$0.63 \cdot 0.63 \approx 0.40$$

– If the LPMAC is from a random function, the attack succeeds when the second collision doesn't exist. For the 2^{68} structures, there are $2^{237}/2^{160} = 2^{77}$ expected collisions in total, so the success probability is about:

$$((1 - \frac{1}{2^{160}})^{2^{34}})^{2^{77}} \approx 1.$$

Therefore, the success rate of the whole attack is about

$$\frac{1}{2} \times 0.40 + \frac{1}{2} \times 1 = 0.70.$$

Note that the success probability can be increased by repeating this attack several times, doubling the size of the structure S or the number of different pairs $(\overline{M_1}, \overline{M_1'})$.

Table 4 illustrates the comparison of our distinguishing attacks on LPMAC-SHA-1 with other attacks on HMAC-SHA-1.

Table 4. Comparison Between the Distinguishing Attacks on MACs with SHA-1

	MAC	steps	data
Kim et al. [8]	HMAC	43	$2^{154.9}$
Rechberger et al. [12]	HMAC	50	$2^{153.5}$
		43	$2^{124.5}$
This paper	LPMAC	50	$2^{136.5}$
		61	$2^{154.5}$

5   Conclusions

A new distinguisher is introduced to recognize the secret-prefix MAC which prepends the message length before hashing. The new distinguisher utilizes a near-collision differential path instead of a collision path, and detects an inner near-collision in the first round. The core of the attack is to capture the mathematical characters of a near-collision differential path which can be utilized to fulfill a birthday attack. Our attack is applicable to some other LPMACs such as LPMAC with reduced SHA-2.

Acknowledgements. We would like to thank Christian Rechberger and three reviewers for their very helpful comments on the paper.

References

1. Bellare, M., Canetti, R., Krawczyk, H.: Keying Hash Functions for Message Authentication. In: Koblitz, N. (ed.) CRYPTO 1996. LNCS, vol. 1109, pp. 1–15. Springer, Heidelberg (1996)
2. Biham, E., Chen, R.: Near-Collisions of SHA-0. In: Franklin, M.K. (ed.) CRYPTO 2004. LNCS, vol. 3152, pp. 290–305. Springer, Heidelberg (2004)
3. Biham, E., Chen, R., Joux, A., Carribault, P., Lemuet, C., Jalby, W.: Collisions of SHA-0 and Reduced SHA-1. In: Cramer, R. (ed.) EUROCRYPT 2005. LNCS, vol. 3494, pp. 36–57. Springer, Heidelberg (2005)
4. Contini, S., Yin, Y.L.: Forgery and Partial Key-Recovery Attacks on HMAC and NMAC Using Hash Collisions. In: Lai, X., Chen, K. (eds.) ASIACRYPT 2006. LNCS, vol. 4284, pp. 37–53. Springer, Heidelberg (2006)
5. den Boer, B., Bosselaers, A.: Collisions for the Compression Function of MD5. In: Helleseth, T. (ed.) EUROCRYPT 1993. LNCS, vol. 765, pp. 293–304. Springer, Heidelberg (1994)
6. Fouque, P.-A., Leurent, G., Nguyen, P.Q.: Full Key-Recovery Attacks on HMAC/NMAC-MD4 and NMAC-MD5. In: Menezes, A. (ed.) CRYPTO 2007. LNCS, vol. 4622, pp. 13–30. Springer, Heidelberg (2007)
7. Galvin, J.M., McCloghrie, K., Davin, J.R.: Secure Management of SNMP Networks. Integrated Network Management II, 703–714 (1991)
8. Kim, J., Biryukov, A., Preneel, B., Hong, S.: On the Security of HMAC and NMAC Based on HAVAL, MD4, MD5, SHA-0, and SHA-1. In: De Prisco, R., Yung, M. (eds.) SCN 2006. LNCS, vol. 4116, pp. 242–256. Springer, Heidelberg (2006)
9. NIST: Secure Hash Standard. Federal Information Processing Standard, FIPS-180-1 (April 1995)
10. Preneel, B., van Oorschot, P.: MDx-MAC and Building Fast MACs from Hash Functions. In: Coppersmith, D. (ed.) CRYPTO 1995. LNCS, vol. 963, pp. 1–14. Springer, Heidelberg (1995)
11. Rechberger, C., Rijmen, V.: On Authentication with HMAC and Non-random Properties. In: Dietrich, S., Dhamija, R. (eds.) FC 2007 and USEC 2007. LNCS, vol. 4886, pp. 119–133. Springer, Heidelberg (2007)
12. Rechberger, C., Rijmen, V.: New Results on NMAC/HMAC when Instantiated with Popular Hash Functions. Journal of Universal Computer Science 14(3), 347–376 (2008)

13. Tsudik, G.: Message Authentication with One-Way Hash Functions. ACM Comput. Commun. Rev. 22(5), 29–38 (1992)
14. Wang, L., Ohta, K., Kunihiro, N.: New Key-Recovery Attacks on HMAC/NMAC-MD4 and NMAC-MD5. In: Smart, N.P. (ed.) EUROCRYPT 2008. LNCS, vol. 4965, pp. 237–253. Springer, Heidelberg (2008)
15. Wang, X., Lai, X., Feng, D., Chen, H., Yu, X.: Cryptanalysis of the Hash Functions MD4 and RIPEMD. In: Cramer, R. (ed.) EUROCRYPT 2005. LNCS, vol. 3494, pp. 1–18. Springer, Heidelberg (2005)
16. Wang, X., Yin, Y.L., Yu, H.: Finding Collisions in the Full SHA-1. In: Shoup, V. (ed.) CRYPTO 2005. LNCS, vol. 3621, pp. 17–36. Springer, Heidelberg (2005)
17. Wang, X., Yu, H.: How to Break MD5 and Other Hash Functions. In: Cramer, R. (ed.) EUROCRYPT 2005. LNCS, vol. 3494, pp. 19–35. Springer, Heidelberg (2005)
18. Wang, X., Yu, H., Wang, W., Zhang, H., Zhan, T.: Cryptanalysis on HMAC/NMAC-MD5 and MD5-MAC. In: Joux, A. (ed.) EUROCRYPT 2009. LNCS, vol. 5479, pp. 121–133. Springer, Heidelberg (2009)
19. Wang, X., Yu, H., Yin, Y.L.: Efficient Collision Search Attacks on SHA-0. In: Shoup, V. (ed.) CRYPTO 2005. LNCS, vol. 3621, pp. 1–16. Springer, Heidelberg (2005)
20. Yu, H., Wang, G., Zhang, G., Wang, X.: The Second-Preimage Attack on MD4. In: Desmedt, Y.G., Wang, H., Mu, Y., Li, Y. (eds.) CANS 2005. LNCS, vol. 3810, pp. 1–12. Springer, Heidelberg (2005)

Fast and Secure CBC-Type MAC Algorithms

Mridul Nandi

National Institute of Standards and Technology
mridul.nandi@gmail.com

Abstract. The CBC-MAC or cipher block chaining message authentication code, is a well-known method to generate message authentication codes. Unfortunately, it is not forgery-secure over an arbitrary domain. There are several secure variants of CBC-MAC, among which OMAC is a widely-used candidate. To authenticate an s-block message, OMAC costs $(s+1)$ block cipher encryptions (one of these is a zero block encryption), and only one block cipher key is used. In this paper, we propose two secure and efficient variants of CBC-MAC: namely, GCBC1 and GCBC2. Our constructions cost only s block cipher encryptions to authenticate an s-block message, for all $s \geq 2$. Moreover, GCBC2 needs only one block cipher encryption for almost all single block messages, and for all other single block messages, it costs two block cipher encryptions. We have also defined a class of generalized CBC-MAC constructions, and proved a sufficient condition for prf-security. In particular, we have provided an unified prf-security analysis of CBC-type constructions, e.g., XCBC, TMAC and our proposals GCBC1 and GCBC2.

Keywords: CBC-MAC, OMAC, padding rule, prf-security.

1 Introduction

In cryptography, a common trend is to design fast and secure algorithms. In this paper, we propose two fast and secure block cipher-based message authentication codes. A message authentication code, or MAC, is useful in those applications where data integrity and authenticity are essential. In terms of security, we want a MAC to be a pseudorandom function, or prf, which means that it is computationally indistinguishable from an ideal random function. Prf-security is a strong security notion, and it also guarantees that the MAC is unforgeable. In this paper, we use the words "secure" and "prf-secure" synonymously. Several secure and fast authentication algorithms are already known. We first broadly classify them into three main categories, based on their underlying building blocks.

HASH-MAC: These are based on hash functions. HMAC [1] is a widely used candidate in this class that has been standardized by the National Institute of Standards and Technology (NIST). The other efficient, popular candidates include the cascaded-PRF [2], sandwich-MAC [21], and KMDP [14].

O. Dunkelman (Ed.): FSE 2009, LNCS 5665, pp. 375–393, 2009.

UNIVERSAL HASH BASED MAC: These MACs use universal hash functions and small domain pseudorandom functions. In software, these are very fast for long messages [11,19]. These generally require field multiplications, key expansions, and invocations of a smaller domain pseudorandom function which may be overhead for short messages. Some popular examples of universal hash based authentications are UMAC [8] and poly1305 MAC [7]. In [17], a 4-round version of AES [12] is used to obtain a universal hash function that eventually produces a very fast MAC computation for long messages (close to two times faster than OMAC). But, this is also slower than OMAC, due to the overhead required for processing short messages.

BLOCK CIPHER BASED MAC: In this paper, we study this category in more detail. These MACs are usually based on several invocations of a block cipher, either in a feedback mode (cipher block chaining or CBC-MAC) or in a parallel mode (e.g., PMAC [9] or XOR-MAC [3]). *A block cipher is a permutation $e_K : \{0,1\}^n \to \{0,1\}^n$, for each key K chosen from the key space $\{0,1\}^k$, where n (the block size) and k (the key size) are positive integers. We fix these parameters throughout the paper.* Intuitively, a block cipher is called pseudorandom permutation or prp-secure if the keyed block cipher family is computationally indistinguishable from an ideal random permutation. CBC-MAC (cipher block chaining message authentication [4]) is the first construction in this category. Given a message $M = (m_1 \| \cdots \| m_\ell) \in (\{0,1\}^n)^\ell$, the CBC-MAC of the message M based on e_K is computed as follows:

$$\text{CBC-MAC}_K(M) = e_K(e_K(\cdots e_K(m_1) \oplus m_2 \cdots) \oplus m_\ell).$$

However, CBC-MAC is not secure for variable length messages due to the length extension attack. Many different modifications of it have been proposed so far, among which OMAC [15] or one-key CBC-MAC[1] is efficient (requires one extra zero block encryption compared to CBC-MAC computation), as well as requiring only one key. Another simple modification, called XCBC-MAC [10] or XCBC, is faster in software, but it needs three keys, which may not be suitable in many applications. These keys may be derived from one key at the cost of few block cipher invocations, which causes slower performance for short messages. The TMAC requires only two keys and it is as efficient as CBC-MAC. If the output of zero block encryption of OMAC is stored as a key (to save one block cipher encryption) then eventually, OMAC and TMAC look almost identical.

1.1 Our Proposals GCBC1 and GCBC2

Let $M = (m_1 \| \cdots \| m_\ell) \in (\{0,1\}^n)^\ell$ with $\ell \geq 2$. The GCBC1-MAC of the message M based on e_K is computed as follows:

$$\text{GCBC1}_K(M) = e_K\big(e_K(\cdots e_K(e_K(m_1) \oplus m_2) \cdots)^{\lll 1} \oplus m_\ell\big).$$

For all other messages M, we pad 10^d (for smallest choice of d) to make sure that the message size is multiple of n and has at least two blocks. Let

---

[1] It is also known as CMAC [13] as recommended by the NIST.

$M' = (m_1\|\cdots\|m_\ell) \in (\{0,1\}^n)^\ell$, $\ell \geq 2$, be the padded message. In this case we compute the tag as follows,

$$\mathsf{GCBC1}_K(M) = e_K\big(e_K(\cdots e_K(e_K(m_1) \oplus m_2) \cdots)^{\lll 2} \oplus m_\ell\big).$$

In other words, we apply one or two left shifts to the last intermediate chaining value[2] of CBC-MAC before xor-ing the last message block of the padded message. This small tweak eventually helps to avoid length extension attack. Moreover we prove that it is prf-secure (see Section 5). Handling the last intermediate input in two different manners depending on the size of the last message block, is very common in MACs and it is used, e.g., in XCBC, OMAC, TMAC, etc.

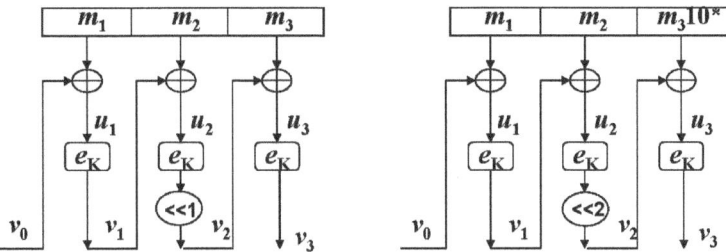

Fig. 1. GCBC1, Generalized CBC, which uses a left shift variation operation, an underlying iterative function e_K (block cipher) and a simple padding rule

Theorem (*Security Bound of GCBC1*)

$$\mathbf{Adv}^{\mathrm{prf}}_{\mathsf{GCBC1}}(q,\sigma) \leq \frac{5(\sigma+q)(\sigma+q-1)}{2^{n-4}} + \mathbf{Adv}^{\mathrm{prp}}_e(\sigma).$$

Our second construction, called GCBC2, authenticates almost all single block messages by using one block cipher encryption. It considers several cases, depending on the message size. For, $x \in \{0,1\}^n$, define $\bar{x} = x$, and if $|x| \leq n-1$ then define $\bar{x} = x10^{n-1-|x|}$. We define $\delta = 2$ if the message size is multiple of n otherwise, we set $\delta = 1$.

1. Let $|M| \leq n-4$, then $\mathsf{GCBC2}_K(M) = e_K(\overline{M})$.
2. Let $n-3 \leq |M| \leq n$, then write $M = m_1 := m'_1 m''_1$ where $|m'_1| = n-3$ and $|m''_1| \leq 3$. We define

$$\mathsf{GCBC2}_K(M) = e_K(e_K(m'_1\|011) \oplus \overline{m''_1}).$$

---

[2] Note that for a single block padded message, the last intermediate chaining value is nothing but the message block and so, left shift operations on it is a predictable operation. So we can not prevent length extension attack for a single block padded message. Hence, we need to make sure that padded message has at least two blocks.

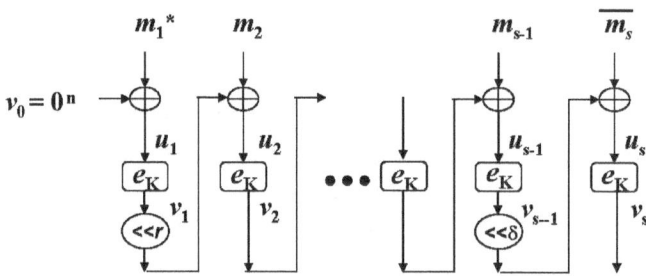

Fig. 2. GCBC2 for more than one message block. Let $M = m_1\|m_2\|\cdots\|m_s$, $m_1 = m'_1\|m''_1$, $|m''_1| = 3$. We denote $m^*_1 = \overline{m}_1$ if $m''_1 = 000$, o.w., $m^*_1 = m_1$. The nonnegative integer r, $0 \le r \le 4$, denotes the amounts of left shift applied to the output of the first encryption (the value of r can be determined in the definition of GCBC2).

3. Let $M = m_1\|m_2$, $m_1 = m'_1 m''_1$, $|m'_1| = n - 3, |m''_1| = 3$ and $|m_2| \le n$. We define

$$\mathsf{GCBC2}_K(M) = \begin{cases} e_K(e_K(m_1)^{\lll\delta+1} \oplus \overline{m_2}) & \text{if } m''_1 \neq 000 \\ e_K(e_K(m'_1\|100)^{\lll\delta-1} \oplus \overline{m_2}) & \text{if } m''_1 = 000 \end{cases}$$

4. Let $M = m'_1 m''_1\|m_2\|m_3\|\cdots\|m_s$ where $|m'_1| = n - 3$, $|m''_1| = 3$ and $|m_2| = \cdots = m_{s-1} = n, |m_s| \le n$. We define

$$\mathsf{GCBC2}_K(M) = e_K\big((e_K(\cdots e_K(e_K(\ e_K(m'_1\|m''_1)^{\lll 4} \oplus m_2) \oplus m_3)\cdots)\big)^{\lll\delta} \oplus \overline{m_s}\big) \text{ if } m''_1 \neq 000$$

$$= e_K\big((e_K(\cdots e_K(e_K(\ e_K(m'_1\|100)^{\lll 5} \oplus m_2) \oplus m_3)\cdots)\big)^{\lll\delta} \oplus \overline{m_s}\big) \text{ if } m''_1 = 000$$

Theorem (*Security Bound of GCBC2*, see Section 5)

$$\mathbf{Adv}^{\mathrm{prf}}_{\mathsf{GCBC2}}(q, \sigma) \le \frac{33(\sigma + q)(\sigma + q - 1)}{2^n} + \mathbf{Adv}^{\mathrm{prp}}_e(\sigma).$$

OMAC vs Our MACs: For all messages having block sizes at least two, both GCBC1 and GCBC2 have similar performance to CBC-MAC, whereas OMAC costs one extra zero block encryption. Zero block encryption can be computed off line, in which case we have to store the output as a key and hence it has similar performance to TMAC (which has two-key and similar to CBC performance). For single block messages (of size less than or equal to n), both OMAC and GCBC1 need two block cipher encryptions (in case of OMAC, one of the encryption is zero block encryption). The GCBC2 costs exactly one block cipher encryption like CBC-MAC for almost all single block messages (except the messages of size in between $n - 3$ and n, and in which case it costs two encryptions). Thus, GCBC2 is a good choice whenever the short messages are authenticated frequently. In some applications, we might know before hand that message sizes are at least $(n + 1)$-bits. In these applications, GCBC1 would be a good choice due to its simplicity and performance (same as CBC-MAC, but secure for arbitrary length messages). In table 1, a comparison of software performances are given.

Table 1. It provides a performance comparison of known CBC-type MACs along with our proposals. The software speed is computed (in the platform Intel(R) Pentium(R) 4 CPU 3.60 GHz, 1GB RAM) with AES-128 as the underlying block cipher. Here, # BC denotes the number of invocations of the block cipher $e : \{0,1\}^k \times \{0,1\}^n \to \{0,1\}^n$ that is used to authenticate an s-block message. Time is computed by taking the average over several executions.

Name of MAC	microsec (1-15 bytes)	microsec (16 bytes)	microsec (17 - 32 bytes)	# BC for s-block	Total Keysize
XCBC [10]	43.7	43.7	78.46	s	$k + 2n$
TMAC [16]	43.98	44.05	78.80	s	$k + n$
OMAC [15]	78.72	78.80	113.80	$s+1$	k
GCBC1	77.9	77.92	77.95	s	k
GCBC2	43.58	78.26	78.37	s	k

A GENERALIZED CBC-MAC: In this paper, we have also provided a general class of CBC-type constructions, called gcbc, which includes almost all popularly known CBC-type constructions, e.g., XCBC, TMAC, OMAC, and our proposals. We have also given a sufficient condition for prf-secure gcbc constructions and we have shown that almost all known constructions such as XCBC, TMAC, and our proposals satisfy the sufficient condition. So, we have provided that, how an unified way of security analysis of CBC-type MACs can be provided.

Organization of the Paper. We first provide basic definitions, and notations in Section 2. In Section 3, we propose a generalized CBC-type message authentication algorithms and also show that most of the CBC-type constructions belong to the class. The security analysis has been made by using decorrelation technique. The detailed security analysis of the generalized CBC-type constructions is given in Section 4. Finally, in Section 5, we specify two fast and secure constructions, called GCBC1 and GCBC2, from the generalized class.

2   Preliminaries

2.1   Definitions and Notations

Given any set S, $S^+ = \cup_{i=1}^\infty S^i$, and $S^* = \cup_{i=0}^\infty S^i = S^+ \cup \{\lambda\}$, where λ is the empty string. For example, $\{0,1\}^+$ is the set of all non-empty finite bit-sequences. Let $|x| = i$ for any $x \in \{0,1\}^i$. Any $X \in S^+$ can be written as $X = (x_1, \cdots, x_i)$ for some $i \geq 1$ and $x_1, \cdots, x_i \in S$. A tuple $Y = (y_1, \cdots, y_j) \in S^*$ is a prefix of X if $j \leq i$ and $y_1 = x_1, \cdots, y_j = x_j$. Trivially, λ is a prefix of any X, and is called a trivial prefix. Any other prefixes are called non-trivial prefixes. Let $x = x_1 x_2 \cdots x_n \in \{0,1\}^n$, $x_i \in \{0,1\}$. Then for any two integers $i \leq j$, the set $\{i, i+1, \cdots, j\}$ is denoted as $[i..j]$, and $x_i x_{i+1} \cdots x_j$ is denoted as $x[i..j]$ whenever $1 \leq i \leq j \leq n$. If $i > j$, $x[i..j]$ is nothing but λ. Let $x[i]$ represent the ith bit of x. The notation $x^{\ll t}$ (or $x^{\gg t}$) is denoted for t-bit left shift (or right shift, respectively) of an n-bit string x. The set $\{0,1\}^n$ is sometimes identified as $GF(2^n)$ by fixing a primitive polynomial $z^n + c_1 z^{n-1} + \cdots + c_{n-1} z + c_n$, where $c_i \in \{0, 1\}$. Let $0^n = \mathbf{0}$ (the additive identity), $0^{n-1}1 = \mathbf{1}$ (the multiplicative identity) and $\alpha = 0^{n-2}10 \in GF(2^n)$ (known as a primitive element). For any element $x \in \{0, 1\}^n$, the field multiplication with α is denoted as $\alpha \cdot x$, and it can be computed as $x^{\ll 1}$ if $x[1] = 0$; otherwise, it is $x^{\ll 1} \oplus c$, where $c = c_1 c_2 \cdots c_n$. We use $x \xleftarrow{*} S$ to mean that x is chosen uniformly from the set S, and it is independently chosen from all other previously described distributions.

Definition 1. (Ideal Random Function and Ideal Random Permutation) ρ *is said to be an ideal random function from* $\mathcal{M}$ *to* $\{0,1\}^n$ *if, for any distinct* $m_1, \cdots, m_q \in \mathcal{M}$, $(\rho(m_1), \cdots, \rho(m_q))$ *is uniformly distributed over* $(\{0,1\}^n)^q$ *for any* $q > 0$. *In other words, for any* q *elements* $y_1, \cdots, y_q \in \{0,1\}^n$,

$$\Pr[\rho(m_1) = y_1, \cdots, \rho(m_q) = y_q] = \frac{1}{2^{nq}}.$$

Similarly, τ *is said to be an ideal random permutation on* $\{0,1\}^n$ *if, for any distinct* $x_1, \cdots, x_q \in \{0,1\}^n$ *and distinct* $y_1, \cdots, y_q \in \{0,1\}^n$,

$$\Pr[\tau(m_1) = y_1, \cdots, \tau(m_q) = y_q] = \frac{1}{2^n(2^n - 1) \cdots (2^n - q + 1)}.$$

When $\mathcal{M}$ is a finite set, there is an alternative way to view an ideal random function. Let $\text{Func}(\mathcal{M}, \{0,1\}^n)$ denote the set of all functions from $\mathcal{M}$ to $\{0,1\}^n$. The set of all functions from $\{0,1\}^n$ to $\{0,1\}^n$ is denoted as $\text{Func}(n, n)$. An ideal random function from $\mathcal{M}$ to $\{0,1\}^n$ is defined as a function chosen at random (uniformly) from $\text{Func}(\mathcal{M}, \{0,1\}^n)$ (this is not possible when $\mathcal{M}$ is an infinite set). This definition is an equivalent to the previous definition, as one can show that

$$\Pr[\rho(m_1) = y_1, \cdots, \rho(m_q) = y_q : \rho \xleftarrow{*} \text{Func}(\mathcal{M}, \{0,1\}^n)] = \frac{1}{2^{nq}}$$

for any distinct $m_1, \cdots, m_q \in \mathcal{M}$, and any $y_1, \cdots, y_q \in \{0,1\}^n$. We sometimes write an ideal random function as a keyed function family rand_ρ, where $\text{rand}_\rho(x) = \rho(x)$ and $\rho \in \text{Func}(\mathcal{M}, \{0,1\}^n)$. A block cipher is a function $e : \{0,1\}^k \times \{0,1\}^n \to \{0,1\}^n$, such that for any key $K \in \{0,1\}^k$, $e_K := e(K, \cdot)$ is a permutation on $\{0,1\}^n$. In this paper, we fix n, and any element $x \in \{0,1\}^i$ is called a block if $1 \le i \le n$. A block is called complete if $i = n$, otherwise, it is called incomplete. For any $x \in \{0,1\}^*$, we denote $\lceil \frac{|x|}{n} \rceil$ as $||x||$ (called the number of blocks of x). Let A be an oracle adversary. We say A is a q-adversary if it makes at most q queries, and we say it is a (q, σ)-adversary if it makes at most q queries, and the total number of blocks in all queries is at most σ. For simplicity, we assume that a q-adversary makes exactly q queries, as there is no loss when making some extra dummy queries. We say q is the number of input queries, whereas σ as the number of block-queries.

Definition 2. (Pseudorandom Function) *Let* $F_{K'}$ *be a keyed function family, where* $K' \in \mathcal{K}'$ *and* $F_{K'} : \mathcal{M} \to \{0,1\}^n$ *for a message space* $\mathcal{M}$. *For any* *probabilistic oracle adversary A, we define the prf-advantage of it over the function family F as*

$$\mathbf{Adv}_F^{\mathrm{prf}}(A) = |\Pr[A^{F_{K'}} = 1 : K' \xleftarrow{*} \mathcal{K}'] - \Pr[A^\rho = 1]|,$$

where ρ is an ideal random function from $\mathcal{M}$ to $\{0,1\}^n$, and the probabilities are computed over the internal randomness of A, the uniform distribution of K' and randomness of the output behavior of ρ. When $\mathcal{M} = \{0,1\}^n$, we can equivalently compute the prf-advantage as

$$\mathbf{Adv}_F^{\mathrm{prf}}(A) = |\Pr[A^{F_{K'}} = 1 : K' \xleftarrow{*} \mathcal{K}] - \Pr[A^{\mathrm{rand}_\rho} = 1 : \rho \xleftarrow{*} \mathrm{Func}(n,n)]|.$$

The prf-advantage of F is defined as $\mathbf{Adv}_F^{\mathrm{prf}}(q,\sigma) = \max_A \mathbf{Adv}_F^{\mathrm{prf}}(A)$, where the maximum is taken over all (q,σ)-adversaries A. When $\mathcal{M} = \{0,1\}^n$, we have $\sigma = q$ and hence, we also write $\mathbf{Adv}_F^{\mathrm{prf}}(\sigma)$. We say that a function family F is a (q,σ,ϵ)-prf (or (σ,ϵ)-prf, for the case that $\mathcal{M} = \{0,1\}^n$) if $\mathbf{Adv}_F^{\mathrm{prf}}(q,\sigma) \leq \epsilon$.

Definition 3. (Pseudorandom Permutation) *The prp-advantage of an oracle adversary A over a block cipher $e : \{0,1\}^k \times \{0,1\}^n \to \{0,1\}^n$ is computed as*

$$\mathbf{Adv}_e^{\mathrm{prp}}(A) = |\Pr[A^{e(K,\cdot)} = 1 : K \xleftarrow{*} \{0,1\}^k] - \Pr[A^\tau = 1]|,$$

where the probabilities are computed over internal randomness of A, uniform distribution of K and randomness of the ideal random permutation τ on $\{0,1\}^n$. The prp-advantage of the block cipher e is defined as $\mathbf{Adv}_e^{\mathrm{prp}}(q) = \max_A \mathbf{Adv}_e^{\mathrm{prp}}(A)$, where the maximum is taken over all q-adversaries A.

Lemma 1. (Switching Lemma) *For any function family $F = (F_K)_{K \in \mathcal{K}}$, $F_K : \{0,1\}^n \to \{0,1\}^n$, we have*

$$\mathbf{Adv}_F^{\mathrm{prf}}(\sigma) \leq \mathbf{Adv}_F^{\mathrm{prp}}(\sigma) + \frac{\sigma(\sigma-1)}{2^{n+1}}.$$

The proof of the switching lemma can be found in [6], for example.

3 Generalized **CBC-MAC** Class

3.1 Building Blocks

Every MAC for a message space $\mathcal{M}$ has two main components, namely a randomized key-generation algorithm and a tag-generation algorithm which may be deterministic or probabilistic. The key-generation algorithm returns a key (K,L) at random from it's key space $\mathcal{K} \times \{0,1\}^\ell$. In this paper, we consider deterministic tag-generation algorithms which consist of three main building blocks as described below.

PADDING RULE. A padding rule pad : $\mathcal{M} \to ([0..t] \times \{0,1\}^n)^+$ which ensures that the padded message is in a desired form. The non-negative integer t is said to be the variation number. Given a message m, the padded message $\mathsf{pad}(m) = X$ will be written as $((\delta_1, x_1), \cdots, (\delta_s, x_s))$ for some positive integer s, $x_i \in \{0,1\}^n$ and $\delta_i \in [0..t]$, where $1 \leq i \leq s$. We denote the set of all possible δ_1 values as

$$\Delta_{\mathsf{pad}} = \{\delta_1 : \exists m \in \mathcal{M}, \ \mathsf{pad}(m) = ((\delta_1, x_1), \cdots)\}.$$

The role of the x_i's is similar to that of CBC message block, whereas the δ_i values are used to tweak the intermediate outputs of the block cipher by using a variation operation (for example, applying δ_i amounts of left shift, etc.). A padding rule pad is said to be *prefix-free* if, for any $m \neq m'$, $\mathsf{pad}(m)$ is not a prefix of $\mathsf{pad}(m')$.

ITERATIVE FUNCTION. An underlying iterative function $f : \{0,1\}^n \to \{0,1\}^n$ which is determined via a key $K \in \mathcal{K}$. A block cipher e_K or an ideal random function rand_ρ for $\rho \xleftarrow{*} \mathrm{Func}(n, n)$ (note, $\mathsf{rand}_\rho(x) = \rho(x)$) are different examples of iterative functions.

VARIATION OPERATION. A t-variate variation operation is a function $h : [0..t] \times \{0,1\}^n \to \{0,1\}^n$, such that $h(0, x) = x$. These operations are defined to be very efficient functions. The variation operation may use a key L (called the *auxiliary key*) and the underlying iterative function f as a subroutine. Thus, the variation operation may be determined by the key K of f. In this case, we say the operation is a secret variation operation. If the operation does not use f and any auxiliary key L, then h is a publicly computable function, and we say that it is a public variation operation. A simple example of a t-variate public variation operation is

$$h(i, x) = x^{\ll i} \text{ for all } 0 \leq i \leq t, \ x \in \{0,1\}^n.$$

It is called a type-I secret variation operation if it only depends on the auxiliary key and not on the underlying iterative function f. All other secret variation operations are called type-II. In this paper, we consider public or type-I secret variation operations when we study the security analysis of the generalized CBC constructions. But, we also see some secure constructions, such as OMAC or CMAC, that use type-II secret variation operations.

3.2   Definition of a Generalized CBC-MAC

We define a class of generalized CBC message authentication algorithms (denoted as $\mathcal{C}_{\mathsf{gcbc}}$). Any authentication algorithm from this class for a message space $\mathcal{M}$ has two main functionalities, namely a randomized key-generation algorithm (or Key-Gen) with a key space $\mathcal{K} \times \{0,1\}^\ell$ and a deterministic tag-generation algorithm $\mathsf{gcbc}^{f,h,\mathsf{pad}}$ (defined below). Thus, we need to specify the key space $\mathcal{K} \times \{0,1\}^\ell$, a message space $\mathcal{M}$, the underlying iterative operation f, a t-variate variation operation h and a padding rule with variation number t. The only randomness of the generalized CBC comes from the key $(K, L) \in \mathcal{K} \times \{0,1\}^\ell$ and hence, we denote the authentication algorithm as $\mathsf{gcbc}_{K,L}$ whenever all the above are clear from the context.

1. Key-Gen : $(K, L) \xleftarrow{*} \mathcal{K} \times \{0,1\}^\ell$, where $\mathcal{K} \times \{0,1\}^\ell$ is the key space. Key-generation is parameterized by the key space only.

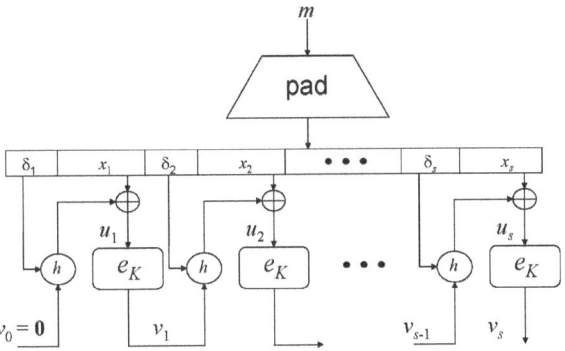

Fig. 3. Generalized CBC which uses variation operation h, an underlying iterative function e_K (block cipher) and a padding rule pad

2. $\text{gcbc}^{f,h,\text{pad}}$: The tag-generation algorithm for a message space $\mathcal{M}$ uses three subroutines viz.,
   - a padding rule $\text{pad} : \mathcal{M} \to ([0..t] \times \{0,1\}^n)^+$, with a variation number $t \geq 0$,
   - a t-variate variation operation (public or secret) $h : [0..t] \times \{0,1\}^n \to \{0,1\}^n$, and
   - an underlying iterative function $f : \{0,1\}^n \to \{0,1\}^n$.

   These subroutines, except for the padding rule pad, are specified by the key (K, L) (the output of Key-Gen), where K is the key for the underlying iterative function f, and L is the auxiliary key that is used for the secret variation operation h. For the public variation operation, $\ell = 0$. For any message m, we define $\text{gcbc}^{f,h,\text{pad}}(m) = v_s$, where v_s is computed as follows (also described in Algorithm 1 and illustrated in Figure 3):

$$v_0 = 0^n, \ u_i = h(\delta_i, v_{i-1}) \oplus x_i, \ v_i = f(u_i), 1 \leq i \leq s \tag{1}$$

   where $\text{pad}(m) = ((\delta_1, x_1), \cdots, (\delta_s, x_s))$, $\delta_i \in [0..t]$, $x_i \in \{0,1\}^n$.

Remark 1. Usually, $x_1, \cdots, x_s$ in Algorithm 1 are all different message blocks and $\delta_1, \cdots, \delta_s$ correspond to tweaking of chaining values. If $\delta_i = 0$ then there is no change in the ith intermediate input (same as CBC chaining for intermediate inputs). For non-zero values of δ_i, we might do left shift operations on intermediate chaining value before xor-ing with the message block. The variation function $h(\delta, \cdot)$ corresponds to these tweaking operations.

Remark 2. An efficiency of the tag-generation algorithm $\text{gcbc}^{f,h,\text{pad}}$ depends on the number of invocations of f, as the underlying iterative function is the most costly operation (we also desire a strong security notion from it, such as it being a pseudorandom function). Note that the number of invocations of f is at least s, and it may be more if we use the type-II secret variation operation.

To keep it small, we should carefully design a padding rule so that the value of s is as small as possible. The padding rule and the variation operations (except the computation of f, which may be used in h) are usually very cheap and hence, we mostly focus on the number of invocations of f when we compare the performance of different constructions.

Algorithm 1. Generalized Cipher Block Chaining Message Authentication

---

Require:

    key. $K\|L \in \mathcal{K} \times \{0,1\}^\ell$. \\ an output of the key generation algorithm Key-Gen,
                                      \\ which is used in the functions f and h

    function. $f : \{0,1\}^n \rightarrow \{0,1\}^n$,
                    $h : [0..t] \times \{0,1\}^n \rightarrow \{0,1\}^n$,
                    pad $: \mathcal{M} \rightarrow ([0..t] \times \{0,1\}^n)^+$ \\$\mathcal{M} = \{0,1\}^*$.

    input. $m \in \mathcal{M}$.

1: $X = \mathsf{pad}(m)$
2: **divide** X as $((\delta_1, m_1) \cdots (\delta_s, x_s))$ where $x_i \in \{0,1\}^n, \delta_i \in [0..t], 1 \le i \le s$
3: $v_0 = 0^n$
4: **for** $j = 1$ to s **do**
5:     $u_j = h(\delta_j, v_{j-1}) \oplus x_j$
6:     $v_j = f(u_j)$
7: **end for**
8: **return** v_s

3.3   Known CBC-Type MACs Are Generalized CBC-MAC

This class is indeed a generalized class, as it contains almost all CBC-type authentication algorithms. We describe more precisely how XCBC, TMAC and OMAC belong to the class. A common choice of the underlying iterative function is a block cipher e_K, $K \in \mathcal{K} = \{0,1\}^k$, and a common choice of padding rule pad is described below. Given a message $m = m_1 \| \cdots \| m_{s-1} \| m_s$ with $m_1, \cdots, m_{s-1} \in \{0,1\}^n$, $m_s \in \{0,1\}^r$, and $1 \le r \le n$, the padding rule (with variation number as two) is defined as

$$\mathsf{pad}(m) = (0, m_1), \cdots, (0, m_{s-1}), (\delta, \overline{m}_s),$$

where $\delta = 1$ if $r < n$; otherwise, we set $\delta = 2$. Also,

$$\overline{m}_s = \begin{cases} m_s \| 10^{n-1-|x|} & \text{if } |m_s| < n \\ m_s & \text{if } |m_s| = n \end{cases}$$

The value of ℓ and the 2-variate variation operations h are described below for each construction. Recall that the key generation algorithm returns a key $(K, L) \xleftarrow{*} \{0,1\}^k \times \{0,1\}^\ell$.

XCBC: Let $\ell = 2n$, and write $L = L_1 \| L_2$, where $L_1, L_2 \in \{0,1\}^n$. Define $h(i,x) = x \oplus L_i \ \forall x \in \{0,1\}^n, i = 1, 2$.

TMAC: Let $\ell = n$, and define $h(i, x) = x \oplus (L \cdot \alpha^{i-1})$ $\forall x \in \{0, 1\}^n$, for $i = 1, 2$, where α is a primitive element and $\alpha^0 = 1$ (multiplicative identity).

OMAC: Let $\ell = 0$, and define $h(0, x) = x$ and $h(i, x) = x \oplus (e_K(\mathbf{0}) \cdot \alpha^i)$ $\forall x \in \{0, 1\}^n, i = 1, 2$.

Remark 3. OMAC is an example where the variation operation is a type-II secret. In the case of the other two examples, the variation operations are type-I secret variation operations. In the next section, we propose two different padding rules and two public variation operations. The use of public variation operations helps to keep the key size as low as possible. We also see that these public variation operations are efficiently computable. We should be careful in the security analysis when we choose a public variation operation, since the security analysis is not straightforward.

4   Security Analysis

4.1   Decorrelation Technique

Vaudeney's Decorrelation Theorem (Lemma 22 of [20][3]) is used in the security analysis. Based on our notations, we state the following version of the Decorrelation Theorem.

Theorem 1. (Decorrelation Theorem)
Let q and σ be two fixed integers, and let $F_{K'} : \mathcal{M} \to \{0, 1\}^n$ be a family of functions indexed by key K' that is chosen uniformly from the key space $\mathcal{K}'$. Suppose that the following holds for a positive real number ϵ;

$$\Pr[F_{K'}(x_1) = y_1, \cdots, F_{K'}(x_q) = y_q : K \xleftarrow{*} \mathcal{K}] \geq 2^{-nq}(1 - \epsilon) \text{ for any}$$
$$(y_1, \cdots, y_q) \in (\{0, 1\}^n)^q \text{ and distinct } m_1, \cdots, m_q \in \mathcal{M}, \sum_{i=1}^q \|m_i\| \leq \sigma.$$

Then for any distinguisher A which asks q queries with σ blocks present in all queries, $\mathbf{Adv}_F^{\mathrm{prf}}(A) \leq \epsilon$.

What does the Decorrelation Theorem mean? The above condition means that the output behavior of the function family is very close to that of an ideal random function for any choices of distinct inputs (note, $2^{-nq} = \Pr[\rho(x_1) = y_1, \cdots, \rho(x_q) = y_q]$). Thus, the prf-advantage of a (q, σ)-adversary should be small, independent of how the adversary works. Note that the adversary (at the end of the query-responses) has a set of inputs and outputs, and he has to distinguish the function family from an ideal random function based on the query-responses. However, the values of ϵ can depend on q and σ.

---

[3] It was mentioned in [20] that the decorrelation theorem was freely adapted from Patarin's coefficient H-techniques [18].

4.2 Security Analysis of Generalized CBC Algorithm

A variation operation can be public or secret. A variation operation is said to be *allowed* if either it is public, or it is a type-I secret (which does not use the underlying iterative as a subroutine and only uses an auxiliary key).

Definition 4. *Let Δ be a subset of $[0..t]$ and ϵ be a nonnegative real number. An allowed t-variate variation operation $h : [0..t] \times \{0,1\}^n \rightarrow \{0,1\}^n$ is said to be a (ϵ, Δ)-xor weak universal operation if, for any $0 \leq \delta \neq \delta' \leq t$, $c \in \{0,1\}^n$, and for all auxiliary key $L_0 \in \{0,1\}^\ell$, the following conditions are satisfied.*

> *W1:* $\Pr[h_{L_0}(i, R) = c : R \xleftarrow{*} \{0,1\}^n] \leq \epsilon$, *for any* $0 \leq i \leq t$,
>
> *W2:* $\Pr[h_L(\delta, 0^n) \oplus h_L(\delta', 0^n) = c : L \xleftarrow{*} \{0,1\}^\ell] \leq \epsilon$ *whenever* $\delta, \delta' \in \Delta$
>
> *W3:* $\Pr[h_L(\delta, R) \oplus h_L(\delta', R) = c : (R, L) \xleftarrow{*} \{0,1\}^n \times \{0,1\}^\ell] \leq \epsilon$.

Note that when Δ is a singleton set, then the condition W2 is vacuously true. We provide a sufficient condition for prf-secure of generalized CBC constructions.

A sufficient condition for prf-secure gcbc

1. Let pad be a prefix-free padding rule with a variation number $t \geq 0$, h be a $(\epsilon, \Delta_{\mathsf{pad}})$-xor weak universal, allowed variation operation, and
2. the underlying iterative function family $(f_K)_{K \in \mathcal{K}}$ is (σ, μ)-prf.

The generalized CBC, based on the above such building blocks, is (q, σ, ϵ')-prf, where $\epsilon' = \sigma'(\sigma' - 1)\epsilon + \mu$ and where σ' denotes the total number of blocks in all queries after padding. (which have been queried by an adversary)[4]. As we are going to apply decorrelation theorem, we want to show the following probability for distinct $m_1, \cdots, m_q \in \mathcal{M}$ and distinct $y_1, \cdots, y_q \in \{0,1\}^n$,

$$p = \Pr_{\rho,L}[\mathsf{gcbc}_{\rho,L}(m_1) = y_1, \cdots, \mathsf{gcbc}_{\rho,L}(m_q) = y_q] \geq \frac{1 - \frac{\sigma'(\sigma'-1)\epsilon}{2}}{2^{nq}} \qquad (2)$$

where the probability is computed over $(\rho, L) \xleftarrow{*} \mathsf{Func}(n, n) \times \{0,1\}^\ell$. Then by applying decorrelation theorem and switching lemma, we know that $\mathsf{gcbc}_{K,L}$ is (q, σ, ϵ')-prf-secure. Now, it remains to show the above equation. We first introduce some notations as given below.

Notations. Let $\mathsf{pad}(m_i) = X_i = ((\delta_{i,1}, x_{i,1}) \cdots, (\delta_{i,\ell_i}, x_{i,s_i}))$, where $x_{i,1}, \cdots, x_{i,s_i} \in \{0,1\}^n$ and $\delta_{i,1}, \cdots, \delta_{i,s_i} \in [0..t]$. Let $X_{i,j} = ((\delta_{i,1}, x_{i,1}) \cdots, (\delta_{i,j}, x_{i,j}))$ for $0 \leq j \leq s_i$, where $X_{i,0} = \lambda$ for any i. For each $1 \leq i \leq q$, we have the following sequences of u_i's and v_i's values.

$$v_i\text{-sequence} : \quad 0^n \xrightarrow{(\delta_{i,1}, x_{i,1})} v_{i,1} \xrightarrow{(\delta_{i,2}, x_{i,2})} v_{i,2} \cdots \xrightarrow{(\delta_{i,s_i}, x_{i,s_i})} v_{i,s_i}$$

$$u_i\text{-sequence} : \quad \lambda \xrightarrow{(\delta_{i,1}, x_{i,1})} u_{i,1} \xrightarrow{(\delta_{i,2}, x_{i,2})} u_{i,2} \cdots \xrightarrow{(\delta_{i,s_i}, x_{i,s_i})} u_{i,s_i}$$

---

[4] Later, we propose two padding rules where the number of block can only increase by at most one for each message and hence, $\sigma' \leq \sigma + q$.

Note that all these variables $u_{i,j}, v_{i,j}$ are random variables, whereas $\delta_{i,j}, x_{i,j}$ are fixed constants. Moreover, $u_{i,j} = h_L(\delta_{i,j}, v_{i,j-1}) \oplus x_{i,j}$ and $\mathsf{rand}_\rho(u_{i,j}) = v_{i,j}$. Thus, $u_{i,j}$ corresponds to the ith input of rand_ρ while computing $\mathsf{gcbc}_{\rho,L}(m_i)$ and $v_{i,j}$ is its corresponding output. The most common approach is to recognize all trivial collisions of inputs (which are because of the fact that some parts of two messages are identical). Then we try to prove that any other input collision has low probability (in other we would provide an upper bound on all other input collisions). Next, we would define admissible tuples which correspond to all intermediate inputs and outputs which do not have any non-trivial collisions.

Lemma 2. *If $X_{i,j} = X_{i',j'}$ then $u_{i,j} = u_{i',j'}$ and $v_{i,j} = v_{i',j'}$ with probability 1.*

Definition 5. *Let $L_0 \in \{0,1\}^\ell, V_{i,j} \in \{0,1\}^n, 1 \le i \le q, 1 \le j \le s_i$ for some fixed integers $s_1, \cdots, s_q$ and q. We say that a tuple $(L_0, V_{i,j})_{i,j}$ is admissible if*
- *$V_{i,j} = V_{i',j'}$ whenever $X_{i,j} = X_{i',j'}$,*
- *$V_{i,0} = 0^n, V_{i,s_i} = y_i$ for all $1 \le i \le q$ and*
- *$h_{L_0}(\delta_{i,j}, V_{i,j-1}) \oplus x_{i,j} \ne h_{L_0}(\delta_{i',j'}, V_{i',j'-1}) \oplus x_{i',j'}$ for all i, j, i', j' such that $X_{i,j} \ne X_{i',j'}$.*

Let σ_1 be the maximum number of all pairs (i, j) with distinct $X_{i,j}$'s. More precisely, $\sigma_1 = |\{X : X = X_{i,j}$ for some $i, j\}|$. Clearly, $\sigma_1 \le \sigma' = \sum_{i=1}^q \|X_i\|$.

Lemma 3. *Given any admissible tuple $(L_0, V_{i,j})_{i,j}$, $\Pr[L = L_0, v_{i,j} = V_{i,j}] = \frac{1}{2^{n(\sigma_1+q)}} \times \frac{1}{2^\ell}$.*

Proof. Given any such admissible tuple, $u_{i,j} = u_{i',j'}$ if and only if $X_{i,j} = X_{i',j'}$ and all $u_{i,j} = U_{i,j} = h_{L_0}(\delta_{i,j}, V_{i,j-1}) \oplus x_{i,j}$ values are fixed. Thus,

$$\Pr[L = L_0, v_{i,j} = V_{i,j}] = \Pr[L = L_0, \rho(U_{i,j}) = V_{i,j} \text{ for all } i, j]$$
$$= \Pr[\rho(U_{i,j}) = V_{i,j} \text{ for all } i, j] \times \Pr[L = L_0]$$
$$= \frac{1}{2^{n(\sigma_1+q)}} \times \frac{1}{2^\ell}.$$

Lemma 4. *The number of admissible tuples is at least $2^{n\sigma_1+\ell}(1 - \frac{\epsilon\sigma_1(\sigma_1-1)}{2})$.*

Proof. We have $2^{n\sigma_1+\ell}$ tuples $(L_0, (V_{i,j})_{i,j})$ such that $V_{i,0} = 0^n, V_{i,s_i} = y_i$ for all $1 \le i \le q$ and $V_{i,j} = V_{i',j'}$ whenever $X_{i,j} = X_{i',j'}$. Now we need to find an estimate of the number of tuples among these such that $h_{L_0}(\delta_{i,j}, V_{i,j-1}) \oplus x_{i,j} \ne h_{L_0}(\delta_{i',j'}, V_{i',j'-1}) \oplus x_{i',j'}$ for all i, j, i', j' such that $X_{i,j} \ne X_{i',j'}$. To do so, we count the complement. Suppose that for some i, j, i', j' with $X_{i,j} \ne X_{i',j'}, h_{L_0}(\delta_{i,j}, V_{i,j-1}) \oplus x_{i,j} = h_{L_0}(\delta_{i',j'}, V_{i',j'-1}) \oplus x_{i',j'}$. The number of such tuples is at most $2^{n\sigma_1+\ell-1}$, since h is a weakly $(\epsilon, \Delta_{\mathsf{pad}_1})$-xor universal variation operation. The total number of possible values of i, j, i', j' such that $X_{i,j} \ne X_{i',j'}$ is $\binom{\sigma_1}{2}$. Subtracting all such non-admissible tuples, we see that there are at least $2^{n\sigma_1+\ell}(1 - \frac{\epsilon\sigma_1(\sigma_1-1)}{2})$ admissible tuples. ∎

Combining the above two lemmas, we can prove the following theorem.

Theorem 2. *Let h be a weakly $(\epsilon, \Delta_{\mathsf{pad}})$-xor universal operation and pad be a prefix-free padding rule. Suppose that the underlying iterative function is an ideal random function $(\mathsf{rand}_\rho)_{\rho \in \mathrm{Func}(n,n)}$, and we denote the corresponding generalized CBC authentication algorithm as $\mathsf{gcbc}_{\rho,L}$. Let σ' be the largest number of blocks after padding q messages having at most σ blocks in total. Then for any distinct $m_1, \cdots, m_q \in \mathcal{M}$ (message space), and any $y_1, \cdots, y_q \in \{0,1\}^n$,*

$$\Pr_{\rho,L}[\mathsf{gcbc}_{\rho,L}(m_1) = y_1, \cdots, \mathsf{gcbc}_{\rho,L}(m_q) = y_q] \geq \frac{1 - \frac{\sigma'(\sigma'-1)\epsilon}{2}}{2^{nq}}.$$

Theorem 3. *Based on all notations defined so far, we have,*

$$\mathbf{Adv}^{\mathrm{prf}}_{\mathsf{gcbc}}(q, \sigma) \leq \frac{\sigma'(\sigma' - 1)\epsilon}{2} + \mathbf{Adv}^{\mathrm{prf}}_f(\sigma')$$

$$\leq \frac{\sigma'(\sigma' - 1)}{2}(\epsilon + \frac{1}{2^n}) + \mathbf{Adv}^{\mathrm{prp}}_f(\sigma')$$

Theorem 4. *The variation operations defined in XCBC and TMAC are weakly $(\frac{1}{2^n}, \Delta_{\mathsf{pad}})$-xor universal operations and $\sigma' = \sigma$. So,*

1. $\mathbf{Adv}^{\mathrm{prf}}_{\mathsf{XCBC}}(q, \sigma) \leq \frac{\sigma(\sigma-1)}{2^n} + \mathbf{Adv}^{\mathrm{prp}}_f(\sigma).$
2. $\mathbf{Adv}^{\mathrm{prf}}_{\mathsf{TMAC}}(q, \sigma) \leq \frac{\sigma(\sigma-1)}{2^n} + \mathbf{Adv}^{\mathrm{prp}}_f(\sigma).$

5   Two New Efficient Generalized **CBC-MAC**: **GCBC1** and **GCBC2**

In this section, we propose two secure, generalized CBC constructions, namely GCBC1 and GCBC2 both have message space $\{0,1\}^*$.

5.1   **GCBC1**

We first define a padding rule $\mathsf{pad}_1 : \{0,1\}^{>n} \to ([0..2] \times \{0,1\}^n)^+$. For any $m = m_1 \cdots m_{s-1} m_s \in \{0,1\}^{>n}$, where $m_1, \cdots, m_{s-1} \in \{0,1\}^n$ and $m_s \in \{0,1\}^r$, $1 \leq r \leq n$, we define the padded message as

$$\mathsf{pad}_1(m) = ((0, m_1), \cdots, (0, m_{s-1}), (\delta, \overline{m}_s)),$$

where $\delta = 1$ if $r < n$; otherwise, $\delta = 2$. We extend the definition of the padding rule to the message space $\{0,1\}^*$ as follows. Let $m_1 \in \{0,1\}^r$, define

$$\mathsf{pad}_1(m_1) = \begin{cases} ((0, \overline{m}_1), (1, \mathbf{0})) & \text{if } r < n \\ ((0, m_1), (1, 10^{n-1})) & \text{if } r = n. \end{cases}$$

Thus, s-block messages have s-block padded messages for all $s \geq 2$, and one-block messages have two-block padded messages. It is also easy to observe that $\Delta_{\mathsf{pad}_1} = \{0\}$. Moreover, the padding rule is prefix-free.

Algorithm 2. GCBC1

Require:

    key. $K \xleftarrow{*} \{0,1\}^k$. \\ block cipher key

    function. $e_K : \{0,1\}^n \rightarrow \{0,1\}^n$. \\ block cipher

    input. $m \in \{0,1\}^*$.

1: **divide** m as $(m_1, \cdots, m_{s-1}, m_s)$

    where $m_1, \cdots, m_{s-1} \in \{0,1\}^n, m_s \in \{0,1\}^r, 1 \le r \le n$.

2: **if** $s = 1$ and $r = n$ **then**

3:    $m_2 = 10^{n-1}, s = 2, r = n - 1$.

4: **else if** $s = 1$ **then**

5:    $m_1 = \overline{m}_1, m_2 = \mathbf{0}, s = 2$.

6: **end if**

7: $v_0 = \mathbf{0}$

8: **for** $j = 1$ to $s - 1$ **do**

9:    $u_j = v_{j-1} \oplus x_j$

10:   $v_j = e_K(u_j)$

11: **end for**

12: **if** $r < n$ **then**

13:   $u_s = v_{s-1}^{\lll 1} \oplus x_s$

14: **else**

15:   $u_s = v_{s-1}^{\lll 2} \oplus x_s$

16: **end if**

17: $v_s = e_K(u_s)$

18: **return** v_s

Proposition 1. *The padding rule* pad_1 *over the message space* $\{0,1\}^*$ *is a prefix-free padding rule.*

Proof. Suppose $m = m_1 \cdots m_{s-1} m_s$ and $m' = m'_1 \cdots m'_{s'-1} m'_{s'}$ where $s \le s'$, $\mathsf{pad}_1(m)$ is a prefix of $\mathsf{pad}_1(m')$ and $m_1, m'_1, \cdots, m_{s-1}, m'_{s'-1} \in \{0,1\}^n, m_s \in \{0,1\}^r, m'_{s'} \in \{0,1\}^{r'}, 1 \le r, r' \le n$.

Case $s \ge 2$: $\mathsf{pad}_1(m) = ((0, m_1), \cdots, (0, m_{s-1}), (\delta, \overline{m}_s))$ is a prefix of $\mathsf{pad}_1(m')$ $= ((0, m'_1), \cdots, (0, m'_{s'-1}), (\delta', \overline{m'}_{s'}))$. Since $\delta, \delta' \ne 0$, $s' = s$ and $\delta = \delta'$. Moreover, $m'_1 = m_1, \cdots, m_{s-1} = m'_{s-1}, \overline{m}_s = \overline{m'}_s$. Now, $\overline{m}_s = \overline{m'}_s$ and $\delta = \delta'$ implies that $m_s = m'_s$. Thus $m = m'$.

Case $s = 1, s' \ge 2$: $\mathsf{pad}_1(m) = ((0, x_1), (1, x_2))$ where $\mathsf{pad}_1(m') = ((0, m'_1),$ $\cdots, (0, m'_{s'-1}), (\delta', \overline{m'}_{s'})$. By comparing δ values of the second pair, we can see that $s' = 2$, $r' < n$ and $\overline{m'}_2 = x_2$. But x_2 is either $\mathbf{0}$ or 10^{n-1} which can't be $\overline{m'}_2$ for any $m'_2 \in \{0,1\}^{r'}, 1 \le r' < n$. So this case does not arise.

Case $s = 1, s' = 1$: Obviously, if $\mathsf{pad}_1(m_1)$ is a prefix of $\mathsf{pad}_1(m'_1)$, then they should be equal. But, it is easy to see that they can be equal only when $m_1 = m'_1$ and hence, $m = m'$. ∎

Now we define a simple public variation operation ls, which has two variations. For any $x \in \{0,1\}^n$ and $0 \le \delta \le 2$, $\mathsf{ls}(\delta, x) = x^{\lll \delta}$.

Proposition 2. *The operation* Is *with 2 variations is a public and weakly $\frac{1}{2^{n-2}}$- xor universal for $\Delta = \Delta_{\mathsf{pad}_1} = \{0\}$. The same operation* Is *with 5 variations is a public and weakly $\frac{1}{2^{n-5}}$-xor universal for $\Delta = \Delta_{\mathsf{pad}_1} = \{0\}$.*

Proof. By definition, $\mathsf{ls}(0, x) = x$. It is easy to see that $x^{\ll i} = c$ has at most 4 solutions of x for any $i \leq 2$ and any constant $c \in \{0, 1\}^n$. So, condition W1 holds. The condition W2 trivially holds, since $\Delta = \{0\}$. To see the condition W3, we first prove that the number of solutions of $x \oplus x^{\ll 1} = c$ is exactly one for any constant $c \in \{0, 1\}^n$. In fact, the solution is $x[n] = c[n], x[n-1] = c[n-1] \oplus c[n-1], \cdots, x[1] = c[1] \oplus \cdots \oplus c[n]$. Similarly, one can see that the number of solutions of x for the equation $x \oplus x^{\ll 2} = c$ is exactly one. Now we want to find the number of solutions of the equation $x^{\ll 1} \oplus x^{\ll 2} = c$. Let $y = x^{\ll 1}$. We have exactly one solution of y and hence, there are exactly two solutions of x. Combining all these observations, we can see that condition W3 is true. Thus, Is is a weakly $\frac{1}{2^{n-2}}$-xor universal. The case for 5 variation operation can be proved similarly. ∎

We define the tag-generation algorithm of GCBC1 as the generalized CBC algorithm (see Figure 1) $\mathsf{gcbc}^{e, \mathsf{ls}, \mathsf{pad}_1}$ with a message space of $\{0, 1\}^*$ and the padding rule pad_1 (see Algorithm 2).

Theorem 5. (Security Bound of GCBC1)

$$\mathbf{Adv}^{\mathrm{prf}}_{GCBC1}(q, \sigma) \leq \frac{5(\sigma + q)(\sigma + q - 1)}{2^n} + \mathbf{Adv}^{\mathrm{prp}}_e(\sigma + q)$$

Proof. We apply the result of the above two propositions to the generalized CBC security bound (see Theorem 3 where $sigma' \leq \sigma + q$). ∎

5.2   GCBC2

We first define a padding rule pad_2 for the message space $\{0, 1\}^*$ with variation number 5. Let $m = m_1 \cdots m_{s-1} m_s \in \{0, 1\}^*$, where $m_1, \cdots, m_{s-1} \in \{0, 1\}^n$ and $m_s \in \{0, 1\}^r$, $0 \leq r \leq n$ ($r = 0$ only when we have an empty message m). Let us denote $\delta = 1$ if $r < n$; otherwise, $\delta = 2$. If $|m| \geq n - 3$ then denote $m_1 = m_1' \| m_1''$, where $m_1' \in \{0, 1\}^{n-3}$ and $m_1'' \in \{0, 1\}^*$. Define $\mathsf{pad}_2(m)$ to depend on s.

Case $s = 1$, $\mathsf{pad}_2(m_1) = \begin{cases} ((0, m_1' \| 011), (0, \overline{m''}_1)) & \text{if } r \geq n - 3 \\ (0, \overline{m}_1) & \text{if } r \leq n - 4 \end{cases}$

Case $s = 2$, $\mathsf{pad}_2(m_1, m_2) = \begin{cases} ((0, m_1), (\delta + 1, \overline{m}_2)) & \text{if } m_1'' \neq 000 \\ ((0, m_1' \| 100), (\delta - 1, \overline{m}_2)) & \text{if } m_1'' = 000 \end{cases}$

For all other cases, i.e., $s \geq 3$,

$\mathsf{pad}_2(m) = \begin{cases} ((0, \overline{m'}_1), (5, m_2), (0, m_3), \cdots, (0, m_{s-1}), (\delta, \overline{m}_s)) & \text{if } m_1'' = 000 \\ ((0, m_1), (4, m_2), (0, m_3), \cdots, (0, m_{s-1}), (\delta, \overline{m}_s)) & \text{if } m_1'' \neq 000 \end{cases}$

Note, $\Delta_{\mathsf{pad}_2} = \{0\}$ and it increases one block only when the message size is in between $n - 3$ and n. All other messages and their padded messages have same number of blocks. Now we prove that it is a prefix-free padding rule.

Proposition 3. pad_2 *is prefix-free padding rule over the message space* $\{0,1\}^*$.

Proof. Suppose that $m = m_1 \cdots m_{s-1} m_s$ and $m' = m'_1 \cdots m'_{s'-1} m'_{s'}$ where $s \leq s'$, $\mathsf{pad}_2(m)$ is a prefix of $\mathsf{pad}_2(m')$ and $m_1, m'_1, \cdots, m_{s-1}, m'_{s'-1} \in \{0,1\}^n$, $m_s \in \{0,1\}^r$, $m'_{s'} \in \{0,1\}^{r'}$, $1 \leq r, r' \leq n$. Let $\mathsf{pad}_2(m) = ((0,x_1),(\delta,x_2),\cdots)$ if $s \geq 2$; otherwise, $\mathsf{pad}_2(m) = (0,x_1)$. Similarly we denote $\mathsf{pad}_2(m') = ((0,x'_1),(\delta',x'_2),\cdots)$ if $s \geq 2$; otherwise, $\mathsf{pad}_2(m') = (0,x'_1)$. When $s = 1$, s' must be 1; otherwise, for any $s' \geq 2$, we always have $(x_1,\delta) \neq (x'_1,\delta')$. It is also easy to see that if $s = s' = 1$, then $m = m'$. Let $s \geq 2$. By comparing the values of δ and δ' we must have $s = s' = 2$ or $s, s' \geq 3$. From the definition of pad_2, one can check that $m = m'$.   ∎

We define the tag-generation algorithm of GCBC2 as the generalized CBC algorithm (see Figure 2) $\mathsf{gcbc}^{e,\mathsf{ls},\mathsf{pad}_2}$ with a message space of $\{0,1\}^*$ and the padding rule pad_2 (see Algorithm 1 for generalized CBC tag generation algorithm or see a complete description of GCBC in introduction). The proof of the following theorem is immediate from Theorem 3 ($sigma' \leq \sigma + q$).

Theorem 6. (Security Bound of GCBC2)

$$\mathbf{Adv}^{\mathrm{prf}}_{GCBC2}(q,\sigma) \leq \frac{33(\sigma+q)(\sigma+q-1)}{2^n} + \mathbf{Adv}^{\mathrm{prp}}_e(\sigma+q).$$

Remark 4. Our bound is of the form $\sigma^2/2^n$, whereas in [5], it had been shown that CBC-MAC is prf-secure for prefix-free messages with the security bound of the form $\ell q^2/2^n$, where ℓ denotes the number of blocks of the longest query among all q queries. Note that GCBC1 can be viewed as a CBC-MAC, where the last message block is modified by the last intermediate chain value. Because of this modification, it seems hard to obtain prefix queries. If it is so, then we can apply the result from [5] to obtain a bound of the form $\ell q^2/2^n$ for GCBC1. This would be our future research work and we leave this as an open problem.

An Efficient Variation Operation. One can choose an efficient, different public variation operation $h = \mathsf{tr}$ for a generalized CBC MAC algorithm, which is defined as follows. Let n' be a divisor of n and $x = x_1 \cdots x_{n'}$, where $x_i \in \{0,1\}^w$. The actual value of w can depend on the underlying block cipher and when using AES, we choose $w = 8$. Define, $\mathsf{tr}(0, x_1 \cdots x_{n'}) = (x_1, \cdots, x_{n'})$, $\mathsf{tr}(1, x_1 \cdots x_{n'}) = \mathsf{tr}(x_1 \cdots x_{n'}) := x_2 \cdots x_{n'} x_1^{\lll 1}$ and inductively define for $i \geq 2$,

$$\mathsf{tr}(i, x_1 \cdots x_{n'}) = \mathsf{tr}(i-1, \mathsf{tr}(1, (x_1 \cdots x_{n'}))) = \mathsf{tr}(i-1, x_2 \cdots x_{n'} x_1^{\lll 1}).$$

In particular, we have $\mathsf{tr}(2, x_1 \cdots x_{n'}) := x_3 \cdots x_{n'} x_1^{\lll 1} x_2^{\lll 1}$ and $\mathsf{tr}(3, x_1 \cdots x_{n'}) := x_4 \cdots x_{n'} x_1^{\lll 1} x_2^{\lll 1} x_3^{\lll 1}$ and so on. This would be very efficient in software when we use a w-bit processor. In case of GCBC2 with the above defined variation operation, it needs at most three 8-bit shift operations. Note, an 8-bit implementation of a single shift on 128 bits needs 16 shift operations. For example, if we use AES, then a single shift on 128 bits (it is partitioned into 16 bytes) requires 16 shift operations and several bitwise-and and bitwise-or operations. The proof is very similar to that of proposition 2 and hence we omit it.

Proposition 4. *The operation* tr *with 5 variations is a public and weakly* $\frac{1}{2^{n-5}}$- *xor universal for* $\Delta = \Delta_{\mathsf{pad}_2} = \{0\}$.

6   Conclusion

In this paper, many popular CBC-type message authentication algorithms are viewed in a unified way. In particular, a wide class of authentication algorithms called generalized CBC algorithms is introduced. This class contains almost all known CBC-type secure authentication algorithms. Moreover, we have proposed two secure constructions GCBC1 and GCBC2 from this class which are optimum in key size and the number of block cipher invocations. These constructions may have significant performance compared to OMAC for short messages. We also characterize the prf-secure generalized CBC constructions. We hope the idea of generalizing CBC constructions can also help us to generalize other similar constructions for different security goals.

Acknowledgement. We would like to thank Elaine Barker, William Burr, Tetsu Iwata and the anonymous referees of FSE-09 who have made several editorial and technical comments. We would also like to thank Liting Zhang who found a small error on the earlier version of GCBC2 and based on his comments we have made a minor modification of GCBC2.

References

1. Bellare, M., Canetti, R., Krawczyk, H.: Keying Hash Functions for Message Authentication. In: Koblitz, N. (ed.) CRYPTO 1996. LNCS, vol. 1109, pp. 1–15. Springer, Heidelberg (1996)
2. Bellare, M., Canetti, R., Krawczyk, H.: Pseudorandom Functions Revisited: The Cascade Construction and Its Concrete Security. In: FOCS, pp. 514–523 (1996)
3. Bellare, M., Guérin, R., Rogaway, P.: XOR MACs: New Methods for Message Authentication Using Finite Pseudorandom Functions. In: Coppersmith, D. (ed.) CRYPTO 1995. LNCS, vol. 963, pp. 15–28. Springer, Heidelberg (1995)
4. Bellare, M., Kilian, J., Rogaway, P.: The Security of the Cipher Block Chaining Message Authentication Code. J. Comput. Syst. Sci. 61(3), 362–399 (2000)
5. Bellare, M., Pietrzak, K., Rogaway, P.: Improved Security Analyses for CBC MACs. In: Shoup, V. (ed.) CRYPTO 2005. LNCS, vol. 3621, pp. 527–545. Springer, Heidelberg (2005)
6. Bellare, M., Rogaway, P.: The Security of Triple Encryption and a Framework for Code-Based Game-Playing Proofs. In: Vaudenay, S. (ed.) EUROCRYPT 2006. LNCS, vol. 4004, pp. 409–426. Springer, Heidelberg (2006)
7. Bernstein, D.J.: The Poly1305-AES Message-Authentication Code. In: Gilbert, H., Handschuh, H. (eds.) FSE 2005. LNCS, vol. 3557, pp. 32–49. Springer, Heidelberg (2005)
8. Black, J., Halevi, S., Krawczyk, H., Krovetz, T., Rogaway, P.: UMAC: Fast and Secure Message Authentication. In: Wiener, M.J. (ed.) CRYPTO 1999. LNCS, vol. 1666, pp. 216–233. Springer, Heidelberg (1999)

9. Black, J., Rogaway, P.: A Block-Cipher Mode of Operation for Parallelizable Message Authentication. In: Knudsen, L.R. (ed.) EUROCRYPT 2002. LNCS, vol. 2332, pp. 384–397. Springer, Heidelberg (2002)
10. Black, J., Rogaway, P.: CBC MACs for Arbitrary-Length Messages: The Three-Key Constructions. J. Cryptology 18(2), 111–131 (2005)
11. Carter, L., Wegman, M.N.: Universal Classes of Hash Functions. J. Comput. Syst. Sci. 18(2), 143–154 (1979)
12. Daemen, J., Rijmen, V.: The Design of Rijndael: AES - The Advanced Encryption Standard (2002),
    http://csrc.nist.gov/CryptoToolkit/aes/rijndael/Rijndael-ammended.pdf
13. Dworkin, M.: Recommendation for Block Cipher Modes of Operation: The CMAC Mode for Authentication,
    http://csrc.nist.gov/publications/nistpubs/index.html#sp800-38B
14. Hirose, S., Park, J.H., Yun, A.: A Simple Variant of the Merkle-Damgård Scheme with a Permutation. In: Kurosawa, K. (ed.) ASIACRYPT 2007. LNCS, vol. 4833, pp. 113–129. Springer, Heidelberg (2007)
15. Iwata, T., Kurosawa, K.: OMAC: One-Key CBC MAC. In: Johansson, T. (ed.) FSE 2003. LNCS, vol. 2887, pp. 129–153. Springer, Heidelberg (2003)
16. Kurosawa, K., Iwata, T.: TMAC: Two-Key CBC MAC. In: Joye, M. (ed.) CT-RSA 2003. LNCS, vol. 2612, pp. 33–49. Springer, Heidelberg (2003)
17. Minematsu, K., Tsunoo, Y.: Provably Secure MACs from Differentially-Uniform Permutations and AES-Based Implementations. In: Robshaw, M.J.B. (ed.) FSE 2006. LNCS, vol. 4047, pp. 226–241. Springer, Heidelberg (2006)
18. Patarin, J.: Etude des Générateurs de Permutations Basés sur le Schéma du D.E.S. Phd Thèsis de Doctorat de l'Université de Paris 6 (1991)
19. Rogaway, P.: Bucket Hashing and Its Application to Fast Message Authentication. J. Cryptology 12(2), 91–115 (1999)
20. Vaudenay, S.: Decorrelation: A Theory for Block Cipher Security, vol. 16, pp. 249–286 (2003)
21. Yasuda, K.: "Sandwich" Is Indeed Secure: How to Authenticate a Message with Just One Hashing. In: Pieprzyk, J., Ghodosi, H., Dawson, E. (eds.) ACISP 2007. LNCS, vol. 4586, pp. 355–369. Springer, Heidelberg (2007)

HBS: A Single-Key Mode of Operation for Deterministic Authenticated Encryption

Tetsu Iwata[1] and Kan Yasuda[2]

[1] Dept. of Computational Science and Engineering, Nagoya University, Japan
iwata@cse.nagoya-u.ac.jp
[2] NTT Information Sharing Platform Laboratories, NTT Corporation, Japan
yasuda.kan@lab.ntt.co.jp

Abstract. We propose the HBS (Hash Block Stealing) mode of operation. This is *the first* single-key mode that provably achieves the goal of providing deterministic authenticated encryption. The authentication part of HBS utilizes a newly-developed, vector-input polynomial hash function. The encryption part uses a blockcipher-based, counter-like mode. These two parts are combined in such a way as the numbers of finite-field multiplications and blockcipher calls are minimized. Specifically, for a header of h blocks and a message of m blocks, the HBS algorithm requires just $h + m + 2$ multiplications in the finite field and $m + 2$ calls to the blockcipher. Although the HBS algorithm is fairly simple, its security proof is rather complicated.

Keywords: Universal hash function, counter mode, SIV, security proof.

1 Introduction

The goals for blockcipher modes of operation are twofold. One is to establish *authenticity*, or data integrity. The other is to preserve *privacy*, or data confidentiality. These two goals are realized by the mechanism of *authenticated encryption*, or AE for short. An AE mode establishes authenticity and preserves privacy concurrently by producing a ciphertext into which both the tag and the encrypted message are embedded.

A crucial aspect of an AE mode is that its security is based on the use of either a randomized salt or a state-dependent value, which has been formalized as *nonce-based* AE [13,14,15]. In fact, many of the modern AE modes, including CCM [19], GCM [9] and OCB [13], are all nonce-based.

The nonce, however, needs to be handled with great care, because the misuse of nonce (i.e., repeating the same value) would generally lead to the complete collapse of systems based on these AE modes. This is due to the fact that the security designs have no concern in what happens when the nonce assumption becomes no longer true.

The problem of nonce misuse has been settled by the introduction of *deterministic authenticated encryption* [16], or DAE for short. A DAE mode provides a deterministic, stateless algorithm that produces a ciphertext from the pair

O. Dunkelman (Ed.): FSE 2009, LNCS 5665, pp. 394–415, 2009.

of a header and a message. A DAE mode can be used also as a conventional nonce-based AE mode by embedding a nonce value into the header. The virtue of a DAE mode is that it still ensures a certain level of security (i.e., all that an adversary can do is to detect a repetition of the same header-message pair) even when the DAE mode is not combined with a nonce element. Hence DAE is more robust than nonce-based AE.

The work [16] also proposes a concrete DAE mode of operation called SIV. The SIV mode is a blockcipher-based scheme, utilizing a vector-input version of the CMAC algorithm [5,11] for its authentication part and the CTR (counter) mode for its encryption part. The SIV mode requires two independent keys for the underlying blockcipher, one being for the CMAC algorithm and another for the CTR mode.

The purpose of the current paper is to improve usability and performance over SIV. For this, we present a new DAE mode of operation called HBS (which stands for Hash Block Stealing), which has the following features.

1. The HBS mode requires just one key. It uses the same key for authentication and for encryption.
2. For authentication, the HBS mode adopts a vector-input polynomial-based universal hash function (rather than the blockcipher-based CMAC).
3. For encryption, the HBS mode uses a CTR-like scheme using a blockcipher, which operates differently from an ordinary CTR mode.
4. The numbers of finite-field multiplications (in hashing) and blockcipher calls are minimized.
5. We provide the HBS mode with concrete proofs of security.

In general, reducing keying material without compromising security reduces the cost of sharing, saving, and updating the secret key. Also, changing from a blockcipher-based MAC to polynomial-based universal hashing increases efficiency on many platforms [3,17], as in changing from CCM [19], which uses a CBC-MAC, to GCM [9], which employs polynomial-based hashing.

It turns out that meeting the above objectives is demanding. We introduce the notion of a vector-input ϵ-almost XOR universal hash function and develop a new polynomial-based hashing that meets our requirements. We also devise a somewhat odd way of incrementation in a CTR mode, which is necessary for the security of the scheme and is non-trivially binded with the new polynomial hash function. As a result, the security proofs become rather involved.

2   Preliminaries

Notation. If x is a finite string, then $|x|$ denotes its length in bits. If x and y are two equal-length strings, then $x \oplus y$ denotes the XOR of x and y. If x and y are finite strings, then $x\|y$ denotes their concatenation. Given finite strings $x_0, x_1, \ldots, x_{m-1}$, we use the notation $x_0\|x_1\|\cdots\|x_{m-1}$ and the vector notation $(x_0, x_1, \ldots, x_{m-1})$ interchangeably. For a positive integer n, $\{0,1\}^n$ is the set of all n-bit strings, and $(\{0,1\}^n)^+$ is the set of all strings whose lengths are positive multiples of n bits. Whenever we write $X = (X[0], X[1], \ldots, X[m-1]) \in (\{0,1\}^n)^+$, we implicitly assume that $m \geq 1$ is an integer and $|X[i]| = n$ for $0 \leq i \leq m - 1$. $\{0,1\}^*$ is the set of all finite strings (including the empty string), and for a positive integer ℓ, $(\{0,1\}^*)^\ell$ is the set of all vectors $(x_0, x_1, \ldots, x_{\ell-1})$, where $x_i \in \{0,1\}^*$ for $0 \leq i \leq \ell - 1$. For positive integers n and j such that $n \leq 2^j - 1$, $\langle j \rangle_n$ is the big-endian n-bit binary representation of j. For a finite string x and a positive integer n such that $|x| \geq n$, $\mathrm{msb}(n, x)$ is the most significant n bits of x. For a positive integer n, 0^n is the n-times repetition of 0's.

We write N for the set of non-negative integers. Given a real number x, the symbol $\lceil x \rceil$ denotes the smallest integer greater than or equal to x. If X is a finite set, then $\#X$ is the cardinality of X, and we let $x \xleftarrow{R} X$ denote the process of selecting an element from X uniformly at random and assigning it to x.

The Finite Field of 2^n Elements. We regard the set $\{0,1\}^n$ as the finite field of 2^n elements (relative to some irreducible polynomial). For $x, y \in \{0,1\}^n$, the symbol $x \cdot y$ denotes the product of x and y. We write $x^2 = x \cdot x$, $x^3 = x \cdot x \cdot x$, and so on. The $\oplus$ operation corresponds with the addition in the field.

Blockciphers and SPRP Adversaries. A blockcipher is a function $E : \mathcal{K} \times \{0,1\}^n \rightarrow \{0,1\}^n$ such that for any $K \in \mathcal{K}$, $E(K, \cdot) = E_K(\cdot)$ is a permutation on $\{0,1\}^n$ (i.e., a permutation family). The positive integer n is the block length, and an n-bit string is called a block. If $\mathcal{K} = \{0,1\}^k$, then k is the key length.

The SPRP notion for blockciphers was introduced in [6] and later made concrete in [1]. Let $\mathrm{Perm}(n)$ denote the set of all permutations on $\{0,1\}^n$. This set can be regarded as a blockcipher by assigning a unique string (a key) to each permutation. We say that P is a random permutation if $P \xleftarrow{R} \mathrm{Perm}(n)$. An adversary is a probabilistic algorithm (a program) with access to one or more oracles. An *SPRP-adversary* A has access to two oracles and returns a bit. The two oracles are either the encryption oracle $E_K(\cdot)$ and the decryption oracle $E_K^{-1}(\cdot)$, or a random permutation oracle $P(\cdot)$ and its inverse oracle $P^{-1}(\cdot)$. We define the advantage $\mathbf{Adv}_E^{\mathrm{sprp}}(A)$ as

$$\left| \Pr(K \xleftarrow{R} \mathcal{K} : A^{E_K(\cdot), E_K^{-1}(\cdot)} = 1) - \Pr(P \xleftarrow{R} \mathrm{Perm}(n) : A^{P(\cdot), P^{-1}(\cdot)} = 1) \right|.$$

For an adversary A, A's running time is denoted by $time(A)$. The running time is its actual running time (relative to some fixed RAM model of computation) and its description size (relative to some standard encoding of algorithms). The details of the big-O notation in a running-time reference depend on the RAM model and the choice of encoding.

3   Specification of HBS

3.1   A Vector-Input Universal Hash Function F

In this section, we define a new vector-input universal hash function F. Let n be a fixed block length, e.g., $n = 128$. Before defining F, we first define a polynomial universal hash function $f : \{0,1\}^n \times (\{0,1\}^n)^+ \to \{0,1\}^n$. Let $L \in \{0,1\}^n$ be a key for f and $X = (X[0], X[1], \ldots, X[m-1]) \in (\{0,1\}^n)^+$ be an input. Define

$$f_L(X) = L^m \oplus L^{m-1} \cdot X[0] \oplus L^{m-2} \cdot X[1] \oplus \cdots \oplus X[m-1].$$

Note that we need $(m-1)$ multiplications to compute $f_L(X)$.

Now we define the vector-input hash function F. It internally uses a padding function $\text{pad}(\cdot) : \{0,1\}^* \to (\{0,1\}^n)^+$, which takes $x \in \{0,1\}^*$ and outputs

$$\text{pad}(x) = \begin{cases} x & \text{if } x \in (\{0,1\}^n)^+, \\ x\|10^i & \text{otherwise,} \end{cases}$$

where i is the smallest non-negative integer such that the total length of $x\|10^i$ in bits becomes a positive multiple of n. F also takes a vector dimension ℓ as a parameter, and we write $F^{(\ell)}$ for F with a parameter ℓ. $F^{(\ell)}$ takes $L \in \{0,1\}^n$ as a key and $\mathcal{X} = (x_0, x_1, \ldots, x_{\ell-1}) \in (\{0,1\}^*)^\ell$ as its input. For $0 \le i \le \ell - 1$, let $X_i = \text{pad}(x_i)$ and $z_i = f_L(X_i)$. Define $F^{(\ell)} : \{0,1\}^n \times (\{0,1\}^*)^\ell \to \{0,1\}^n$ as

$$\begin{aligned} F_L^{(\ell)}(\mathcal{X}) &= F_L^{(\ell)}(x_0, x_1, \ldots, x_{\ell-1}) \\ &= L \cdot z_0^\ell \cdot \langle c_0 \rangle_n \oplus L^2 \cdot z_1^\ell \cdot \langle c_1 \rangle_n \oplus \cdots \oplus L^\ell \cdot z_{\ell-1}^\ell \cdot \langle c_{\ell-1} \rangle_n, \end{aligned} \quad (1)$$

where, for $0 \le i \le \ell - 1$, we set $z_i = f_L(\text{pad}(x_i))$ and

$$c_i = \begin{cases} 1 & \text{if } x_i \notin (\{0,1\}^n)^+, \\ 2 & \text{otherwise.} \end{cases}$$

The degree of $F_L^{(\ell)}(\mathcal{X})$ as a polynomial in L is $\mu(\mathcal{X})$, which is defined as follows. For $\mathcal{X} = (x_0, \ldots, x_{\ell-1}) \in (\{0,1\}^*)^\ell$, define $\mu : (\{0,1\}^*)^\ell \to N$ as

$$\mu(\mathcal{X}) = \max_{0 \le i \le \ell-1} \{i + 1 + \ell m_i\},$$

where $m_i = \lceil |x_i|/n \rceil$ for $0 \le i \le \ell - 1$. The value m_i is the length of x_i in blocks, where a partial block counts for one block. Note that each z_i is a polynomial in L of degree m_i.

We remark that the multiplication by the constant $\langle 2 \rangle_n$ can be implemented efficiently. For example, for $n = 128$ we can choose $\mathbf{x}^{128} + \mathbf{x}^7 + \mathbf{x}^2 + \mathbf{x} + 1$, which is one of the irreducible polynomials having the minimum number of non-zero coefficients. Then, for $x \in \{0,1\}^{128}$, we can compute the product $x \cdot \langle 2 \rangle_{128}$ as $x \cdot \langle 2 \rangle_{128} = x \cdot 0^{126}10 = x \ll 1$ if $\text{msb}(1, x) = 0$, or as $x \cdot \langle 2 \rangle_{128} = x \cdot 0^{126}10 = (x \ll 1) \oplus 0^{120}10000111$ otherwise, where $x \ll 1$ is the left shift of x by one bit (The first bit of x disappears and a zero comes into the last bit). See [13] for details.

3.2   Vector-Input ϵ-Almost XOR Universal Hash Function

We next define the notion of a vector-input ϵ-almost XOR universal (VI-ϵ-AXU for short) hash function. This plays an important role in our security analysis.

Algorithm HBS.Enc$_K(H, M)$	**Algorithm** CTR.Enc$_K(S, M)$
100 $L \leftarrow E_K(0^n)$	300 **for** $i \leftarrow 0$ **to** $\lceil |M|/n \rceil - 1$ **do**
101 $S \leftarrow F_L^{(2)}(H, M)$	301 $R_i \leftarrow E_K(S \oplus \langle i+1 \rangle_n)$
102 $C \leftarrow$ CTR.Enc$_K(S, M)$	302 $R \leftarrow (R_0, R_1, \ldots, R_{\lceil |M|/n \rceil - 1})$
103 $T \leftarrow E_K(S)$	303 $C \leftarrow M \oplus \mathrm{msb}(|M|, R)$
104 **return** (T, C)	304 **return** C
Algorithm HBS.Dec$_K(H, (T, C))$	**Algorithm** CTR.Dec$_K(S, C)$
200 $L \leftarrow E_K(0^n)$	400 **for** $i \leftarrow 0$ **to** $\lceil |C|/n \rceil - 1$ **do**
201 $S \leftarrow E_K^{-1}(T)$	401 $R_i \leftarrow E_K(S \oplus \langle i+1 \rangle_n)$
202 $M \leftarrow$ CTR.Dec$_K(S, C)$	402 $R \leftarrow (R_0, R_1, \ldots, R_{\lceil |C|/n \rceil - 1})$
203 $S' \leftarrow F_L^{(2)}(H, M)$	403 $M \leftarrow C \oplus \mathrm{msb}(|C|, R)$
204 **if** $S \neq S'$ **then return** $\bot$	404 **return** M
205 **else return** M	

Fig. 1. Definition of the encryption algorithm HBS.Enc (left top) and the decryption algorithm HBS.Dec (left bottom). CTR.Enc (right top) is used in HBS.Enc, and CTR.Dec (right bottom) is used in HBS.Dec.

Definition 1 (VI-ϵ-AXU Hash Function). *Let ℓ be an integer and $F^{(\ell)}$: $\{0,1\}^n \times (\{0,1\}^*)^\ell \to \{0,1\}^n$ be a vector-input function keyed by $L \in \{0,1\}^n$. $F^{(\ell)}$ is said to be VI-ϵ-AXU if, for any two distinct inputs $\mathcal{X}, \hat{\mathcal{X}} \in (\{0,1\}^*)^\ell$ and for any $Y \in \{0,1\}^n$,*

$$\Pr(L \xleftarrow{R} \{0,1\}^n : F_L^{(\ell)}(\mathcal{X}) \oplus F_L^{(\ell)}(\hat{\mathcal{X}}) = Y) \leq \epsilon.$$

We show that the polynomial hash function $F^{(\ell)}$ defined in Sect. 3.1 is VI-ϵ-AXU for a small ϵ. A proof is given in Appendix A.

Theorem 1. *Let ℓ be an integer and $F^{(\ell)}$: $\{0,1\}^n \times (\{0,1\}^*)^\ell \to \{0,1\}^n$ be the polynomial hash function defined in Sect. 3.1, keyed by $L \in \{0,1\}^n$. Let $\mathcal{X} = (x_0, \ldots, x_{\ell-1})$, $\hat{\mathcal{X}} = (\hat{x}_0, \ldots, \hat{x}_{\ell-1}) \in (\{0,1\}^*)^\ell$ be any two distinct inputs to $F^{(\ell)}$. Then for any $Y \in \{0,1\}^n$,*

$$\Pr(L \xleftarrow{R} \{0,1\}^n : F_L^{(\ell)}(\mathcal{X}) \oplus F_L^{(\ell)}(\hat{\mathcal{X}}) = Y) \leq \frac{\max\{\mu(\mathcal{X}), \mu(\hat{\mathcal{X}})\}}{2^n}.$$

3.3   HBS: Hash Block Stealing

We present our HBS, Hash Block Stealing. It takes a blockcipher E as a parameter and uses $F^{(\ell)}$ defined in Sect. 3.1 with $\ell = 2$.

Fix a blockcipher $E : \{0,1\}^k \times \{0,1\}^n \to \{0,1\}^n$. HBS consists of two algorithms, the encryption algorithm (HBS.Enc) and the decryption algorithm (HBS.Dec). These algorithms are defined in Fig. 1. See Fig. 2 for a picture illustrating HBS.Enc (See also Fig. 3 for $F_L^{(2)}(H, M)$).

The encryption algorithm HBS.Enc takes a key $K \in \{0,1\}^k$, a header $H \in \{0,1\}^*$, and a plaintext $M \in \{0,1\}^*$ to return a ciphertext $(T, C) \in \{0,1\}^n \times \{0,1\}^*$, where $|C| = |M|$. We write $(T, C) \leftarrow$ HBS.Enc$_K(H, M)$. The decryption

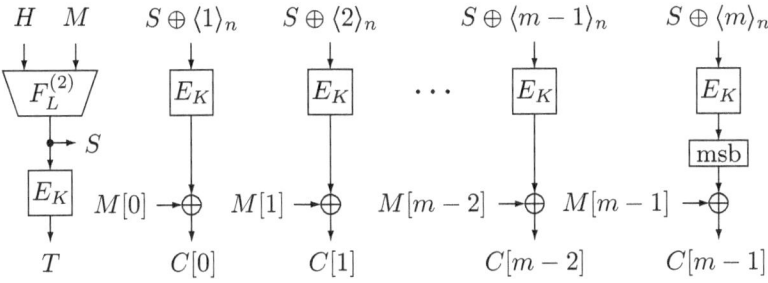

Fig. 2. Illustration of the encryption algorithm $(T, C) \leftarrow \text{HBS.Enc}_K(H, M)$. In the figure, $M = (M[0], M[1], \ldots, M[m-1])$, where $|M[0]| = \cdots = |M[m-2]| = n$ and $|M[m-1]| \leq n$. $L = E_K(0^n)$, and the output of "msb" is the most significant $|M[m-1]|$ bits of $E_K(S \oplus \langle m \rangle_n)$.

algorithm takes K, H, and (T, C) as its inputs to return the corresponding plaintext M or a special symbol $\perp$. The symbol $\perp$ indicates that the given inputs are invalid. We write $M \leftarrow \text{HBS.Dec}_K(H, (T, C))$. We have $\text{HBS.Dec}_K(H, (T, C)) = \perp$ when the decryption process fails.

HBS.Enc and HBS.Dec call subroutines CTR.Enc and CTR.Dec, respectively, which are the encryption and the decryption of the CTR mode using S as its initial counter value. By specification we restrict the message length to $2^{n/2} - 1$ blocks, as the security of the HBS mode becomes vacuous beyond this point. This restriction allows us to write $\langle i \rangle_n = 0^{n/2} \| \langle i \rangle_{n/2}$ for an integer $0 \leq i \leq 2^{n/2} - 1$ (We assume that the block length n is an even integer), which implies that the incrementation of the counter in CTR.Enc and CTR.Dec can be done by adding 1 modulo $2^{n/2}$ rather than modulo 2^n.

$F_L^{(2)}(H, M)$ can be implemented using $h + m$ multiplications and two squaring operations, as in Fig. 3. See also Fig. 4 for pseudocode. In the figures, we let $\lceil |H|/n \rceil = h$ and $\lceil |M|/n \rceil = m$, so that $\text{pad}(H) = (\bar{H}[0], \ldots, \bar{H}[h-1])$ and

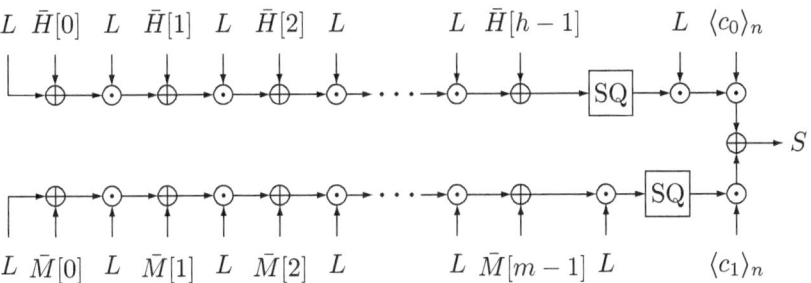

Fig. 3. Illustration of $F_L^{(2)}(H, M)$. In the figure, $\text{pad}(H) = (\bar{H}[0], \ldots, \bar{H}[h-1])$ and $\text{pad}(M) = (\bar{M}[0], \ldots, \bar{M}[m-1])$. $L = E_K(0^n)$, "SQ" outputs the square of its input, "$\odot$" is the multiplication, and $c_0, c_1 \in \{1, 2\}$ are constants.

$$\boxed{\begin{aligned}
&\textbf{Function } F_L^{(2)}(H, M)\\
&100 \quad z_H \leftarrow L \oplus \bar{H}[0]\\
&101 \quad \textbf{for } i = 1 \textbf{ to } h - 1 \textbf{ do}\\
&102 \quad\quad z_H \leftarrow L \cdot z_H \oplus \bar{H}[i]\\
&103 \quad z_M \leftarrow L \oplus \bar{M}[0]\\
&104 \quad \textbf{for } i = 1 \textbf{ to } m - 1 \textbf{ do}\\
&105 \quad\quad z_M \leftarrow L \cdot z_M \oplus \bar{M}[i]\\
&106 \quad \textbf{return } L \cdot z_H^2 \cdot \langle c_0 \rangle_n \oplus (L \cdot z_M)^2 \cdot \langle c_1 \rangle_n
\end{aligned}}$$

Fig. 4. Implementation example of $F_L^{(2)}(H, M)$, where $\mathrm{pad}(H) = (\bar{H}[0], \ldots, \bar{H}[h-1])$ and $\mathrm{pad}(M) = (\bar{M}[0], \ldots, \bar{M}[m-1])$. z_H and z_M are n-bit variables, and $c_0, c_1 \in \{1, 2\}$.

$\mathrm{pad}(M) = (\bar{M}[0], \ldots, \bar{M}[m-1])$. Also, let $c_0 = 1$ if $H \notin (\{0,1\}^n)^+$ and $c_0 = 2$ otherwise, and $c_1 = 1$ if $M \notin (\{0,1\}^n)^+$ and $c_1 = 2$ otherwise. Then $F_L^{(2)}(H, M)$ can be written as

$$L \cdot \left(f_L(\bar{H}[0], \ldots, \bar{H}[h-1]) \right)^2 \cdot \langle c_0 \rangle_n \oplus L^2 \cdot \left(f_L(\bar{M}[0], \ldots, \bar{M}[m-1]) \right)^2 \cdot \langle c_1 \rangle_n$$

$$= L \cdot \left(L^h \oplus L^{h-1} \cdot \bar{H}[0] \oplus L^{h-2} \cdot \bar{H}[1] \oplus \cdots \oplus \bar{H}[h-1] \right)^2 \cdot \langle c_0 \rangle_n$$

$$\oplus \left(L^{m+1} \oplus L^m \cdot \bar{M}[0] \oplus L^{m-1} \cdot \bar{M}[1] \oplus \cdots \oplus L \cdot \bar{M}[m-1] \right)^2 \cdot \langle c_1 \rangle_n.$$

Recall that the multiplication by $\langle 2 \rangle_n$ can be implemented using a left shift and a conditional XOR, which is a small overhead compared to the multiplication by L.

Lastly, we make a remark about handling a message with no header. For such a message, simply ignore the computation of z_H and define the value $(L \cdot z_M)^2 \cdot \langle c_1 \rangle_n$ as the output of the hash function $F_L^{(2)}(\cdot, M)$. Note that this should be differentiated from the case $H = \varepsilon$ (the null string), where $\mathrm{pad}(H) = 10^{n-1}$.

4   Security Analysis of HBS

HBS is a mode of operation for deterministic authenticated encryption. Before presenting our security results, we define the security of the mode. Our security definition follows the one given by Rogaway and Shrimpton in [16].

4.1   Security Definition

Let $E : \{0,1\}^k \times \{0,1\}^n \to \{0,1\}^n$ be a blockcipher. For notational simplicity, we write $\mathcal{E}_K(\cdot, \cdot)$ for $\mathrm{HBS.Enc}_K(\cdot, \cdot)$, and $\mathcal{D}_K(\cdot, \cdot)$ for $\mathrm{HBS.Dec}_K(\cdot, \cdot)$. Let $\mathrm{HBS}[E] = (\mathrm{HBS.Enc}_K, \mathrm{HBS.Dec}_K) = (\mathcal{E}_K, \mathcal{D}_K)$ be defined as in Sect. 3.3.

An adversary A is given access to either a pair of $\mathcal{E}_K(\cdot, \cdot)$ and $\mathcal{D}_K(\cdot, \cdot)$ oracles or a pair of $\mathcal{R}(\cdot, \cdot)$ and $\perp(\cdot, \cdot)$ oracles. On query $(H, M) \in (\{0,1\}^*)^2$, $\mathcal{E}_K(\cdot, \cdot)$ returns $(C, T) \leftarrow \mathcal{E}_K(H, M)$, and the random-bits oracle $\mathcal{R}(\cdot, \cdot)$ returns a random string of $n + |M|$ bits. On query $(H, (T, C)) \in \{0,1\}^* \times (\{0,1\}^n \times \{0,1\}^*)$, $\mathcal{D}_K(\cdot, \cdot)$ returns $\perp$ or $M \leftarrow \mathcal{D}_K(H, (T, C))$, and the $\perp(\cdot, \cdot)$ oracle returns $\perp$ on every input. Queries can be made adaptively, and A's goal is to distinguish between these pairs.

Definition 2. *The DAE-advantage* $\mathbf{Adv}^{\mathrm{dae}}_{\mathrm{HBS}[E]}(A)$ *of an adversary* A *in break-ing* $\mathrm{HBS}[E] = (\mathcal{E}_K, \mathcal{D}_K)$ *is defined as*

$$\mathbf{Adv}^{\mathrm{dae}}_{\mathrm{HBS}[E]}(A) = \left| \Pr(A^{\mathcal{E}_K(\cdot,\cdot),\mathcal{D}_K(\cdot,\cdot)} = 1) - \Pr(A^{\mathcal{R}(\cdot,\cdot),\perp(\cdot,\cdot)} = 1) \right|.$$

In what follows, the first oracle refers to $\mathcal{E}_K(\cdot, \cdot)$ or $\mathcal{R}(\cdot, \cdot)$, and the second oracle refers to $\mathcal{D}_K(\cdot, \cdot)$ or $\perp(\cdot, \cdot)$. We make the following assumptions about A.

- A does not make a query (H, M) to its first oracle if the second oracle has returned M in response to some previous query $(H, (T, C))$.
- A does not make a query $(H, (T, C))$ to its second oracle if the first oracle has returned (T, C) in response to some previous query (H, M).
- A does not repeat a query.

The last assumption is without loss of generality, and the first two assumptions are to prevent trivial wins.

We also consider the security of HBS in terms of privacy. An adversary B is given access to either $\mathcal{E}_K(\cdot, \cdot)$ or $\mathcal{R}(\cdot, \cdot)$, and B's goal is to distinguish between these oracles.

Definition 3. *The PRIV-advantage* $\mathbf{Adv}^{\mathrm{priv}}_{\mathrm{HBS}[E]}(B)$ *of an adversary* B *in break-ing the privacy of* $\mathrm{HBS}[E] = (\mathcal{E}_K, \mathcal{D}_K)$ *is defined as*

$$\mathbf{Adv}^{\mathrm{priv}}_{\mathrm{HBS}[E]}(B) = \left| \Pr(B^{\mathcal{E}_K(\cdot,\cdot)} = 1) - \Pr(B^{\mathcal{R}(\cdot,\cdot)} = 1) \right|.$$

We deal with $\mathrm{HBS}[\mathrm{Perm}(n)] = (\mathcal{E}_P, \mathcal{D}_P)$, where a random permutation $P \xleftarrow{R} \mathrm{Perm}(n)$ is used as the underlying blockcipher. That is, $\mathrm{HBS}[\mathrm{Perm}(n)]$ is defined by replacing E_K and E_K^{-1} in Fig. 1 by P and P^{-1}, respectively.

4.2 Security Theorem

Consider the DAE-advantage of an adversary A attacking $\mathrm{HBS}[\mathrm{Perm}(n)]$. We assume A makes a total of at most q queries. For $0 \le i \le q-1$, A makes a query of the form (H_i, M_i) to the first oracle, or $(H_i, (T_i, C_i))$ to the second oracle. We assume that $\lceil |H_i|/n \rceil \le h_{\max}$, $\lceil |M_i|/n \rceil \le m_{\max}$, and $\lceil |C_i|/n \rceil \le m_{\max}$ for some $h_{\max}, m_{\max} \ge 1$. That is, $h_{\max}$ is the maximum length of H_i, and $m_{\max}$ is the maximum length of M_i and C_i in blocks. The following theorem is our main result on the security of our HBS mode. It shows that HBS provides standard birthday-bound security.

Theorem 2. *Let* A *be an adversary described as above. Then*

$$\mathbf{Adv}^{\mathrm{dae}}_{\mathrm{HBS}[\mathrm{Perm}(n)]}(A) \le \frac{19q^2(1 + h_{\max} + 2m_{\max})^2}{2^n}.$$

Given Theorem 2, it is standard to pass to its complexity-theoretic bound, by replacing $P \xleftarrow{R} \mathrm{Perm}(n)$ with E_K (and hence P^{-1} with E_K^{-1}).

Lemma 1. *Let $E : \{0,1\}^k \times \{0,1\}^n \to \{0,1\}^n$ be a blockcipher, and A be an adversary attacking $\mathrm{HBS}[E]$. Then there exists A' attacking E such that*

$$\mathbf{Adv}^{\mathrm{dae}}_{\mathrm{HBS}[E]}(A) \le \mathbf{Adv}^{\mathrm{sprp}}_E(A') + \frac{19q^2(1 + h_{\max} + 2m_{\max})^2}{2^n},$$

where A' makes at most $1 + q(1 + m_{\max})$ queries, and $\mathrm{time}(A') = \mathrm{time}(A) + O(nq(h_{\max} + m_{\max}))$.

The proof of Lemma 1 is standard and omitted (For example, see [1]). Now we turn to Theorem 2. The proof is based on the following lemma.

Lemma 2. *Let A be an adversary as in Theorem 2. Then there exists an adversary B such that*

$$\mathbf{Adv}^{\mathrm{dae}}_{\mathrm{HBS}[\mathrm{Perm}(n)]}(A) \le 2\mathbf{Adv}^{\mathrm{priv}}_{\mathrm{HBS}[\mathrm{Perm}(n)]}(B) + \frac{11q^2(1 + h_{\max} + 2m_{\max})^2}{2^n} + \frac{q}{2^n},$$

where B makes at most q queries, and it is subject to the same restriction as A on the length of its queries.

We give a proof sketch of Lemma 2, leaving a complete proof in Appendix B. We see that $\mathbf{Adv}^{\mathrm{dae}}_{\mathrm{HBS}[\mathrm{Perm}(n)]}(A)$ is at most $p_0 + p_1$, where we define p_0 and p_1 as

$$\begin{cases} p_0 = \left| \Pr\left(A^{\mathcal{E}_P(\cdot,\cdot), \perp(\cdot,\cdot)} = 1 \right) - \Pr\left(A^{\mathcal{R}(\cdot,\cdot), \perp(\cdot,\cdot)} = 1 \right) \right|, \\ p_1 = \left| \Pr\left(A^{\mathcal{E}_P(\cdot,\cdot), \mathcal{D}_P(\cdot,\cdot)} = 1 \right) - \Pr\left(A^{\mathcal{E}_P(\cdot,\cdot), \perp(\cdot,\cdot)} = 1 \right) \right|. \end{cases}$$

We can view A in p_0 as attacking the privacy of $\mathrm{HBS}[\mathrm{Perm}(n)]$, as the second oracle always returns $\perp$. So there exists B such that $p_0 \le \mathbf{Adv}^{\mathrm{priv}}_{\mathrm{HBS}[\mathrm{Perm}(n)]}(B)$.

To derive the upper bound on p_1, we observe that the oracles are identical unless the $\mathcal{D}_P(\cdot,\cdot)$ oracle returns something other than $\perp$. We thus have

$$p_1 \le \Pr(A^{\mathcal{E}_P(\cdot,\cdot), \mathcal{D}_P(\cdot,\cdot)} \text{ forges}),$$

where $A^{\mathcal{E}_P(\cdot,\cdot), \mathcal{D}_P(\cdot,\cdot)}$ forges if A makes a query $((H,M),T)$ to its second oracle such that $\mathcal{D}_P((H,M),T)) \ne \perp$. To derive the upper bound on the forgery probability, we introduce an oracle $\mathcal{V}_P(\cdot,\cdot)$ that takes $((H,M),T)$ as its input and returns T if $P(F_L^{(2)}(H,M)) = T$ or returns $\perp$ otherwise. We can show that

$$\Pr(A^{\mathcal{E}_P(\cdot,\cdot), \mathcal{D}_P(\cdot,\cdot)} \text{ forges}) \le \Pr(A^{\mathcal{E}_P(\cdot,\cdot), \mathcal{V}_P(\cdot,\cdot)} \text{ forges}) + \frac{11q^2(1 + h_{\max} + 2m_{\max})^2}{2^n}$$

and $\Pr(A^{\mathcal{E}_P(\cdot,\cdot), \mathcal{V}_P(\cdot,\cdot)} \text{ forges}) \le \mathbf{Adv}^{\mathrm{priv}}_{\mathrm{HBS}[\mathrm{Perm}(n)]}(B) + q/2^n$. The proof for the first claim is rather complicated but the second one follows from the standard argument that distinguishing is easier than forging, and Lemma 2 follows.

Now to prove Theorem 2, it suffices to bound $\mathbf{Adv}^{\mathrm{priv}}_{\mathrm{HBS}[\mathrm{Perm}(n)]}(B)$. Recall that B is an adversary attacking the privacy of $\mathrm{HBS}[\mathrm{Perm}(n)]$, where B makes a query of the form (H_i, M_i) for $0 \le i \le q - 1$. We assume that H_i is at most $h_{\max}$ blocks, and M_i is at most $m_{\max}$ blocks. We have the following result.

Lemma 3. *Let B be an adversary as described above. Then*

$$\mathbf{Adv}^{\mathrm{priv}}_{\mathrm{HBS}[\mathrm{Perm}(n)]}(B) \leq \frac{3q^2(1 + h_{\max} + 2m_{\max})^2}{2^n} + \frac{(1 + q + qm_{\max})^2}{2^{n+1}}.$$

Let us explain the intuitive reasoning behind the bound, leaving a complete proof in Appendix C. Suppose B makes q queries $(H_0, M_0), \ldots, (H_{q-1}, M_{q-1})$, where M_i is m_i blocks. Let $S_i = F_L^{(2)}(H_i, M_i)$, which corresponds to the initial counter value, and we consider the set $\mathbb{S}_i = \{S_i, S_i \oplus \langle 1 \rangle_n, S_i \oplus \langle 2 \rangle_n, \ldots, S_i \oplus \langle m_i \rangle_n\}$. Each element in $\mathbb{S}_i$ is the input value to P at the i-th query. Notice that we never have a collision within $\mathbb{S}_i$, but we may have $0^n \in \mathbb{S}_i$, or $\mathbb{S}_i \cap \mathbb{S}_j \neq \emptyset$. Both events may be useful for B to mount an distinguishing attack. Now for the first event, we know that $F_L^{(2)}(H_i, M_i)$ is a non-zero polynomial in L, and hence, for $0 \leq s \leq m_i$, the probability that $F_L^{(2)}(H_i, M_i) = \langle s \rangle_n$ is small. For the second event to occur, we need

$$F_L^{(2)}(H_i, M_i) \oplus \langle s \rangle_n = F_L^{(2)}(H_j, M_j) \oplus \langle t \rangle_n,$$

where $0 \leq s \leq m_i$ and $0 \leq t \leq m_j$. Theorem 1 ensures that the equality holds with only a small probability. It turns out that during the attack, with a high probability, we have $0^n \notin \mathbb{S}_i$ for $0 \leq i \leq q - 1$, and $\mathbb{S}_i \cap \mathbb{S}_j = \emptyset$ for $0 \leq i < j \leq q - 1$. Under this assumption, what B learns is the output values of P for distinct input values, and hence B cannot distinguish it from a truly random string.

Given Lemma 2 and Lemma 3, we find the proof of Theorem 2 straightforward, as we have

$$\begin{aligned}
\mathbf{Adv}^{\mathrm{dae}}_{\mathrm{HBS}[\mathrm{Perm}(n)]}(A) &\leq 2\mathbf{Adv}^{\mathrm{priv}}_{\mathrm{HBS}[\mathrm{Perm}(n)]}(B) + \frac{11q^2(1 + h_{\max} + 2m_{\max})^2}{2^n} + \frac{q}{2^n} \\
&\leq \frac{6q^2(1 + h_{\max} + 2m_{\max})^2}{2^n} + \frac{(1 + q + qm_{\max})^2}{2^n} \\
&\qquad + \frac{11q^2(1 + h_{\max} + 2m_{\max})^2}{2^n} + \frac{q}{2^n} \\
&\leq \frac{19q^2(1 + h_{\max} + 2m_{\max})^2}{2^n},
\end{aligned}$$

where the first inequality follows from Lemma 2, the next one from Lemma 3, and the last one from easy simplification.

5   Rationale of f_L and $F_L^{(\ell)}$ and Comparison with SIV

Rationale of f_L and $F_L^{(\ell)}$. We adopt $f_L(X[0], \ldots, X[m-1]) = L^m \oplus L^{m-1} \cdot X[0] \oplus L^{m-2} \cdot X[1] \oplus \cdots \oplus X[m-1]$ as the polynomial hash in HBS, rather than the usual $g_L(X[0], \ldots, X[m-1]) = L^{m-1} \cdot X[0] \oplus L^{m-2} \cdot X[1] \oplus \cdots \oplus X[m-1]$ as in [18,9]. The choice of f_L enables us to handle variable-length inputs without increasing the number of multiplications. We note that the degree of our polynomial f_L is higher than that of g_L. However, the difference is only one, and the decrease in the security bound due to the increase in the degree should be negligible in practice.

$F_L^{(\ell)}$ is designed to handle an ℓ-dimensional vector efficiently. For an input $(x_0, \ldots, x_{\ell-1}) \in (\{0,1\}^*)^\ell$, we need just $\lceil |x_0|/n \rceil + \cdots + \lceil |x_0|/n \rceil + \ell$ multiplications. We may need additional ℓ shifts and XOR's depending on the length of each component, but the cost for these operations is fairly low. Moreover, $F_L^{(2)}$ allows us to reuse z_H or z_M in Fig. 4 if H or M stays fixed, and the same is true for $F_L^{(\ell)}$ with $\ell \geq 3$.

In contrast, the polynomial hash in GCM was designed to accept only two-dimensional vectors (H, M), and the lengths of H and M were encoded into a block as $\langle |H| \rangle_{n/2} \| \langle |M| \rangle_{n/2}$. One could handle, say, three-dimensional vectors by encoding the lengths into a block, but such an approach would severely limit the maximum length of each component, which is solved in our hash function $F_L^{(\ell)}$.

A drawback of $F_L^{(\ell)}$, as compared to the vectorized CMAC [16], is that the (maximum) dimension ℓ needs to be fixed in advance. In order to lift this restriction, we can start with the hash function $F_L^{(2)}$ for two-dimensional vectors and then "increase" the dimension by utilizing the values $\sqrt{L}$, $\sqrt[4]{L}$, $\sqrt[8]{L}$, ..., upon receiving 3-, 4-, 5-, ..., dimensional vectors. Although we can efficiently compute the square root $\sqrt{L}$ of an element $L \in GF(2^n)$, we admit that this solution results in rather complicated algorithms.

As with the polynomial hash in GCM [9], $L = 0^n$ is a weak key for $F_L^{(\ell)}$ [4]. We may let $L \leftarrow \mathrm{msb}(n-1, E_K(0^n)) \| 1$ to avoid the problem at the cost of slight decrease in the security bound. But since $L = 0^n$ only occurs with a probability $1/2^n$, we do not adopt this approach. The birthday attack on GCM described in [4] can be made also on our HBS mode. Therefore, as with GCM, one needs to update the secret key well before processing $2^{n/2}$ blocks.

Comparison with SIV. HBS basically follows the design of SIV [16], but there are important differences. HBS works with a single key, and SIV [16] uses two separate keys, one for encryption and the other for MAC. SIV works as follows. First, let $IV \leftarrow \mathrm{CMAC}^*_{K_1}(H, M)$ and $C \leftarrow \mathrm{CTR.Enc}_{K_2}(IV, M)$. Then the output is (IV, C), where CMAC^* is constructed from CMAC [5,11] to handle a vector input (H, M).

In Table 1, we make a brief comparison between SIV and HBS. HBS reduces the keying material and replaces $\lceil |M|/n \rceil + \lceil |H|/n \rceil$ blockcipher calls by $\lceil |M|/n \rceil + \lceil |H|/n \rceil + 2$ multiplications at the cost of a stronger assumption about the blockcipher. The SPRP assumption is needed as we "steal" the result of hash computation (hash block) and use it as the initial counter value. However, we argue that this does not seem to make a substantial difference in practice, because many of the modern blockciphers, such as AES, seem to satisfy the strong property of SPRP. Rather, we admit that it may be disadvantageous of HBS to require the inverse cipher. This is especially true for blockciphers of asymmetric encryption/decryption design, such as AES.

Table 1. Comparison between SIV and HBS

	SIV	HBS
Keying material	two blockcipher keys	single blockcipher key
Number of blockcipher calls	$2\lceil|M|/n\rceil + \lceil|H|/n\rceil + 2$	$\lceil|M|/n\rceil + 2$
Number of multiplications	0	$\lceil|M|/n\rceil + \lceil|H|/n\rceil + 2$
Assumption about blockcipher	PRP	SPRP
Security bound	$O(\sigma^2/2^n)$	$O(q^2(h_{\max} + m_{\max})^2/2^n)$

We note that, out of $2\lceil|M|/n\rceil + \lceil|H|/n\rceil + 2$ blockcipher calls in SIV, two calls can be done during idle time (without H or M). Similarly, out of $\lceil|M|/n\rceil + 2$ blockcipher calls in HBS, one blockcipher call (for $L = E_K(0^n)$) does not need H or M. We also note that there is a subtle difference in the security bounds. In SIV, the bound is $O(\sigma^2/2^n)$, while in HBS, the bound is $O(q^2(h_{\max} + m_{\max})^2/2^n)$, where σ is the total block length of H and M. It remains open if our analysis of HBS can be improved to give an $O(\sigma^2/2^n)$ security bound.

6   Further Discussion: Beyond the Birthday Bound

HBS delivers fine performance and provides security up to the birthday bound. It might be beneficial to come up with a DAE construction whose security is *beyond* the birthday bound. Such a construction would most likely be less efficient than HBS but might be desirable for some cases, as explained below.

Recall that DAE demands that the message space be of high entropy. An important example is the *key wrap* [10], since a key space obviously has high entropy. Although a key length is usually very short and a query complexity exceeding the birthday bound is hard to imagine in such a scenario, the highest security possible is desired for key-wrap applications. Therefore, highly secure constructions stand as suitable candidates for such systems, even if their performance is relatively modest.

We could give a beyond-the-birthday-bound construction as follows. First construct a $2n$-to-$2n$-bit blockcipher $E' : \{0,1\}^{2n} \rightarrow \{0,1\}^{2n}$, which has a certain key space, from a blockcipher $E_K : \{0,1\}^n \rightarrow \{0,1\}^n$ with $K \in \{0,1\}^k$ via the sum construction [7] and the Feistel network of six rounds [12]. Then construct the HBS mode with the block size of $2n$ bits, using E' as its underlying blockcipher. This construction ensures security beyond the $2^{n/2}$ bound but has the following three problems.

1. It is inefficient and impractical. One call to E' requires twelve calls to E.
2. It requires more than one key. The key space of E' is large.
3. It produces a long ciphertext. The tag size is $2n$-bit rather than n-bit.

The last point might be contrasted with the double-pipe hash [8], which has $2n$-bit intermediate values but outputs n-bit hash values. It remains open to provide a beyond-the-birthday-bound construction which resolves these problems.

Acknowledgments

The authors would like to express their thanks to the anonymous reviewers of FSE 2009 for many insightful comments.

References

1. Bellare, M., Kilian, J., Rogaway, P.: The security of the cipher block chaining message authentication code. J. Comput. Syst. Sci. 61(3), 362–399 (2000)
2. Bellare, M., Rogaway, P.: The security of triple encryption and a framework for code-based game-playing proofs. In: Vaudenay, S. (ed.) EUROCRYPT 2006. LNCS, vol. 4004, pp. 409–426. Springer, Heidelberg (2006)
3. Gladman, B.: AES and combined encryption/authentication modes (2006), http://www.gladman.me.uk/
4. Handschuh, H., Preneel, B.: Key-recovery attacks on universal hash function based MAC algorithms. In: Wagner, D. (ed.) CRYPTO 2008. LNCS, vol. 5157, pp. 144–161. Springer, Heidelberg (2008)
5. Iwata, T., Kurosawa, K.: OMAC: One-key CBC MAC. In: Johansson, T. (ed.) FSE 2003. LNCS, vol. 2887, pp. 129–153. Springer, Heidelberg (2003)
6. Luby, M., Rackoff, C.: How to construct pseudorandom permutations from pseudorandom functions. SIAM J. Comput. 17(2), 373–386 (1988)
7. Lucks, S.: The sum of PRPs is a secure PRF. In: Preneel, B. (ed.) EUROCRYPT 2000. LNCS, vol. 1807, pp. 470–484. Springer, Heidelberg (2000)
8. Lucks, S.: A failure-friendly design principle for hash functions. In: Roy, B.K. (ed.) ASIACRYPT 2005. LNCS, vol. 3788, pp. 474–494. Springer, Heidelberg (2005)
9. McGrew, D.A., Viega, J.: The security and performance of the Galois/counter mode (GCM) of operation. In: Canteaut, A., Viswanathan, K. (eds.) INDOCRYPT 2004. LNCS, vol. 3348, pp. 343–355. Springer, Heidelberg (2004)
10. NIST: AES key wrap specification (2001)
11. NIST: Recommendation for block cipher modes of operation: The CMAC mode for authentication (2005)
12. Patarin, J.: Security of random Feistel schemes with 5 or more rounds. In: Franklin, M.K. (ed.) CRYPTO 2004. LNCS, vol. 3152, pp. 106–122. Springer, Heidelberg (2004)
13. Rogaway, P., Bellare, M., Black, J., Krovetz, T.: OCB: A block-cipher mode of operation for efficient authenticated encryption. In: ACM CCS, pp. 196–205. ACM Press, New York (2001)
14. Rogaway, P.: Authenticated-encryption with associated-data. In: Atluri, V. (ed.) ACM CCS, pp. 98–107. ACM Press, New York (2002)
15. Rogaway, P.: Nonce-based symmetric encryption. In: Roy, B.K., Meier, W. (eds.) FSE 2004. LNCS, vol. 3017, pp. 348–359. Springer, Heidelberg (2004)
16. Rogaway, P., Shrimpton, T.: A provable-security treatment of the key-wrap problem. In: Vaudenay, S. (ed.) EUROCRYPT 2006. LNCS, vol. 4004, pp. 373–390. Springer, Heidelberg (2006)
17. Satoh, A.: High-speed hardware architectures for authenticated encryption mode GCM. In: Friedman, E.G., Theodoridis, S. (eds.) IEEE ISCAS 2006, pp. 4831–4844. IEEE Press, Los Alamitos (2006)
18. Wegman, M.N., Carter, L.: New hash functions and their use in authentication and set equality. J. Comput. Syst. Sci. 22(3), 265–279 (1981)
19. Whiting, D., Housley, R., Ferguson, N.: Counter with CBC-MAC (CCM). Submission to NIST (2002), http://csrc.nist.gov/groups/ST/toolkit/BCM/index.html

A Proof of Theorem 1

We have 2^n possible values of $L \in \{0,1\}^n$. We show that

$$\#\{L \mid F_L^{(\ell)}(\mathcal{X}) \oplus F_L^{(\ell)}(\hat{\mathcal{X}}) = Y\} \le \max\{\mu(\mathcal{X}), \mu(\hat{\mathcal{X}})\}.$$

It suffices to show that $F_L^{(\ell)}(\mathcal{X}) \oplus F_L^{(\ell)}(\hat{\mathcal{X}}) = Y$ is a non-trivial equation in L of degree at most $\max\{\mu(\mathcal{X}), \mu(\hat{\mathcal{X}})\}$. Let $x_i = (x_i[0], \ldots, x_i[m_i - 1])$, $X_i = \mathrm{pad}(x_i) = (X_i[0], \ldots, X_i[m_i-1])$, $z_i = f_L(X_i)$. Also, let $c_i = 1$ if $x_i \notin (\{0,1\}^n)^+$, and $c_i = 2$ if $x_i \in (\{0,1\}^n)^+$. We use $\hat{x}_i$, $\hat{X}_i$, $\hat{z}_i$ and $\hat{c}_i$ in the same way.

Observe that, from (1), $F_L^{(\ell)}(\mathcal{X}) \oplus F_L^{(\ell)}(\hat{\mathcal{X}})$ can be written as

$$F_L^{(\ell)}(\mathcal{X}) \oplus F_L^{(\ell)}(\hat{\mathcal{X}}) = \bigoplus_{0 \le i \le \ell-1} L^{i+1} \cdot (z_i^\ell \cdot \langle c_i \rangle_n \oplus \hat{z}_i^\ell \cdot \langle \hat{c}_i \rangle_n). \tag{2}$$

We show (2) is a non-zero, non-constant polynomial in the following three cases.

Case 1: $|X_i| \ne |\hat{X}_i|$ for some $0 \le i \le \ell - 1$,
Case 2: $|X_i| = |\hat{X}_i|$ for all $0 \le i \le \ell-1$, but $X_i[j] \ne \hat{X}_i[j]$ for some $0 \le i \le \ell-1$
    and $0 \le j \le m_i - 1$, and
Case 3: $|X_i| = |\hat{X}_i|$ for all $0 \le i \le \ell - 1$, and $X_i[j] = \hat{X}_i[j]$ for all $0 \le i \le \ell - 1$
    and $0 \le j \le m_i - 1$.

For the first case, if $m_i > \hat{m}_i$, then $(z_i^\ell \cdot \langle c_i \rangle_n \oplus \hat{z}_i^\ell \cdot \langle \hat{c}_i \rangle_n)$ is a polynomial in L of degree ℓm_i. To see this, we note that the polynomial is equivalent to

$$\begin{aligned}
&\left(L^{m_i} \oplus L^{m_i-1} \cdot X_i[0] \oplus L^{m_i-2} \cdot X_i[1] \oplus \cdots \oplus X_i[m_i - 1]\right)^\ell \cdot \langle c_i \rangle_n \\
&\oplus \left(L^{\hat{m}_i} \oplus L^{\hat{m}_i-1} \cdot \hat{X}_i[0] \oplus L^{\hat{m}_i-2} \cdot \hat{X}_i[1] \oplus \cdots \oplus \hat{X}_i[\hat{m}_i - 1]\right)^\ell \cdot \langle \hat{c}_i \rangle_n
\end{aligned} \tag{3}$$

and therefore the coefficient of $L^{\ell m_i}$ is $\langle c_i \rangle_n \ne 0^n$. This implies $F_L^{(\ell)}(\mathcal{X}) \oplus F_L^{(\ell)}(\hat{\mathcal{X}}) = Y$ is a non-trivial equation of degree at most $\max\{\mu(\mathcal{X}), \mu(\hat{\mathcal{X}})\}$. Similarly, if $m_i < \hat{m}_i$, then $(z_i^\ell \cdot \langle c_i \rangle_n \oplus \hat{z}_i^\ell \cdot \langle \hat{c}_i \rangle_n)$ is a polynomial in L of degree $\ell \hat{m}_i$, and therefore $F_L^{(\ell)}(\mathcal{X}) \oplus F_L^{(\ell)}(\hat{\mathcal{X}}) = Y$ is a non-trivial equation.

For the second case, consider i and j such that $0 \le i \le \ell - 1$, $0 \le j \le m_i - 1$, and $X_i[j] \ne \hat{X}_i[j]$. Now $(z_i^\ell \cdot \langle c_i \rangle_n \oplus \hat{z}_i^\ell \cdot \langle \hat{c}_i \rangle_n)$ can be written as (3), and since we have $m_i = \hat{m}_i$, a simplification shows that the coefficient of $L^{\ell m_i}$ is $\langle c_i \rangle_n \oplus \langle \hat{c}_i \rangle_n$, and the coefficient of $L^{\ell(m_i-j-1)}$ is $(X_i[j])^\ell \cdot \langle c_i \rangle_n \oplus (\hat{X}_i[j])^\ell \cdot \langle \hat{c}_i \rangle_n$. If $\langle c_i \rangle_n \ne \langle \hat{c}_i \rangle_n$, then the coefficient of $L^{\ell m_i}$ is non-zero. Otherwise the coefficient of $L^{\ell(m_i-j-1)}$ is non-zero since

$$(X_i[j])^\ell \cdot \langle c_i \rangle_n \oplus (\hat{X}_i[j])^\ell \cdot \langle \hat{c}_i \rangle_n = (X_i[j] \oplus \hat{X}_i[j])^\ell \cdot \langle c_i \rangle_n,$$

and the right hand side is zero if and only if $X_i[j] = \hat{X}_i[j]$.

Finally, we consider the third case. In this case, we claim that we must have $x_i \in (\{0,1\}^n)^+$ and $\hat{x}_i \notin (\{0,1\}^n)^+$ (or $x_i \notin (\{0,1\}^n)^+$ and $\hat{x}_i \notin (\{0,1\}^n)^+$) for some $0 \le i \le \ell-1$. Indeed, if $x_i, \hat{x}_i \in (\{0,1\}^n)^+$ holds for all $0 \le i \le \ell-1$, then we have $x_i = \hat{x}_i$ for all $0 \le i \le \ell - 1$, since $X_i = \mathrm{pad}(x_i) = x_i$ and $\hat{X}_i = \mathrm{pad}(\hat{x}_i) = \hat{x}_i$. This contradicts the condition $\mathcal{X} \ne \hat{\mathcal{X}}$. Similarly, if $x_i, \hat{x}_i \notin (\{0,1\}^n)^+$ holds for all $0 \le i \le \ell - 1$, then again it contradicts the fact $\mathcal{X} \ne \hat{\mathcal{X}}$. Now without loss of generality, we may assume that $x_i \in (\{0,1\}^n)^+$ and $\hat{x}_i \notin (\{0,1\}^n)^+$ for some $0 \le i \le \ell - 1$, which implies $\langle c_i \rangle_n \ne \langle \hat{c}_i \rangle_n$. Then $(z_i^\ell \cdot \langle c_i \rangle_n \oplus \hat{z}_i^\ell \cdot \langle \hat{c}_i \rangle_n)$ can be written as (3), and since we have $m_i = \hat{m}_i$, the coefficient of $L^{\ell m_i}$ is $\langle c_i \rangle_n \oplus \langle \hat{c}_i \rangle_n \ne 0^n$. $\qquad\square$

B   Proof of Lemma 2

Let p_0 and p_1 be as defined in Sect. 4.2. We have already shown that $p_0 \le \mathbf{Adv}^{\mathrm{priv}}_{\mathrm{HBS[Perm}(n)]}(B)$ and $p_1 \le \Pr(A^{\mathcal{E}_P(\cdot,\cdot),\mathcal{D}_P(\cdot,\cdot)}$ forges$)$. In the rest of this section we evaluate the last probability.

We introduce two oracles $\mathcal{O}_P(\cdot,\cdot)$ and $\mathcal{V}_P(\cdot,\cdot)$. The $\mathcal{O}_P(\cdot,\cdot)$ oracle takes $(T,s) \in \{0,1\}^n \times \mathbf{N}$ $(s < 2^n)$ as its input and returns $P(P^{-1}(T) \oplus \langle s \rangle_n)$. The $\mathcal{V}_P(\cdot,\cdot)$ oracle takes $((H,M),T)$ as its input and returns T if $P(F_L^{(2)}(H,M)) = T$ or returns $\perp$ otherwise. We observe that the $\mathcal{D}_P(\cdot,\cdot)$ oracle can be perfectly simulated by using these two oracles $\mathcal{O}_P(\cdot,\cdot)$ and $\mathcal{V}_P(\cdot,\cdot)$. Therefore, there exists an adversary A_1 such that

$$\Pr(A^{\mathcal{E}_P(\cdot,\cdot),\mathcal{D}_P(\cdot,\cdot)} \text{ forges}) \le \Pr(A_1^{\mathcal{E}_P(\cdot,\cdot),\mathcal{O}_P(\cdot,\cdot),\mathcal{V}_P(\cdot,\cdot)} \text{ forges}).$$

We use the following system of notation. Consider all queries to the $\mathcal{E}_P(\cdot,\cdot)$ oracle. Also consider those queries to the $\mathcal{V}_P(\cdot,\cdot)$ oracle which make the oracle return T (i.e., forgery). Enumerate these queries, in the order of being made, as (H_1, M_1), (H_2, M_2), ... and set $S_i = F_L^{(2)}(H_i, M_i)$. Define sets $\mathbb{S}_i = \{S_i, S_i \oplus \langle 1 \rangle_n, S_i \oplus \langle 2 \rangle_n, \ldots, S_i \oplus \langle m_i \rangle_n\}$. Define vectors $\mathbb{Y}_i$ as follows. In the case of $\mathcal{E}_P(\cdot,\cdot)$-query, define $\mathbb{Y}_i = (P(S_i), P(S_i \oplus \langle 1 \rangle_n), \ldots, P(S_i \oplus \langle m_i \rangle_n))$ so that $\mathbb{Y}_i[j] = P(S_i \oplus \langle j \rangle_n)$. In the case of $\mathcal{V}_P(\cdot,\cdot)$-query, define $\mathbb{Y}_i = (P(S_i))$ (A 1-dimensional vector). With abuse of notation we identify $\mathbb{Y}_i$ with the set $\{\mathbb{Y}_i[0], \ldots, \mathbb{Y}_i[m_i]\}$.

Consider all queries to the $\mathcal{O}_P(\cdot,\cdot)$ oracle. Let $(T_1, s_1), (T_2, s_2), \ldots$ be these queries and $Y_1, Y_2, \ldots$ the values returned by the oracle so that we have $Y_i = \mathcal{O}_P(T_i, s_i)$. We classify the queries to the $\mathcal{O}_P(\cdot,\cdot)$ oracle into three categories: *root* queries, *chain* queries, and *extension* queries.

Root. We say that (T_i, s_i) is a *root query* if $T_i \ne T_j$, $T_i \ne Y_j$ for all $j < i$ and $T_i \notin \mathbb{Y}_k$ for each previous k-th query to the $\mathcal{E}_P(\cdot,\cdot)$ / $\mathcal{V}_P(\cdot,\cdot)$ oracle.

Chain. We say that (T_i, s_i) is a *chain query* if there exists a $j < i$ such that the j-th query (T_j, s_j) was a root query and either $T_i = T_j$ or $T_i = Y_j$. Recursively, we call (T_i, s_i) a chain query also if there exists a $j < i$ such that the j-th query (T_j, s_j) was a chain query and $T_i = Y_j$.

Extension. We say that (T_i, s_i) is an *extension query* if there exists some previous k-th query to the $\mathcal{E}_P(\cdot,\cdot)$ / $\mathcal{V}_P(\cdot,\cdot)$ oracle such that $T_i \in \mathbb{Y}_k$. Recursively, we call (T_i, s_i) an extension query also if there exists a $j < i$ such that the j-the query (T_j, s_j) was an extension query and $T_i = Y_j$.

Note that some of the chain/extension queries to the $\mathcal{O}_P(\cdot,\cdot)$ oracle, even if the query itself is new, may be *trivial* in the sense that the adversary A_1 already knows the to-be-returned value. For example, if $P(T \oplus \langle s_1 \rangle_n) = Y_1$ and $P(T \oplus \langle s_2 \rangle_n) = Y_2$, then we know that $Y_2 = \mathcal{O}_P(Y_1, \langle s_1 \rangle_n \oplus \langle s_2 \rangle_n)$ (The second argument is treated as an integer). Whenever possible, we implicitly exclude these trivial queries from our consideration.

Now we are ready to introduce a modified oracle $\mathcal{O}_{P,\tilde{P}}(\cdot,\cdot)$ for A_1, where $\tilde{P} \overset{R}{\leftarrow} \mathrm{Perm}(n)$ is a random permutation independent of P. The oracle is defined in the style of lazy sampling for $\tilde{P}$. The oracle keeps a record of domain points $\tilde{S}_1, \tilde{S}_2, \ldots$ and that of range points $\tilde{Y}_1, \tilde{Y}_2, \ldots$ in a linked way. For this, the oracle needs to observe the vectors $\mathbb{Y}_i$ output by the $\mathcal{E}_P(\cdot,\cdot)$ / $\mathcal{V}_P(\cdot,\cdot)$ oracle. Let (T, s) be the current query made to the $\mathcal{O}_{P,\tilde{P}}(\cdot,\cdot)$ oracle.

1. If (T, s) is a trivial query, then the oracle simply returns the expected value.
2. If (T, s) is a root query, the oracle adds the point T to the set of range points $\{\tilde{Y}_1, \tilde{Y}_2, \ldots\}$. Then the oracle picks $\tilde{S} \overset{R}{\leftarrow} \{0,1\}^n \setminus (\{0^n\} \cup \bigcup_i \{\tilde{S}_i\})$, where i runs over already-defined domain points. The oracle updates the record $\{\tilde{S}_1, \tilde{S}_2, \ldots\}$ by adding the new domain point $\tilde{S}$ (The oracle establishes a link $\tilde{P}(\tilde{S}) = T$). If $s = 0$, then the oracle returns T. If $s \geq 1$, then the oracle checks if $\tilde{S} \oplus \langle s \rangle_n \in \{\tilde{Y}_1, \tilde{Y}_2, \ldots\}$. If so, then the oracle returns $\tilde{P}(\tilde{S} \oplus \langle s \rangle_n)$. If not, then the oracle adds the point $\tilde{S} \oplus \langle s \rangle_n$ to the set of domain points, picks $\tilde{Y} \overset{R}{\leftarrow} \{0,1\}^n \setminus (\bigcup_i \mathbb{Y}_i \cup \bigcup_j \{\tilde{Y}_j\})$, adds $\tilde{Y}$ to the set of range points and returns $\tilde{Y}$.
3. If (T, s) is a chain query, then it means that $\tilde{S} = \tilde{P}^{-1}(T)$ is in the set $\{\tilde{S}_1, \tilde{S}_2, \ldots\}$ but $\tilde{S} \oplus \langle s \rangle_n$ is not. So the oracle adds the point to the set $\{\tilde{S}_1, \tilde{S}_2, \ldots\}$, picks $\tilde{Y} \overset{R}{\leftarrow} \{0,1\}^n \setminus (\bigcup_i \mathbb{Y}_i \cup \bigcup_j \{\tilde{Y}_j\})$ and adds $\tilde{Y}$ to the set of range points. This establishes a link $\tilde{P}(\tilde{S} \oplus \langle s \rangle_n) = \tilde{Y}$, and the value $\tilde{Y}$ is returned.
4. Finally, if (T, s) is an extension query, then it means that the oracle has located a vector $\mathbb{Y}_i$ and an integer t such that $T = \mathcal{O}_{P,\tilde{P}}(\mathbb{Y}_i[0], t)$. The task is to return a value corresponding to $\mathcal{O}_{P,\tilde{P}}(\mathbb{Y}_i[0], \langle t \rangle_n \oplus \langle s \rangle_n)$. For this, the oracle picks $\tilde{Y} \overset{R}{\leftarrow} \{0,1\}^n \setminus (\bigcup_i \mathbb{Y}_i \cup \bigcup_j \{\tilde{Y}_j\})$, adds $\tilde{Y}$ to the set of range points and returns $\tilde{Y}$. Note that no domain point is selected; no domain point is linked to the range point $\tilde{Y}$. However, the range point $\tilde{Y}$ is linked to the vector $\mathbb{Y}_i$ through the "distance" $\langle t \rangle_n \oplus \langle s \rangle_n$.

We want to evaluate the quantity

$$p_2 = \left| \mathrm{Pr}\left(A_1^{\mathcal{E}_P(\cdot,\cdot),\mathcal{O}_P(\cdot,\cdot),\mathcal{V}_P(\cdot,\cdot)} \text{ forges}\right) - \mathrm{Pr}\left(A_1^{\mathcal{E}_P(\cdot,\cdot),\mathcal{O}_{P,\tilde{P}}(\cdot,\cdot),\mathcal{V}_P(\cdot,\cdot)} \text{ forges}\right) \right|. \quad (4)$$

It turns out that the evaluation of p_2 requires a fair amount of work. So we defer the treatment of p_2 until we finish evaluating the second forgery probability in (4). Observe that there exists an adversary A_2 such that

$$\mathrm{Pr}\left(A_1^{\mathcal{E}_P(\cdot,\cdot),\mathcal{O}_{P,\tilde{P}}(\cdot,\cdot),\mathcal{V}_P(\cdot,\cdot)} \text{ forges}\right) \leq \mathrm{Pr}\left(A_2^{\mathcal{E}_P(\cdot,\cdot),\mathcal{V}_P(\cdot,\cdot)} \text{ forges}\right),$$

because the adversary A_2 can perfectly simulate the $\mathcal{O}_{P,\tilde{P}}(\cdot,\cdot)$ oracle by observing the vectors $\mathbb{Y}_i$ and performing lazy sampling for $\tilde{P}$.

We apply the standard argument that distinguishing is easier than forging. Namely, there exists an adversary B such that

$$\Pr(A_2^{\mathcal{E}_P(\cdot,\cdot),\mathcal{V}_P(\cdot,\cdot)} \text{ forges}) \leq \mathbf{Adv}_{\text{HBS}[\text{Perm}(n)]}^{\text{priv}}(B) + \frac{q}{2^n}.$$

The adversary B simulates the two oracles for A_2 by using its own oracle and outputs 1 as soon as A_2 forges. If B's oracle is $\mathcal{E}_P(\cdot,\cdot)$, the simulation is perfect and hence $\Pr\left(B^{\mathcal{E}_P(\cdot,\cdot)} = 1\right) = \Pr(A_2^{\mathcal{E}_P(\cdot,\cdot),\mathcal{V}_P(\cdot,\cdot)} \text{ forges})$. If B's oracle is $\mathcal{R}(\cdot,\cdot)$, then each time A_2 makes a query to the second oracle, the probability of forgery is $1/2^n$ (prior to the execution of the game) and so $\Pr\left(B^{\mathcal{R}(\cdot,\cdot)} = 1\right) \leq q/2^n$.

Now we come back to the evaluation of p_2. Consider those queries to the $\mathcal{V}_P(\cdot,\cdot)$ oracle for which the oracle returns $\bot$. Enumerate these queries, in the order of being made, as $((H_1, M_1), T_1), ((H_2, M_2), T_2), \ldots$. Set $U_i = F_L^{(2)}(H_i, M_i)$ and $Z_i = P(U_i)$.

For $S \in \{0,1\}^n$, define $\mathcal{N}(S) = \bigcup_{s_1,\ldots,s_q}\{S \oplus \langle s_1 \rangle_n \oplus \cdots \oplus \langle s_q \rangle_n\}$, where each s_i runs over $0 \leq s_i \leq m_{\max}$. This set is not so large, as we have $\mathcal{N}(S) \subset \{S, S \oplus \langle 1 \rangle_n, \ldots, S \oplus \langle 2m_{\max} \rangle_n\}$. Put $\mathbb{S}_i^* = \mathcal{N}(S_i)$, $\tilde{\mathbb{S}}_i^* = \mathcal{N}(\tilde{S}_i)$ and $\mathbb{U}_i^* = \mathcal{N}(U_i)$.

We consider the following bad events in the latter game in (4). (i) $\mathbb{S}_i^* \ni 0^n$ or $\mathbb{U}_i^* \ni 0^n$ for some i. (ii) $\mathbb{S}_i^* \cap \mathbb{S}_j^* \neq \emptyset$, $\mathbb{U}_i^* \cap \mathbb{U}_j^* \neq \emptyset$ or $\mathbb{S}_i^* \cap \mathbb{U}_j^* \neq \emptyset$ for some i and j such that $(H_i, M_i) \neq (H_j, M_j)$. (iii) $\mathbb{S}_i^* \cap \tilde{\mathbb{S}}_j^* \neq \emptyset$ or $\mathbb{U}_i^* \cap \tilde{\mathbb{S}}_j^* \neq \emptyset$ for some i and j. (iv) $\tilde{\mathbb{S}}_i^* \ni 0^n$ for some i. (v) $\mathbb{Y}_i \ni \tilde{Y}_j$ for some i and j. (vi) $Z_i = \tilde{Y}_j$ for some i and j. (vii) $L = \tilde{Y}_i$ for some i.

We argue that the two games in (4) (referred to as the *first* and the *second* games) proceed exactly the same as long as none of (i)–(vii) bad events occurs. This means that we have

$$p_2 \leq \Pr\left(A_1^{\mathcal{E}_P(\cdot,\cdot),\mathcal{O}_{P,\tilde{P}}(\cdot,\cdot),\mathcal{V}_P(\cdot,\cdot)} \text{ causes one of (i)–(vii)}\right).$$

To see this, it is helpful to consider the permutation P as lazily sampled. We check this one by one.

1. Consider a query (say the i-th query) either to the $\mathcal{E}_P(\cdot,\cdot)$ oracle or to the $\mathcal{V}_P(\cdot,\cdot)$ oracle (returning 1). Thanks to (i), (ii) and (iii), it is guaranteed that the set $\mathbb{S}_i$ consists of entirely fresh domain points for P (unless a query with (H_i, M_i) has been made to the $\mathcal{V}_P(\cdot,\cdot)$ oracle returning $\bot$). This results in lazy sampling of points for $\mathbb{Y}_i$. In the first game, these points are sampled from $\{0,1\}^n \setminus (\{L\} \cup \bigcup_j \mathbb{Y}_j \cup \bigcup_j \{\tilde{Y}_j\} \cup \bigcup_j \{Z_j\})$, where j runs over already-defined points. In the second game, the points are sampled from $\{0,1\}^n \setminus (\{L\} \cup \bigcup_j \mathbb{Y}_j \cup \bigcup_j \{Z_j\})$. The two distributions remain the same due to (v).
2. Consider a root query (T, s) to the $\mathcal{O}_P(\cdot,\cdot)$ or $\mathcal{O}_{P,\tilde{P}}(\cdot,\cdot)$ oracle. Note that (vii) eliminates the possibility $T = L$. In the first game, a domain point S is sampled for $P^{-1}(T)$ from $\{0,1\}^n \setminus (\{0^n\} \cup \bigcup_i \mathbb{S}_i \cup \bigcup_i \{\tilde{S}_i\} \cup \bigcup_i \{U_i\})$. In the second game, a domain point S is sampled (for $\tilde{P}^{-1}(T)$) from $\{0,1\}^n \setminus (\{0^n\} \cup$

$\bigcup_i\{\tilde{S}_i\}$). These are the same distribution due to (iii). In both the games, the shifted domain point $S \oplus \langle s \rangle_n$ is not fresh under the same condition that there exists some i such that $S \oplus \langle s \rangle_n = \tilde{S}_i$. This is due to (iii) and (iv). If it is fresh, then in the first game a range point $\tilde{Y}$ is sampled from $\{0,1\}^n \setminus (\{L\} \cup \bigcup_i \mathbb{Y}_i \cup \bigcup_i \{\tilde{Y}_i\} \cup \bigcup_i \{Z_i\})$, whereas in the second game a range point $\tilde{Y}$ is sampled from $\{0,1\}^n \setminus (\bigcup_i \mathbb{Y}_i \cup \bigcup_i \{\tilde{Y}_i\})$. These are the same owing to (vi) and (vii).

3. Consider a chain query to the $\mathcal{O}_P(\cdot,\cdot)$ or $\mathcal{O}_{P,\tilde{P}}(\cdot,\cdot)$ oracle. Thanks to (i), (ii) and (iii), the query results in a fresh domain point $P^{-1}(T) \oplus \langle s \rangle_n$ in the first game and $\tilde{P}^{-1}(T) \oplus \langle s \rangle_n$ in the second. In the first game, the corresponding range point is sampled from $\{0,1\}^n \setminus (\{L\} \cup \bigcup_i \mathbb{Y}_i \cup \bigcup_i \{\tilde{Y}_i\} \cup \bigcup_i \{Z_i\})$. In the second game, the range point is sampled from $\{0,1\}^n \setminus (\bigcup_i \mathbb{Y}_i \cup \bigcup_i \{\tilde{Y}_i\})$. These two yield the same distribution because of (vi) and (vii).

4. Consider an extension query (T,s) to the $\mathcal{O}_P(\cdot,\cdot)$ or $\mathcal{O}_{P,\tilde{P}}(\cdot,\cdot)$ oracle. Thanks to (i), (ii) and (iii), the query results in a fresh domain point $P^{-1}(T) \oplus \langle s \rangle_n$. In the first game, the corresponding range point is sampled from $\{0,1\}^n \setminus (\{L\} \cup \bigcup_i \mathbb{Y}_i \cup \bigcup_i \{\tilde{Y}_i\} \cup \bigcup_i \{Z_i\})$. In the second game, the range point is sampled from $\{0,1\}^n \setminus (\bigcup_i \mathbb{Y}_i \cup \bigcup_i \{\tilde{Y}_i\})$. These two sampling processes yield the same distribution due to (vi) and (vii).

5. Consider a query (say the i-th query) to the $\mathcal{V}_P(\cdot,\cdot)$ oracle returning $\perp$. If a query with (H_i, M_i) has been made to one of the oracles, then no sampling is performed. Otherwise, owing to (i), (ii) and (iii), the point U_i is a fresh domain point. This results in lazy sampling of Z_i. In the first game, Z_i is sampled from $\{0,1\}^n \setminus (\{L\} \cup \bigcup_j \mathbb{Y}_j \cup \bigcup_j \{\tilde{Y}_j\} \cup \bigcup_j \{Z_j\})$. In the second game, it is sampled from $\{0,1\}^n \setminus (\{L\} \cup \bigcup_j \mathbb{Y}_j \cup \bigcup_j \{Z_j\})$. These two distributions remain the same because of (vi).

Now we replace P with a random function $R \xleftarrow{R} \mathrm{Func}(n)$ (The set of all functions on $\{0,1\}^n$). This is possible because the inverse cipher P^{-1} never appears in the definitions of the three oracles. By the PRP/PRF switching lemma [2], we get

$$\left| \Pr\left(A_1^{\mathcal{E}_P(\cdot,\cdot),\mathcal{O}_{P,\tilde{P}}(\cdot,\cdot),\mathcal{V}_P(\cdot,\cdot)} \text{ causes one of (i)-(vii)}\right) \right.$$
$$\left. - \Pr\left(A_1^{\mathcal{E}_R(\cdot,\cdot),\mathcal{O}_{R,\tilde{P}}(\cdot,\cdot),\mathcal{V}_R(\cdot,\cdot)} \text{ causes one of (i)-(vii)}\right) \right| \leq \frac{(1+q+qm_{\max})^2}{2^{n+1}}.$$

We introduce a modified oracle $\mathcal{E}_R^\bullet(\cdot,\cdot)$. This oracle behaves just like the $\mathcal{E}_R(\cdot,\cdot)$ oracle, computing the values for $\mathbb{S}_i$ using $L = R(0^n)$, except that at the end of query process, even if a bad event has occurred, the oracle returns a random string, which defines the value of $\mathbb{Y}_i$ (unless a query with (H_i, M_i) has been made to one of the oracles, in which case those already-defined values are used). The $\mathcal{V}_R^\bullet(\cdot,\cdot)$ oracle is defined similarly. That is, unless a query with (H_i, M_i) has been already made, the oracle computes the value $S_i \ / \ U_i$ using $L = R(0^n)$ and, regardless of a bad event, picks a random point in $\{0,1\}^n$. If this point happens to be the same as T, then the oracle returns T. Otherwise, the random point is set to Z_i, and the oracle returns $\perp$. The modified oracles are identical to the original ones until a bad event occurs, so we have

$$\Pr\left(A_1^{\mathcal{E}_R(\cdot,\cdot),\mathcal{O}_{R,\bar{P}}(\cdot,\cdot),\mathcal{V}_R(\cdot,\cdot)} \text{ causes one of (i)–(vii)}\right)$$
$$= \Pr\left(A_1^{\mathcal{E}_R^{\bullet}(\cdot,\cdot),\mathcal{O}_{R,\bar{P}}(\cdot,\cdot),\mathcal{V}_R^{\bullet}(\cdot,\cdot)} \text{ causes one of (i)–(vii)}\right).$$

Now observe that the values returned by the three oracles are independent of the value $L = R(0^n)$. So we are ready to evaluate the probabilities of (i)–(vii). Without loss of generality we assume that the bad events are disjoint. This can be done by defining a bad event to occur prior to any other bad event. So, for example, when we say "the event (i) occurs," we implicitly mean that no other bad event has occurred.

An event of type (i) occurs upon a query (H_i, M_i), made to the $\mathcal{E}_R^{\bullet}(\cdot,\cdot)$ oracle or to $\mathcal{V}_R^{\bullet}(\cdot,\cdot)$, which satisfies an equation $F_L^{(2)}(H_i, M_i) = \langle s \rangle_n$ for some $0 \leq s \leq 2m_{\max}$. Here, we have $1 \leq i \leq q$, and the degree of the equation is at most $2(1 + h_{\max} + m_{\max})$. Therefore, we have

$$\Pr\left(A_1^{\mathcal{E}_R^{\bullet}(\cdot,\cdot),\mathcal{O}_{R,\bar{P}}(\cdot,\cdot),\mathcal{V}_R^{\bullet}(\cdot,\cdot)} \text{ causes (i)}\right) \leq \frac{2q(1 + 2m_{\max})(1 + h_{\max} + m_{\max})}{2^n}.$$

Event (ii) corresponds to queries (H_i, M_i) and (H_j, M_j) $(1 \leq i < j \leq q)$, made to the $\mathcal{E}_R^{\bullet}(\cdot,\cdot)$ oracle or to $\mathcal{V}_R^{\bullet}(\cdot,\cdot)$, which satisfy an equation $F_L^{(2)}(H_i, M_i) \oplus F_L^{(2)}(H_j, M_j) = \langle s \rangle_n$ with $0 \leq s \leq 2m_{\max}$. We have

$$\Pr\left(A_1^{\mathcal{E}_R^{\bullet}(\cdot,\cdot),\mathcal{O}_{R,\bar{P}}(\cdot,\cdot),\mathcal{V}_R^{\bullet}(\cdot,\cdot)} \text{ causes (ii)}\right) \leq \frac{q^2(1 + 2m_{\max})(1 + h_{\max} + m_{\max})}{2^n}.$$

Event (iii) corresponds to a query (H_i, M_i) and a root query (T_j, s_j) (either one may be made before the other) such that $F_L^{(2)}(H_i, M_i) = \tilde{S}_j \oplus \langle s \rangle_n$ with $0 \leq s \leq 2m_{\max}$. Note that A_1 makes at most q-many root queries (rather than $q(1 + m_{\max})$-many). So we have

$$\Pr\left(A_1^{\mathcal{E}_R^{\bullet}(\cdot,\cdot),\mathcal{O}_{R,\bar{P}}(\cdot,\cdot),\mathcal{V}_R^{\bullet}(\cdot,\cdot)} \text{ causes (iii)}\right) \leq \frac{2q^2(1 + 2m_{\max})(1 + h_{\max} + m_{\max})}{2^n}.$$

An event of type (iv) occurs upon a root query (T_i, s_i) satisfying $\tilde{S}_j = \langle s \rangle_n$ for some $0 \leq s \leq 2m_{\max}$. The sampling is performed from the space $\{0,1\}^n \setminus \{0^n\}$ for at most q-many distinct points, so we have

$$\Pr\left(A_1^{\mathcal{E}_R^{\bullet}(\cdot,\cdot),\mathcal{O}_{R,\bar{P}}(\cdot,\cdot),\mathcal{V}_R^{\bullet}(\cdot,\cdot)} \text{ causes (iv)}\right) \leq \frac{1 + 2m_{\max} - 1 + 1}{2^n} = \frac{1 + 2m_{\max}}{2^n}.$$

An event of type (v) occurs upon a new query (H_i, M_i) to the $\mathcal{E}_R^{\bullet}(\cdot,\cdot)$ oracle or to the $\mathcal{V}_R^{\bullet}(\cdot,\cdot)$ oracle returning T. Such an query causes random sampling of the vector $\mathbb{Y}_i$. The values $\tilde{Y}_1, \tilde{Y}_2, \ldots$ consist of root points selected by A_1 plus additional "semi-random" points. The number of random sampling for the vectors $\mathbb{Y}_i$ is at most $q(1 + m_{\max})$, and the size of $\{\tilde{Y}_1, \tilde{Y}_2, \ldots\}$ is at most $q(1 + m_{\max})$. So we obtain

$$\Pr\left(A_1^{\mathcal{E}_P^{\bullet}(\cdot,\cdot),\mathcal{O}_{P,\bar{P}}(\cdot,\cdot),\mathcal{V}_P^{\bullet}(\cdot,\cdot)} \text{ causes (v)}\right) \leq \frac{q^2(1 + m_{\max})^2}{2^n}.$$

For event (vi), there are three cases. The first case is upon the sampling of Z_i. This case can be treated similarly to event (v), and the probability is at most $q \cdot q(1+m_{max})/2^n = q^2(1+m_{max})/2^n$. The second case is upon a root query (T, s) such that $T = Z_i$ for some i. The values $Z_1, Z_2, \ldots$ are simply random points, the number of which is at most q. There are at most q-many root queries, so the probability is at most $q^2/2^n$. The third case is upon the sampling of a range point $\tilde{Y}_j$. Note that such a sampling is performed for at most $q(1+m_{max})$ times. So the probability is at most $q^2(1+m_{max})/2^n$. Overall, the probability for event (vi) is at most

$$\frac{q^2(1+m_{max})}{2^n} + \frac{q^2}{2^n} + \frac{q^2(1+m_{max})}{2^n} \leq \frac{2q^2(1+2m_{max})}{2^n}.$$

For event (vii), observe that the number of points $\tilde{Y}_1, \tilde{Y}_2, \ldots$ is at most $q(1+m_{max})$. So we simply have

$$\Pr\left(A_1^{\mathcal{E}_{\tilde{P}}^\bullet(\cdot,\cdot), \mathcal{O}_{P,\tilde{P}}(\cdot,\cdot), \mathcal{V}_P^\bullet(\cdot,\cdot)} \text{ causes (vii)}\right) \leq \frac{q(1+m_{max})}{2^n}.$$

Finally we sum up the terms. This yields

$$p_2 \leq (1+2+1+2+1+1+2+1)\frac{q^2(1+h_{max}+2m_{max})^2}{2^n}$$
$$= \frac{11q^2(1+h_{max}+2m_{max})^2}{2^n}.$$

$\square$

C   Proof of Lemma 3

We consider the encryption algorithm $\mathcal{E}_R(\cdot, \cdot)$ of "HBS[Func(n)]," where Func(n) is the set of all functions over $\{0,1\}^n$, and a random function $R \xleftarrow{R} \text{Func}(n)$ is used as the underlying blockcipher (Obviously the decryption algorithm cannot be defined). We see that

$$\mathbf{Adv}^{priv}_{HBS[Perm(n)]}(B) \leq \mathbf{Adv}^{priv}_{\text{"HBS[Func}(n)]\text{"}}(B) + \frac{(1+q+qm_{max})^2}{2^{n+1}} \quad (5)$$

from the PRP/PRF switching lemma [2]. We then use the following lemma to prove Lemma 3.

Lemma 4. *Let $h_0, \ldots, h_{q-1}, m_0, \ldots, m_{q-1}$ be integers such that $h_i \leq h_{max}$ and $m_i \leq m_{max}$ for $0 \leq i \leq q-1$. Also, let $H_0, \ldots, H_{q-1}, M_0, \ldots, M_{q-1}, T_0, \ldots, T_{q-1},$ and $C_0, \ldots, C_{q-1}$ be bit strings that satisfy the following conditions for $0 \leq i \leq q-1$. $H_i \in (\{0,1\}^n)^{h_i}, M_i \in (\{0,1\}^n)^{m_i}, T_i \in \{0,1\}^n,$ and $C_i \in (\{0,1\}^n)^{m_i}$. Furthermore, assume $(H_i, M_i) \neq (H_j, M_j)$ holds for $0 \leq i < j \leq q-1$. Then we have*

$$\frac{p_3}{p_4} \geq 1 - \frac{3q^2(1+h_{max}+2m_{max})^2}{2^n}, \quad (6)$$

where we define p_3 and p_4 as $p_3 \overset{\text{def}}{=} \Pr(\mathcal{E}_R(H_i, M_i) = (T_i, C_i))$ *for* $0 \le i \le q - 1$ *and* $p_4 \overset{\text{def}}{=} \Pr(\mathcal{R}(H_i, M_i) = (T_i, C_i))$ *for* $0 \le i \le q - 1$.

Proof. The proof is based on a counting argument. We count the number of functions $R \in \text{Func}(n)$ that satisfy $\mathcal{E}_R(H_i, M_i) = (T_i, C_i)$ for $0 \le i \le q - 1$. We first count the number of $L = R(0^n)$ and then count the rest. Let $S_i = F_L^{(2)}(H_i, M_i)$. S_i corresponds to the initial counter value. Consider the set $\mathbb{S}_i = \{S_i, S_i \oplus \langle 1 \rangle_n, S_i \oplus \langle 2 \rangle_n, \ldots, S_i \oplus \langle m_i \rangle_n\}$. Now we claim that the number of $L \in \{0, 1\}^n$ such that

$$\{0^n\} \cap \mathbb{S}_i = \emptyset \text{ for } 0 \le i \le q - 1 \text{ and } \mathbb{S}_i \cap \mathbb{S}_j = \emptyset \text{ for } 0 \le i < j \le q - 1 \quad (7)$$

is at least $2^n - 2q(1 + m_{\max})(1 + h_{\max} + m_{\max}) - q^2(1 + 2m_{\max})(1 + h_{\max} + m_{\max})$.

Fix $0 \le i \le q - 1$ and $0 \le s \le m_i$. Consider the equation $F_L^{(2)}(H_i, M_i) = \langle s \rangle_n$. As we have seen in Sect. 3.1, $F_L^{(2)}(H_i, M_i)$ is a non-zero polynomial in L of degree $\mu(H_i, M_i) = \max\{1 + 2h_i, 2 + 2m_i\}$. We therefore have $\#\{L \mid F_L^{(2)}(H_i, M_i) = \langle s \rangle_n\} \le \max\{1 + 2h_i, 2 + 2m_i\} \le 2(1 + h_{\max} + m_{\max})$, which implies

$$\#\{L \mid \{0^n\} \cap \mathbb{S}_i \ne \emptyset \text{ for some } 0 \le i \le q - 1\} \le 2q(1 + m_{\max})(1 + h_{\max} + m_{\max}).$$

Next, fix $0 \le i < j \le q - 1$ and consider the equation $F_L^{(2)}(H_i, M_i) \oplus \langle s \rangle_n = F_L^{(2)}(H_j, M_j) \oplus \langle t \rangle_n$, where $0 \le s \le m_i$ and $0 \le t \le m_j$. Theorem 1 implies that this equation has at most $\max\{\mu(H_i, M_i), \mu(H_j, M_j)\} = \max\{1 + 2h_i, 2 + 2m_i, 1 + 2h_j, 2 + 2m_j\}$ solutions. Observe that the equation is equivalent to $F_L^{(2)}(H_i, M_i) \oplus F_L^{(2)}(H_j, M_j) = \langle s \rangle_n \oplus \langle t \rangle_n$, and the right hand side takes at most $(1 + m_i + m_j)$ values (rather than $(1 + m_i)(1 + m_j)$ values). We thus have

$$\#\{L \mid \mathbb{S}_i \cap \mathbb{S}_j \ne \emptyset\} \le (1 + m_i + m_j) \max\{1 + 2h_i, 2 + 2m_i, 1 + 2h_j, 2 + 2m_j\}$$
$$\le 2(1 + 2m_{\max})(1 + h_{\max} + m_{\max}),$$

which implies $\#\{L \mid \mathbb{S}_i \cap \mathbb{S}_j \ne \emptyset \text{ for some } 0 \le i < j \le q - 1\}$ is at most

$$q^2(1 + 2m_{\max})(1 + h_{\max} + m_{\max}).$$

Once we fix any L that satisfies (7), the inputs to R, $\{0^n\} \cup \mathbb{S}_0 \cup \cdots \cup \mathbb{S}_{q-1}$, are all distinct. Therefore, the left hand side of (6) is at least

$$\frac{2^n - 2q(1 + m_{\max})(1 + h_{\max} + m_{\max}) - q^2(1 + 2m_{\max})(1 + h_{\max} + m_{\max})}{2^n},$$

which is at least $1 - (3q^2(1 + h_{\max} + 2m_{\max})^2)/2^n$. $\qquad\square$

Proof (of Lemma 3). Without loss of generality, we assume that B makes exactly q oracle queries. Also, since B is computationally unbounded, we assume that B is deterministic. Now we can regard B as a function $f_B : (\{0, 1\}^n)^{q(1+m_{\max})} \to \{0, 1\}$. To see this, let $Y \in (\{0, 1\}^n)^{q(1+m_{\max})}$ be an arbitrary bit string of

$q(1 + m_{\max})$ blocks. The first query, (H_0, M_0), is determined by B. If we return the first $n + |M_0|$ bits of Y, then the next query, (H_1, M_1), is determined. Similarly, if we return the next $n + |M_1|$ bits of Y, then the next query, (H_2, M_2), is determined. By continuing the procedure, the output of B, either 0 or 1, is determined. Therefore, the output of B and the value of q queries, $(H_0, M_0), \ldots, (H_{q-1}, M_{q-1})$, are all determined by fixing Y. Let $\mathbf{v}_{\text{one}} = \{Y \mid f_B(Y) = 1\}$, and $P_{\mathcal{R}} \stackrel{\text{def}}{=} \Pr(B^{\mathcal{R}(\cdot,\cdot)} = 1)$. Then we have

$$P_{\mathcal{R}} = \sum_{Y \in \mathbf{v}_{\text{one}}} p_4. \tag{8}$$

On the other hand, let $P_{\text{HBS}} \stackrel{\text{def}}{=} \Pr(A^{\mathcal{E}_R(\cdot,\cdot)} = 1)$ and observe

$$P_{\text{HBS}} = \sum_{Y \in \mathbf{v}_{\text{one}}} p_3 \geq \left(1 - \frac{3q^2(1 + h_{\max} + 2m_{\max})^2}{2^n}\right) \sum_{Y \in \mathbf{v}_{\text{one}}} p_4,$$

where the last inequality follows from Lemma 4. Then P_{HBS} is at least

$$\left(1 - \frac{3q^2(1 + h_{\max} + 2m_{\max})^2}{2^n}\right) P_{\mathcal{R}} \geq P_{\mathcal{R}} - \frac{3q^2(1 + h_{\max} + 2m_{\max})^2}{2^n}$$

from (8). Now, we have $P_{\text{HBS}} \geq P_{\mathcal{R}} - (3q^2(1 + h_{\max} + 2m_{\max})^2)/2^n$, and by applying the same argument to $1 - P_{\text{HBS}}$ and $1 - P_{\mathcal{R}}$, we have $1 - P_{\text{HBS}} \geq 1 - P_{\mathcal{R}} - (3q^2(1 + h_{\max} + 2m_{\max})^2)/2^n$. Finally, from (5), we obtain the claimed bound.   □

Author Index